LESSONS LEARNED IN REGULATING SMALL MODULAR REACTORS

The following States are Members of the International Atomic Energy Agency:

AFGHANISTAN	GERMANY	PALAU
ALBANIA	GHANA	PANAMA
ALGERIA	GREECE	PAPUA NEW GUINEA
ANGOLA	GRENADA	PARAGUAY
ANTIGUA AND BARBUDA	GUATEMALA	PERU
ARGENTINA	GUYANA	PHILIPPINES
ARMENIA	HAITI	POLAND
AUSTRALIA	HOLY SEE	PORTUGAL
AUSTRIA	HONDURAS	QATAR
AZERBAIJAN	HUNGARY	REPUBLIC OF MOLDOVA
BAHAMAS	ICELAND	ROMANIA
BAHRAIN	INDIA	RUSSIAN FEDERATION
BANGLADESH	INDONESIA	RWANDA
BARBADOS	IRAN, ISLAMIC REPUBLIC OF	SAINT KITTS AND NEVIS
BELARUS	IRAQ	SAINT LUCIA
BELGIUM	IRELAND	SAINT VINCENT AND
BELIZE	ISRAEL	THE GRENADINES
BENIN	ITALY	SAMOA
BOLIVIA, PLURINATIONAL	JAMAICA	SAN MARINO
STATE OF	JAPAN	SAUDI ARABIA
BOSNIA AND HERZEGOVINA	JORDAN	SENEGAL
BOTSWANA	KAZAKHSTAN	SERBIA
BRAZIL	KENYA	SEYCHELLES
BRUNEI DARUSSALAM	KOREA, REPUBLIC OF	SIERRA LEONE
BULGARIA	KUWAIT	SINGAPORE
BURKINA FASO	KYRGYZSTAN	SLOVAKIA
BURUNDI	LAO PEOPLE'S DEMOCRATIC	SLOVENIA
CAMBODIA	REPUBLIC	SOUTH AFRICA
CAMEROON	LATVIA	SPAIN
CANADA	LEBANON	SRI LANKA
CENTRAL AFRICAN	LESOTHO	SUDAN
REPUBLIC	LIBERIA	SWEDEN
CHAD	LIBYA	SWITZERLAND
CHILE	LIECHTENSTEIN	SYRIAN ARAB REPUBLIC
CHINA	LITHUANIA	TAJIKISTAN
COLOMBIA	LUXEMBOURG	THAILAND
COMOROS	MADAGASCAR	TOGO
CONGO	MALAWI	TONGA
COSTA RICA	MALAYSIA	TRINIDAD AND TOBAGO
CÔTE D'IVOIRE	MALI	TUNISIA
CROATIA	MALTA	TÜRKİYE
CUBA	MARSHALL ISLANDS	TURKMENISTAN
CYPRUS	MAURITANIA	UGANDA
CZECH REPUBLIC	MAURITIUS	UKRAINE
DEMOCRATIC REPUBLIC	MEXICO	UNITED ARAB EMIRATES
OF THE CONGO	MONACO	UNITED KINGDOM OF
DENMARK	MONGOLIA	GREAT BRITAIN AND
DJIBOUTI	MONTENEGRO	NORTHERN IRELAND
DOMINICA	MOROCCO	UNITED REPUBLIC
DOMINICAN REPUBLIC	MOZAMBIQUE	OF TANZANIA
ECUADOR	MYANMAR	UNITED STATES OF AMERICA
EGYPT	NAMIBIA	URUGUAY
EL SALVADOR	NEPAL	UZBEKISTAN
ERITREA	NETHERLANDS	VANUATU
ESTONIA	NEW ZEALAND	VENEZUELA, BOLIVARIAN
ESWATINI	NICARAGUA	REPUBLIC OF
ETHIOPIA	NIGER	VIET NAM
FIJI	NIGERIA	YEMEN
FINLAND	NORTH MACEDONIA	ZAMBIA
FRANCE	NORWAY	ZIMBABWE
GABON	OMAN	
GEORGIA	PAKISTAN	

The Agency's Statute was approved on 23 October 1956 by the Conference on the Statute of the IAEA held at United Nations Headquarters, New York; it entered into force on 29 July 1957. The Headquarters of the Agency are situated in Vienna. Its principal objective is "to accelerate and enlarge the contribution of atomic energy to peace, health and prosperity throughout the world".

IAEA-TECDOC-2003

LESSONS LEARNED IN REGULATING SMALL MODULAR REACTORS

CHALLENGES, RESOLUTIONS AND INSIGHTS

INTERNATIONAL ATOMIC ENERGY AGENCY
VIENNA, 2022

COPYRIGHT NOTICE

For further information on this publication, please contact:

Regulatory Activities Section
International Atomic Energy Agency
Vienna International Centre
PO Box 100
1400 Vienna, Austria
Email: Official.Mail@iaea.org

© IAEA, 2022
Printed by the IAEA in Austria
June 2022

IAEA Library Cataloguing in Publication Data

Names: International Atomic Energy Agency.
Title: Lessons learned in regulating small modular reactors / International Atomic Energy Agency.
Description: Vienna : International Atomic Energy Agency, 2022. | Series: IAEA TECDOC series, ISSN 1011–4289 ; no. 2003 | Includes bibliographical references.
Identifiers: IAEAL 22-01516 | ISBN 978–92–0–125022–3 (paperback : alk. paper) | ISBN 978–92–0–124922–7 (pdf)
Subjects: LCSH: Nuclear reactors — Safety measures. | Nuclear reactors — Security measures. | Nuclear reactors — Nuclear energy.

FOREWORD

Many Member States have shown an increasing interest in the design and deployment of small modular reactors (SMRs). However, there is limited international regulatory experience in this field. To date, very few nuclear regulatory authorities have issued a construction licence for such reactors and few other authorities have conducted regulatory reviews of novel reactor designs prior to approving the construction of units using SMR technologies.

A wide range of approaches to the regulation of SMRs have been identified and uncertainties exist concerning the best approach to address the various novel features of these reactors. The majority of Member State regulatory bodies with experience in SMR regulation meet at the SMR Regulators' Forum, which offers them an opportunity to share valuable information on these topics. To complement the work of the Forum, the IAEA embarked on the task of documenting experience in regulating SMRs on various technical topics that go beyond the discussions of the Forum. Member States facing challenges with SMR regulation, responded to questionnaires on aspects such as legal and regulatory framework, safety design and analysis, and other regulatory challenges, to gather information that could be useful to the regulatory body of any Member State intending to deploy SMRs.

The primary objective of this publication is to document the experience gained by Member State regulatory bodies over the past twenty years on the regulation of SMRs, including licensing and compliance verification. This publication also focuses on challenges encountered and their resolutions.

The IAEA expresses its thanks to all those involved in the preparation of this publication. The IAEA officers responsible for this publication were M. Santini and M. Santos of the Division of Nuclear Installation Safety.

CONTENTS

1. INTRODUCTION

1.1. BACKGROUND

There is currently very limited international experience in regulating and licensing small modular reactors (SMRs). Very few regulatory bodies have issued a construction or an operating licence.

There also appears to be a wide range of approaches to the regulation of SMRs and significant uncertainty arises as a result of the various novel features of these reactors. It is expected that sharing existing regulatory experiences will help all Member States preparing for SMR deployment and will also support the development of common regulatory strategies based on the current state of practice.

Many countries, including embarking countries, are considering SMRs as the main option for the deployment of new nuclear power generation. It is expected that this publication will support those countries in improving or establishing a regulatory framework and infrastructure suitable for the regulation of SMRs.

Practical experience of SMR regulation by Member States has been gathered through a survey using a detailed questionnaire covering potential challenges in regulating SMRs. This was subsequently supplemented by incorporating responses to follow up questions, and the experiences shared in documents issued by the SMR Regulators' Forum Working Groups [1].

1.2. OBJECTIVE

The primary objective of this TECDOC is to document existing experience gained by regulatory bodies on the regulation of SMRs, including licensing and compliance assurance, with particular focus on challenges encountered, their resolutions, and insights to future issues.

This TECDOC identifies key regulatory challenges and lessons learned that have emerged in regulatory decision making related to SMRs in Member States. This publication also presents the early challenges and lessons learned by regulatory bodies in preparation for review of an application for an SMR licence. The document also provides some forward-looking insights on how regulatory bodies expect to address the challenges in the near future. It is expected that the TECDOC will help enhance the effectiveness of regulating SMRs deployed in the short and medium term.

This TECDOC is intended for use by regulatory bodies responsible for the regulation of SMRs.

1.3. SCOPE

For the purpose of this TECDOC, SMRs are defined as Nuclear Power Plants (NPPs) that typically have the following features:

— Nuclear reactors of power of typically <300 MW(e) or <1000 MW(th) per reactor.
— Reactors designed for commercial use (including prototypes or demonstration plants), i.e. electricity production, desalination, process heat (as opposed to research and test reactors).
— Reactors designed to allow addition of multiple modules in close proximity to the same infrastructure (modular reactors).

— Novel designs that have not been widely analysed or licensed by regulatory bodies. Technologies considered include water-cooled, high temperature gas, liquid metal cooled and molten salt reactors.

— Reactors that may be underwater, land-based or floating nuclear power plants (FNPPs).

Specific features of small modular reactors

The specific features of SMRs that were considered by the SMR Regulators' Forum Working Groups have been grouped into four categories: 'facility size', 'use of novel technologies', 'modular design' and 'deployment'. These categories are not mutually exclusive. They simply provide a useful framework for identifying important SMR specific features.

The key SMR specific features and claims made by designers and vendors are listed below:

(a) Facility size

(1) Smaller plant footprint (as compared to a conventional nuclear power plant (NPP)).
(2) Smaller power of the core:
— Reduced decay heat load;
— Increased core stability (in some designs);
— Smaller inventory of radionuclides.

(b) Use of novel technologies

(1) Passive cooling mechanisms:
— Natural circulation;
— Gravity driven injection.
(2) Integral design (incorporation of primary system components into a single vessel).
(3) Non-traditional barriers to fission product release.
(4) Unique fuel designs (e.g. ceramic materials, molten salt fuel).
(5) Passive safety systems.

(c) Modular design:

(1) Compact and simplified designs:
— Practical elimination of some severe accident scenarios.
— Inherent safety features (e.g. longer grace periods).
— Fewer structures, systems, and components (SSCs):
    i. Elimination of some traditional initiating events (IEs).
— Introduction of new events:
    i. Internal to a single module;
    ii. Module to module interactions;
    iii. New construction techniques;
    iv. New human factors.
(2) Manufacturing, assembly and testing in factory.
(3) Multi-module facilities:
— Control room staffing;
— Sharing of SSCs among modules;
— Modules' dependence/independence;

— Multi-module failure in hazards conditions.

(d) Deployment (siting and transportation)

    (1) Siting:
       — On ground;
       — Underground;
       — Floating;
       — Underwater;
       — Movable;
       — In regions lacking essential infrastructure (e.g. electrical grid, cooling water).
    (2) Module transportation:
       — During construction;
       — During the operation of other modules;
       — For refuelling purposes in some designs.

All these features may influence the legal and regulatory aspects that are analysed in detail in Sections 2, 3 and 4.

1.4. STRUCTURE

This publication consists of four sections. Section 1 covers the background, objectives, scope, and structure of the publication. Section 2 elaborates on aspects related to legal framework, regulation and licensing. Section 3 discusses aspects related to safety design and analysis, and Section 4 describes other regulatory challenges in regulating SMRs. Finally, the annex documents each of the detailed responses to the questionnaire and follow up questions submitted by the Member State participants.

The regulatory expectations, practical examples/challenges and forward-looking approaches based on the responses provided by the Member State participants are described for each of the topics addressed in this publication.

2. ASPECTS RELATED TO LEGAL AND REGULATORY FRAMEWORK

The legal and regulatory frameworks for the licensing of nuclear installations generally vary across countries at both the fundamental principles level (i.e. performance based versus prescriptive) and in their specific goals, expectations and requirements.

It is widely acknowledged that the IAEA safety standards and the legal framework in Member States were developed and established in the context of deploying large pressurised water reactors (PWRs) as the dominant technology. As many countries, including embarking countries, consider the deployment of SMRs using innovative features, concepts, technology and deployment models, it is anticipated that there may be challenges and changes needed to the legal and regulatory frameworks to enable effective regulation.

In this section, the practical experience from Member State participants regarding the principles of their respective (1) legal framework, (2) regulations and guidance, (3) licensing process, and (4) regulatory approaches are discussed. Practical experiences and recommendations arising from a review of documents issued by the SMR Regulators' Forum Working Groups (WGs) are also discussed [1].

For each topic, an overview of the legal/regulatory expectations for Member State participants is presented, followed by a summary of the key practical experiences and challenges and how they were resolved. In addition, for each topic, a subsection titled 'Looking ahead' discusses proposed approaches to resolve the key challenges in regulating SMRs and the areas where further work is still needed from the perspective of the regulatory bodies. The full set of questions, follow up questions and the responses for each topic are documented in the Annex.

2.1. LEGAL FRAMEWORK

This subsection documents the key points from the responses to the questionnaire concerning the legal framework (Question 1): challenges/practical experience and forward-looking activities reported.

It is widely acknowledged that there are generally two broad types of approaches, with some variations, to regulate the safety and security of nuclear installations across Member States: a prescriptive approach based on legal requirements to comply with specific rules by means that are specified in laws and regulations, and goal-setting (non-prescriptive or performance based) legal frameworks in which the legal requirement is broadly set as 'safety goals' to be achieved by the applicant/licensee. The latter leaves room and options for the applicant/licensee to decide how to achieve the goals whilst nevertheless putting the responsibility on them to demonstrate that the goal has been achieved.

2.1.1. Regulatory expectations

Based upon the questionnaire responses received, regulatory bodies expect that the legal framework for nuclear installations will be adequate for the projects involving SMRs, or else expect that necessary modifications to the framework will be made in advance of the regulatory activities. These regulatory activities may involve the development and/or modification of regulations and guidance.

2.1.2. Practical experiences and challenges

The following considerations represent the key experiences/challenges at the legal framework level identified in the questionnaire responses:

(a) In general, Member State participants operating under goal-setting legal frameworks reported that no or very limited changes were needed to enable the regulation of SMRs. This is generally expected when the goals are expressed on a broadly technology-neutral basis and was reported to be the case in Canada and the United Kingdom (UK).

(b) Member State participants using rule-based frameworks and approaches to regulation (which are not technology neutral) reported that changes have been made or will be necessary to enable the regulation of new technology (high temperature gas cooled reactors (HTGR) and FNPPs). These changes were reported as needed at the secondary legislation levels such as regulations, requirements and standards. As an example, the Czech Republic responded that their legal framework contains a list of specific equipment and parameters (including maximum operating pressure and temperature) and a nominal diameter to be met. Another example from the Czech Republic is the legal provisions pertaining to the implementation of the defence in depth concept (DiD). For nuclear installations with a nuclear reactor, the function of physical safety barriers has to be ensured by independent SSCs.

(c)   Limited or no need for changes at the legal framework and regulations level were reported when the requirements and regulations already cover or were specifically developed to cover the technology under consideration (e.g. LWR technology when considering LWR-SMR projects) as reported by Argentina and the United States.

(d)   Member State participants have identified gaps or challenges to their existing legal frameworks when these were used to develop the regulation of a specific technology type (e.g. LWRs) and a broad set of novel technologies need to be considered. In this case, changes may be necessary in existing regulatory guidance to identify and resolve areas that may not be technology neutral (or to interpret the expectation in the context of new technologies). For example, in the Czech Republic the currently used technology is LWR, but a challenge could be presented if SMR designs based on different technologies (e.g. Generation IV technologies) are considered.

(e)   Member State participants have recognised that, in the area of nuclear security laws (and regulations), regulation has tended to be of a more prescriptive nature and revisions are underway to recognise that new technologies and approaches can offer opportunities to better protect nuclear reactors and nuclear security infrastructure against threats. Therefore, it appears that there has been a shift towards goal setting/performance-based and technology inclusive approaches. Goal setting, technology inclusive security expectations are applied, for example, in the UK, as documented in its Security Assessment Principles (SyAPs). The SyAPs provide the essential foundation for the introduction of outcome focussed regulation for all constituent security disciplines: physical, personnel, transport, and cyber security and information assurance. This regulatory philosophy is aligned with the UK non-prescriptive nuclear safety regime and provides applicant/licensees with a coherent regulatory approach across the UK civil nuclear industry.

Further information can be found in the Section A–1 of the Annex, in the responses to Question 1.

2.1.3. Looking Ahead

The review of information provided in the questionnaire responses identified the following considerations for future activities and initiatives at the legal framework level.

There is recognition that national legislative frameworks that are technology specific need to be adjusted to reflect distinct technologies and/or be formulated in a more technology neutral manner to facilitate the deployment of SMRs, for example:

(a)   In the consideration of new technologies (for instance HTGR) according to their laws, some Member State participants have invited public comments on the scientific and technical documents as part of the application process (as implemented in Japan).

(b)   Other Member States may wish to implement similar approaches that may involve changes at the legal framework level or the introduction of processes for public consultation.

(c)   Canada is using a risk informed approach to identify changes and maintain flexibility in the regulatory approach in a way that considers the potential radiological consequences and health impacts.

(d)   The regulatory approaches reported by the Russian Federation indicate the need for further work:

(i) In accordance with the regulatory approach adopted by the Russian Federation, the BREST-300 facility does not belong in the SMR class. Hence it is subject to the licensing process established for conventional NPPs.

(ii) The licensing procedure for the construction of a power unit with BREST-300 was not complete at the time of writing this TECDOC. BREST-300 is innovative in its design, but it has few features inherent in other designs of SMRs that are being considered around the world. The regulatory body is using the existing regulatory and legal framework to regulate the construction, considering the specifics of the fuel and coolant to be used.

See Section A–1 of the Annex (responses to Question 1) for more details.

2.2. REGULATIONS AND GUIDANCE

This subsection documents the key points from the questionnaire responses concerning regulations and guidance (Question 2): regulatory expectations, practical experiences/challenges and forward looking activities reported.

Requirement 32 of IAEA Safety Standards Series No. GSR Part 1 (Rev. 1), Governmental, Legal and Regulatory Framework for Safety [2] states:

"The regulatory body shall establish or adopt regulations and guides to specify the principles, requirements and associated criteria for safety upon which its regulatory judgements, decisions and actions are based."

Requirement 33 of GSR Part 1 (Rev. 1) [2] states:

"Regulations and guides shall be reviewed and revised as necessary to keep them up to date, with due consideration of relevant international safety standards and technical standards and of relevant experience gained."

Member State participants with mature and generally goal-setting regimes noted that regulations typically do not specify detailed criteria for use in assessing licence applications or judging compliance.

Member State participants with practical experience using the regulatory framework in the licensing of SMRs, e.g. HTGR, are involved in the development of hold-points, expectations and guidance to enable staged licensing submissions that would scrutinise an application's readiness for the next phase. This practice, in the case of South Africa, was linked to the evolution and development of the safety case by the applicant/licensee.

2.2.1. Regulatory expectations

Regulatory bodies understand that existing regulations and guidance applicable to nuclear reactors are also applicable to SMRs. However, for particular situations where modifications are necessary because regulations and guidance are not as technology neutral as they need to be, those changes are expected to be implemented in advance of the SMR projects.

2.2.2. Practical experiences and challenges

The following considerations represent the key experiences/challenges at the level of regulations and guidance that were identified in the questionnaire responses:

(a) Member State participants with largely technology-neutral and goal setting regulatory regimes generally concluded that there is limited need for changes or new regulations to regulate and license an SMR. However, additional guidance or interpretations may be needed to consider specific innovative features of the proposed SMRs. This was reported to be the case in Argentina, Canada and the UK.

(b) Regulatory bodies reported that one of the challenging aspects was the provision of guidance to the applicant and the designer on the processes that needed to be undertaken to demonstrate compliance with regulatory requirements as the review and interactions progressed. The need for this guidance was identified because of requests from industry or where the opinion of regulatory bodies was that further engagement was needed. Regulatory requirements and guidance ideally need to be in place in advance of a licence application to inform about the licensing process and the development of the safety case.

(c) Regulatory bodies also reported challenges associated with developing regulatory requirements for 'new' types of reactors when there is limited operating experience available and international safety standards have either not been developed or are in an early stage.

(d) Argentina reported that:

> (i) There is no need to change the regulations for the CAREM 25 reactor prototype. However, in parallel with the regulatory reviews, the regulatory body began a process for reviewing the country's regulatory requirements.
>
> (ii) To fill in gaps in the national requirements or guidelines, the regulatory body reported that international standards were used as a guide to expand the scope of the country's regulatory requirements.

(e) Canada and the UK reported that they periodically update guidance and regulatory documents to consider lessons learned, good practices and new knowledge. Generally, SMR proposals are expected to demonstrate, with suitable information, that they meet regulatory requirements and expectations. In these countries, the requirements and guidance for reactor facilities are generally articulated to be technology neutral and favour the use of a graded approach.

(f) China reported a number of challenges and areas of experience as follows:

> (i) Since all NPPs previously built in China were pressurized water reactors (PWRs), the regulatory requirements formulated by the regulatory body are mainly applicable to this type of reactor. Therefore, if an SMR uses PWR technology, such as the ACP100, most of the regulatory requirements are applicable.

(ii)   For FNPPs, although relevant regulations have not yet been modified, additional research has been carried out and changes might be proposed in the following aspects:

— Construction;
— Internal and external hazard;
— Safety systems and design extension conditions (DEC) for multi-unit NPPs;
— Reactor core control;
— Design for loss of off-site power;
— Air conditioning and ventilation systems, etc.

(iii)   The nuclear safety principles available were based on land-based thermal neutron reactors and, therefore, for SMRs such as FNPPs, the base regulations and accompanying guides needed to be improved or supplemented.

(iv)   The regulations established to define safety goals for SMRs provide for a higher level of protection of the public than those expected from a large NPP based on light water technology. The long term goal for nuclear safety as established in 2020 is to eliminate the possibility of a large radioactive release and therefore a probabilistic safety assessment (PSA) goal would have to be deleted.

(g)   The Czech Republic reported that if an SMR of a very distinct design were deployed in the country, a more detailed analysis of the secondary legislation would be necessary. However, since there are no plans for deployment of an SMR (of a particular design) in the Czech Republic, no specific challenges could be identified.

(h)   France reported that it has not identified any need to change their regulations as a result of a future project involving SMRs. However, to deal with specific issues, they could enact resolutions to enforce new requirements on licensees as this has been an approach that has been taken in the past. Regarding regulatory guides, the regulatory body has not identified any need for the creation of new guidance specific to SMRs.

(i)   Japan reported that it has introduced new regulatory requirements to address lessons learned from the Fukushima Daiichi NPP accident. Under such circumstances, the regulatory body have conducted a review of the High Temperature Engineering Test Reactor (HTTR), with is considered a research reactor. On reviewing HTTR, Japan reported the following challenges and measures:

(i)   In order to comply with the new regulatory requirements, portable power supply equipment has been installed in HTTR for monitoring of its shutdown state, in case of loss of the commercial or the emergency power supply, which is different from NPPs. In addition, HTTR is designed for passive decay heat dissipation (utilizing natural convection and radiation), therefore emergency power is not required for reactor cooling through active engineered means in a power outage scenario.

(ii)   Due to the inherent safety features of the HTTR design, the integrity of fuel assembly and coolant pressure boundaries can be maintained even when the cooling function (such as blowers, heat exchangers, etc.) is lost. For that reason, with regards to measures against internal fires, the design for fire protection is not necessarily based on typical measures or requirements stipulated in the regulations and guides for the following:

— The self-extinguishing nature of fire-retardant cables;
— Equipping different types of fire detection system;
— Activating stationary fire extinguishing equipment from the central control room;
— Separation of each fire area.

(iii) The regulatory body reviewed and confirmed that the required safety levels were met, considering the specific characteristics of the HTTR design.

(j) The Russian Federation reported that their regulations only cover FNPPs. For land based SMRs, the regulatory body is planning to carry out research to adapt the existing regulatory requirements and assess the need to develop new regulatory requirements for the case of serial production of modular reactors.

(k) South Africa reported the development of technology neutral safety requirements based on a graded approach according to the scale of the hazard, risk and defined dose criteria. These technology neutral requirements need to be underpinned by the development of many specific regulatory requirements and guidance documents to support them and enable the review of the safety case as presented by the applicant.

(l) UK report included in the item e, above.

(m) The United States of America (USA) reported the following initiatives and changes to regulations and associated guidance documents for the licensing of SMRs:

(i) Changes to address population-related siting considerations for advanced reactors;

(ii) Changes to address emergency planning issues for future SMRs and other new technologies, including non-LWRs;

(iii) A potential inequity between the insurance requirements for facilities of different sizes was identified;

(iv) A limited-scope revision to the regulations and guidance related to physical security for advanced reactors;

(v) Three potential licensing structure alternatives for multi-module facilities were reviewed and it was determined that the licensing of each module individually was preferred;

(vi) An equitable assessment of annual fees for SMRs was proposed;

(vii) The criteria to ensure appropriate treatment of important risk insights related to multi-module design and operation were documented;

(viii) Changes to allow applicants to submit a site-specific estimate of decommissioning costs with a supporting analysis and adequate justification for an exemption to the minimum funding requirements for large LWRs in the NRC regulations.

(n) The SMR Regulators' Forum considers the following challenge:

As the concept of SMR 'module' is not equivalent to the 'unit' or 'plant' concept for large reactors, the safety principles developed for the 'multi-units' issue cannot be transposed to 'multi-modules' in SMR facilities. Therefore, the forum recommends that principles and requirements for the safety assessment of a 'multi-module' SMR need to be developed. For additional challenges/considerations by the SMR Regulators Forum please refer to [1], where the reports can be found.

Further information can be found in Section A–2 of the Annex, in the responses to Question 2.

2.2.3. Looking ahead

The review of information provided in the questionnaire responses identified the following considerations for future activities and initiatives in regulatory expectations and guidance:

(a) Some Member State participants operating in technology neutral, goal-setting regulatory regimes reported that they are proactively undertaking a review of existent guidance for compatibility with SMRs. This aims to address lessons learned from the experiences of other Member State participants who developed processes, guidance and requirements in parallel with licensing and found difficulties.

(b) There is generally an expectation and approach to develop and implement relevant guidance to the assessment and licensing of SMRs by seeking references from (and contributing to) IAEA safety standards and guidance development activities. Some Member State participants have, for example, referred to the work undertaken in relation to the applicability of the requirements of IAEA Safety Standards Series no. SSR-2/1 (Rev. 1), Safety of Nuclear Power Plants: Design [3] to ensure compatibility with HTGRs.

(c) The Czech Republic reported that their secondary legislation is relatively specific and prescriptive: it reflects currently used technology and therefore is not technology neutral. It is acknowledged that changes would be needed if it is to reflect distinct technologies and facilitate their deployment, but they await the specific technological solution that would be deployed prior to engaging in changes of such legislation.

2.3. LICENSING PROCESS

This subsection documents the key points from the questionnaire responses concerning the licensing process (Question 3): regulatory expectations, practical experience/challenges and forward-looking activities are reported.

The Member State arrangements for the authorization process is the principal means by which regulatory bodies apply the legal and regulatory framework to the applicant/licensee willing to deploy and operate a nuclear reactor. Requirement 23 of GSR Part 1 (Rev. 1) [2] requires that these processes are documented. The processes can provide for the granting of instruments such as licences, registrations, permits, certificates, written statements of no objection, letters/memoranda or other documents that provide written permission for a licensee to proceed beyond a hold-point or for agreeing to the activity proposed by the licensee.

It is generally the case that prior to granting an authorization, the regulatory bodies require the applicant/licensee to submit a detailed demonstration that the proposed facilities will operate within safety bounds, meet regulatory criteria of safety and security and environmental protection goals. This is generally reflected in Requirement 24 of GSR Part 1 (Rev.1) [2]. The regulatory bodies, usually by assessment, consider whether the safety and security of the activity will be assured, and whether the safety objectives, principles and regulatory criteria will be satisfied for the lifetime of the facility. The regulatory bodies' process and assessment are usually supported by guidance on the format and content of the documents and information to be submitted.

The SMR Regulators Forum WGs have reported key considerations in the lifecycle steps for the licensing of SMRs versus large NPPs, and these are introduced here for context.

Figure 1 shows the high-level stages of activities in the authorization process of an NPP, as defined in IAEA Safety Standards Series No. SSG-12, Licensing Process for Nuclear Installations [4].

In Figure 1, the arrows represent potential hold points (or key regulatory interventions) along the process (for illustrative purposes only).

In addition to novel design features and approaches, SMR projects may introduce several differences to a new-build project, ranging from factory manufacturing and testing tasks, to new construction and commissioning methods, and to new programmes for long term operation and maintenance. These, in turn, may impact the potential stages (as defined in SSG-12 [4]) for SMR authorization process.

Figure 2 shows the potential stages for the lifecycle of an SMR. In Figure 2, the arrows represent potential hold points (or key regulatory interventions) along the process (for illustrative purposes only).

Figure 2 shows that there could be two new stages: manufacturing and transport. All stages could introduce changes in the associated activities that may introduce safety and regulatory challenges and an increasing need for regulatory activities and oversight as discussed in detail below.

Additionally, the responsibilities and activities of the applicant/licensee increase in the above context. The applicant/licensee needs to have influence over the design and procurement of an SMR so as to ensure safety, including those aspects of safety ensured by design and quality insurance in the period of first supply and assembly.

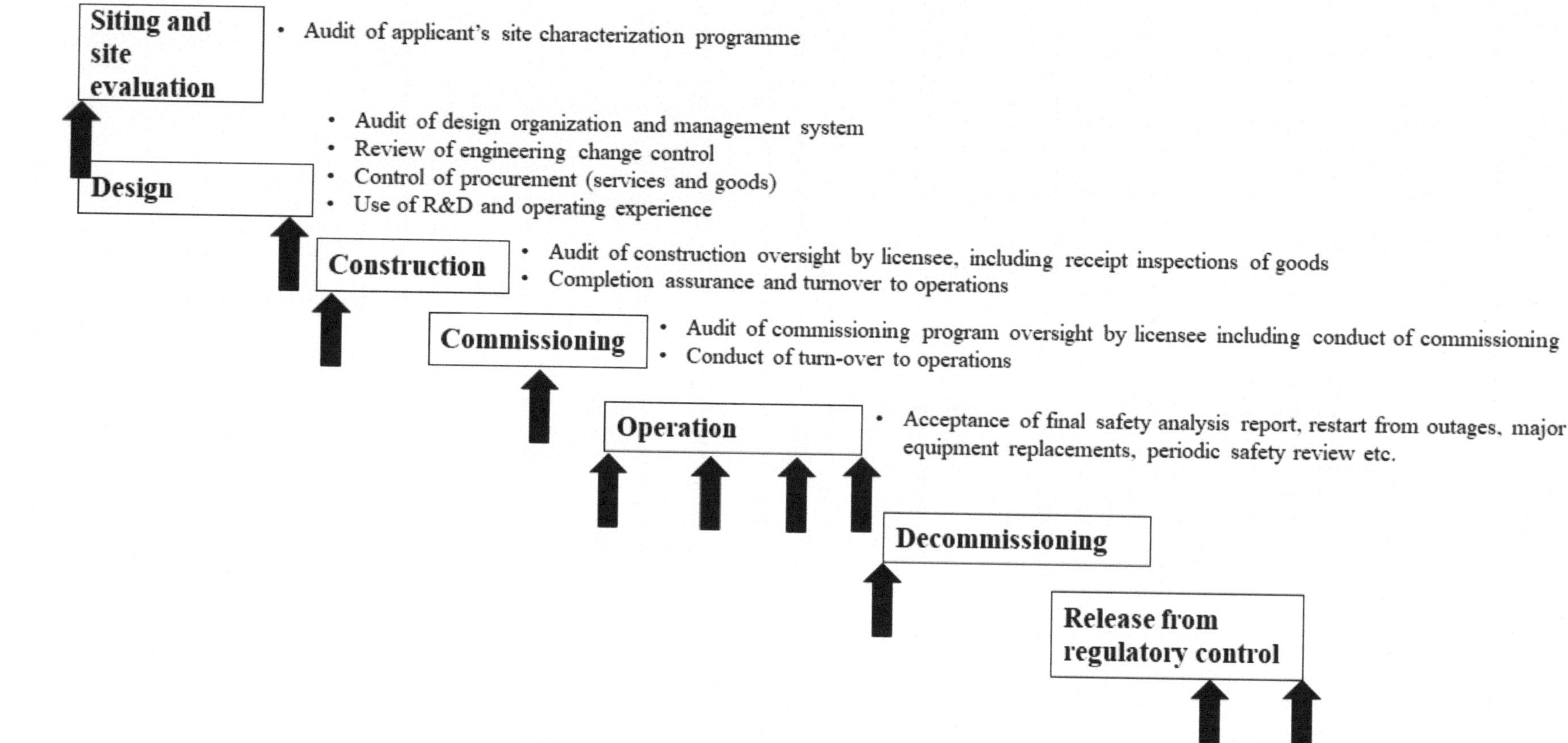

FIG. 1. High-level stages of activities in the authorization process of an NPP (adapted from SSG-12 [4]). The arrows represent potential hold points (or key regulatory interventions) along the process (for illustrative purposes only).

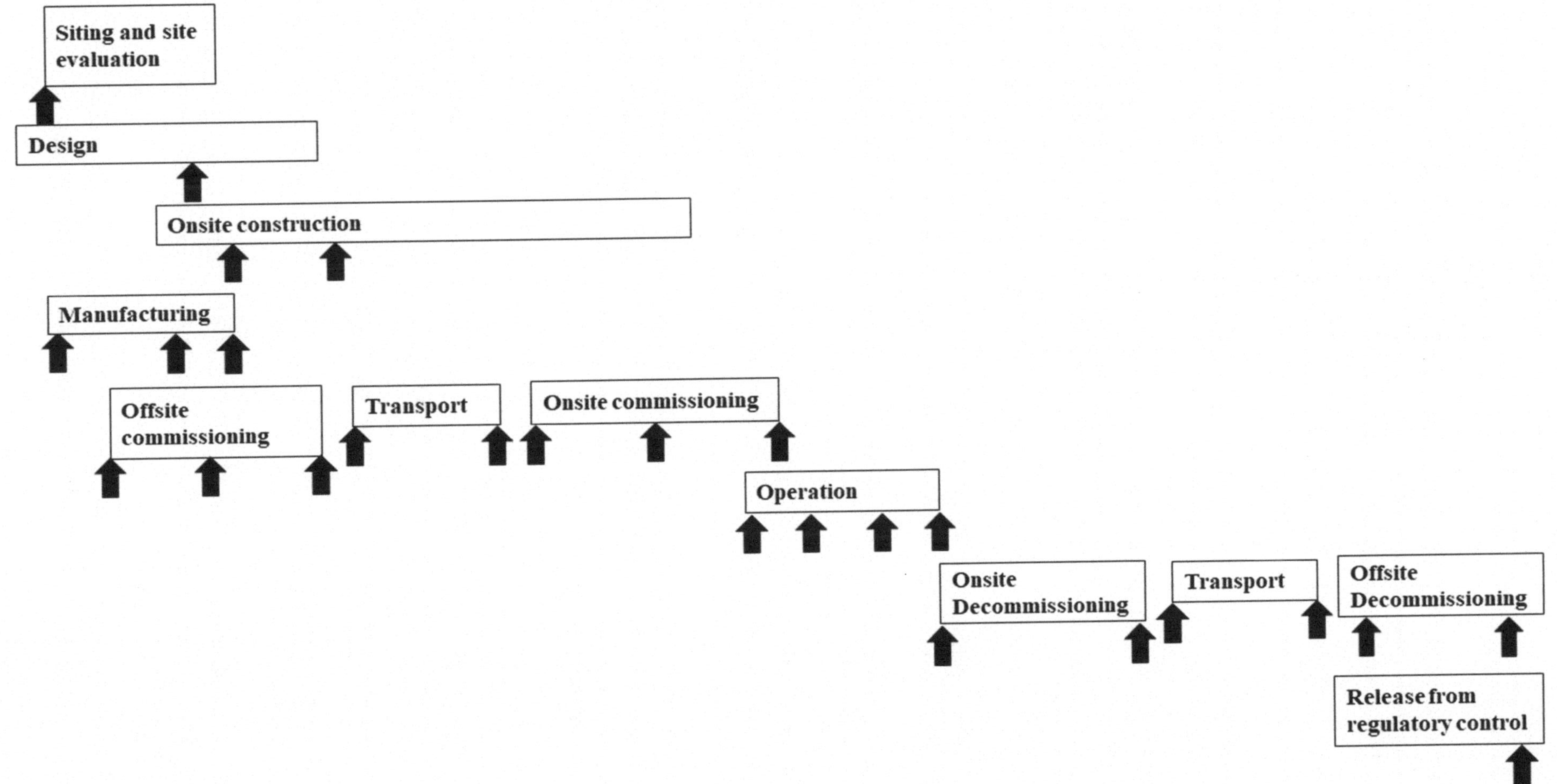

FIG. 2. Potential stages for the lifecycle of an SMR]). The arrows represent potential hold points (or key regulatory interventions) along the process (for illustrative purposes only).

(a)   Siting and site evaluation

All activities associated with the SMR proposal, including the impact of construction and operation of multiple modules (or units) on a single site, have to be considered in the licence application. This would be particularly important for cases such as:

(i)   Multiple-module SMRs where an applicant proposes to place only a few modules into service to begin with, with an option to install and operate further units in the future.

(ii)   Spent modules that may be removed and replaced with newer modules, which could differ technically from the original unit.

The impact of adjacent units planned on a site, along with the proximity to population has to be also considered. In some cases, more than one licensee may be present on a site, and any possible interactions would need to be considered by all current or potential licensees.

Similar considerations apply to FNPPs (for the facility where the reactor is fuelled).

Considering SMR specific features (see Section 1.3.1), the selected site characteristics could be an important challenge with regard to the implementation of DiD.

The design needs to take due account of site-specific conditions to determine the maximum delay time by which off-site services need to be available.

Siting aspects may have an important influence on SMR safety design and different DiD levels due to the possible range of suitable sites for SMR installations, including underground, underwater or floating on water.

New site configurations may involve the evaluation of vulnerability to additional specific external hazards and environmental phenomena. For multi-unit/module plant sites, designs have to take due account of the potential for specific hazards giving rise to simultaneous impacts on several units/modules on the site.

(b)   Design

There are no fundamental changes in the design review process for an SMR compared to a large-scale NPP design. However, due consideration is expected to be given during the review of a first-of-a-kind (FOAK) design since this type of facility will differ in the type of evidence and operating experience available to support the safety case.

In addition, regulatory bodies may need to adopt new guidelines/approaches adapted to SMRs in order to meet the underlying regulatory requirements.

Another challenge that arises with SMRs is the level of maturity of design organizations, some of them being industry newcomers with little or no experience in nuclear safety, nuclear security or safeguards.

Regulatory bodies also need to seek assurance that processes are in place to ensure efficient and effective knowledge transfer from the designer to the licence applicant.

The trend towards global standardization and certification of SMR designs desired by some designers may be challenging for current licensees and regulatory bodies as it may require significant changes in the national licensing process.

Finally, the design of onsite supporting systems (especially if they are shared in multi modules) needs to be evaluated in terms of overall plant safety.

(c)   Manufacturing

The goal of many SMR designs is to manufacture SMR modules off the site and then transport them to a site for installation and use.

The engineered modules could be manufactured serially in a controlled factory environment. The premise is that factory manufacturing results in high-quality construction, short manufacturing times, and economies of scale.

These engineered modules would be delivered from production factories to be assembled on the deployment site, with the assumption that construction time would be significantly reduced. Significant additional issues arise in the case of transportable SMRs, e.g. safeguards issues.

It is also claimed that some of the commissioning work could be done during manufacturing, reducing the onsite time to bring the plant to commercial operation. This concept has been proven in the shipbuilding and aerospace industries. Traditional reactor construction has already utilized this approach; however, some proposed SMR designs would use it on a wider scale, with some proposing to build, fuel and commission reactors before delivering them to site.

With the manufacturing model described above, there are three major differences compared to traditional reactors in that:

(i)   Assembly is mostly performed at the factory;

(ii)   Manufacturing and assembly may take place before the future licensee has decided to build the facility, i.e. prior to the beginning of any licensing process;

(iii) Manufacturing processes, if not implemented correctly for safety-significant modules, could result in latent safety issues.

Configuration management and stability needs to be verified from the FOAK manufactured SMR to the 'nth-of-a-kind' (NOAK) manufactured SMR (including situations that involve changes in manufacturing facilities or vendors).

SMR manufacturers will be expected to demonstrate the capacity and capability to address all safety requirements.

Site construction and commissioning of SMRs is a licensee activity. Licensees need to exercise oversight of in-factory manufacturing and testing to achieve an assembled SMR that is safe and meets all regulatory requirements.

There is a need to establish regulatory oversight for safety systems and components important to safety that are built and assembled in a factory, and the availability of onsite system inspections needs to be considered. The scope of regulatory oversight may be limited to the licensee's procurement process for systems that are easily verified after onsite installation.

Additional consideration is expected to be given to a manufacturing facility involved in fuel loading; these considerations have to recognize safety, nuclear security and safeguards aspects. Factory fuelled and sealed transportable reactor modules represent a unique issue to regulation that will need further discussion about the role of the 'factory' licensee versus the 'site' licensee during the manufacturing, testing, delivery/installation and commissioning phase. Some questions to be addressed include:

(i)   When the module is being assembled (and possibly tested) at the factory, what is the role of the deployment site licensee?

(ii)   The factory would need an operating licence to load fuel into each reactor module, perform any testing and store the module prior to deployment in a guaranteed shutdown state. The operating licence for such activities would likely begin with the requirement applicable to NPPs (and a safety case) but a graded approach is likely to be applied commensurate with the scope of activities. When construction of site structures is in progress under a construction licence, it is for the purpose of future installation and operation of the reactor module. What is therefore the role of the site licensee in the factory-based activities involving the reactor? Is any factory testing part of commissioning?

Finally, an additional challenge arises for components or modules that have been manufactured abroad. Regulatory bodies may consider developing processes for approval of components whose manufacturing they (or the applicant/licensee) have not been able to oversee. It is likely to be the case that regulatory bodies will not be able to inspect factories based outside their jurisdiction unless an existing or prospective licensee within the jurisdiction of the regulatory body is in place and facilitates that regulatory oversight abroad. It is also likely that export controls of intellectual property across jurisdictions may pose barriers to the transfer of knowledge needed to achieve regulatory assurance.

(d)   Construction

Construction time is expected to be shorter for SMRs than for traditional NPPs. This is due to their smaller footprint, and the possibility that many key components might be manufactured offsite and then transported to the site for final inspection and installation (greater use of modularization).

For sites with multiple modules, the simultaneous construction of modules in parallel with operation of other modules needs to be considered. Any construction activity could pose an additional hazard to existing units.

(e)   Off-site commissioning

For some SMRs, off-site commissioning consists of the commissioning tests that are performed on a module (or other equipment) before it leaves the off-site assembly facility.

Some off-site commissioning might also represent the last opportunity a licensee or regulatory body has to inspect some portions of a module. Off-site commissioning can replace some onsite commissioning tests.

Off-site commissioning plans have also to take into consideration possible damage during transportation, and the time between testing and module use. If there is a prolonged period between a piece of equipment or module being tested and its use, an appropriate asset care programme is expected to be developed and implemented.

Regulatory bodies may expect the applicant's personnel to conduct or supervise off-site commissioning to ensure that appropriate commissioning standards are being adhered to, hence ensuring proper transfer of knowledge and responsibility to the applicant.

As for manufacturing, additional challenges may arise when off-site commissioning has been completed earlier, meaning that the regulatory body and the applicant would not have had the opportunity to observe the tests.

The licensee needs to demonstrate how the commissioning programme takes into account any uncertainties due to the lack of operating experience.

The licensee will need to justify the representativeness of full-scale replica tests results and first-plant-only-tests (FPOT) if wanting to take credit for those tests in the commissioning phase and detail the commissioning tests to be performed on the licensed plant to check their full applicability.

(f)   Transportation

Some SMR concepts consider using a compact nuclear core vessel that would either be entirely replaceable or that would have its entire fuel inventory replaced in a manner like a fuel cartridge.

Using this approach, licensees intend to reduce or even eliminate lengthy refuelling operations at the deployment site and possibly facilitate quicker removal from the deployment site. The spent fuel inventory might then either be stored onsite or shipped to another location for refurbishment or disposal. Transporting reactor vessels is especially challenging, as there is no certified packaging that is large enough for most (or all) reactor cores.

For regulating fuelling at the factory and transport of reactor modules that contain fuel, it is recognized that many safety, nuclear security, safeguards and legal challenges arise.

(g)   On-site commissioning

Continuity from off-site commissioning to onsite commissioning needs to be ensured.

Integration testing of all modules and systems also has to be considered.

With on-site commissioning, difficulties may be introduced as new modules are added. If multiple units/modules are shared in one facility or some units/modules will be added later on, the following challenges are likely to arise:

(i) There will be shared SSCs that may necessitate certain commissioning activities to take place as the first modules are installed and placed into service.

(ii) Due consideration needs to be given to the performance of shared systems when adding units or modules and whether additional or new or repeated commissioning tests may be needed (a shared plant HVAC system, for example, is important to environmental qualification).

(iii) Commissioning may have the objective to demonstrate/verify the compatibility with the existing plant.

(h) Operation

Some SMR designs propose to have multiple smaller reactors operating on a single site. The multiple reactor modules may have services that are shared between modules, such as common electrical systems, compressed air systems or civil structures. For facilities with multiple modules, additional consideration needs to be given to the impact of activities involving each module on the operation of the other modules.

When licensing an SMR site or facility, regulatory bodies also need to consider:

(i) That some novel designs may need additional regulatory controls for operation;

(ii) That many operating concepts can be different from traditional reactors:

— Remotely operated facilities with no operators on the site;
— Multiple modules operated from a common control room by the same operating personnel;
— Different companies undertaking different actions (refuelling, maintenance).

(iii) The security arrangements of remote sites;

(iv) The accident response needed at remote sites;

(v) That there may be multiple operating organizations on one site;

(vi) The length of the operating licence, and the interval between periodic safety reviews.

(i) On-site decommissioning

Some SMR facilities may have plans for sequenced decommissioning, i.e. some modules may still be operating while some are decommissioned. This could lead to decommissioning personnel working near operating modules. Safety and nuclear security issues are expected to be considered under these circumstances.

(j) Off-site decommissioning

Unique regulatory perspectives for off-site commissioning can include disposal considerations for unconventional fuels, and various reuse or refurbishment possibilities for modules. For example, a reactor module may just need to be refuelled and key components inspected before redeploying at the same or different site where it came from.

Decommissioning in an off-site facility would likely be more controlled than traditional onsite decommissioning activities.

(k)   Release from regulatory control

This stage is unlikely to present any significant difference from traditional NPPs.

The above themes touch upon numerous considerations and challenges that are discussed in Sections 3 and 4.

2.3.1. Regulatory expectations

Considering the questionnaire responses received, regulatory bodies generally expect that the licensing process for NPPs has to be adequate for projects involving SMRs, or that the necessary modifications to the licensing process have to be made in advance of the regulatory activities.

2.3.2. Practical experiences and challenges

The key practical experience and challenges in applying regulatory processes to the licensing of SMRs that were reported in the questionnaire responses were as follows:

(a)   The absence of a specific licensing process that has been tested or applied to SMRs has been considered as a key challenge by Member State participants, therefore resulting in the need to develop and implement a process once the license application is received and begins to be scrutinised by the regulatory body. The licensing process needs to be sufficiently flexible for developmental projects as was the case with the Pebble Bed Modular Reactor (PBMR) project in South Africa.

(b)   Member State participants currently constructing SMRs, such as the CAREM 25 prototype in Argentina, reported changes to the licensing process, mainly at the beginning of the construction stage. These were due to the reactor being considered as a prototype. The licensing scheme included the hold-point for the start of construction and, in order to issue the construction authorization, the regulatory body established additional mandatory documentation requirements compared to the traditional NPP licensing. The fulfilment of these regulatory requirements conditioned the beginning of the construction of the nuclear module of the reactor, for example, on the resolution of findings of the safety analysis report. The findings were related to engineering, and which have an impact on civil works, whether due to structural functions, confinement or shielding.

(c)   Some regulatory bodies offer a pre-licensing process such as GDA in the UK, VDR in Canada, and the Memorandum of Understanding (MOU) in Argentina and China. These generally intend to increase flexibility and better adapt to the differing levels of maturity and development of SMR vendors and their technologies, whilst remaining consistent with previous approaches.

   (i)   The UK, for example, allows applicants to make use of existing submissions (e.g. to other regulatory bodies) – supplemented to meet UK expectations where necessary. There is greater emphasis on early engagement and agreement of scope/submissions throughout the process. The UK reports that a new key feature was the introduction of additional outputs (GDA statements) to show that the design (or a meaningful assessment scope from the design) has shown alignment with UK regulatory expectations through the various stages. The acceptability of partial scopes and new GDA statement outputs is summarised in the Annex. The UK also reported that a new suite of guidance for RPs including detailed Technical Guidance [5] with key expectations and lessons learnt were published in 2019.

(ii)   Canada reported flexibility in pre-licensing activities. VDRs can take place in parallel with a licence application; for example, a vendor may decide to engage in the VDR Phase 2 or 3 in parallel with the regulatory body's review of an application for a licence to prepare the site.

(iii)   Argentina reported the establishment in the MOU of the regulatory requirements and expectations in terms of licensing process and safety level that has to be fulfilled by the design of the proposed plant and demonstrated through the safety analysis to be further submitted to the regulatory body.

(iv)   China reported that the regulatory body has carried out pre-licensing activities with vendors before the formal licensing process. In accordance with the design characteristics of SMRs, the reviewers will intervene in the safety review of SMR in advance and carry out corresponding technical exchanges with designers.

(d)   Member State participants licensing FNPPs, such as China, reported the need for adjustments to the authorization process as described earlier. Others, such as the Russian Federation, reported experiences in the licensing of an FNPP, in cooperation with the maritime regulatory body, and which involve a transfer of responsibility for the safety of vessels when the acceptance documents for the vessel with a nuclear installation are signed. There is therefore a permit for building and commissioning ships and other vessels as nuclear installations which is received by the shipbuilding organizations, and then followed by a permit to operate which is received by the operating organizations. In the Russian Federation, the conduct of comprehensive tests to confirm the declared characteristics, and the signing of a state acceptance report of the facility is needed prior to issuing the operation license. The change in licensee is overseen as follows:

(i)   The licensee submits a set of documents for obtaining a licence to operate a nuclear installation – a vessel with a nuclear reactor(s).

(ii)   The set of documents includes a safety analysis report developed by the designers of the reactor and the vessel and adjusted with due regard for the changes in the initial design of the vessel made as a result of its construction. Thus, a safety review of the nuclear installation is carried out for the licensee, but on the basis of the documents of the parent design organization and the reactor designer.

(iii)   The safety review has to be completed before the signing of the state acceptance report of the facility. If the safety review reveals significant inconsistencies with the requirements of federal regulations and rules in the field of the use of atomic energy, which leads to the refusal to grant the licensee a license to operate the vessel, the state acceptance report is not signed, the commissioning stage is not considered completed, and the licensee does not become responsible for the safety of the facility.

(e)   China reported the following practical experiences:

    (i)   Designers and vendors were encouraged to contact and communicate safety design and safety regulatory issues with the regulatory body before the license application, especially for innovative designs.

    (ii)   The licensing process for FNPPs built and loaded in the factory was adjusted. If the factory has already built similar FNPPs, it can directly apply for a construction permit, instead of having to first conduct a site safety assessment.

(f)   In France, the licensing process establishes that the set-up of multi-units can be authorized by the same approval if they are operated by the same licensee, on the same site. Firstly, before requesting an authorization to set up a nuclear installation, an applicant/licensee can ask to the regulatory body the opinion about its project's safety options. This step is not mandatory.

(g)   Japan reported that to install and operate a new reactor, it is necessary to (1) obtain the permit for reactor installation ('Reactor Installation Permit') for the specific design; (2) obtain the approval of plan for constructing the specific design ('Construction Plan') and (3) carry out construction work; and finally, (4) obtain the approval of Operational Safety Programmes prior to start of operation. The regulatory body noted that it had invited public comment from a scientific and technical view on the draft review documents. This was reported as the first time in which public comments were invited for research reactor (as SMRs are designated in Japan) plans.

(h)   The Russian Federation reported no experience in licensing land based SMRs. However, it reported that the existing licensing process for the NPPs could be applied to land based SMRs.

(i)   South Africa reported the following practical experiences:

    (i)   The differences between the regulatory approach applied to the licensing of a twin unit large NPP based on PWR technology as a turnkey project, compared to a developmental project like the PBMR (and HTGR SMR) for which the licensing basis (and therefore) the safety case was in development, have been considered. The regulatory body reported that there was no broad consensus on general design criteria and design rules for HTGRs at that time, and a key lesson learned was to concentrate efforts on this upfront, and to include the designers, applicant and the regulatory body.

    (ii)   Effort was dedicated to the development and implementation of a process for the structured development and assessment of the safety cases, taking into account the limited design criteria and rules. This provides clarity of expectations and a logical link between the various steps of the design process, the safety assessment and the development of operational support programme.

    (iii)   Engagement with designers is a key point (and a framework that allows for direct engagement with the designer/architect engineer with involvement of a potential client and eventual licensee needs to be in place). This is akin to the necessary transfer of knowledge and safety case between the applicant/licensee in the UK's generic design assessment (GDA) process and Canada's approach to vendor design reviews (VDR) and licensing.

(iv) The expectations for engagement with designers involve the identification and agreement on key safety issues early so that the focus is on the proposed technical resolution that is adequately developed with a stable design before development of the safety case and licensing engagements.

(j) In the USA, an application for a design certification, combined licence, design approval, or manufacturing licence, respectively, must include the principal design criteria for a proposed facility, whether it is a LWR or a SMR. The principal design criteria establish the necessary design, fabrication, construction, testing, and performance requirements for SSCs important to safety; that is, SSCs that provide reasonable assurance that the facility can be operated without undue risk to the health and safety of the public. The current General Design Criteria (GDC) establish minimum requirements for the principal design criteria for water-cooled NPPs similar in design and location to plants for which construction permits have been issued by the US Nuclear Regulatory Commission. The General Design Criteria are also considered to be generally applicable to other types of nuclear power units and are intended to provide guidance in establishing the principal design criteria for such other units. Many of the existing GDC are applicable to varying degrees to SMRs currently under review by the US Nuclear Regulatory Commission.

The SMR Regulators Forum WGs also provided practical experience/challenges for licensing of multi-unit's sites:

(a) It will be a challenge for one authorized nuclear operating organization to operate multiple-module/multiple-unit facilities. This would include, for example, an ageing management programme for 'common services' features that are shared between modules – including civil structures, common electrical systems, and compressed air systems.

(b) For a proposal for a multiple-module licence to construct or operate a facility, it is important for the applicant to consider the facility's ultimate total capacity over its life and the timelines for deploying the modules. This will affect, for example the environmental assessment (study of potential adverse impacts to the environment) as well as the safety analyses that will support the facility's safety case. In the license application, the regulatory body expects the applicant's programmes and processes to describe how multiple-unit activities will be managed under all safety and control areas.

(c) The whole-site risk is not expressed as a single number but rather as an informed judgement based on a broad range of qualitative and quantitative information.

(d) It needs to be noted that PSA methodologies and metrics may need some development for application to SMRs, due to novel designs.

(e) Although the current regulatory experience is relevant and applicable to multi-unit/multi-module SMR facilities, the novelty of most SMR designs including their deployment strategy (e.g. replaceable reactor modules, different reactor designs on the same site, multiple reactors operated by one operating organization), substantiation of passive and inherent safety features, quality management system and the supply chain control for multiple design developer may pose additional challenge for future regulatory reviews and licensing.

The above challenges are also addressed in Sections 3 and 4 where the relevant questionnaire topics are covered.

Further information can be found in Section A–3 of the Annex, in the responses to Question 3.

2.3.3. Looking ahead

The following activities and prospective changes to the assessment and licensing processes for SMRs were reported:

(a) Most Member State participants operating in technology-neutral, goal-setting regulatory regimes, reported that they have not been engaged in changing the licensing or regulatory permitting work associated with the deployment of an SMR at the time of receiving the questionnaire, but it could be a near future task.

(b) Other Member State participants, such as the USA, reported that the licensing process addressing SMRs has to be adjusted as needed, but always taking into account the core specific elements already existing in their licensing process, including environmental assessments, public engagement, etc.

(c) Pre-licensing activities are considered important for some of the Member State participants in the licensing process of SMRs.

2.4. REGULATORY APPROACH

This subsection documents the key points from the questionnaire responses concerning the regulatory approach (Question 4): practical experience/challenges and forward-looking activities.

Review and assessment activities are undertaken by regulatory bodies to determine whether the applicant's safety cases adequately demonstrate the safety of the facility and that the regulatory body's safety objectives, principles and criteria have been satisfied. The overall objectives from these activities generally entail ensuring that:

(a) The available information provided by the applicant demonstrates the safety of the proposed activity;

(b) The information contained in the safety case is accurate and sufficient to enable confirmation of compliance with the expectations and requirements of the regulatory body;

(c) The technical solutions, in particular novel ones, have been proven or qualified by experience or testing or both, and are capable of achieving the required level of safety.

Effective regulatory bodies define and document the processes for reviewing and assessing safety cases and manage the process to ensure that the review and assessment is completed prior to issuing an authorization. The questionnaire responses documented in Section A–4 of the Annex provide an overview of the processes and assessments undertaken by each of the Member State participants, including the principles, criteria and guidance used to make judgements and decisions, as well as the possible format and content of the documents and information to be submitted by the applicant.

2.4.1. Regulatory expectations

Considering the questionnaire responses received, regulatory bodies generally expect that the applicant/licensee will submit a detailed demonstration that the proposed SMR facility will operate within safety bounds, and meet regulatory criteria of safety, nuclear security, safeguards and environmental protection goals.

The regulatory bodies' assessment process is expected to be documented, for example in regulatory guidance and in the management system of the regulatory body. This may include the regulatory expectations on the content and quality of the documents and information to be submitted.

2.4.2. Practical experiences and challenges

The key practical experience/challenges with regard to regulatory assessment reported in the questionnaire responses were as follows:

(a) Member State participants reported limited changes to the approach for reviewing documentation as part of the authorization process. This was, for example, the case of Japan and Argentina. In the case of Japan, the changes reported were limited to the regulatory requirements introduced after the Fukushima Daiichi NPP accident. Argentina emphasized the benefit of early engagement in the goal-setting regulatory approach, as effective interactions between the regulatory body and the applicant allowed the early detection of misalignments with regulatory requirements avoiding issues in later phases (manufacturing or construction). Similar emphasis on early engagement is reported by Canada, South Africa and the UK.

(b) In the context of the PBMR, South Africa emphasised the importance of independent safety assessments to ensure that the safety case submitted by the applicant complies with the licensing requirements. South Africa reported that the traditional safety assessment process was adapted to take into account the developmental nature of the project without reducing the margin of safety required for a new design. Particular attention was given to events that were unique to the design to ensure that the amount of margin against safety limits was clearly understood. It recommended that representative bounding cases were selected, to ensure that the most severe consequences are covered by the analyses and emphasis was placed on the calculation routes as the goals to be met included amongst others numeric targets.

(c) Canada reported changes and enhancements of the review reports that document the results from carrying out the individual work instructions to provide reviewers with a single document that contains or references practically all information relevant to their review. This was reported to benefit future licence application assessments as early identification of how and where the submitted documentations during the design review have changed since they were submitted allows for a targeted review of the changes instead of the entire submission. A similar approach is reported in the transition between the GDA safety case and site-specific licensing in the UK.

(d) In Japan's legislation the definition for research and test reactors is 'a reactor that is not categorized as a power reactor'. As the Japanese HTTR is for test and research purposes and does not have a power generation function, it is categorized and regulated as research and test reactor accordingly.

(e) In the Russian Federation, the regulatory approach to the review and assessment process for floating SMRs was reported as not needing any changes. It was also reported:

(i)   The main difference between the floating SMRs and land-based ones was the absence of a separate 'siting' stage. As a result, there was no need to obtain a siting license for an atomic energy facility. The assessment of the possibility of operating a vessel at a specific site can be carried out both at the stage of its construction and at the stage of operation and performed in each specific case. At the same time, in accordance with the existing regulations, the design of a floating SMR needs to formulate the requirements that the site has to satisfy.

(ii)   During the operation of a floating SMR, a vessel may occupy several sites; the list of seaports of the Russian Federation where vessels and other floating crafts with nuclear power installations are allowed is regulated by the Government.

(iii) In addition, the set of documents to demonstrate nuclear safety and radiation protection during the operation of nuclear installations of vessels and other floating equipment is expected to contain a certificate of classification and examination of the vessel in the Russian Maritime Register of Shipping. Also, when licensing floating plants, unlike stationary installations, the set of demonstrating documents does not include the results of observations of buildings and structures.

(iv)   The review processes are the same for all nuclear installations that, according to the IAEA terminology, can be classified as SMRs.

(f) The UK reported no changes to the approach and that regulatory capability (training and knowledge management), research activities and proactive review of guidance (safety assessment principles (SAPs) and technical assessment guides (TAGs)) were some of the activities the regulatory body embarked upon early on to ensure that it was ready to regulate innovative designs and technology. These activities are covered in separate questions. Regarding whether 'additional' analysis, checks are expected, the overarching regulatory requirement is for the applicant/licensee to demonstrate that the risks have been reduced to as low as reasonably practicable (ALARP), considering engineering, operations and the management of safety. Therefore, the regulatory approach has not changed.

(g) France and the USA reported no changes in their regulatory approach for the introduction of SMRs.

(h) China reported that at present, the SAR review of the floating reactor is being carried out, and some specific issues encountered in the review process will be the subject of special discussions. The problems and experiences in the SMR review will be collected and summarized to support the further review process, and relevant regulations, guidelines and technical documents will be modified accordingly. The documents under consideration include the safety regulatory approach, emergency management and security requirements.

Further information can be found in Section A–4 of the Annex, in the responses to Question 4.

2.4.3. Looking ahead

Some Member State participants currently in the early phases of assessing license applications or not yet assessing licensing applications, safety cases or SMR designs, generally report a drive to undertake readiness activities, proactively reviewing guidance, processes, and approaches. For example:

(a)   Canada reported that work is ongoing to enhance the collective understanding of regulatory requirements and guidance regarding management systems and of an 'intelligent customer' — particularly in regard to the capabilities of an applicant with regards to scrutinizing a technology vendor or the applicant's contractors and sub-contractors.

(b)   As reported in Section 2.3 (Question 3), the UK has undertaken GDA modernization processes and issued guidance to increase flexibility and better adapt to the differing levels of maturity and development of SMR vendors and their technologies, whilst remaining consistent with previous approaches. This included the production of technical guidance across all technical disciplines in the context of previous GDAs, as well as introducing interim regulatory assessment statements providing clarity on the level of alignment with regulatory expectations as the design assessment progresses through the steps. The UK has engaged and continues to engage with reactor vendors to communicate relevant regulatory expectations and learn about their design proposals ahead or formal regulatory activities.

3. ASPECTS RELATED TO SAFETY DESIGN AND ANALYSIS

As part of the regulatory process, the regulatory body needs to review and assess the installation's design and analyse the documentation provided by the applicant/licensee. Paragraph 2.9 of SSR-2/1 (Rev. 1) [3] states:

"…a comprehensive safety assessment of the design is required to be carried out. Its objective is to identify all possible sources of radiation and to evaluate possible doses that could be received by workers at the installation and by members of the public, as well as possible effects on the environment, as a result of operation of the plant. The safety assessment is required in order to examine: (i) normal operation of the plant; (ii) the performance of the plant in anticipated operational occurrences; and (iii) accident conditions. On the basis of this analysis, the capability of the design to withstand postulated initiating events (PIEs) and accidents can be established, the effectiveness of the items important to safety can be demonstrated and the inputs (prerequisites) for emergency planning can be established."

Practical experiences shared by Member State participants have shown that challenges relating to specific topics and areas of regulatory review have already emerged. Topics addressed in this section relate to safety functions, safety analysis and safety in design, based on the responses to the questionnaire (see Section A–4 of the Annex). For each topic, an overview of the regulatory expectations, followed by a summary of the practical experiences/challenges and the forward-looking activities presented by the Member State participants is given. The forward-looking subsections discuss approaches proposed to resolve the key challenges in regulating SMRs and the areas where further work is still needed from the perspective of the regulatory bodies.

These topics have been selected by a panel of experts based on their experience and challenges in regulating SMRs. Therefore, it is not intended to be a comprehensive set of topics for the areas discussed herein.

In the area of safety functions, the topics considered are:

(a) Reactivity control function;
(b) Heat removal function;
(c) Confinement function.

In the area of safety analysis, the topics considered are:

(a) Initiating events;
(b) External events;
(c) Defence in depth;
(d) Core damage and severe accidents;
(e) Sharing of safety systems and features;
(f) Safety objectives for multi-units facilities;
(g) Accident source term;
(h) Computer codes.

In the area of safety design, the issues considered are:

(a) Safety classification of SSCs;
(b) Novel/innovative design features;
(c) Qualification of SSCs;
(d) Industry codes and standards.

It is noted that these topics have also links to topics addressed in Sections 2 and 4.

3.1. SAFETY FUNCTIONS

Requirements 46, 51 and 55 of SSR-2/1 [3] establish safety requirements for: (a) reactor shutdown; (b) removal of residual heat from the reactor core; and (c) control of radioactive releases from the containment, all of which have to be fulfilled by the design of an NPP. These safety functions are explained further in the IAEA Safety Glossary [7], as follows:

"(a) The capability to safely shut down the reactor and maintain it in a safe shutdown condition during and after appropriate operational states and accident conditions;

(b) The capability to remove residual heat from the reactor core, the reactor and nuclear fuel in storage after shutdown, and during and after appropriate operational states and accident conditions;

(c) The capability to reduce the potential for the release of radioactive material and to ensure that any releases are within prescribed limits during and after operational states and within acceptable limits during and after design basis accidents."

This is commonly expressed as the three 'fundamental safety functions' for NPPs (see Requirement 4 of SSR-2/1 (Rev. 1) [3]), as follows:

(a) Control of reactivity;
(b) Removal of heat;
(c) Confinement of radioactive material.

The safety functions are fulfilled using engineered safety systems and their associated components and support systems. Safety systems may include active and/or passive components.

This subsection provides information on regulatory expectations, Member State participants experience and challenges, and forward-looking considerations on the three fundamental safety functions.

3.1.1. Reactivity control function

This subsection documents the key points from the questionnaire responses concerning the reactivity control function (Question 18): regulatory expectations, practical experience/challenges and forward-looking activities reported.

Member State participants reported that they have requirements relating to the reactivity control function, and in some cases for the design of the reactivity control systems.

The regulatory body has to review and assess the information provided by the applicant/licensee related to the reactivity control function to determine whether the facility design or activity is in compliance with the relevant safety requirements and regulatory requirements.

Member State participants consider the adequacy of IAEA safety requirements for shutdown and control systems in land-based facilities [3] and recognise that these requirements might be potentially reformulated from a functional perspective for all SMRs including floating types.

3.1.1.1. Regulatory expectations

Considering the questionnaire responses received, regulatory bodies generally expect that the applicant/licensee will submit a detailed demonstration that the reactivity control function of the proposed SMR meets the following safety requirements:

(a) Requirements for reactivity control: Operational safety limits are to be established from safety analyses. Specific variables are measured by the reactor protection system and the shutdown systems are activated if predetermined set points are reached. Reactivity control methods meet the seismic requirements.

(b) Shutdown capability: it is generally expected, as required by para. 6.9 of SSR-2/1 (Rev.1) [3], that at least two diverse and independent systems for shutdown with adequate safety margin are provided; one of the systems is expected to be fast acting and the other needs to provide adequate reliability in DEC.

These safety requirements have been identified by a panel of experts based on their experience and challenges in regulating SMRs and on the responses to the questionnaire provided by Member State participants. It is not intended to be a comprehensive set of requirements applicable to the reactivity control function.

3.1.1.2. Practical experience/challenges

The following considerations represent the key experience/challenges that were identified in the questionnaire responses.

(a) In general, regulations for the design of NPPs are applicable to SMRs.

(b) Some vendors claim that their proposed designs are: simpler; incorporate enhanced engineered safety features, passive features and inherent safety during normal operation, in the event of malfunctions and in accidental conditions; and are able to demonstrate fulfilment of safety functions by diverse means.

(c) For the reactivity control function, the claims of vendors are often related to a significant negative power coefficient of reactivity, significant thermal inertia, and sometimes claiming that some shutdown systems do not need to be classified as safety systems that are necessary to manage design basis accidents (DBAs).

(d) For lead-cooled reactors, corrosion has been identified as a key issue for shutdown systems relying on control rods, as it may hinder their insertion. Some Member State participants, such as the UK, expect that reliability of control rods in lead-cooled reactors might be an order of magnitude lower than current LWR solutions. Therefore, they expect a robust demonstration of reliability, and an independent and diverse shutdown system.

(e) When estimating the effectiveness of the proposed reactivity control or shutdown functions in their submissions (for example, shutdown margins values), some Member State participants, such as Canada, indicated that vendors are not expected to over-rely on computer modelling. Claims are to be fully justified in the context of a specific design and proven by a rigorous test and qualification programme.

(f) The novel approaches for the design of reactivity control systems need to be provided with quality assured and credible information that is supported by research and development in order to demonstrate claims of safety. Supporting information and data need to be demonstrated to be relevant and validated. In some cases, information from historical prototype experimental reactors may not have the necessary pedigree to support modern FOAK reactors. In some other cases, data extrapolation may produce unquantified uncertainties that will need to be addressed by additional research and development work supported by modern quality assurance practices. For example, in Argentina, the regulatory body focused attention on the following three main issues when assessing the development of hydraulic reactivity control and shutdown systems for the CAREM 25 reactor:

(i) Structural and functional verification in seismic conditions;
(ii) Validation and qualification of the position measurement system;
(iii) Validation and qualification of the mechanisms and drive system.

(g) In the helium-cooled and graphite-moderated HTGR, some Member State participants such as South Africa and the UK have considered that a rupture of the primary piping would not result in significant safety consequences. This is because such a rupture is a design basis event. In general, in such a loss of coolant accident (LOCA), the reactor would be shut down by means of control rods, and the decay heat would be removed passively. However, their respective regulatory bodies concluded that the potential for an anticipated transient without action of the primary shutdown system needs to be considered. Reactor vendors tend to rely on the negative reactivity feedback of the fuel and argue that if air can be excluded, fuel damage can be avoided. A major concern is graphite oxidation damage to the fuel and core if a major air ingress takes place through the breached primary pressure boundary. In the case of fuel operating in an oxidizing environment, maintenance of fuel integrity would involve operation at temperatures significantly below normal. The maintenance of the temperature below design limits it is a crucial factor in this event and is likely to need diverse shutdown means for the reactor, together with sufficient diversity and redundancy of the cooling systems.

(h) Regarding the challenge of the design of an SMR-LWR reactivity control system, the USA reported on the effects of boron volatility and redistribution during a passive cooling operation in long term shutdown. During this mode of operation, boron free steam will enter the downcomer and containment, which can potentially challenge the reactor core shutdown margin. The loss of boron from the reactor coolant can potentially lead to a return to power following a postulated accident or anticipated operational occurrence. This challenge was ultimately resolved using analytical methods and current design guidance.

Further information can be found in the Section A–18 of the Annex, in the responses to Question 18.

3.1.1.3. *Looking ahead*

The following are the forward-looking considerations related to the adequacy of reactivity control function under normal operation and accident conditions:

(a) Some designers and license applicants claim that shutdown or reactivity control by natural physical phenomenon such as convection, gravity and reactivity feedback cannot fail and do not need testing. However, natural phenomena are not a guarantee to operability as channels can get blocked and gravity driven control/shutdown rods can get stuck. Therefore, a robust demonstration of the effectiveness of passive systems is necessary, along with an extensive verification of the reliability claims and a robust demonstration of conservative sub-criticality.

(b) Additional analyses of the reactivity control function are needed in respect of the significant negative coefficient of reactivity with power. In this case, reactor shutdown might be a temporary situation; a reactor cooling down without xenon could lead to criticality at a later stage.

(c) In relation to the shutdown function, the initial expectation remains, considering that two independent and diverse shutdown systems are required to be provided.

(d) To demonstrate the safety margin, an adequate categorization of Initiating Events related to reactivity control, the use of models, codes and software to perform verification and validation (V&V[1]), and a quantification of uncertainties in core behaviour and computer code predictions are needed.

3.1.2. Heat removal function

This subsection documents the key points from the questionnaire responses concerning the heat removal function (Question 19): regulatory expectations, practical experience/challenges and forward-looking activities reported.

In general, Member State participants reported that they have specific expectations on the design of reactor cooling systems to ensure that the heat removal function will continue to be fulfilled under normal operation, anticipated operational occurrences (AOO) and accident conditions.

Regulatory bodies generally review and assess the information provided by the applicant/licensee relating to nuclear heat removal function to judge whether the facility or activity is in compliance with the relevant safety requirements and regulatory requirements.

---

[1] V&V include methodologies to evaluate the accuracy and the reliability of the models, codes and software used in safety assessments of designs of reactors.

3.1.2.1. *Regulatory expectations*

Considering the questionnaire responses received, regulatory bodies generally expect that the applicant/licensee will submit a detailed demonstration that the heat removal function of the proposed SMR facility includes the following regulatory requirements:

(a) A coherent heat removal pathway for all PIEs considered for the design that will ensure that the maximum acceptable fuel temperatures are not exceeded. In addition, for events that are considered DEC, the intent is that adequate heat removal criteria will be available such that maximum acceptable fuel temperatures are not exceeded.

(b) Sufficient cooling has to be provided for all parts of the core to remove heat such that the temperature limits for the fuel and structural components are not exceeded.

(c) Monitoring systems and instrumentation of the reactor core and the cooling circuits has to be provided such that it can reliably identify the cooling conditions inside the reactor core.

These requirements have been identified by a panel of experts based on their experience and challenges in regulating SMRs and on the responses to the questionnaire provided by Member State participants. It is not intended to be a comprehensive set of requirements applicable to the heat removal function.

3.1.2.2. *Practical experience/challenges*

The following considerations represent the key experience/challenges that were identified in the questionnaire responses:

(a) The UK reported that some of the requirements established in SSR-2/1 (Rev.1) [3] might not be fully applicable to SMRs: some may require a technology specific interpretation, and, in some cases, additional requirements may be needed. For example, for HTGRs, the isolation of leaks is more important for preventing long term oxidation of the core (air ingress scenario) or reactivity control (water ingress scenario), than for maintaining the coolant for heat removal purposes.

(b) The UK also reported that some SMR designers claim not to require forced circulation for heat removal during normal operations, AOO or accident conditions (and thus include no pumps in the primary circuit). Some vendors claim to have effectively eliminated the possibility of certain LOCAs, for example, by using an integrated approach, guard vessel, or shared pool within which the reactor sits. Other designers claim that sufficient decay heat removal can be provided by natural circulation, even following a LOCA.

(c) South Africa reported that regulatory assessments identified the following as issues to be resolved:

    (i) The important case of heat removal subject to chemical reactions (In certain conditions associated with air ingress, carbon can oxidise. The reactions are complex and are both exothermic and endothermic. The net effect is an exothermic reaction that, if unconstrained, could lead to very high temperatures in the core that might lead to additional fuel failures and a release of radioactivity).

    (ii) Temperature limits for the reactor core as well as core structure ceramics need to be adequately justified.

(d)   Argentina reported that it is building the CAREM 25 prototype, for the experimental demonstration of the prediction of codes and models, and the robustness and reliability of the safety systems and safety features including specially those related to the heat removal function.

Further information can be found in the Section A–19 of the Annex, in the responses to Question 19.

3.1.2.3.   Looking ahead

The following are the forward-looking considerations related to the adequacy of nuclear heat removal function under normal operation, AOO and accident conditions:

(a)   The adequacy of a FOAK design for passive cooling, and the need for testing and qualification of passive systems.

(b)   The reliability of cooling systems based on natural circulation;

(c)   Determination of driving force for natural circulation to effectively cool the fuel and prevent dry out following accidents resulting in the reactor pressure vessel's depressurization. Additionally, a certain temperature gradient is needed to ensure the export of heat. In this condition, the core could keep very high temperature for a long time. This does not comply with the current LWR design requirements in some Member States, so a reasonable demonstration is needed.

(d)   Determination of thermal hydraulic instabilities in the cooling systems under different conditions.

(e)   The applicability of computer code and models and V&V.

(f)   Modelling of passive valves.

(g)   Diversity and redundancy of important in the cooling systems.

(h)   The implementation of the single failure criterion for passive and semi-passive cooling systems.

(i)   Operational challenges with novel steam generators such as helical-coil steam generators, and the effects of flow oscillations in the secondary side on the primary side coolant.

(j)   Whether there is sufficient redundancy against failures of these systems or for performance of periodic maintenance and in-service inspections.

(k)   The capacity and long-term performance of such systems, such as whether they are sized only for decay heat removal or can begin performing their cooling functions at higher reactor power (e.g. scenarios involving a failure to shut down the reactor).

(l)   For some SMR designs, the use of one system (passive or active) to mitigate consequences of failure of more than one level of DiD.

(m)   For some SMR designs, the robustness of performance of SSCs against combinations of internal and external events, particularly in some reactor states during DEC when barriers to releases may have been compromised.

(n)   For some SMR designs, the use of non-nuclear design standards.

(o)   The applicability of obsolete or cancelled design standards for SMRs.

3.1.3. Confinement function

This subsection documents the key points from the questionnaire responses concerning the confinement function (Question 17): regulatory expectations, practical experience/challenges and forward-looking activities reported.

Traditionally NPP designs used a high-pressure retaining, 'leak-tight' containment structure and confinement systems that meet the following three safety objectives:

(a)   Confinement of radioactive substances in operational states and in accident conditions;
(b)   Protection of the reactor against natural external events and human induced events;
(c)   Radiation shielding in operational states and in accident conditions.

Some SMR designs (and technological trends) include different technical solutions to confine radioactive substances and control radioactive releases to the environment during accident conditions.

Regulatory bodies generally review and assess the information provided by the applicant/licensee relating to the confinement function to judge whether the facility or activity is in compliance with the relevant safety requirements and regulatory requirements.

3.1.3.1.   *Regulatory expectations*

Considering the questionnaire responses received, regulatory bodies generally expect that the applicant/licensee will submit a detailed demonstration that for any SMR proposal:

(a)   The confinement function is fulfilled in all plant states with sufficient reliability taking into account the need for in-service inspections, reliability testing and maintenance;

(b)   The failure of the containment system would not impact the ability of the reactivity control and heat removal systems to perform their functions;

(c)   Sufficient DiD is maintained (see Section 3.2.3 for more details);

(d)   The fuel qualification programme activities and other tests will provide results that will support the credibility of containment and confinement performance (if applicable).

These regulatory requirements have been identified by a panel of experts based on their experience and challenges in regulating SMRs and on the responses to the questionnaire provided by Member State participants. It is not intended to be a comprehensive set of requirements applicable to the confinement function.

3.1.3.2. Practical experience/challenges

The following considerations represent the key experience/challenges that were identified in the questionnaire responses:

(a) Canada reported that the following overarching technological trends have been found within proposals made by SMR vendors related to the confinement function:

   (i) The concept of 'functional containment', where radionuclides are retained within multiple barriers with an emphasis on retention at their source (in the TRISO fuel) with the reactor vessel pressure boundary serving to complement the confinement function (HTGR design);

   (ii) The concept of 'low-leakage containment', where there is continuous leakage from the containment during normal operation (liquid cooled fast reactor technologies);

   (iii) A traditional concrete containment structure is argued as not necessary to fulfil confinement safety functions and meet acceptance criteria for accident conditions;

   (iv) The containment isolation is provided by loop seals;

   (v) The containment requirements could be minimized given that over-pressurization accidents would not be credible;

   (vi) The containment design follows the passive safety design principles and therefore, no engineered automatic actuation of containment isolation devices would be proposed;

   (vii) There are potentially novel safety classification and code classification approaches that challenge traditional views that the containment is a safety system.

(b) Canada also reported (refer to the responses to Question 11) that a safety 'means' does not necessarily mean a dedicated system or structure. For example, containment was traditionally understood to mean a hardened concrete/steel structure, whereas now, 'means of containment' is being interpreted more broadly to include all factors that contribute to confinement of radionuclides and containment of releases. For molten salt concepts, this may include retention in fuel salts coupled with inherent temperature behaviours and physical barriers including the reactor 'pressure boundary' and surrounding civil structures working together. Canada also mentioned that different considerations need to be made for solid fuels and for liquid fuels. For solid fuels, the integrity of the fuel is typically the acceptance criteria in the overall confinement scheme. For liquid fuels, such as molten salt fuels, the criteria may be the thermal physical condition and radio nucleus release limits.

(c) The USA reported that for some SMR designs, the containment is a small, high pressure, ASME Section III, Class 1 Ref. [6] that is more comparable to typical reactor pressure vessels in design and dimensions than to typical containment structures. The post-accident containment atmospheric pressure is very high, and the containment volume is relatively small compared to the containments for large light water reactors (LLWRs). This causes the allowable leakage to be extremely low, making it extremely difficult to be able to accurately measure the leakage rate from the containment.

(d) China reported that:

(i) For the confinement function, the most important barrier is the coating layer of the TRISO coated particle and the integrity of TRISO coated particles.

(ii) The most serious accident condition to challenge the TRISO coated particle is a depressurization accident, which might increase the fuel temperature to the limit.

(iii) For the air ingress accident, considered to be a beyond design basis accident, or at worst, classified as DEC (i.e. to be taken into consideration in the design), it is claimed that there is enough grace time (e.g. 72 hours) to take action to mitigate the air ingress accident limiting the failure rate of coated particles and, hence the release of fission products.

(iv) In this context, the break of primary pressure boundary, the opening of the reactor building, does not result directly in a large radioactive release, although the integrity of primary boundary or reactor building can reduce the release.

(v) Currently the acceptance criterion for the confinement function is associated with the accumulated frequency for accident scenarios whose release can result in radiation exposures in excess of 50 mSv. Within the site boundary, this frequency has to be less than 1E-6 per reactor year.

(vi) The common mode factors for the DiD levels and multiple barriers need more investigation, although no obvious factor has been found.

Further information can be found in Section A–17 of the Annex, in the responses to Question 17.

3.1.3.3. *Looking ahead*

The following are the forward-looking considerations related to the adequacy of the confinement function under normal operation and accidental conditions:

(a) Considering that regulations require the containment (in a new nuclear reactor) to be a leak tight civil engineering structure, the proposals made by SMR vendors mentioned above may challenge this requirement. The proposals also provide challenges in terms of interpreting guidance on the adequacy of the confinement function, as well as involve a greater interpretation of technical standards used in detailed design activities and an enhanced need for supporting research and development work to establish the reliability of SSCs to fulfil the confinement function.

3.2. SAFETY ANALYSIS

This subsection provides information on the regulatory expectations, Member State participants experience and challenges, and forward-looking considerations regarding the following safety analysis issues:

(a) Initiating events;
(b) External events;
(c) Defence in depth;
(d) Core damage and severe accidents;

(e) Sharing of safety systems and features;
(f) Safety objectives for multi-unit facilities;
(g) Accident source term;
(h) Computer codes.

The topics listed above have been identified by a panel of experts based on their experience and challenges in regulating SMRs and on the responses to the questionnaire provided by Member State participants. It is not intended to be a comprehensive set of topics associated with safety analysis.

3.2.1. Initiating events

This subsection documents the key points from the questionnaire responses concerning IEs – Question 8: practical experience/challenges and forward-looking activities reported.

Paragraph 3.19 of IAEA Safety Standards Series No. SSG-2 (Rev. 1), Deterministic Safety Analysis for Nuclear Power Plants [3] states:

> "The set of postulated initiating events should be identified in a systematic way. This should include a structured approach to the identification of the postulated initiating events, such as:
>
> (a) Use of analytical methods such as hazard and operability analysis, failure modes and effects analysis, engineering judgement and master logic diagrams;
>
> (b) Comparison with the list of postulated initiating events developed for safety analysis of similar plants (ensuring that previously identified deficiencies are not propagated);
>
> (c) Analysis of operating experience data for similar plants;
>
> (d) Use of insights and results from probabilistic safety analysis."

PIEs include only those failures (either initial or consequential) that directly lead to the challenging of safety functions and ultimately to threatening the integrity of barriers to the release of radioactive material. Therefore hazards, either internal or external (natural or human induced), do not need to be considered as PIEs by themselves. However, the loads associated with these hazards need to be considered a potential cause of PIEs, including multiple failures resulting from these hazards.

For sequential deployment or maintenance of units, it needs to be ensured that a hazard in unit/module under construction, in or maintenance or in operation would not have any safety consequences for a neighbouring operating unit/module, or else such safety consequences need to be carefully considered.

The set of PIEs needs to be reviewed as the design and safety assessment proceed, as part of an iterative process between these two activities.

3.2.1.1.   Regulatory expectations

Considering the questionnaire responses received, regulatory bodies generally expect that the applicant/licensee will submit a detailed demonstration that it has developed and applied a systematic approach (such as master logic diagram (MLD), hazard and operability study (HAZOP), failure mode and effects analysis (FMEA) and expert judgment), for identifying PIEs in the proposed SMR facilities, considering elements that are specific of the design, covering common cause events and all credible failures and hazards, and taking into account all plant states.

3.2.1.2. *Practical experience/challenges*

The following considerations represent the key experience /challenges that were identified in the questionnaire responses:

(a) Regulatory bodies of Member State participants consider that the main challenges associated with the identification of IEs for SMRs is the lack of operating experience and the need for a well-established list of faults derived from international experience such is the case for LWRs.

(b) Some Member State participants (e.g. China, the Russian Federation) reported that the selection of IEs is based on existing designs and supplemented by either fault tree and/or hazard and operability analyses.

(c) Other Member State participants (e.g. Argentina, South Africa and UK) reported that it is expected that a systematic methodology will be used to derive a complete list of IEs from hazard and operability analyses, fault tree analyses, engineering judgment, deterministic and probabilistic assessments and informed by past applications and experience.

(d) The Member State participants responding to this question did not primarily address external events. Canada, Japan and South Africa indicated the need for their consideration and for the inclusion of common mode events.

(e) Canada reported that identification and classification of IEs affecting multiple units and some IEs happening in one unit might have non-immediate cascading effect on other units during the accident evolution as a possible challenge that needs more attention.

(f) South Africa further required the selection and categorisation of licensing basis events which are PIEs analysed in the safety demonstration.

(g) France reported that a triggering event can be 'excluded' if the licensee demonstrates that it is physically impossible or extremely unlikely to occur with a high confidence level regarding the safety objectives. Sufficient design and construction provisions completed by operating provisions have to be implemented to justify this exclusion.

(h) The USA reported with regard to practical experience/challenges that:

    (i) For the SMR design, some challenges existed because the assumed IE frequencies contain large uncertainties, as plant-specific operating experience and associated data are not available to inform design-specific IE frequency estimates.

    (ii) The design, in conjunction with the use of simplifying assumptions, allows the potential accident sequences to be reasonably represented by design initiators. This was possible because the design uses fail-safe features, passive core cooling, and heat removal capabilities, thereby relying less on active systems than a traditional LLWR (i.e. PWRs).

(i)   China reported with regard to practical experience that:

    (i)   The review process of PIEs and classification of high temperature gas cooled reactor (HTR) is complicated. The general principle is to refer to the existing guidelines which are based on PWRs, and to undertake analysis case by case for HTR designs. For example, PSA is used to re-evaluate the condition classification of the IEs, and the risk-inform method is used to review the safety classification of systems and equipment.

    (ii)   A review of the integrity of IEs is generally conducted in the following ways:

        — Review of the list of previous IEs;
        — Identification of specific IEs resulting from the failure or mis-operation of the frontier system or supporting system;
        — Reference to the list of IEs for NPPs of the same type;
        — Use the main logic graph method for deductive analysis.

    (iii) The follow list of PIEs was applied for the high temperature gas-cooled reactor pebble-bed module (HTR-PM) safety analysis:

      — AOOs and DBAs:

          • One control rod spuriously withdrawn during power operation (100% and 50%);
          • Main helium blower spurious speed-up;
          • Loss of off-site power supply;
          • Loss of normal feed water flow;
          • Double-end rupture of the large primary piping;
          • Double-end rupture of the DN10 mm instrument piping;
          • Double-end rupture of the steam generator tube.

      — BDBA (DEC):

          • Anticipated transient without scram (ATWS) during loss of the off-site power supply;
          • ATWS during loss of the normal feed water flow;
          • ATWS during the case of one control rod spuriously withdrawn;
          • ATWS during operational basis earthquake;
          • Loss of feed water flow overlapped with the failure of the blower isolation valve;
          • Steam generator tube rupture (SGTR) overlapped with the failure of the steam generator discharge system;
          • Rupture of the steam generator tube plate;
          • Air ingress accident (chimney effect);
          • Loss of the passive heat removal system.

The SMR Regulators' Forum WGs adds:

(a)   The selection of IEs is impacted by the set of internal hazards and external hazards identified for the design. It is expected that, for sites with more than one unit, IEs will include those that can simultaneously affect more than one unit (e.g. loss of off-site power) or events that can arise in one unit and lead to an IE in another unit (e.g. a strike from a missile generated by disintegration of a turbine in an adjacent unit). The selection of IEs also needs to consider faults originating in SSCs used by more than one reactor, such as fuel handling equipment.

(b)   Most SMR designs claim the use of inherent and passive safety features, which might reduce their vulnerabilities to some PIEs and external hazards which impact the whole site. However, given that a significant number of SMR designs envision multiple modules or units on the site, they may use shared SSCs, thus it is expected that the importance of some internal IEs and external events for safety may increase and they may need to be adequately addressed in the design (e.g. support system faults).

(c)   Because SMRs may be located remotely or in many different environments, a detailed analysis of all possible hazards and associated risks for SMRs needs to be performed for each specific SMR application. The lessons learned after the Fukushima Daiichi NPP accident are also expected to be extensively used in the design of SMRs regarding the risks of external hazards.

(d)   Criteria for exclusion of IEs need to be established.

(e)   Common mode events due to internal hazards and their influence on the independence of DiD levels need to be considered, taking into account SMR design specifics (e.g. modules, compact design and multi units/modules aspects).

Further information can be found in Section A–8 of the Annex, in the responses to Question 8.

3.2.1.3.   *Looking ahead*

The following activities and prospective changes for the assessment of IEs were reported by Member State participants:

(a)   Regulatory bodies faced with the authorization of new reactor designs need to ensure that their standards for identification and classification of IEs are included in the regulatory framework and that a systematic process that includes engineering judgement, deterministic assessments and probabilistic assessments is necessary.

(b)   Criteria for exclusion of events have to be established.

3.2.2. External events

This subsection documents the key points from the questionnaire responses concerning external events (Question 9): regulatory expectations, practical experience/challenges and forward-looking activities reported.

The regulatory body needs to review and assess the information on the facility provided by the applicant/licensee to determine whether the facility design is in compliance with the relevant safety requirements and regulatory requirements, and, in particular, review information on external events from other facilities coupled to the proposed SMR.

For sites with multiple facilities or multiple activities, account has to be taken in the safety assessment of the effects of external events on all facilities and activities, including the possibility of concurrent events affecting different facilities and activities, and of the potential hazards presented by each facility or activity to the others.

For facilities on a site that would share resources under accident conditions, the safety assessment needs to demonstrate that the required safety functions can be fulfilled at each facility under any operational state and in accident conditions in the other facilities.

3.2.2.1. *Regulatory expectations*

Considering the questionnaire responses received, regulatory bodies generally expect that the applicant/licensee will submit a detailed demonstration that it has developed and applied a systematic approach, for identifying and analysing all natural and human induced events, in the proposed SMR facilities, considering hazards associated with other facilities on the site, and the combination of and/or consequential events.

Accounting for external natural and human-induced impacts for SMRs, including hazards associated with facilities on-site, needs to be done in accordance with the existing regulatory requirements.

3.2.2.2. *Practical experience/challenges*

The following considerations represent the key experience/challenges that were identified in the questionnaire responses:

(a) South Africa reported that the NPPs design has to consider all hazards including those from events from other facilities on the site, including other units as well as industrial facilities that the reactor facility is directly connected to.

(b) Canada, China, France, the Russian Federation, the UK and the USA published specific regulatory documents that deal with external events.

(c) South Africa identified the need for regulatory guidance that specifically addresses issues such as proximity (hazard), separation, possible coupling, feedback effects, etc. of other facilities connected with the reactor systems.

(d) Some Member State participants, such as South Africa and UK, indicated that the combination of, and/or consequential events need to be considered and included in the analyses.

(e) Some Member State participants, such as Canada and South Africa, require the identification of potential cliff edge effects for DEC events to be considered when analysing external hazards.

(f) The USA reported that any applicant referencing a design under its regulation, will address the locations and distances from the plant of nearby industrial, military, and transportation facilities, and such data need to agree with data obtained from other sources, when available.

(g) The Russian Federation reported that the list of external events for floating SMRs includes:

    (i) Ship accidents, which include stranding, collision with a ship (pier), and water ingress into the power and auxiliary compartments, capsizing, flooding in the shallows, flooding in deep water;

(ii)   Shock waves caused by explosions on board the ship, human activity while the ship is in port, fire in the main control room (MCR), power compartment, engine room, electrical compartment, reactor compartment, and rooms with the equipment of the Integrated Marine Automation Systems;

(iii) Helicopter crash, including that on the premises of the NPP and on the hull structures of the vessel containing potentially dangerous equipment (equipment working under pressure, equipment filled with hydrogen, oxygen, aviation fuel);

(iv)   Loss of cooling water.

Further information can be found in Section A–9 of the Annex, in the responses to Question 9.

3.2.2.3.   *Looking ahead*

The following activities and prospective changes to the assessment of external events need to be considered by SMR embarking countries:

(a) Evaluating the adequacy of the applicable regulatory requirements for assessment of external hazards emanating from other facilities coupled to the SMR.

(b) Defining and addressing in the SMR design, the list of external events for floating SMRs.

3.2.3. Defence-in-depth

This subsection documents the key points from the questionnaire responses concerning DiD – Question 11: regulatory expectations, practical experience/challenges and forward-looking activities reported.

Paragraph 2.14 of SSR-2/1 (Rev. 1) [3] states:

> "A relevant aspect of the implementation of defence in depth for a nuclear power plant is the provision in the design of a series of physical barriers, as well as a combination of active, passive and inherent safety features that contribute to the effectiveness of the physical barriers in confining radioactive material at specified locations. The number of barriers that will be necessary will depend upon the initial source term in terms of the amount and isotopic composition of radionuclides, the effectiveness of the individual barriers, the possible internal and external hazards, and the potential consequences of failures."

In addition, Requirement 7 of SSR-2/1 (Rev.1) [3] states that: "The design of a nuclear power plant shall incorporate defence in depth. The levels of defence in depth shall be independent as far as is practicable."

The application of the DiD concept focuses on preventing accidents in a NPP by providing different levels of defence and mitigating the consequences of accidents if they do occur.

Paragraph 2.12 of SSR-2/1 (Rev.1) [3] requires that the DiD concept is applied to all safety related activities (including organizational, behavioural and design related) in both normal operation and different shutdown states. The independent layers of defence need to ensure that if a failure were to occur, appropriate SSCs would detect it and compensate for or correct it by appropriate measures.

The SMR Regulators' Forum WGs added that:

"PSAs could be used to check that DiD principles have been properly applied. PSA results could reflect the reliability of the features implemented at each DiD level and the sufficient independence of the levels. PSAs could also be used for the identification of so-called complex DEC sequences and for the assessment of the risks induced by multi-modules."

3.2.3.1. *Regulatory expectations*

Considering the questionnaire responses received, regulatory bodies generally expect that the applicant/licensee will submit a detailed demonstration that five levels of DiD are incorporated in the design and operation of the proposed SMR facilities, including physical barriers to prevent uncontrolled releases of radioactive material to the environment. The different levels of DiD are expected to be independent to the extent practicable.

3.2.3.2. *Practical experience/challenges*

The key practical experience and challenges in DiD reported in the questionnaire responses were as follows:

(a) The regulatory requirements in the Member State participants have typically evolved consistently with IAEA safety standards to include multiple failure events as DEC, as was reported by Argentina and Japan.

(b) Requirements for independence of levels of DiD are being challenged through the introduction of new multiple function features. That is, some developers are proposing specific designs for SSCs that perform multiple safety functions and, in some cases, provisions are being claimed to satisfy multiple levels of DiD. Incomplete supporting information on behaviours of passive and inherent features increases uncertainties in safety claims. The regulatory bodies are challenging vendors to justify with validated data, their proposals that include challenges to the independence of levels of DiD.

(c) Argentina reported practical experience with the enhanced implementation of the DiD concept. Challenges have been experienced with the implementation of the DiD principles and requirements specifically related to some design concepts. As an example, the scope of regulatory activities was enlarged for the purpose of analysing the inclusion of design aspects to fulfil safety functions for events occurring in sub-level 3B of DiD.

(d) China reported with regard to practical experience that:

    (i) The safety review principles for HTR-PM still preserve the five levels of DiD adopted for PWRs and BWRs. However, considering the safety characteristics of HTR-PM, the design consideration of each level could be different.

    (ii) For multiple barriers, the coating layer of TRISO coated particle is especially important for both normal operation and accident conditions and can be considered as another feature in DiD.

(e) Canada reported that the target of Level 4 is to mitigate the consequence of DEC according to the ALARA principle. There are challenges by designers who claim that level 5 DiD is not necessary.

(f) The UK reported with regard to practical experience that:

(i) As part of a proactive review of guidance for compatibility with advanced nuclear technology, the regulatory body is developing examples of aspects to be considered when assessing the adequacy of DiD provisions for SMR designs (specifically sodium-cooled fast reactors (SFRs), lead-cooled fast reactors (LFRs), high temperature gas-cooled reactors (HTGRs) and molten salt reactors (MSRs)) through multi-disciplinary workshops.

(ii) Key conclusions from the workshops have been that the concept of DiD, and related expectations as laid out in the SAPs, remain fully applicable. Although the implementation of DiD in some designs may differ, the regulatory expectations, methodology and key considerations for assessment are expected to remain largely unchanged from assessment of a mature reactor technology. In the short term, the regulatory body is consolidating the outcome of the multidisciplinary reports in technical notes. The regulatory body may develop additional guidance to inspectors to provide further clarity on those expectations in the context of advanced technologies.

(g) The UK reported as potential challenges arising from the technology design and safety approaches:

(i) The sharing of safety systems between reactors modules;

(ii) Reduction of protection provided at level 4 of DiD on the basis of enhanced passive safety features;

(iii) Absence of severe accident analysis;

(iv) Reduced independence between levels of DiD;

(v) Absence of safety classified instrumentation and control (I&C) and increased reliance on arguments of 'practical elimination'.

(h) The SMR Regulators' Forum WGs added:

(i) Regarding the application of DiD for multi-unit NPPs, historically, the safety assessment and safety demonstration for large reactors are typically based on single-unit safety concept. For the majority of participating countries in the WGs, a license is given for a single unit without specific regulatory requirements for multi-unit issues. However, in many countries (e.g. Canada, UK and USA) there are requirements related to the sharing of SSCs important to safety among nuclear units. It needs to be demonstrated that such sharing will not significantly impair each unit's ability to perform its safety functions. Additionally, shared SSCs may be a challenge for the regulatory bodies because it may introduce risk significant vulnerabilities into the design.

(ii) General design safety requirements include those related to the application of DiD.

(iii) The practical elimination concept could not be used to justify omission of a complete DiD level. For example, it could not be used to justify absence of severe accident management arrangements and capabilities that are expected at DiD level 4 or, in the absence of off-site emergency response, at level 5.

Further information can be found in Section A–11 of the Annex, in the responses to Question 11.

3.2.3.3. *Looking ahead*

The following activities and prospective changes for the assessment on DiD were reported by Member States participants:

(a) Because of the broad scope and different discipline applications of DiD provisions (such as application of the barrier concept, margins, redundancy, diversity, and independence), regulatory bodies need to consider providing guidance on these topics to make implementation of DiD useable for the design engineers.

(b) The implementation of DiD for SMRs involves robust regulatory scrutiny, particularly if there is significant departure from the traditional DiD concept.

3.2.4. Core damage and severe accidents

This subsection documents the key points from the questionnaire responses concerning core damage and severe accidents (Question 21): regulatory expectations, practical experience/challenges and forward-looking activities reported.

Core damage and other criteria have traditionally been used in certain countries as a metric to evaluate the adequacy of reactor designs. Core damage is traditionally thought of as damage to the fuel and reactor core components (e.g. piping, supports, instrumentation). There are certain SMR designs where the concept of core damage may not be readily applicable, such as the case of liquid fuel reactors or using fuels with very high temperature failure tolerance (TRISO fuel).

Paragraph 7.56 of SSG-2 (Rev. 1) [8] states:

> "The analysis of severe accidents should identify the bounding plant parameters resulting from the postulated core melting sequences, and demonstrate that:
>
> (a) The plant can be brought into a state in which the containment functions can be maintained in the long term;
>
> (b) The plant SSCs (e.g. the containment) and procedures can prevent a large radioactive release or an early radioactive release, including containment bypass;
>
> (c) Control locations remain habitable to allow performance of required staff actions;
>
> (d) Planned severe accident management measures are effective."

Paragraph 7.57 of SSG-2 (Rev. 1) [8] states:

> "The safety analysis of severe accidents should demonstrate that compliance with the acceptance criteria is achieved by features implemented in the design, combined with implementation of procedures or guidelines for accident management."

The regulatory body may establish more specific rules or requirements describing acceptable ways to demonstrate 'practical elimination' of the possibility of conditions arising that could lead to an early radioactive release or a large radioactive release.

3.2.4.1. *Regulatory expectations*

Considering the questionnaire responses received, regulatory bodies generally expect that the applicant/licensee will submit a detailed demonstration about core damage and severe accidents in the safety analyses. This needs to include an analysis of DEC to further improve the safety of the SMR design. The release sequences are expected to be analysed against defined criteria: for sequences that are not excluded by the inherent safety of the reactor design, organisational accident management measures need to be implemented.

3.2.4.2. *Practical experience/challenges*

The following considerations represent the key experience/challenges that were identified in the questionnaire responses:

(a)   Challenges are being experienced where certain quantitative criteria have been defined in regulatory requirements relating to core damage and radioactive releases that are sometimes only applicable to technologies already licensed in the respective country.

(b)   Some Member State participants, such as Canada and Czech Republic, reported that the concept of 'practical elimination' has been included in the regulatory requirements relating to large early releases.

(c)   Canada reported that some designers claim that there is a need for SMRs to have extensive use of passive features. According to these claims, these features prevent some or most of the traditional accident events and scenarios from causing any core damage or releasing radioactive material to the environment. Applicants' concerns were that the proposed use of passive features is challenging to model in traditional safety analyses and alternate approaches need to be recognized as applicable to or acceptable for safety cases. Canada also noted that uncertainties presented by alternative and innovative features can affect the confidence on the outcomes of safety analyses.

(d)   In Argentina, South Africa and UK, the safety goals are technology neutral and relate to the ultimate objective to protect the public. However, in Canada and South Africa, the definition of core damage, as defined in their regulations, and the associated safety goals may not be applicable, as written, to all reactor designs.

(e)   In South Africa and UK, the term core degradation instead of core damage is used acknowledging the need to define the state of the core as part of the release model. The UK identified as a potential challenge the definition of an appropriate degraded plant condition as the starting point for severe accident analysis, and the selection of an appropriate metric for PSA studies.

(f)   South Africa reported some indication of the challenges associated with interpretation or definition of core damage and severe accidents in the context of the PBMR, as follows:

(i)   The PSA for a PBMR is fundamentally different to that for an LWR. The concept of core damage frequency cannot be used. Extensive use of passive features has to be modelled. Events of very low frequency have to be addressed.

(ii)   The PSA for a PBMR seeks to achieve the same overall goal as for an LWR, but is structurally different. This follows directly from the differences in the design and safety philosophy:

— The concept of core damage and large early release end states is not considered;

— The PSA is fundamentally a challenge-response analysis of the fission product barrier — the fuel particle coating;

— Extensive use is made of passive systems for which failure probabilities are correspondingly small and therefore difficult to justify from operational or test data.

(iii)   The following issues therefore need to be addressed:

— The reliability of passive systems in particular for their long term response;

— The modelling of fuel degradation as a function of time and temperature;

— The urgency for developing the PSA depends upon the intention to employ risk-informed methodologies and the way in which these interact with the design process;

— Since early PBMR PSA proposals excluded the most severe accidents, the PSA's representativity of the actual risk associated with the PBMR was questioned. As a result, the regulatory body required assessments of the probability of occurrence of severe core damage states and the risk associated with large off-site releases.

(g)   Canada reported that design requirements also permit the proposal of other surrogates under the condition that the underlying objectives continue to be met. For example, for sodium reactors, a high temperature operational constraint (e.g. 800°C) with a sufficient degree of conservatism has been proposed as a surrogate to a formal definition of core damage.

(h)   Japan reported that for SMRs considered as research and test reactor facilities, there are no specific limits for the amount of radioactive material released. However, to prevent excessive exposure of the public, it is required that the dose evaluation value for DEC does not exceed 5 mSv per accident.

(i)   China reported that current value for the possible failure rate of TRISO coated particles in modular HTGRs is of the order of 1E-4 per year, even for most serious accidents of depressurization. Therefore, it claimed that there is no core damage and no severe accident for modular HTGRs.

(j)   The SMR Regulators' Forum WGs added:

(i)   The progression of faults/accidents has to be analysed assuming failure or degradation of the primary barriers (levels 1–3 of DiD) to fission product release in order to establish a facility's vulnerability to severe accidents. All areas of a facility having the potential for severe accidents have to be assessed.

(ii)   Where claims are being made that severe accidents will be precluded by design provisions, such conclusions need to document how accidents based on unmitigated consequences associated with a fault have been characterised and analysed. Any assumptions with respect to the maintenance of barrier integrity should be robustly justified.

Further information can be found in Section A–21 of the Annex, in the responses to Question 21.

3.2.4.3.   *Looking ahead*

For the accident sequence modelling, regulatory bodies report that the following forward-looking considerations are needed:

(a)   To delineate single and multi-unit accident sequences;

(b)   To account for multi-unit common cause and causal dependencies, including functional, human, and spatial dependencies;

(c)   To consider adverse impacts of a single reactor/facility accident on other units, thus creating additional multi-unit accident scenarios;

(d)   To consider how operator actions may be adversely affected by multi-unit interactions;

(e)   To consider the timing of releases from different units;

(f)   To consider how radiological contamination of the site may inhibit operator actions and accident management measures.

In addition, Member State participants reported the need for the development of internal guidance on the appropriate selection of a degraded plant condition for severe accident analysis of advanced reactor technologies.

3.2.5. Sharing of safety systems and features

This subsection documents the key points from the questionnaire responses concerning sharing of safety systems and features (Question 16): regulatory expectations, practical experience/challenges and forward looking activities reported.

The SMR module concept lends itself for multiple reactor modules to be deployed near each other and to share safety and supporting systems.

Requirement 33 of SSR-2/1 (Rev.1) [3] states that: "Each unit of a multiple unit nuclear power plant shall have its own safety systems and shall have its own safety features for design extension conditions".

Paragraph 5.63 of SSR-2/1 (Rev.1) [3] states that: "To further enhance safety, means allowing interconnections between units of a multiple unit nuclear power plant shall be considered in the design".

The regulatory body needs to review and assess the information provided by the applicant/licensee to determine whether the facility or activity is in compliance with the relevant safety requirements and regulatory requirements.

The SMR Regulators' Forum WGs considers:

(a) Typically, the current requirements and guidance limit the sharing of SSCs important to safety between reactors. In exceptional cases sharing of SSCs important to safety is permitted if it can be demonstrated that it is not in detrimental to nuclear safety. As such, if sharing of SSCs between reactors is arranged, safety requirements need to be met for each reactor for all plant states. Also, in the event of an accident involving one of the reactors, orderly shutdown, cool down and removal of residual heat have to be achievable for the other reactors.

(b) For SMR designs that share SSCs, the safety assessment needs to consider all relevant safety implications, in recognition that such sharing may introduce risk significant vulnerabilities in the design.

3.2.5.1. *Regulatory expectations*

Considering the questionnaire responses received, regulatory bodies generally expect that the applicant/licensee will submit a detailed demonstration in the safety analyses that the sharing of systems and features is not detrimental to safety or it further improves the safety of the SMR. It also needs to determine that sharing of systems or features will not lead to the violation of safety requirements for any module or for the facility as a whole.

3.2.5.2. *Practical experience/challenges*

The following considerations represent the key experience/challenges that were identified in the questionnaire responses:

(a) Challenges that may be encountered relate to the potentially conflicting requirement of applying the single failure criterion for DBAs and how this will be implemented for DEC. There is international experience in Canada, China and the Russian Federation with shared safety systems. For example, shared SSCs features are designed to supplement unit specific DiD. Some of these features include:

  (i) One MCR with dedicated space allocated for each operating unit panels, including unit 0 and fuel handling;

  (ii) Common containment, including one vacuum building;

  (iii) Common emergency coolant injection system functions;

  (iv) Emergency power system to supplement unit-specific electrical supply architecture;

  (v) Emergency service water to supplement unit specific water-cooling systems.

(b)   The UK regulatory body has documented its expectations in this regard in line with the IAEA safety standards and expects that facilities have their own dedicated safety systems to protect against design basis faults and that such safety systems are not shared between facilities. This arises from the strong design basis expectation that very high reliability safety systems are needed to protect against high consequence faults to demonstrate the adequacy of Level 3 DiD. Equipment designed to assist with controlling or mitigating accidents may be shared where this is justified to be in the interests of safety (e.g. if this provides a diverse, alternative means of restoring a lost safety function). Where equipment is shared, the UK regulatory body expects the safety case to demonstrate that the sharing does not increase either the likelihood or the consequences of an accident at any of the facilities.

(c)   The Argentina regulatory body has also documented requirements in this regard in line with the IAEA safety standards and do not allow sharing of safety systems for DBAs.

(d)   Canada reported with regards to sharing of SSCs in multi-unit facilities, that it is possible to have shared SSCs, but the applicants/licensees have to demonstrate that the safety functions can be achieved even with failure of an SSC in another unit. Thus, there is sharing of SSCs, but each unit has to be able to operate independently of the others. More information and details is given in the answer to Question 7 and next subsection.

(e)   The USA reported that SSCs important to safety are not to be shared among nuclear power units unless it can be shown that such sharing will not significantly impair the ability to perform their safety functions, including an accident in one unit, an orderly shutdown and cooldown of the remaining units.

(f)   The Russian Federation reported that:

  (i)   Regarding FNPPs, the multipurpose use of safety systems and their components to perform safety functions as well as normal operation functions is allowed if the process of operation of the system and its functions are not affected, and the safety functions has priority over normal operation functions. For example, the primary circuit cleaning and cooling system can be used for both cooling down during normal operation of the reactor and cooling down in the case of accidents.

  (ii)   In the limited space of a vessel, some systems or their components can be used for two reactors. Sharing means that the system has a channel design with redundant components within the channel. At the same time, there is a backup channel that can replace the failed safety channel of any of the reactors. Also, with appropriate demonstration, it is allowed to operate the channel of the system of one reactor for the other reactor or for both at once. If this is the case, it needs to demonstrate that the operation of one channel is sufficient to perform the safety functions in the event of accidents on two reactors simultaneously.

  (iii)   For land based NPPs, the multipurpose use of safety systems and their elements needs to be justified. Combining safety functions with normal operation functions are not expected to lead to a violation of safety requirements or a decrease in the performance of safety functions. The safety systems of one unit of a multi-unit NPP have to be independent of the safety systems of another unit of the same NPP. Research work to assess new requirements for the use of combined systems in multi-reactor plants is planned.

Further information can be found in Section A–16 of the Annex, in the responses to Question 16.

3.2.5.3. *Looking ahead*

The following activities and prospective changes for the assessment of sharing of safety systems and features were reported by Member State participants:

(a) For SMR designs that include sharing of safety systems and features, the safety assessment needs to consider all relevant safety implications, in recognition that such sharing may introduce risk significant vulnerabilities in the design.

(b) Research work to assess new regulatory requirements for the use of combined systems in multi-reactor plants is planned in some countries.

3.2.6. Safety objectives for multi-unit facilities

This subsection documents the key points from the questionnaire responses concerning multi-unit facilities (Question 7): regulatory expectations, practical experience/challenges and forward-looking activities reported.

Multi-unit/module SMRs may use shared systems to a greater extent that multi-unit NPPs because of their compact configuration and proximity, and this may impact among others, the selection of IEs, internal and external hazards, the approach to shared systems, DiD, human factors, engineering and risk assessment. Governments and/or regulatory bodies across Member State participants have well-established dose limits for occupational exposure, and control public exposure through limits on the discharges of radioactive material to the environment. These limits are communicated to the applicant/licensee during the authorization process.

3.2.6.1. *Regulatory expectations*

Considering the questionnaire responses received, regulatory bodies generally expect that the applicant/licensee will submit a demonstration that the safety objectives for multi-unit SMR facilities are met. This includes consideration of the potential for hazards originating from one nuclear module/installation to affect other nuclear module/installation located on the same site or on adjacent sites. The assessment needs to include comparison with safety objectives and numerical targets, e.g. dose limits, dose constraints, authorized limits on discharges, and PSA goals.

3.2.6.2. *Practical experience/challenges*

The following considerations represent the key experience/challenges that were identified in the questionnaire responses:

(a) Regulatory bodies understand the safety and regulatory implications of sharing SSCs and/or infrastructure. In these cases, it is important to ensure that the safety of the SMR is not negatively impacted by the adoption of a modular reactor deployment and recognize that multi-unit/multi-module SMR designs may have certain potential operational and safety benefits, such as interconnections between units/modules to strengthen the availability and reliability support services (i.e. electric power, compressed air, water) or qualified personnel. The responses indicate that no changes to safety objectives and numerical targets (i.e. dose limits, dose constraints, authorized limits on discharges, PSA goals) were introduced or needed to address the proposed SMRs.

(b)   Some Member State participants such as Canada and the UK indicate that the safety goals and numerical targets apply on a site-wide basis, others such as Japan and the Russian Federation apply these metrics to single unit basis; others, such as Argentina, apply these metric on both a site and single unit basis.

(c)   The USA reported that the regulatory body determined that the applicant's design considers the risk and safety effects of the multi-module plant operation with shared systems to ensure the independence and protection of the safety systems of each unit during all operational modes. Staff also determined that the applicant's multi-module evaluation was adequate for the design certification (DC), since the applicant considered potential system interactions with other reactor modules.

(d)   The SMR Regulators' Forum WGs added about the licensing of multiple module/unit facilities:

(i)   Ownership model: SMRs offer a potential for having several designs and several operators of different reactor cores on a single site. This may lead to additional challenges regarding emergency planning and response, strict liability, the use of common services and a conflict of support groups (shared maintenance…).

(ii)   Considerations of a single license for several cores/modules/units: some countries are pondering the benefits of issuing a single licence for several reactor cores/modules/units, acknowledging the necessity of incrementally bringing units/modules/cores online. The impact on the license of replacing an SMR once the fuel is spent (for relevant designs) is also worth considering.

(iii)   Shared systems important to safety: some designs propose that systems important to safety be shared between several SMRs. It is worth examining how this specificity may have an impact on safety and on the type of licence that could be issued to the SMR. Special attention should be paid to not overlap with the work of the Design and Safety Analysis working group.

(iv)   Shared personnel (exclusion control rooms): several teams, such as maintenance, emergency response, or training, are expected in some situations (design and country dependent) to be shared. It could be a maintenance team owned by a separate organization, or an emergency response team common to several owners on a same site. Each of these unique situations might raise a new challenge and has to be examined from a licensing perspective.

(v)   Control room: some designers propose control rooms shared between several SMRs, a reduced number of operators in the control room per reactor core compared to large-scale plants, or even remotely operated facilities. Since these specificities are new for most countries, challenges arise when it comes to their acceptability.

(e)   The SMR Regulators' Forum WGs added about risk assessment for multi-unit/multi-module sites:

(i)   The Fukushima Daiichi NPP accident demonstrated the possibility of accidents involving nearly concurrent core damage at multiple reactor units and spent fuel pools. It was recognized that the accident progression was influenced by complex interactions involving operator actions to protect each facility, as well as interactions and dependencies among the facilities.

(ii)   In this context, there is a need for the evaluation of site risk in an integrated way, which includes consideration of the potential for accidents involving multiple installations concurrently.

(iii)   In a multi-unit PSA (MUPSA), it is necessary to consider multi-unit accidents either of a causal nature, in which a single-reactor accident may propagate to affect other units, or as a result of a common cause event that affects multiple units or radiological sources concurrently.

(iv)   It would be beneficial for both designers and regulatory bodies to think beyond the single unit mindset. This might involve extending their considerations to whole site risk including developing methods of aggregating risk from differing on-site sources (e.g. new and old reactors, spent fuel pools).

(v)   Even if the SMR concept is based on a modular design with small unique power on multi modules/units sites, the SMR design needs to take due account of the potential consequences of several – or even all – units failing simultaneously due to external hazards. It may affect the methodology for emergency planning zone (EPZ) assessment.

Further information can be found in Section A–7 of the Annex, in the responses to Question 7.

3.2.6.3.   *Looking ahead*

The following activities and prospective changes for the safety assessment related to multi-module units were reported by Member State participants:

(a)   For SMRs, PSAs (or PRAs) need to explicitly consider multiple units or modules. The precise content and scope of the PSA are likely to be dependent on specific design details, such as interactions or dependencies between units or modules. For example, the current practice in Canada is that the PSA has to reflect the station design, not just the unit design. This includes the following considerations:

(i)   How do the units interact with each other in different station states?

(ii)   How are shared/common systems divided up in the PSA to reflect individual unit safety?

(iii)   Site based PSA versus unit based PSA?

      — External events (human-induced or natural);
      — Common-cause failures;
      — What is the modelled release size and inventory?

(b) Consideration of multi-unit PSA has been an ongoing topic of research in some Member States. In the UK it is expected that assessment of a new build NPP against the numerical targets of PSA would consider the results of Level 1, 2 and 3 PSAs. The UK has reported contributions to the OECD/NEA Working Group on Risk Assessment (WGRISK) task on-site level PSA and the development of an IAEA Safety Report on multi-unit PSA. The UK also reported that it has conducted research on the effects of multiple releases from multiple units on the same site on Level3 PSA consequences. These activities have been considered in the recent updates to PSA guidance.

(c) Specific safety aspects, relevant to multi-units/multi-modules, identified by the SMR Regulators' Forum WGs include the following:

(i) The potential for interactions among the modules.

(ii) The potential for sharing safety systems and features.

(iii) Multi-module failure in hazard conditions.

(iv) Modules dependence/independence.

(v) Human factors engineering, including aspects related to:

— Main control room;
— Supplementary control and other emergency response facilities and locations;
— Maintenance of the multiple modules;
— Potential remote control of the main control room;
— Minimum shift complement;
— Training.

(vi) Emergency preparedness and response.

(vii) Capacity for the addition of future modules.

3.2.7. Accident source term

This subsection documents the key points from the questionnaire responses concerning the accident source term (Question 22): regulatory expectations, practical experience/challenges and forward-looking activities reported.

The source term is the amount and isotopic composition of radioactive material released (or postulated to be released) from a facility and it is used in modelling releases of radionuclides to the environment, in the context of accidents at nuclear installations [7].

Under accident conditions, source term evaluation involves simulation codes capable of predicting the release of fission products from fuel elements, transport through the primary system and containment or spent fuel pool building, the related chemistry affecting this transport and the characteristics in which the radioactive material would be released.

Paragraph 2.18 of SSG-2 (Rev. 1) [8] states:

"The source term is evaluated for operational states and accident conditions for the following reasons:

(a)   To confirm that the design is optimized so that the source term is reduced to a level that is as low as reasonably achievable in all plant states;

(b)   To support the demonstration that the possibility of certain conditions arising that could lead to an early radioactive release or a large radioactive release can be considered to have been 'practically eliminated'.

(c)   To demonstrate that the design ensures that requirements for radiation protection, including restrictions on doses, are met;

(d)   To provide a basis for the emergency arrangements that are required to protect human life, health, property and the environment in case of an emergency at the nuclear power plant;

(e)   To support specification of the conditions for the qualification of the equipment required to withstand accident conditions;

(f)   To provide data for training activities regarding emergency arrangements;

(g)   To support the design of safety features for the mitigation of the consequences of severe accidents".

Determining the accident source term depends on: fuel and fission product inventories; fuel behaviour under accident conditions; release mechanisms; energies, temperatures, pressure and timing involved in the response of facility safety systems and in the behaviour of facility containment/confinement systems.

The regulatory body needs to review and assess the information related to the source term provided by the applicant/licensee to determine whether the facility or activity is in compliance with the relevant safety requirements and regulatory requirements.

3.2.7.1. *Regulatory expectations*

Considering the questionnaire responses received, regulatory bodies generally expect that the applicant/licensee will submit detailed information about the source term, which needs to be included. For example, the radionuclide inventories in the fuel elements associated with the SMR, the fission product release from fuel elements in operational states and in accident conditions, its transport through the primary system and the containment or spent fuel pool building, the related chemistry affecting this transport and the chemical and physical form in which the radioactive material would be released.

3.2.7.2. *Practical experience/challenges*

The following considerations represent the key experience/challenges that were identified in the questionnaire responses:

(a)   The UK reported that the challenges in this topic are:

   (i)   Limited operating experience;

   (ii)   Limited maturity/analysis of source term for normal operation and faults;

(iii) Mobility of source term (dust), particularly for HTGRs;

(iv) Fission product release rates from novel fuel;

(v) Plate out and clean up rates of activation and fission products from coolant in novel designs;

(vi) Performance of novel heating, ventilation and air conditioning (HVAC) systems and containment/confinement systems in removing airborne activity.

(b) The USA reported as practical experience:

(i) Some SMR designs did not include large piping of reactor cooling system (RCS); therefore, the accident scenario that would result in a fission product release to containment would not be the same as for the LLWR LOCA.

(ii) The SMR applicant proposed a methodology to develop a core damage source term based on several severe accident scenarios taken from the design-specific probabilistic risk assessment (PRA).

(c) Member State participants, such as South Africa, have identified challenges with regards to HTGRs which include:

(i) The behaviour of TRISO fuel;

(ii) The extent of graphite dust formation and suspension;

(iii) Fission product behaviour in TRISO fuels;

(iv) Fission product behaviour in purification systems;

(v) Interaction between steam and fission products.

(d) With regards to molten salt reactors, the following impact on determination of the accident source term have been identified for Canada:

(i) Establishing the appropriate compositions and solubility behaviour of fuel salt mixtures as part of setting the safe operating envelope;

(ii) Uncertainties in fission product behaviour in the fuel matrix, and in purification and storage systems;

(iii) Compatibility of the molten salt with structural materials and corrosion control to minimize barrier degradation.

(e) The SMR Regulators' Forum WGs added on issues and challenges of accident progression and source term characterization:

(i) Existing severe accident models that are limited to single-reactor accidents will have to be enhanced to treat multi-unit and fuel storage accidents.

(ii) There is a need to define new release categories that adequately describe the releases from multi-unit accidents; this includes release magnitudes, energies, and timing from reactor units, spent fuel storage and other radiological sources.

Further information can be found in Section A–22 of the Annex, in the responses to Question 22.

3.2.7.3. *Looking ahead*

The following activities and prospective changes for the assessment of the accident source term were reported by Member State participants: research and development work is needed in the areas of fuel and fission product behaviour during accident conditions and in the behaviour of safety systems and containment/confinement systems in order to have an adequate confidence in postulated accident source terms for some designs of SMRs.

3.2.8. Computer codes

This subsection documents the key points from the questionnaire responses concerning computer codes (Question 13): regulatory expectations, practical experience/challenges and forward-looking activities reported.

The models and methods used in available computer codes for deterministic safety analysis have to be appropriate and adequate for the purpose. The extent of their V&V and the means for achieving it depend on the type of application and the purpose of their analysis.

Validation against test data is the primary means of validation. However, in cases where no means to achieve appropriate data are available for test cases, it is possible to enhance confidence in the results by means of code-to-code comparisons or using bounding engineering judgement to compensate for limitations in the full validation. The approach taken to validation and the use of the code will need to be justified.

Paragraph 4.60 of IAEA Safety Standards Series No. GSR Part 4 (Rev. 1), Safety Assessment for Facilities and Activities [8] states:

> "Any calculational methods and computer codes used in the safety analysis shall undergo verification and validation to a sufficient degree. Model verification is the process of determining that a computational model correctly implements the intended conceptual model or mathematical model; that is, whether the controlling physical equations and data have been correctly translated into the computer codes. System code verification is the review of source coding in relation to its description in the system code documentation. Model validation is the process of determining whether a mathematical model is an adequate representation of the real system being modelled, by comparing the predictions of the model with observations of the real system or with experimental data. System code validation is the assessment of the accuracy of values predicted by the system code against relevant experimental data for the important phenomena expected to occur. The uncertainties, approximations made in the models, and shortcomings in the models and the underlying basis of data, and how these are to be taken into account in the safety analysis, shall all be identified and specified in the validation process. In addition, it shall be ensured that users of the code have sufficient experience in the application of the code to the type of facility or activity to be analysed."

3.2.8.1. *Regulatory expectations*

56

Considering the questionnaire responses received, regulatory bodies generally expect that the applicant/licensee will submit detailed information on the calculational methods and computer codes used in the safety analysis of SMRs, as well as on the V&V process applied to them for the different scenarios during accident conditions.

3.2.8.2.   Practical experience/challenges

The following considerations represent the key experience/challenges that were identified in the questionnaire responses:

(a)   The use of legacy and commercial off the shelf (COTS) codes presented challenges in the licensing of some of the projects such as PBMR. Special attention is needed to be paid to limitations in the QA status of legacy codes and the status of experimental and analytical verification and validation of them. It is also necessary that the design of the test facilities is such that they are directly able to support the validation of the computer codes for the phenomena modelled.

(b)   Member State participants, such as Canada, South Africa and the UK, have regulatory requirements on the V&V of computer models in line with the IAEA safety standards, while other Member State participants such as the Russian Federation have experience in the use of various software tools for comprehensive analyses of various safety aspects, providing regulatory expectations regarding users of the codes, quality assurance, and representatively of data used in validation and modelling.

(c)   Some Member State participants have guides for the V&V and for the regulatory assessment of computer codes. For example:

   (i)   Canada reported the use of a risk informed approach to determine the level of regulatory review depending on the importance of the code to the safety case. The regulatory body will examine the information submitted by the vendor/licensee and take into account the assumptions, available information, uncertainties and estimated margins to safety when establishing the level of review. Due to limited computer codes that are validated for safety demonstration of SMRs, the regulatory body expects a vendor/licensee to generate quality data for code validation via their research program.

   (ii)   France reported that the recommendations for qualification of scientific computing tools used in the nuclear safety case, deals with:

   — The intended scope of utilization of the scientific computing tools in the safety case, which needs to be defined before the process of verification, validation and transposition;
   — The process of verification and validation of the scientific computing tool (SCT), which lies at the heart of qualification;
   — The process of transposition of the validation cases to the intended scope of utilization;
   — The declaration of qualification;
   — Several points concerning certain software (pre- and post-processing, coupling, etc.) and specialized uses of the SCT, like neutronics, thermohydraulics, mechanics etc.

(iii) South Africa reported that:

— The regulatory assessment of the PBMR Safety Case requires a comprehensive set of sophisticated independent analytical tools that use a variety of techniques. Independent computer codes are used both to replicate calculations performed by PBMR and to provide additional calculations deemed necessary by the assessors. All areas of importance to the safety case that may require the use of computer codes for assessment have been identified and listed and suitable techniques for assessment of computer codes and/or models were identified. It is also important to consider interfaces and data flow between analysis areas.

— The regulatory body interests in a licence applicant's V&V activities is not limited to safety analyses but also includes design analyses to the extent that design analyses have nuclear safety implications.

— Lack of validated codes can have an impact on the licensing process and schedule and may require very conservative assumptions and analyses due to model uncertainties, as was experienced during the PBMR project.

(iv) China indicated that V&V of computer codes presented challenges such as limited availability of experimental data.

(d) The SMR Regulators' Forum WGs added the following about the models used to evaluate radiological consequences:

(i) Models to evaluate radiological consequences need to consider how to model releases from multi-unit and multi-facility accidents; this includes consideration of different points of release from the plant, possible differences in time of release and release energies for plume rise considerations.

(ii) The method for decoupling models for radiological consequences from radioactive inventories needs revision for spent fuel accidents.

Further information can be found in Section A–13 of the Annex in the responses to Question 13.

3.2.8.3. *Looking ahead*

The following activity and prospective changes for the assessment on computer codes were reported by Member State participants: the availability of validated computer codes and models for safety demonstration of SMRs may be limited and will necessitate extra effort and early engagement by vendors/licensee with regulatory bodies to understand regulatory requirements and expectations.

3.3. SAFETY IN DESIGN

This subsection provides information on the regulatory expectations, Member State participants' experience and challenges, and forward-looking considerations regarding:

(a) Safety classification of SSCs;
(b) Novel/innovative design features;
(c) Qualification of SSCs;
(d) Industry codes and standards.

The areas listed above have been identified by a panel of experts based on their experience and challenges in regulating SMRs and on the responses to the questionnaire provided by Member State participants. It is not intended to be a comprehensive set of areas associated with safety in design.

3.3.1. Safety classification of structure, systems and components

This subsection documents the key points from the questionnaire responses concerning safety classification of SSCs (Question 10): regulatory expectations, practical experience/challenges and forward-looking activities reported.

Paragraph 4.4 of IAEA Safety Standards Series No. SSG-30, Safety Classification of Structures, Systems and Components in Nuclear Power Plants [10]states that:

> "A complete set of engineering design rules is expected to ensure that the SSCs will be designed, manufactured, constructed, installed, commissioned, operated, tested, inspected and maintained to appropriate quality standards. To achieve this, designers and applicants/licensees generally apply design rules that identify appropriate levels of capability, reliability (dependability) and robustness."

The design rules have also to take due account of regulatory requirements relevant to safety classification of SSCs, as documented in the relevant regulations, standards and guidance of the Member State participants.

Paragraph 4.2 of SSG-30 [10] states:

> "The engineering design rules should be chosen so that the plant design meets the objective that the most frequent postulated initiating events yield little or no adverse consequences, while more extreme events (those having the potential for the greatest consequences) have a very low probability of occurrence".

3.3.1.1.   Regulatory expectations

Considering the questionnaire responses received, regulatory bodies generally expect that the applicant/licensee will submit detailed information on the classification of SSCs important to the safety of the SMR.

The correspondence between the safety class, and the associated engineering design and manufacturing rules, including the codes and/or standards that apply to each SSC, need to be justified.

3.3.1.2.   Practical experience/challenges

The key practical experience and challenges in safety classification of SSCs reported in the questionnaire responses were as follows:

(a)   Regulatory bodies have developed the basis, requirements, and guidance for assessing the methods and engineering design rules defined by the designers and applicants/licensees. Some of them have requirements to establish rules and acceptance criteria related to capability, reliability, and robustness.

(b)   Regulators have specific regulations to establish and evaluate requirements on the safety classification of SSCs. For example, Canada and China reported that their regulations are being aligned with SSG-30 [10]; other Member State participants, such as Argentina, are in the process of updating their regulations, using SSG-30 [10] as a reference.

(c)   Argentina's regulatory body mentioned that the 'integral' assessment of the safety case of a new reactor design was one of the most challenging aspects of the licensing process for the CAREM 25 prototype reactor. This 'integral' assessment links the demonstration of safety and the safety classification of SSCs. From the safety classification of SSCs, engineering requirements were derived and compliance with these had to be demonstrated in the safety analysis, in accordance with the acceptance criteria set by the regulatory body.

(d)   Canada reported that the challenges are more likely to arise when assessing the adequacy of safety classification bases, which rely on outputs from the safety analysis and research and development programmes. For SMRs incorporating more passive and inherent safety features, the influencing phenomena will need to be properly understood before they can inform the safety classification of SSCs. Examples of relevant information include mission times and reliability targets for the shutdown means and containment isolation.

(e)   The Czech Republic reported that one of the more challenging aspects of SMR licensing would be the classification of the SSCs, as the current system of classification and conformity assessment is mainly adapted to the LWRs and thus the most detailed requirements are very specific and focused on this technology.

(f)   The UK noted that it would be useful to develop guidance to define the requirements for the classification of electrical power supply systems as part of the plant safety assessment of SMRs. This is because there is a lack of consistency between the IAEA (and the International Electrotechnical Commission (IEC)) guidance and the Institute of Electrical and Electronics Engineers (IEEE) guidance for example. It would also be beneficial if the terminology used in future guidance was aligned with the IAEA safety glossary [7] definitions.

(g)   The practical experience/challenges reported for China was given in Section 3.2.1.2.

Further information can be found in Section A–10 of the Annex in the response to Question 10.

3.3.1.3.   *Looking ahead*

The following activities and prospective changes for the assessment on safety classification of SSC were reported by Member State participants:

(a)   Most of the Member State participants stated that this area does not present extra challenges.

(b)   In general, all the Member State participants agree on the importance of applying robust approaches to the safety classification of SSCs to enhance safety by design.

3.3.2.   Qualification of structures, systems and components

This subsection documents the key points from the questionnaire responses concerning qualification of SSCs (Question 15): regulatory expectations, practical experience/challenges and forward-looking activities reported.

Requirement 30 of SSR2/1 (Rev. 1) states that the qualification of SSCs is required to "verify that items important to safety at a nuclear power plant are capable of performing their intended functions when necessary" and to ensure that environmental conditions are duly considered throughout their design life.

Typically, qualification conditions include the entire range of operating conditions from normal operation of the plant to accident conditions derived from DBAs. Qualification programmes are required to be developed for items important to safety also considering ageing effects and natural external events.

Design requirements for SSCs should reflect specific qualification requirements: SSCs are then designed, manufactured, constructed, installed, commissioned, operated, tested, inspected, and maintained in accordance with established processes that ensure design specifications and the expected level of safety performance is achieved.

3.3.2.1. *Regulatory expectations*

Considering the questionnaire responses received, regulatory bodies generally expect that the applicant/licensee will submit detailed information on the qualification of SSCs important to safety of the SMR, to confirm compliance with regulatory requirements and with any conditions specified in the authorization.

3.3.2.2. *Practical experience/challenges*

The following considerations represent the key experience/challenges that were identified in the questionnaire responses on this subject:

(a) In general, the Member State participants that had practical experience in the qualification of SSCs (e.g. Argentina and Canada) established guidelines and defined conditions so that the licensee can develop a qualification programme to demonstrate that they meet safety requirements in the early design stages. The objective of this programme is defining the items within its scope, establishing the qualification methods, and establishing the measures for maintaining the qualification.

(b) Several Member States participants (e.g. Argentina, Canada, China, the Russian Federation) put an emphasis on their national safety regulations and on the quality assurance programmes of the operating organizations (extended to suppliers), in order to ensure the safety of NPPs. This includes site selection, design, manufacturing, construction, commissioning, operation, and decommissioning stages, which have to be completed by qualified personnel.

(c) China reported that the safety analysis and environmental qualification requirements for SSCs in a floating NPP have to reflect the marine environmental conditions. As the review of the information for the floating reactor has not been completed, there is no specific experience.

(d) The Russian Federation reported that the rules used for assessment of compliance of items important to safety, do not apply to ships and other vessels with nuclear installations. In this case, the regulatory body established a temporary procedure for assessment of the conformity of SSCs and materials by other organizations allowed to perform this task.

(e)   The UK reported that at the time of writing this TECDOC, a meaningful assessment of the qualification process for SSCs in some proposed SMR designs cannot be made because of the limited level of design maturity and development. It means that there is limited information on the criterion that will be used to support the qualification of SSCs. The UK response also recognizes that qualifying equipment for operation up to 60 years is challenging, particularly in the context of advanced reactors for which there is limited experience.

(f)   Argentina reported that the CAREM 25 project developed a comprehensive and detailed methodology for establishing the engineering requirements for SSCs. Its objective was to establish an SSC safety classification system (based on the safety analysis) in order to set the design, manufacturing, assembly, testing, inspections, operation, and QA requirements among others, that will apply to each SSC, demonstrating the functional safety of the design.

(g)   Canada reported that one challenge could be whether a vendor submits creditable and sufficient information to support the qualification claims due to lack of operating experience on the proposed SMR design. The information provided by vendors were often studies from decades ago, but the technology was not mature enough to be built and used at the time. SSCs qualification continues to be work in progress. Other challenges are related to the introduction of novel items in the design.

(h)   France reported that qualification durability can be ensured by provisions regarding studies, building, testing, controls and maintenance.

Further information can be found in Section A–15 of the Annex in the response to Question 15.

3.3.2.3.   Looking ahead

The following activities and prospective changes for the assessment on qualification of SSC were reported by Member State participants:

(a)   There are challenges related to the introduction of novel items in the design. The demonstration of fulfilling their safety function for operational states and accident conditions might necessitate additional testing, modelling, and greater margins. In the construction stage, new approaches may be used which could result in varying quality of construction in new builds.

(b)   There are potential challenges with long lead items and the supply chain through the introduction of new suppliers who may have limited experience in addressing regulatory requirements. In some cases, engineering standards might not exist, or they may be utilized differently.

(c)   In novel technologies, identifying key commissioning activities is needed to provide greater confidence that they can be leveraged for future projects.

3.3.3. Novel/innovative design features

This subsection documents the key points from the questionnaire responses concerning novel/innovative design features (Question 14): regulatory expectations, practical experience/ challenges and forward-looking activities reported.

New reactor designs such as SMRs typically include novel and innovative design and safety features (see Section 1.3.1). For such facilities, the experts involved in this publication agree that the review and assessment by the regulatory body need to confirm that the applicant/licensee has performed a suitable and sufficient safety analysis of the SSCs important to safety and has used the results to demonstrate that the regulatory requirements are met and are reflected in operational procedures.

In performing the review and assessment of the SMR design, the regulatory body has to acquire an understanding of the design of the facility and its SSCs, the concepts on which the safety of the design is based and the operating principles proposed by the applicant/licensee to satisfy itself that operational and technical provisions — in particular for any novel features — have been proved or qualified by an appropriate programme of research, analysis and testing, complemented by a subsequent programme of monitoring during operation.

3.3.3.1. *Regulatory expectations*

Considering the questionnaire responses received, regulatory bodies generally expect that the applicant/licensee provides information of the innovative and novel design features that can have an impact on safety, security, safeguards and environmental protection, particularly when they depart from established practices. The consideration of these novel features needs to be supported by appropriate programmes of research, analysis and testing to demonstrate that they do not adversely affect safety, nuclear security, safeguards or environmental protection. This needs to be complemented by a subsequent programme of monitoring during operation.

3.3.3.2. *Practical experience/challenges*

The following considerations represent the key experience/challenges that were identified in the questionnaire responses about novel/innovative design features:

(a) It is recognized that the regulatory decision related to novel/innovative design features needs to consider various factors and the safety demonstration needs to be commensurate with the novelty, complexity and hazards associated with the facility and be based on research, analysis and testing, as well as computer modelling and simulations using V&V codes and past experience.

(b) Some Member State participants with an active SMR programme, such as Argentina, reported that vendors and applicants followed the route of a prototype and test facilities to inform the design and safety demonstration of new reactor designs.

(c) Other Member State participants, such as Canada, reported that vendors and applicants encountered challenges with new designs with innovative design features submitted for regulatory review and assessment in the absence of prototype facilities and/or relevant test and research programmes to support certain reliability and safety claims impacting on the project schedule.

(d) China reported that relevant test verification for a new design or a new equipment has to be carried out. For example, test verifications of a new fuel assembly and safety facilities for an SMR were carried out. The regulatory body needs to judge whether the new design or equipment meets the safety requirements and whether this is adequately demonstrated by the test verification scheme, the verification process and the verification result.

(e)   The USA reported that:

(i)   New designs, by their nature, challenge the regulatory body's review because they employ some features and methodologies that have not previously been evaluated. Much of the technology is not 'off the shelf' and the technical staff assigned to the review may not be familiar with it. The regulatory body has established and refined a request for information (RAI) process for dealing with questions that arise during the review. The US regulatory body found that multiple rounds of RAIs are inefficient; therefore it instituted procedures (e.g. detailed teleconferences explaining the intention of RAIs, when necessary, prior to issuing the final RAI) that have been effective in limiting the number of second round RAIs that they need to issue.

(ii)   Resolving issues with complicated new designs may not, however, be suited solely to the RAI process. Questions can be too intricate and tend to build one on the other. On-site audits and face to face meetings are often necessary to resolve complex problems. In conjunction with the audits, RAIs are commonly generated to get documented responses to resolve those complex problems.

(iii)   When new computer codes are used, the problems can become extremely complicated. benchmarking, validation, and verification can pose challenges.

The SMR Regulators' Forum WGs added:

(1)   Requirements and guidance are necessary for qualification programmes of new materials and features applicable to SMR designs including the extent and scale of the testing, verification and validation of models, and fabrication processes.

Further information can be found in Section A–14 of the Annex in the response to Question 14.

3.3.3.3.   Looking ahead

The following activities and prospective changes for the assessment of novel/innovative design features by Member State participants were reported:

(a)   Any new reactor design, such as SMRs involving novel and innovative design features, needs to be supported by a systematic approach based on sound engineering principles and past experiences, with the aim of demonstrating through robust research, tests and an adequate qualification programme that the regulatory requirements are met and that the design will withstand all postulated transient conditions and accident conditions.

(b)   New designs are expected to be tested before being brought into service and need to be monitored in service to verify that the expected behaviour is achieved.

3.3.4. Industry codes and standards

This subsection documents the key points from the questionnaire responses concerning industry codes and standards (Question 12): regulatory expectations, practical experience/challenges and forward-looking activities reported.

The experts involved in this review agreed that design and construction standards provide a framework to develop engineering rules and tools for designers, applicants, licensees. In addition, they are tools to 'assess' the fulfilment of the safety (engineering) requirements for SSCs. In SSG-30 [9], engineering rules are seen as related to the following characteristics:

(a)   Capability: to perform the required function;

(b)   Reliability: to perform the function with an acceptably low failure rate;

(c)   Robustness: to ensure that the operational loads of the demanding sequence, do not affect the performance.

3.3.4.1.   Regulatory expectations

Considering the questionnaire responses received, regulatory bodies generally expect that the applicant/licensee identifies and implements the applicable industry codes and standards relevant to the SMR.

Codes and standards applicable to SSCs have to be identified and their use needs to be in accordance with their classification. If different codes and standards are used for different types of items (e.g. for piping and for electrical systems), consistency between them should be demonstrated.

In the case of SSCs for which there are no appropriate established codes or standards, an approach derived from existing codes or standards for similar equipment may be applied, or in the absence of such codes and standards, the results of experience, tests, analysis, or a combination of these may be applied.

3.3.4.2.   Practical experience/challenges

The key practical experience and challenges in industry codes and standards reported in the questionnaire responses were as follows:

(a)   Member State participants, such as Argentina, Canada, China and South Africa, emphasize that when industry codes and standards maybe not fully applicable, other internationally recognized standards and codes, combined with good practices, may be used in order to meet regulatory requirements for equipment qualification. They also emphasize that the SSCs have to be designed according to the latest or currently applicable approved codes and standards, and be consistent with the plant reliability goals necessary for safety. In South Africa, the regulations require that where an unproven design or feature is introduced, a complete analysis supporting the design and the codes and standards used need to be provided. An example of a technical area that might pose challenges in this respect is high-temperature materials; however, the same approach has to be applied for all codes and standards to be selected.

(b)   Member State participants that have goal-setting regulatory approaches, such as the UK, do not prescribe specific design solutions, codes, and standards; vendors can propose alternative approaches in demonstrating the fulfilment of safety functions.

(c)   In South Africa, the PBMR design differs significantly from current LWRs. Therefore, the existing rules for the choice of design codes, standards, guidelines, and regulations cannot simply be applied. The problem with a blanket application of the current LWR code selection rules, is that it could result in the choice of an inappropriate code, resulting in a deficient design, or irreconcilable inconsistencies in the subsequent code choices.

(d)   China reported that some equipment for HTR-PM and ACP100 are designed and manufactured according to ASME III code Ref. [6].

(e)   Argentina reported that a challenge for the regulatory body was the monitoring and reviewing of activities related to the manufacture of the reactor pressure vessel of the CAREM 25 under the ASME code Ref. [6], taking into account the limited experience in the application of industry codes from other countries and that CAREM 25 is a FOAK case. Canada reported similar challenges.

(f)   The Russian Federation reported, in relation to land based SMRs, that preliminary analysis showed the need to develop new regulations based on the existing regulations for floating NPPs, taking into account the materials and alloys used, as well as design features and manufacturing methods of the reactor equipment.

(g)   The USA reported that one area that raised additional challenges and needed additional evaluation was related to the applicable code for small diameter (<5 cm) nuclear components.

Further information can be found in Section A–12 of the Annex in the response to Question 12.

3.3.4.3.   Looking ahead

The following activities and prospective changes for the assessment on industry codes and standards by Member State participants were reported:

(a)   Innovative approaches (e.g. in the deployment model, the safety analysis, etc.), materials and design features are being proposed for SMRs; however, there is sometimes an absence of relevant standards or codes that are applicable to them. In addition, there may also be limited experience in some Member State participants on the application of industry codes and standards from other Member States. This is a common theme in the responses from Argentina, Canada, South Africa and the UK.

(b)   Regulatory acceptance of innovative features or approaches is expected to depend on several factors such as the safety significance of the component, the availability of other components and other proven measures to fulfil the safety function, the level of analysis and evidence available to demonstrate the fitness for use, and the conservatism built into the analysis and design. As operating experience is gained, it is likely that standardised approaches may arise and regulatory bodies may evolve their regulatory positions and expectations on specific cases.

4. OTHER REGULATORY CHALLENGES

This section provides information on the regulatory expectations, Member State participants' experience and challenges, and forward-looking considerations regarding:

(a) Regulatory approach for suppliers;
(b) On-site inspections;

(c) Inspection of reactor internals, civil structures, and structures, systems and components;
(d) Emergency planning zone;
(e) Staffing levels of multi-unit plants;
(f) Occupational exposure;
(g) Safeguards;
(h) Nuclear security.

It is important to note that these topics have also links to topics addressed in Sections 2 and 3. The full set of questions for each topic in turn and the responses is documented in the Annex.

4.1. REGULATORY APPROACH TO SUPPLIERS

This subsection documents the key points from the questionnaire responses concerning regulatory approach to suppliers: regulatory expectations, practical experience/challenges and forward-looking activities reported.

The applicant/licensee is responsible for safety, regardless of the use of contractors or subcontractors. Regulatory approaches to verify compliance with requirements by suppliers vary from Member State to Member State.

4.1.1. Regulatory expectations

Considering the questionnaire responses received, regulatory bodies need to be conferred with the legal authority to obtain all necessary safety related information from SMR suppliers, even if this information is proprietary, and as well as the right to make regulatory inspections on the suppliers' site.

Whenever the applicant/licensee makes use of the safety related services or products of a contractor, the regulatory body expects that the contractor will be supervised by the applicant/licensee as part of its inspection programme in all steps of the authorization process. This may comprise inspection of the design and manufacturing of components, including, where appropriate, activities performed in other countries.

4.1.2. Practical experience and challenges

The following considerations represent the key experience/challenges identified by Member State participants in the questionnaire responses:

(a) Since there will be an increasing role of the manufacturer/supplier, i.e. in assembling equipment and modules in factory conditions, regulatory inspections performed in the factory are particularly important and new guidance for procedures for such inspections may need to be developed.

(b) Member State participants already have specific regulatory requirements for suppliers, and agree that these do not need to be modified to request information and inspections of suppliers.

(c) In general, Member State participants also agree on the importance of safety management programmes during design, manufacturing, construction, commissioning, operation, and decommissioning stages of the facility. to confirm compliance with regulatory requirements.

(d)   Regarding the experience of the UK, in line with UK law, the regulatory body can ask for information from suppliers to the nuclear industry, and powers are therefore given to inspectors to ask for this information for the purposes of nuclear safety. The law applies equally to suppliers to SMRs as to suppliers of existing nuclear installations, and suppliers to any new facilities currently being built. These arrangements include inspection of suppliers, irrespective of whether they supply existing facilities, new build, or (potentially) SMRs.

(e)   Argentina noted that their regulatory body only conducts inspections and audits to the licensee, not to vendors or suppliers. It is the responsibility of the licensee to oversee that the supplier management programme is implemented and the defined design criteria are accomplished. The regulatory body reviews the documentation presented, in order to define and plan the inspection tasks and audits related to the licensee oversight task.

(f)   Canada, the Czech Republic, and the USA reported that the regulatory body is authorized to inspect suppliers as well as licensees, and that accredited persons could verify the regulatory requirements.

(g)   The USA also reported that the regulatory body has modified its regulatory approach. The application of the current design certification differs from those received in the past and special circumstances resulted in the use of vendors or suppliers with specialized capabilities, some of whom may not be well known to the regulatory body. While regulations allow applicants to use these alternate vendors or suppliers, the regulatory body has to assess the vendors' or suppliers' technical and financial qualifications. To determine such qualifications the regulatory body has dispatched multi-disciplinary teams of technical specialist to conduct audits and find answers to questions as follows:

(i)   Has the applicant adequately assessed the ability of the vendor or supplier (including their subcontractors) to provide the necessary information?
(ii)   Is there reasonable assurance that the process employed by the applicant was adequate to identify all information needed?
(iii)   Is there reasonable assurance that the vendors or suppliers and their subcontractors will be able to assume the duties assigned?
(iv)   Do the vendors or suppliers have the expertise and technical competence to manage and control design changes and support the licensing process?
(v)   Do the vendors or suppliers have adequate financial resources to provide the services needed for the duration of the project?

Further information can be found in Section A–6 of the Annex in the response to Question 6.

4.1.3. Looking ahead

The following activities and prospective changes for the assessment on the regulatory approach for suppliers by Member State participants were reported:

(a)   The impact of ineffective management of contractors (as demonstrated by several adverse events across various high-risk industries) has been noted by some Member State participants. This highlights the need for organizations (licensees, vendors, designers) to retain the ability to understand, specify, oversee, and accept contractors' technical and physical work undertaken on its behalf.

In accordance with the characteristics of each type of SMR, and the discussions in Section 2.3, modifications may be necessary in some stages of the licensing process to the programme of regulatory inspections of suppliers. See for example Sections: off-site commissioning, transportation and off-site decommissioning.

4.2. ON-SITE INSPECTIONS

This subsection documents the key points from the questionnaire responses concerning on-site inspections: regulatory expectations, practical experience/challenges and forward-looking activities reported.

Paragraph 4.50 of GSR Part 1 (Rev. 1) [2] states:

> "The regulatory body shall develop and implement a programme of inspection of facilities and activities, to confirm compliance with regulatory requirements and with any conditions specified in the authorization. In this programme, it shall specify the types of regulatory inspection and shall stipulate the frequency of inspections and the areas and programmes to be inspected, in accordance with a graded approach."

The inspectors from regulatory bodies review documents provided by the applicant/licensee and perform site inspections, which could include observation of activities, interviews of personnel and reviews of records. The regulatory body also performs audits of the management system processes of the applicant/license.

4.2.1. Regulatory expectations

Considering the questionnaire responses received, regulatory bodies need to develop and implement a programme of inspection of SMR facilities and activities, to confirm the compliance with regulatory requirements and with any condition specified in the authorization.

4.2.2. Practical experience and challenges

The following considerations represent the key experience/challenges identified by Member State participants in the questionnaire responses:

(a) In general, Member State participants have specific regulations for organizing and conducting on-site inspections, and use international experience to improve the inspection activities. They do not need to modify their regulatory approach in order to manage on-site inspections of SMRs.

(b) In general, Member State participants agree on the importance of on-site inspections to confirm compliance with regulatory requirements.

(c) Regarding the experience of Argentina in the CAREM 25 prototype, the regulatory body established an inspection plan for SSCs with the highest nuclear safety qualification, during the construction stage. For these SSCs, the compliance with the safety requirements is verified, in a complete (exhaustive) manner, at the commissioning stage. As the acceptance criteria for the tests are essentially functional, the inspections during the functional tests in all the safety systems is carried out in a complete way, which in turn need to be full scope. Another example of experience is the managing of knowledge acquired in other national projects (completion of construction, commissioning and operation of the Atucha II NPP, refurbishment of the Embalse NPP, construction of RA-10 RR). It is important that this experience and knowledge could be applicable to next stages of CAREM 25 and will be not lost over time due to the possible generational turnover of inspection personnel.

(d) China reported that has developed special on-site inspection programmes and procedures for HTGR, taking into account that the systems and components are different to PWRs.

Further information can be found in Section A–5 of the Annex in the response to Question 5.

4.2.3. Looking ahead

The following activities and prospective changes for on-site inspections were reported by Member State participants:

(a) In accordance with the characteristics of each type of SMR, and the discussions in Section 2.3, modifications may be necessary to the on-site inspection programme, in different stages of the licensing process (e.g. construction, on-site commissioning and operation).

4.3. INSPECTION OF REACTOR INTERNALS, CIVIL STRUCTURES, AND STRUCTURES, SYSTEMS AND COMPONENTS

This subsection documents the key points from the questionnaire responses concerning: inspection of reactor internals, civil structures and SSCs: regulatory expectations, practical experience/challenges and forward-looking activities reported.

Inspection of reactor internals, civil structures and SSCs need to cover:

(a) All types of maintenance performed on SSCs, including maintenance performed in the physical and radiological conditions at the facility;

(b) Testing, including the conduct of all surveillance testing activities, all in-service inspection (ISI) and calibration of instruments, equipment operability tests and other special tests.

In accordance with Requirement 14 of IAEA Safety Standards Series No. SSR-2/2 (Rev. 1), Safety of Nuclear Power Plants: Commissioning and Operation [11], the applicant/licensee is required to establish and maintain an ageing management programme (AMP) that includes the following:

(a) Documented methods and criteria for identifying SSCs covered by the AMP;

(b) A list of SSCs covered by the AMP and records that provide information for use in the management of ageing;

(c) An evaluation of and documentation of potential ageing related degradation that may affect the safety functions of SSCs;

(d) Details of the extent of understanding of the dominant mechanisms of ageing for SSCs;

(e) Details of the programme for the timely detection and mitigation of ageing processes and/or ageing effects;

(f) Acceptance criteria and required safety margins for SSCs;

(g) Awareness of the physical condition of SSCs, including actual safety margins.

Surveillance programmes using representative material samples (e.g. material specimens for surveillance of the reactor pressure vessel (RPV), cable samples and corrosion coupons) need to be reviewed and extended or supplemented for ageing management purposes within the period of long term operation, if necessary.

4.3.1. Regulatory expectations

Considering the questionnaire responses received, regulatory bodies need to review and assess the information provided by the applicant/licensee in relation to inspection of reactor internals, civil structures and SSCs, including the case of innovative inspection approaches for SMRs, to confirm compliance with regulatory requirements and with any conditions specified in the authorization.

4.3.2. Practical experience and challenges

The following considerations represent the key experience/challenges identified by Member State participants in the questionnaire responses:

(a) In general, Member State participants reported that they have specific regulations to establish and evaluate requirements related to monitoring, testing, sampling and inspection of SSCs important to safety. These requirements establish rules and acceptance criteria and consider the impact of ageing and the effect of degradation mechanisms.

(b) China reported that the reactor internals and SSCs of small power reactors have to be fully inspected before putting into operation. However, the accessibility needed for the ISI of some SSCs cannot be met due to the integrated compact design adopted in the design of the small reactor. The licensee could apply for an exemption for the ISI of these SSCs, on the basis of the reliability of the equipment and the safety impact after the defect occurs.

(c) In South Africa, it was noted that the traditional LWR approach to inspection and testing requirements did not always produce coherent surveillance programmes commensurate with the SSCs importance to safety. Such NPP surveillance requirements are generally focused on LWR practices and may not be applicable to the PBMR. The adequacy of modern techniques and practices for inspection and testing in long term projects needs to be considered. South Africa reported additionally (for the PBMR):

   (i)   Online refuelling would possibly introduce dynamic loads and stresses on the core structure ceramics, introducing the need for routine inspections of the structures to determine the integrity of these internal structures;

   (ii) Maintenance outages were planned for every five years. As such, the internal structures need to maintain their integrity over long operating periods;

   (iii) It was necessary for the NPP to have adequate provision for monitoring and inspections as well as for instrumentation;

   (iv) The dust source term as well as activation products would present challenges in terms of occupational exposure during inspection activities. The contamination of the turbine due to the direct cycle also introduced concerns about occupational exposure during outages.

(d)   Some Member State participants reported as a challenge the deficiency in the accessibility for ISI due to some of the characteristics of SMR technologies (e.g. in designs with sealed cores or civil engineering structures). Canada reported that SMRs may present a number of challenges associated with the regulatory assessment of the provision for inspection of reactor internals, civil structures, and other SSCs. Vendors are opting for alternative methods and/or conservative safety margins in lieu of a typical periodic testing regime. Innovative inspection approaches have also been proposed by existing licensees in Canada, so this challenge is not necessarily unique to SMRs.

(e)   Other challenges were highlighted by the Czech Republic and the UK regarding the use of novel materials and technologies. The main challenges associated with the selected equipment and other SSCs are mainly to be expected if novel technologies are used in a situation where there is no or very limited experience with their application in the nuclear sector. For example, different coolant properties may necessitate novel inspection technologies.

Further information can be found in Section A–23 of the Annex in the response to Question 23.

4.3.3. Looking ahead

Member State participants reported the following challenges related to the inspection of reactor internals, civil structures and SSCs which may result in the need for innovative inspection approaches by regulatory bodies:

(a)   The inaccessibility for ISI due to some characteristics of SMR technologies (i.e. integrated compact design);

(b)   The dust source term as well as activation products present occupational exposure challenges during inspection activities at the PMBR;

(c)   Use of novel materials and technologies;

(d)   The use of novel technologies in a situation when there is no or very limited experience of their application in the nuclear sector.

4.4. EMERGENCY PLANNING ZONE

This subsection documents the key points from the questionnaire responses concerning the EPZ: regulatory expectations, practical experience/challenges and forward-looking activities reported.

Requirements for EPZs for nuclear installations are established in IAEA Safety Standards Series No. GSR Part 7, Preparedness and Response for a Nuclear or Radiological Emergency [12].

Some Member State participants prescribe the size of the EPZ, while others take a performance-based approach. For the prescriptive approach, the applicants (and/or vendors) are expected to demonstrate that the projected distance for which emergency measures would be required meets prescribed regulatory requirements for the size of EPZs. In the performance-based approach, applicants (and/or vendors) can use a mechanistic source term analysis, along with jurisdiction-specific considerations, to support their proposal for the sizes of EPZs.

The following considerations need to be addressed in demonstrating the size of the EPZ (in the performance-based approach), or in demonstrating that the impacted area meets the regulatory requirements in the prescriptive approach:

(a)   The rationale for the choice of the accident(s) (more than one accident may have to be selected depending on the jurisdiction).

(b)   The scenario of the selected accident(s). The radionuclide composition and activity of the radioactive release at the early phase of the accident, starting with its onset (until confinement (elimination) of the source term).

(c)   Analysis of the radiological consequences of the selected accident(s) and the predicted doses in terms of external and internal exposure.

(d)   The potential number of persons who might need medical assistance, considering the severity and type of exposure.

(e)   Assessment of the radiological situation within the facility (including the MCR and the emergency control room), within the site boundary, and beyond the site boundary.

(f)   The key calculation data for the EPZs, their boundaries, and characteristics (contamination zones, zones for protective action planning, zones for planning the actions on mandatory evacuation of the personnel, contamination levels at the zone boundaries) under assumed meteorological conditions.

(g)   The key measures to protect persons, based on the calculation of consequences predicted for the selected accident(s).

(h)   The calculated sizes of the EPZs in case of a nuclear or radiological emergency, including the prevailing winds ('wind rose'), average speed of the surface wind, and average air temperature.

4.4.1. Regulatory expectations

Considering the questionnaire responses received, regulatory bodies need to review and assess the information provided by the applicant/licensee related to the EPZ associated with the SMR project, to confirm compliance with regulatory requirements and with any condition specified in the authorization.

The arrangements for emergency preparedness and response, including the EPZ need to be developed based on the results of a hazard assessment, taking into account events of very low probability and events not considered in the design.

High uncertainties and the need for urgent response actions may persist for SMRs, hence the need for an emergency classification system and pre-established response plans.

4.4.2. Practical experience and challenges

The following considerations represent the key experience/challenges identified by Member State participants in the questionnaire responses:

(a)   In general, Member State participants have not made changes to their approach for establishing EPZs for SMRs. Some Member State participants, such as Canada and the Russian Federation, prescribe the size of EPZ, while others, such as South Africa and the UK, take a performance-based approach.

(b)   In Canada, to date there is insufficient information on accidents, malfunctions, and accident releases to determine the size of EPZs for SMRs (see information regarding in Section A–22 of the Annex, on accident source term).

(c)   China response's mentioned that the HTR-PM and ACP100 are located in the same site as other large commercial PWRs: consequently, the size and supervision of the EPZ has already been covered by that of large commercial PWRs.

(d)   Japan reported that the EPZ is set for each nuclear facility taking into consideration the risks inherent in each nuclear facility and potential degree of impact in the event of an accident. As an example, for a specific NPP facility a 'Precautionary Action Zone' (PAZ: within a radius of approximately 5 km) and an 'Urgent Protective Action Planning Zone' (UPZ: within a radius of approximately 30 km) are set. The size of the EPZ of the test and research reactor has been set based on its thermal power as well as the IAEA standard. The classification levels of thermal power until 2016 was different from the IAEA standard, but the same levels are used in the current guideline.

(e)   South Africa reported that any potential challenges with regard to demonstrating an acceptable EPZ for the PBMR were masked by the main challenge of an insufficiently mature safety analysis to identify such challenges.

(f)   The UK reported some key challenges relevant to SMRs and provided complementary information answering a follow-up question. If the regulations are deemed to apply, then the licensee has to conduct a hazard evaluation and consequence assessment, including sensitivity studies. Information provided by some vendors includes claims that no postulated credible events would result in radiological consequences large enough to require offsite emergency planning. Some vendors explicitly identify the goal to limit any detailed EPZ to within the site boundary.

(g)   The USA reported that the regulatory body's existing emergency preparedness programme for NPPs is focused on LLWRs. Based on the challenges of the proposed SMR design, the regulatory body is proposing to amend its regulations to create an alternate emergency framework for SMRs and other new technologies.

(h)   The Russian Federation reported that there have been no difficulties in reviewing the emergency response plans of the FNPP due to the experience gained by the regulatory body in reviewing the emergency plans for nuclear and radiological accidents (including those on vessels with nuclear reactors). Moreover, there is no intention to make amendments in the existing regulations regarding prevention and management of emergencies.

(i)   The SMR Regulators' Forum WGs [1] added:

   (i)   For SMRs without on-site refuelling capability, there is a need to consider the establishment of an EPZ at any intermediate location and land-based maintenance facility used for the handling and the storage of the fuel assemblies.

(ii) The existing requirements and guidance for emergency preparedness for multiple units emphasize the use of available means and/or support from other units, provided that their safe operation is not compromised. Proper consideration needs to be given to the operating mode (e.g. operation/shutdown/maintenance) of all unaffected units on the site and the limitations of non-standard equipment (e.g. cross-ties of electrical or heat removal systems) that might be shared between the units. The size of the EPZ may be impacted by the number of reactor modules/units postulated to be built at the site (in a simultaneous or sequential deployment); therefore, these aspects need to be adequately addressed at the design stage. Most SMR technology developers claim that their passive and inherent safety features, simpler operation and smaller source terms, result in a need for very limited arrangements for emergency preparedness and response and render the size of EPZ significantly smaller than that of large NPPs.

(iii) A pre-application process may be considered to discuss the requirements and standards for siting and determining EPZs with potential applicants.

(iv) The presence of multiple modules/units at the site could exacerbate challenges that the plant personnel would face during an accident. The events and consequences of an accident at one unit may affect the accident progression or hamper accident management activities at the neighbouring unit; available resources (personnel, equipment and consumable resources) would need to be shared among several units. These challenges need to be identified, and the available resources and mitigation strategies shown to be adequate.

Further information can be found in Section A–24 of the Annex in the response to Question 24.

4.4.3. Looking ahead

The following activities and prospective changes to the EPZ were reported by Member State participants:

(a) In view of the experience, challenges and considerations noted for the accident source term (see Section A–22 of the Annex), it may be a challenge for applicants (and/or vendors) to support their proposals regarding the size of EPZs. A conservative approach may have to be considered to compensate for uncertainties (or incomplete information) regarding the behaviour of:

    (i) Fuel and fission products;
    (ii) Safety systems;
    (iii) Containment/confinement structures.

(b) In addition, regulatory bodies may need to consider if their approach, be it prescriptive or performance-based, is suitable for emerging technologies, and changes such as changes in intervention levels, are acceptable to stakeholders.

(c) The USA reported that the alternative regulatory requirements and implementing guidance would adopt a risk-informed, performance-based, technology-inclusive, and consequence-oriented approach. The alternative requirements would include a scalable approach for determining the size of the EPZ around each facility[2]. The regulatory body is interested in addressing specific emergency preparedness policy issues such as:

(i)   How planning activities apply to the performance-based approach;

(ii)   How hazard analysis is applied to the performance-based approach;

(iii)   What specific factors or technical considerations are needed when applying the scalable EPZ approach.

4.5. STAFFING LEVELS OF MULTI-UNIT PLANTS

This subsection documents the key points from the questionnaire responses concerning staffing levels of multi-unit plants: regulatory expectations, practical experience/challenges and forward-looking activities reported.

The approach to the regulatory assessment of the adequacy of the staffing levels can be described as follows:

(a)   Licensees are expected to conduct and maintain a systematic analysis to determine the basis of the minimum staff complement, while considering:

(i)   Human performance and reliability;

(ii)   Staffing strategies under all operating conditions, including normal operation, AOO, DBA and DEC, taking into account the multi-unit facility configuration;

(iii)   The actions to be performed by operating personnel;

(iv)   The interactions among personnel;

(v)   Staff numbers, competencies, qualifications and workload demands associated with the tasks to be performed.

(b)   Licensees are expected to demonstrate safe operation and response to the most resource-intensive conditions (including events that affect more than unit) under all operating states including normal operations, AOO, DBA and DEC.

Anticipated areas of specific interest for SMRs include:

(a)   The impact of inherently safe/passively safe engineering;

(b)   The impact of a significant reduction of active safety systems;

---

[2] Recently the NRC reviewed an exception request by the Tennessee Valley Authority to scale down the EPZ for small modular reactors from the standard requirement. The NRC granted TVA its exemption from the 10-mile EPZ for future combined construction and operating licence applications for which radioactive source term is bounded by the conditions established by the NRC. Further, The NRC proposed to amend its regulations to include new alternative emergency preparedness (EP) requirements for SMRs and other new technologies to acknowledge technological advancements and other differences from large LWRs that are inherent in SMRs and other new technologies. The enactment of the proposed amendment is pending.

(c)   Single control room operation of multi-units;

(d)   Off-site control rooms;

(e)   Mission (load-following, or district heating or desalination instead of or addition to electrical power generation);

(f)   Unmanned operation;

(g)   Proposed significantly reduced staffing levels;

(h)   Changes to refuelling concepts (return to factory or none);

(i)   Automated manufacture;

(j)   Management of phased deployment models;

(k)   Off-site emergency response.

These aspects may change the role, responsibility, capability requirements and number of operating personnel. In the design of small reactors, measures for accident prevention and human factors engineering are used to reduce the task burden of control room operators, and the active safety system is not relied on excessively for normal, abnormal or accident operation. In addition, the number of systems is far less than that of LLWRs. This can significantly extend the response time available for personnel to take action, thus reducing the necessary actions of operating personnel.

4.5.1. Regulatory expectations

Considering the questionnaire responses received, regulatory bodies need to review and assess the information provided by the applicant/licensee related to staffing levels of multi-unit SMR facilities, to confirm the compliance with regulatory requirements and with any conditions specified in the authorization.

Licensees need to ensure staffing levels for nuclear installations are sufficient to meet the requirements arising from safety, nuclear security, and emergency preparedness and response.

4.5.2. Practical experience and challenges

The following considerations represent the key experience/challenges identified by Member State participants in the questionnaire responses:

(a)   In general, Member State participants have clear expectations regarding staffing levels for NPPs. Member State participants with operating multi-unit facilities, such as Canada, have specific expectations for staffing levels at these facilities.

(b)   For multiple reactor units, task analysis needs to consider operating multiple units in different operating modes. Not only do the actions necessary to operate the unit need to be defined, but also the interaction with other maintenance and support organizations of multiple units has to be carried out in the accident analysis. This situation will bring more complexity to the role of safety supervisors, who need to understand the whole design, the operation concept of each reactor type and the role and responsibility of operating personnel in the accident. Applicants need to establish staffing guidelines to better define the scope of tasks that operating personnel need to perform in a comprehensive accident analysis.

(c)   Canada reported that SMR applicants are required to carry out task analysis for all DBAs, identify appropriate staffing, and determine the functions assigned to the control room operator. The introduction of advanced reactor design and increased use of automatic control systems can have a significant impact on accident analysis, and ultimately affect the role, responsibility, composition and scale of the staff needed to control plant operations. Because of the design differences between SMRs and LWRs, the SMR may need fewer operating personnel to perform the same tasks.

(d)   Argentina reported that in case of the licensing of staff of CAREM 25, which is a FOAK reactor, more working experience of the personnel in the area of operation could be needed and may be a challenge.

(e)   France reported that in French NPPs, some personnel can be shared between two different units in the same plant.

(f)   South Africa reported that higher levels of automation and FOAK systems may influence the number of staff and the knowledge, skills and abilities necessary for control and mitigation actions. The use of FOAK systems does not, in itself, provide justification for less staff, and any claims will need to be supported by a comprehensive staffing analysis.

(g)   The USA reported that an SMR having multiple units/modules coupled with advances in control technologies brought about challenges in assessing the proposal to minimize staffing requirements.

Further information can be found in Section A–25 of the Annex in the response to Question 25.

4.5.3. Looking ahead

The following activities and prospective changes for the assessment of the staffing levels of multi-units plants were reported by Member State participants:

(a)   In view of the experience, challenges, and considerations noted for the accident source term (see Section A–22 of the Annex), and to better define reactor core behaviour under normal operating conditions, it may be a challenge for applicants (and/or vendors) to support their proposals regarding staffing levels. A conservative approach may have to be considered to compensate for uncertainties (or incomplete information) regarding the behaviour of:

(i)   Fuel and fission products;
(ii)   Safety systems;
(iii) Containment/confinement structures.

(b)   Comprehensive analyses will have to be undertaken regarding:

(i)   On-site emergency response capacity and capability (including to support fire protection);
(ii)   Staffing levels to support nuclear security (e.g. including considerations for security by design).

4.6. WORKER DOSE AND PUBLIC EXPOSURE

This subsection documents the key points from the questionnaire responses concerning occupational exposure: regulatory expectations, challenges/practical experience and forward-looking activities reported.

The assessment and review of radiation risks in normal operation is directed towards the determination of occupational exposures and radioactive discharges to the environment. These data will be compared with the safety objectives, requirements, constrains and limits approved by the regulatory body, including application of the principle of optimization of protection and safety.

In the regulatory review and assessment, particular attention needs to be devoted to those aspects that influence the protection of people and the environment in normal operation, which include:

(a) The occupational radiation protection programme and other matters relating to radiation protection of workers;

(b) Radiation protection of the public, with all exposure pathways taken into account;

(c) Discharge, dilution and dispersion of radioactive effluents.

In considering these aspects, the regulatory body has to satisfy itself that radiation doses to workers and the public and radioactive releases to the environment are below relevant limits, are as low as reasonably achievable and that the relevant dose constraints are taken into consideration. Specifically, review and assessment needs to verify that:

(a) The operational limits and conditions and the bases for these have been determined;

(b) The radiation risks associated with operation at these limits have been considered;

(c) Arrangements (including operating procedures) are in place to ensure that protection and safety is optimized.

4.6.1. Regulatory expectations

Considering the questionnaire responses received, regulatory bodies need to review and assess the information provided by the applicant/licensee related to occupational exposure, to confirm compliance with regulatory requirements and with any conditions specified in the authorization.

Regulatory body expectations for occupational exposure are similar for SMRs compared to other technology. It is required that the design of the nuclear installation duly considers occupational exposure during operational states and accident conditions and includes features such as adequate shielding and ventilation systems for radiation protection: see Requirements 5 and 81 of SSR-2/1 (Rev. 1) [3].

Regulatory bodies need to review how the ALARA principle was applied during the design process to identify and describe design features and specifications intended to optimize occupational exposure. This includes occupational exposure from operating modules of the SMR to workers constructing or installing additional modules, and radiation exposure to workers in operating modules.

4.6.2. Practical experience and challenges

The following considerations represent the key experience/challenges identified by Member State participants in the questionnaire responses:

(a) There are several challenges in the context of occupational exposure and public exposure for SMRs, which are in some cases common to many disciplines and technical areas. These have been stated as follows:

    (i) Lack of operating experience (applies also to question 22: accident source term);

    (ii) Lack of analysis of source term for normal operation and faults (also question 22);

    (iii) Unique radiation sources and pathways;

    (iv) Fuel handling, shielding along fuel routes and maintenance of remote and difficult to access plants;

    (v) Coolant activation and the unavailability of data on achievable coolant impurity levels, corrosion of surfaces, mobility and resultant coolant activity levels;

    (vi) Mobility of source term (airborne material) in the case of HTGRs;

    (vii) Structural activation of components near the reactor core in compact designs;

    (viii) Fission product release rates from novel fuel (also question 22);

    (ix) Plate out and cleanup rates of activation and fission products from coolant in novel designs (also question 22);

    (x) Performance of novel HVAC and containment/confinement systems in removing airborne activity (also question 22);

    (xi) Reliance on automation for operation and maintenance.

(b) South Africa reported that the consideration of potential occupational exposure (worker dose) led to limitations on design options. For example events that need early operator actions or actions involving high levels of occupational exposure are not acceptable. Additionally, the lack of progress with fuel qualification was of concern as fuel qualification and performance is the basis of the PBMR safety case and directly related to the source term, confinement issues and consequently the exposure of workers and the public.

Further information can be found in Section A–20 of the Annex in the response to Question 20.

4.6.3. Looking ahead

The following activities and prospective changes in relation to occupational exposure were reported by Member State participants:

(a) Lack of operating experience;

(b) Challenges related to the source term;

(c)   The performance of novel HVAC and containment/confinement systems in removing airborne activity;

(d)   Reliance on automation for operation and maintenance.

4.7. SAFEGUARDS APPROACH

This subsection documents the key points from the questionnaire responses concerning the approach to safeguards: challenges/ practical experience and forward-looking activities reported.

Under a comprehensive safeguards agreement (CSA)[3], the IAEA has the right and obligation to apply safeguards on all nuclear material in all peaceful nuclear activities within the territory of the State, under its jurisdiction or carried out under its control anywhere, for the exclusive purpose of verifying that such material is not diverted to nuclear weapons or other nuclear explosive devices. The IAEA and the State are required to cooperate to facilitate the implementation of safeguards provided for in the CSA.

To ensure the effective implementation of safeguards, the State has the obligation under the CSA to establish and maintain a national system of accounting for and control of nuclear material and to provide the IAEA with information concerning nuclear material subject to safeguards and the features of facilities (i.e. design information) relevant to safeguarding such material.

Safeguards requirements have to be considered by States in the design of NPPs. Integration of safety measures and safeguards measures will help to ensure that neither compromise the other.

4.7.1. Safeguards obligations

First, the provision of early design information for new facilities is required for all States with a CSA. Preliminary design information for new facilities should be provided to the IAEA as soon as the decision to construct or to authorize construction has been taken, whichever is earlier.

Further information on designs of new facilities should be provided to the IAEA as the designs are developed. Such information, which can be submitted in the form of a preliminary design information questionnaire (DIQ), includes, the physical location, preliminary design drawing or plant process layouts. The completed DIQ of a new facility is to be submitted to the IAEA, based on preliminary construction plans, as early as possible, but not later than 180 days prior to the start of the construction of the facility. Completed DIQs for new facilities, based on 'as-built' designs, should be provided to the IAEA as early as possible, and in any event not later than 180 days before the first receipt of nuclear material at the facility. Significant changes to facility design relevant for safeguards purposes are to be provided to the IAEA for examination sufficiently in advance for the safeguards approach and procedures to be adjusted when necessary. The IAEA has the right to examine and verify design information throughout the lifetime of the facility.

The provision of such information for planned facilities enables the State and the IAEA to cooperate in a timely manner to prepare for safeguards implementation at such facilities by, for example, discussing safeguards by design at a very early stage of the facility planning, to facilitate

---

[3] Based on "The Structure and Content of Agreements between the Agency and States Required in Connection with the Treaty on the Non-Proliferation of Nuclear Weapons" (INFCIRC/153 (Corrected)) [13].

the effective and efficient application of safeguards over the lifetime of the facility (see Section 4.7.2 for additional information on safeguards by design).

Furthermore, a CSA concluded between a State and the IAEA requires that State to establish and maintain a State system of accounting for control of NM (SSAC) within its territory or under its jurisdiction or control elsewhere. The SSAC is normally established by the State authority with responsibility for safeguards implementation, the Safeguards Regulatory Authority(SRA), and has the responsibility to account and control NM in all nuclear activities the State.

The CSA also requires the SSAC to be based on a structure of material balance areas, and to make provision as appropriate and specified in the Subsidiary Arrangements for the establishment of such measures as: (a) measurement system for the determination of the quantities of nuclear material received, produced, shipped, lost or otherwise removed from inventory, and the quantities on inventory; (b) the evaluation of precision and accuracy of measurements and the estimation of measurement uncertainty;(c) procedures for identifying, reviewing and evaluating differences in shipper/receiver measurement;(d) procedures for taking a physical inventory; (e) procedures for the evaluation of accumulations of unmeasured inventory and unmeasured losses; (f) a system of records and reports showing, for each material balance area, the inventory of nuclear material and the changes in that inventory including receipts into and transfers out of the material balance area;(g) provisions to ensure that the accounting procedures and arrangements are being operated correctly; and, (h) procedures for the provisions of reports to the Agency in accordance with the provisions on reports.

An SSAC comprised of all of the elements that enable the SRA to carry out its nuclear material accounting and reporting responsibilities. These elements include information systems (computerized or paper-based); nuclear material accounting systems that produce the accounting data at facilities and other locations; various processes, procedures and administrative controls (such as license requirements including import and export; collection and submittal of design information); quality checks; and oversight activities conducted by the SRA to ensure that safeguards requirements are satisfactorily met.

Further information on States' safeguards obligations under a CSA are provided in Ref. [13].

4.7.2. Safeguards by design

Safeguards by design (SBD) refers to the inclusion of safeguards considerations early in the design process, at any point in a facility's lifetime and at any stage of a State's nuclear fuel cycle for which safeguards are applicable. SBD is fundamentally a voluntary best practice, and should not be confused with a State's obligation for the early provision of design information to the IAEA, as described in Section 4.7.1. As a concept, SBD is consistent with the early provision of design information by a State under its safeguards obligations, but it is not limited to the timeframe associated with these obligations. For example, the State regulatory authorities may wish to encourage SBD by including safeguards considerations in pre-licensing review applications, a process that typically occurs 'upstream' of the requirements for provision of design information for safeguards purposes.

SBD promotes the efficient, effective implementation of safeguards, with potential benefits for all stakeholders (designer, vendor, regulatory body, State authority for safeguards, operating organization, contractors, and the IAEA). Guidance on the principles of safeguards by design is provided in Refs [14, 15, 16].

4.7.3. Practical experience and challenges

The following considerations represent the key experience/challenges by Member State participants in the questionnaire responses:

(a)   In general, Member State participants agree that the safeguards measures applied are based on the design and operation of the facilities. It is likely that a different safeguards approach will be agreed for each of the reactor types as more detailed interactions proceed and as the facility design becomes more definitive.

(b)   The UK highlighted that the safeguards approach will be discussed with all relevant stakeholders, as necessary, during the evolution of that particular reactor type (including vendor/operating organization, regulatory body for safeguards (and security and safety colleagues) and IAEA) during planning, design, construction and commissioning. Direct discussions with the vendor/operating organization will go through a number of stages as the project matures.

(c)   To ensure that the vendor/operating organization understands safeguards requirements and how they will impact the facility (and any associated services), the UK and Canada noted that the regulatory body has engaged in discussions with SMR vendors who have provided design information at an early stage. This has allowed the regulatory bodies to consult with the IAEA at an early stage to allow the vendors to incorporate safeguards requirements into their design and construction plans without any unexpected requirements or retrofits being necessary.

(d)   The Czech Republic reported that any SMR, regardless of its design, is considered as a facility, as defined in a CSA, and is therefore subject to all relevant safeguards legal requirements under a CSA.

(e)   Some Member State participants (e.g. China, Canada) mentioned the approach to SBD. In the case of China, the safeguards approach was discussed with the IAEA from the design stage, and an approach based on the characteristics of the SMR has been agreed upon and applied, in particular in relation to the design of equipment with regard to safeguards.

(f)   South Africa mentioned that the main challenges in a pebble bed reactor with online refuelling relates to fuel accountancy.

(g)   According to Section 2.4, additional challenges in safeguards could appear, at least, in manufacturing facilities involved in fuel loading and nuclear commissioning tests.

Further information can be found in Section A–26 of the Annex in the response to Question 26.

4.7.4. Looking ahead

The following activities and prospective changes to the safeguards approach were reported by Member State participants:

(a)   SBD is an approach whereby early consideration of safeguards is included in the design process of a nuclear installation, allowing optimized design choices that take into account economic, operational, safety, and security factors, in addition to safeguards.

(b)   Canada reported that one challenge with the SBD approach is that early design concepts can change a lot during the design process and therefore an iterative approach is necessary.

(c)   Some SMR deployment scenarios will present specific technical and logistical challenges to safeguards inspectors, for example fleets of smaller SMR facilities distributed across a large and possibly remote geographical region making physical inspections complex from a travel perspective. As a result, alternative but equally rigorous safeguards approaches for such deployment scenarios may need to be developed to facilitate efficient and effective inspections.

(d)   According to Section 2.4, additional challenges in safeguards could appear, at least, in case of manufacturing facilities involved in fuel loading and nuclear commissioning.

4.8. SECURITY APPROACH

This subsection documents the key points from the questionnaire responses concerning security approach: regulatory expectations, practical experience/challenges and forward-looking activities reported.

The overall objective of a State's nuclear security regime is to protect persons, property, society, and the environment from malicious acts involving nuclear material and other radioactive material. The objectives of the State's physical protection regime, which is an essential component of the State's nuclear security regime, are:

(a)   To protect against theft and other unlawful taking of NM;

(b)   To locate and recover missing NM:

(c)   To ensure the implementation of rapid and comprehensive measures to locate and, where appropriate, recover missing or stolen NM;

(d)   To protect nuclear material and nuclear facilities against sabotage;

(e)   To mitigate the radiological consequences of sabotage.

The State's physical protection regime for NM needs to achieve these objectives through:

(a)   Prevention of a malicious act by means of deterrence and by protection of sensitive information;

(b)   Management of an attempted malicious act or a malicious act by an integrated system of detection, delay, and response.

The objectives mentioned above need to be addressed in an integrated and coordinated manner taking into account the different risks covered by nuclear security.

The nuclear material accounting and control system and the physical protection system are two distinct systems that have to complement one another in achieving the nuclear security objective of deterrence and timely detection of unauthorized removal of nuclear material. Each system has its own set of requirements and objectives, and both are important to nuclear security. The lifetime of a nuclear installation extends from the earliest planning stages through to its decommissioning. It is important to consider nuclear security early in the design of new facilities and during partial redesigns or modifications, as it can result in nuclear security for these facilities that is more efficient, more effective and better integrated with safety, safeguards, operational and other measures. Nuclear security measures are also important during commissioning and operation; they are not to cease at decommissioning, as they are important in addressing the protection of the remaining quantities of NM or other radioactive material, which has accumulated during operation.

4.8.1. Regulatory expectations

Considering IAEA safety standards and nuclear security guidance publications, and the questionnaire responses received, regulatory bodies need to review and assess the information provided by the applicant/licensee related to nuclear security, to confirm compliance with regulatory or other national requirements and with any conditions specified in the authorization.

4.8.2. Practical experience and challenges

The following considerations represent the key experience/challenges identified by Member State participants in the questionnaire responses:

(a) The introduction of SyAPs in the UK was a significant move away from a prescriptive security regulatory approach towards a more outcome-focused approach. This new regulatory approach applies to all duty holders and those who wish to build new facilities whatever the design. In assessing SMRs, and especially advanced and novel designs, the regulatory body has started to take an approach based on experience within GDA. In terms of the approach to nuclear security, these technologies and builds offer opportunities to reduce the security risk and also may present new risks. The adoption of SyAPs enables a flexible and risk-based approach that is applicable to SMRs. Using the GDA security framework for reviewing and assessing, changes to the approach in the context of SMRs can be identified.

(b) Argentina reported that for the regulatory body to grant the authorization to introduce NM to the site, the licensee has to comply with the regulatory requirements and procedures for physical protection.

(c) Canada reported that the Nuclear Security Regulations (NSR) and associated regulatory documents define the nuclear security requirements and guidance for the licensing, construction and operation of nuclear installations (including high-security installations), and for the production, use, transport and/or storage of NM. In addition, the NSRs ensure that Canada continues to achieve conformity with measures of control and international obligations related to nuclear security to which Canada has agreed. Developers of SMR technologies are seeking alternative approaches to security, such as security by design, in order to reduce the need for security personnel. One of the concerns is that current security requirements are not sufficiently flexible to address design approaches that could allow for a reduction in security personnel.

(d) The Russian Federation reported that there are several physical protection aspects in relation to transportation of an FNPP with a fully loaded core, including the need for providing patrol guards along the entire route of its movement to the site. However, the regulations of the Russian Federation establish standard approaches to physical protection, which are applicable to any nuclear installation. The specific features of physical protection for FNPPs are also established in the federal regulations.

(e) The USA reported that some design aspects, such as the below-grade installation of near term SMRs, provide additional security benefits, such as minimizing aircraft impact, limiting access to vital areas and limiting the communication ability of adversaries. These same features may provide an excellent means of enhancing the effectiveness of security systems against sabotage. The application of the traditional multi-layered defensive approach of deterrence, detection, assessment, delay, and interdiction can be used effectively for physical protection of SMRs.

(f) China reported that, as the HTR-PM and ACP100 is located in the same site as other large commercial PWRs, the approach to nuclear security is covered by that of large commercial PWRs.

(g) The Czech Republic reported that no changes in the security approach have been made as no SMR has been deployed in Czech Republic. For the same reason, the Czech Republic has not identified challenges associated with nuclear security. At the same time, there are certain aspects that could be highlighted in the current legislative system governing nuclear security. The provisions of the Atomic Act and the complementary Decree on security of nuclear installation and NM that are relevant to nuclear security are a combination of prescriptive and performance-based approaches. Some of these provisions are very detailed, including very specific description of the requirements. Most probably the biggest challenges could be envisaged in the area of delineation and physical demarcation of guarded areas, protected areas, inner areas or vital areas, as certain provisions, especially in the implementing decree, are quite specific. In includes, for example, a detailed description of the technical measures for delineation (height of fences, CCTV, etc.).

(h) According to Section 2.4, additional challenges in nuclear security could appear, at least, in case of manufacturing facilities involved in fuel loading and nuclear commissioning.

Further information can be found in Section A–27 of the Annex in the response to Question 27.

4.8.3. Looking ahead

The following activities and prospective changes to the security approach were reported by Member State participants:

(a) Some Member State participants, such as Canada, mentioned that developers of SMR technologies are seeking alternative approaches to nuclear security early in the design, in order to reduce the need for security personnel. One of the concerns of Member State participants is that their current security requirements are not sufficiently flexible to address design approaches that could allow for a reduction in security personnel. Other relevant challenges identified were the integration of safety, nuclear security and site design.

(b) Some challenges identified by Member State participants are related to new technologies, including the introduction of new vulnerabilities such as those posed by cybersecurity threats. At the same time, new technologies provide opportunities to better protect nuclear material and facilities, including nuclear security infrastructure, against threats.

(c) According to Section 2.4, additional challenges in nuclear security could appear, at least, in case of manufacturing facilities involved in fuel loading and nuclear commissioning and during their transportation.

REFERENCES

[1]   INTERNATIONAL ATOMIC ENERGY AGENCY, Small Modular Reactor (SMR) Regulators' Forum (2021), https://www.iaea.org/topics/small-modular-reactors/smr-regulators-forum.

[2]   INTERNATIONAL ATOMIC ENERGY AGENCY, Governmental, Legal and Regulatory Framework for Safety, IAEA Safety Standards Series No. GSR Part 1 (Rev. 1), IAEA, Vienna (2016).

[3]   INTERNATIONAL ATOMIC ENERGY AGENCY, Safety of Nuclear Power Plants: Design, IAEA Safety Standards Series No. SSR-2/1 (Rev. 1), IAEA, Vienna (2016).

[4]   INTERNATIONAL ATOMIC ENERGY AGENCY, Licensing Process for Nuclear Installations, IAEA Safety Standards Series No. SSG-12, IAEA, Vienna (2010).

[5]   OFFICE FOR NUCLEAR REGULATION, Technical Assessment Guides, ONR (2021), https://www.onr.org.uk/operational/tech_asst_guides/index.htm.

[6]   AMERICAN SOCIETY OF MECHANICAL ENGINEERS, ASME Boiler and Pressure Vessel Code (BPV Code), ASME, New York (2019).

[7]   INTERNATIONAL ATOMIC ENERGY AGENCY, Safety Glossary: Terminology Used in Nuclear Safety and Radiation Protection, 2018 Edition, IAEA, Vienna (2019).

[8]   INTERNATIONAL ATOMIC ENERGY AGENCY, IAEA Safety Standards Series No. SSG-2 (Rev. 1), IAEA, Vienna (2019).

[9]   INTERNATIONAL ATOMIC ENERGY AGENCY, Safety Assessment for Facilities and Activities, IAEA Safety Standards Series No. GSR Part 4 (Rev. 1), IAEA, Vienna (2016).

[10]   INTERNATIONAL ATOMIC ENERGY AGENCY, Safety Classification of Structures, Systems and Components in Nuclear Power Plants, IAEA Safety Standards Series No. SSG-30, IAEA, Vienna (2014).

[11]   INTERNATIONAL ATOMIC ENERGY AGENCY, Safety of Nuclear Power Plants: Commissioning and Operation, IAEA Safety Standards Series No. SSR-2/2 (Rev. 1), IAEA, Vienna (2016).

[12]   FOOD AND AGRICULTURE ORGANIZATION OF THE UNITED NATIONS, INTERNATIONAL ATOMIC ENERGY AGENCY, INTERNATIONAL CIVIL AVIATION ORGANIZATION, INTERNATIONAL LABOUR ORGANIZATION, INTERNATIONAL MARITIME ORGANIZATION, INTERPOL, OECD NUCLEAR ENERGY AGENCY, PAN AMERICAN HEALTH ORGANIZATION, PREPARATORY COMMISSION FOR THE COMPREHENSIVE NUCLEAR-TEST-BAN TREATY ORGANIZATION, UNITED NATIONS ENVIRONMENT PROGRAMME, UNITED NATIONS OFFICE FOR THE COORDINATION OF HUMANITARIAN AFFAIRS, WORLD HEALTH ORGANIZATION, WORLD METEOROLOGICAL ORGANIZATION, Preparedness and Response for a Nuclear or Radiological Emergency, IAEA Safety Standards Series No. GSR Part 7, IAEA, Vienna (2015).

[13]   INTERNATIONAL ATOMIC ENERGY AGENCY, The Structure and Content of Agreements Between the Agency and States Required in Connection with the Treaty on the Non-Proliferation of Nuclear Weapons, IAEA INFCIRC/153(Corrected), IAEA, Vienna (1972).

[14]   INTERNATIONAL ATOMIC ENERGY AGENCY, Safeguards Implementation Practices Guide on Provision of Information to the IAEA, IAEA Services Series 33, IAEA, Vienna (2016).

[15]   INTERNATIONAL ATOMIC ENERGY AGENCY, International Safeguards in Nuclear Facility Design and Construction, IAEA Nuclear Energy Series no. NP-T-2.8, IAEA, Vienna (2013).

[16]   INTERNATIONAL ATOMIC ENERGY AGENCY, International Safeguards in the Design of Nuclear Reactors, IAEA Nuclear Energy Serie no. NP-T-2.9, IAEA, Vienna (2014).

RESPONSES BY MEMBER STATES TO THE QUESTIONNAIRES

This Annex presents the detailed responses to the 27 topical questions submitted by the Member States with experience on regulating SMRs consulted for this TECDOC. The Annex is structured sequentially, one section per topic with all MS responses on the same topic. When there was a need to expand the details of the responses through follow up questions, the responses were included in the same topical section under the heading of the country submitting the expanded details.

The list of Member States consulted follows:

- Argentina – National Regulatory Authority (ARN).
- Canada – Canadian Nuclear Safety Commission (CNSC).
- China – National Nuclear Safety Administration (NNSA).
- Czech – Republic State Office for Nuclear Safety (SÚJB).
- France – French Nuclear Safety Authority (ASN).
- Japan – Nuclear Regulation Authority (NRA).
- Russian – Federation Rostechnadzor.
- South Africa National Nuclear Regulator (NNR).
- United Kingdom – Office for Nuclear Regulation (ONR).
- United States of America – Nuclear Regulatory Commission (NRC)

As normal practice, the editing of the content of the responses of the MSs in this Annex has been reduced to the minimum.

A–1. LEGAL FRAMEWORK

This Annex presents the responses provided by the following Member State regulatory bodies to Question 1: "Was there a need to change the legal framework (including the Authorization process) in your country as a result of proposed projects involving SMRs?"

A–1.1. ARGENTINA–ARN

A–1.1.1. Question

Was there a need to change the legal framework (including the Authorization process) in your country as a result of proposed projects involving SMRs?

A–1.1.2. Response

No, it was not necessary to change the legal framework to license the prototype of the CAREM 25 reactor, but some laws and decrees were issued to facilitate the development of the project.

The national policy applicable to nuclear activities with peaceful uses is integrated by the provisions of the National Constitution and the legislation adopted by the National Congress by Law No. 24,804 enacted in 1997 [A–1] and it was not modified. The latter rules the Nuclear Activity along with Law No. 24,776 [A–2] which approved the Convention on Nuclear Safety in 1997, and different laws related to the nuclear activity in accordance with treaties, conventions, agreements, and international conventions.

In Argentina, the national nuclear policy was initially established by Decree No. 10,936 enacted in 1950 [A–3], which created the National Atomic Energy Commission (CNEA) with the objective of developing and handling nuclear technology. The control of the safety aspects of all nuclear activities performed in the country until the year 1994 was performed by the CNEA through its regulatory division, according to Law No. 14,467 [A–4] and Decree No. 842/58 [A–5].

In 1994, the National Government assigned the exclusive performance of these duties to an independent state agency with federal competence. This was implemented in the frame of the Decree No. 1,540/94 [A–6] that reorganized the activities from the nuclear sector and divided them into three entities; Nuclear National Regulatory Body (ENREN), Nucleoeléctrica Argentina Sociedad Anónima (NA-SA), and the National Atomic Energy Commission (CNEA), respectively responsible for the regulation, operation of facilities, and for research and development of the sector. Before that division, all these activities were developed by CNEA. The abovementioned decree was then formally substituted by the Federal Law No. 24,084 [A–1] known as the 'National Law of Nuclear Activity' sanctioned by the Argentine National Congress in 1997 and later complemented by the ruling Decree No. 1,390/1998 [A–7]. The Nuclear Regulatory Authority (ARN) was created, as the successor of the ENREN. Within this context, Law No. 24,804 [A–1] is the current legal framework for the peaceful uses of nuclear energy in Argentina. Article 1 of the mentioned Law establishes that concerning nuclear matters the National Government, through the National Atomic Energy Commission (CNEA) and the Nuclear Regulatory Authority (ARN), shall define the state policy.

In 2006, from Decree 1107/06 [A–8] of the Executive Power, the construction and commissioning project of the CAREM Reactor Prototype was declared of National Interest.

In 2009, Law No. 26,566 [A–9] was passed, declaring 'of national interest' and 'entrusting the CNEA with the design, execution, and commissioning of the CAREM Reactor Prototype'. This law established important benefits from the state that facilitate the development of the project. For example, an exemption from all national taxes, authorization for the creation of an escrow to integrate the different sources of resources, facilities for the expropriation of real estate necessary for the works, a special customs control regime for the entry into the country of elements related to the project, mechanisms to facilitate and encourage the hiring of national companies participating in the project, among other measures.

The nuclear policy shall meet all the obligations assumed by the Argentine Republic as a party to the Treaty for the Prohibition of Nuclear Weapons in Latin America and the Caribbean (Tlatelolco Treaty), the Treaty on Non-Proliferation of Nuclear Weapons (NPT), the Agreement for the Application of Safeguards involving the Argentine Republic, the Federative Republic of Brazil, the Brazilian-Argentine Agency for Accounting and Control of Nuclear Materials (ABACC) and the International Atomic Energy Agency (IAEA), in addition to the commitments assumed by Argentina as a member of the Nuclear Suppliers Group and the National Regime for the Control of Sensitive Exports (Decree No. 603/92 [A–10]).

A–1.1.3. Follow-up Question

Please provide a summary of the key features of the regulatory framework (general) and line of sight to the law with regards to the responsibility of the operating organisation.

A–1.1.4. Response

The Argentine Regulatory Standards are based on a set of fundamental concepts, which are part of the Performance Approach philosophy, sustained by the regulatory system, concerning radiological and nuclear safety, safeguards and security.

Regulatory Standards are not prescriptive but of compliance with safety objectives (performance). The compliance of these objectives has to be demonstrated by the licensee by sound procedures within mandatory documents than can be objectively assessed by the Regulatory Body.

Regarding the adoption of a performance based regulatory approach, some advantages, learnt by the verified application experience are the following:

- The nature of the interaction between the Regulatory Body and the Licensee contributes to an early detection of possible non-compliances or deficient compliance with regulatory requirements (in early design stages), avoiding the increase in time and efforts in fulfilling such requirements in later phases of a project (fabrication or construction).

- The design solutions to comply with regulatory requirements come, in general, from the supplier (nuclear vendor) through the Licensee, that know in detail the installation and the system involved in.

- The establishment of safety objectives keeping openness to different design solutions, helps to manage projects from different vendors, i.e. Nuclear reactors with different safety approaches, while keeping coherence on the need of objective (factual) demonstration of the compliance with regulatory requirements.

According to what is stated in Standard AR 0.0.1. 'Licensing of Class 1 Facilities' [A–11], "(…) the Responsible Entity (holder of the licenses of a Class I facility) is the organization responsible for the radiological and nuclear safety (also safeguards and security) of a Class I facility. This responsibility also implies that The Responsible Entity must do everything reasonable and compatible with its possibilities in favour of the safety of the Class I installation, complying, as a minimum, with the standards and requirements of the Regulatory Authority. This responsibility extends to the development of the Class I facility, comprising the design, construction, commissioning, operation and decommissioning stages".

A–1.2. CANADA–CNSC

A–1.2.1. Question

Was there a need to change the legal framework (including the Authorization process) in your country as a result of proposed projects involving SMRs?

A–1.2.2. Response

The Canadian Nuclear Safety Commission (CNSC) operates within a modern and robust legislative and regulatory framework. For more details regarding these frameworks, please see Article 7 of the seventh Canadian National Report for the Convention on Nuclear Safety [A–12]. The Nuclear Safety and Control Act (NSCA) [A–13] is the enabling legislation for the regulatory framework. Under the NSCA, the CNSC has a legislated mandate to regulate the use of nuclear energy and materials in order to protect health, safety, security and the environment.

The conduct of activities associated with site preparation, construction, operation, decommissioning and ultimately making a decision regarding release of the licensee from regulatory control are subject to the federal NSCA [A–13] as well as all other applicable federal, provincial and/or territorial legislation.

As stated in many fora and as documented in DIS-16-04 Small Modular Reactors: Regulatory Strategy [A–14] and the Strategy for Readiness to Regulate Advanced Reactor Technologies [A–15], the current regulatory framework including the Act, Regulations, and regulatory documents are adequate for the licensing of projects involving SMRs. As part of its readiness reviews, changes were identified as beneficial to two regulations, namely the Nuclear Security Regulations (NSR) [A–16] and the regulations related to the Nuclear Liability and Compensation Act (NLCA) [A–17]. The federal government also recently amended its environmental regulations that took into considerations the licensing of 'smaller' reactors.

In accordance with the CNSC's current regulatory framework, the CNSC requires that the environmental effects of all nuclear facilities or activities be considered and evaluated when licensing decisions are made. All licence applications that demonstrate potential interactions with the environment are subject to an environmental review, commensurate with the scale and complexity of the environmental risks associated with the facility or activity.

Environmental reviews are carried out either under the environmental protection provisions under the NSCA or under other applicable federal, provincial and/or territorial legislation such as the Impact Assessment Act (IAA) [A–18], the former Canadian Environmental Assessment Act, 2012 (CEAA 2012) [A–19], and northern environmental assessment regimes. This means that science-based environmental technical reviews are performed throughout the life cycle of a nuclear facility, including new reactor facilities.

The IAA [A–18] came into force in August 2019, replacing CEAA 2012 [A–19]. The IAA broadens the scope of assessments to include environmental, health, social and economic effects, both positive and negative, of a proposed project. Under the IAA and its Physical Activities Regulations, impact assessments will be conducted on projects identified as having the greatest potential for adverse environmental effects in areas of federal jurisdiction. The Physical Activities Regulations is the schedule which lists the physical activities that are considered designated projects (e.g. within scope of conducting a full impact assessment) under the IAA. Each physical activity listed in the schedule, regardless of its relationship to the nuclear industry, includes a description and in most cases a corresponding threshold. Considerations were taken at the time of promulgation of the new Act and Regulations on assessments of smaller reactors (under CEAA 2012 [A–19], all reactors irrespective of their size were treated equally). The clause related to SMRs is provided below:

> "27 The site preparation for, and the construction, operation and decommissioning of, one or more new nuclear fission or fusion reactors if
>
> (a) that activity is located within the licensed boundaries of an existing Class IA nuclear facility and the new reactors have a combined thermal capacity of more than 900 MWth; or
>
> (b) that activity is not located within the licensed boundaries of an existing Class IA nuclear facility and the new reactors have a combined thermal capacity of more than 200 MWth."

Should a project not be captured under the IAA Physical Activities Regulations, it would still undergo an assessment of its environmental impact under the NSCA (part of CNSC mandate is to ensure protection of the environment over activities it regulates).

While the CNSC has not considered it necessary to change the NSCA as a result of proposed projects involving SMRs, amendments to the NLCA [A–17] are being considered. The NLCA, which is administered by Natural Resources Canada (NRCan), was amended in 2017. It increased the amount of compensation available to address civil nuclear damage from $75 million to $1 billion, broadened the number of categories for which compensation may be sought and improved the procedures for delivering compensation. The NLCA has undergone some scrutiny as, in its current form proponents of Small Nuclear Power Reactor Facilities would be subject to a $1 billion dollar liability amount regardless of their size or the risk that the facility represents. Canada-wide consultation on SMR [1] facilitated by NRCan recommended that the NLCA undergo a revision as in its current form.

It is recognized that most licensees would eventually meet their liability obligations, in whole or in part, via a private-sector nuclear liability insurance provider, who would independently determine the applicable insurance premiums that would be charged. Licensees can only buy nuclear insurance from insurers that have been approved by NRCan.

While the liability limit is not the only input that goes into calculating the nuclear liability insurance premium, it is a major factor. NRCan, leveraging technical advice provided by the CNSC, is currently reviewing and considering whether amendment recommendations to the NLCA [A–17] are required to ensure the risks posed by SMRs are appropriately captured.

The Nuclear Security Regulations [A–16] define security-related requirements for certain nuclear facilities, including high-security sites. The regulations ensure that Canada continues to fulfil its international obligations for the security of nuclear and radioactive materials, both in Canada and internationally.

A revision of the Nuclear Security Regulations is underway. The last major revision to the regulations was completed in 2006. Since then, security threats, operational experience and technological advancements have evolved and there is a need to keep up with updated international recommendations, their guidance and best practices.

Technology, which is embedded in many SMR designs, continues to have a major impact on nuclear security. New technology can present new challenges for the security of nuclear facilities, such as cyber-security threats, as well as opportunities to better protect nuclear security infrastructure against threats. Examples of new technology that could improve security include thermal imaging, night vision and infrared cameras, digital fingerprint screening, computed tomography, and advanced imaging and screening technology to detect firearms and explosive substances.

While the revision of the NSRs is still in progress, the CNSC recognizes the need to have a flexible regulatory approach that considers the potential radiological consequences and health impacts.

On November 2019, during the World Institute of Nuclear Security (WINS) Workshop in Ottawa, the President of the CNSC emphasised the importance of integrating security considerations in the

---

[1] Pan Canadian SMR Roadmap report can be found at https://smrroadmap.ca/

design phases of SMRs. It was also confirmed the CNSC's readiness to regulate these new technologies and highlighted the recent memorandum of cooperation with the U.S. Nuclear Regulatory Commission (U.S. NRC) to modernize the regulation of SMRs.

The audience was issued with the following four challenges:

(a)   Set the path forward toward the effective integration of safety, security and safeguards requirements for SMRs.
(b)   Drive the evolution of prescriptive security requirements to a goal-oriented, graded approach commensurate with the risks of SMRs.
(c)   Imagine the best next steps in international harmonisation.
(d)   Develop concrete recommendations toward modern security requirements.

A–1.2.3. Follow-up Question

A revision of the Nuclear Security Regulations is said to be underway and it is stated that the last major revision to the regulations was completed in 2006. Since then, security threats, operational experience and technological advancements have evolved and there is a need to keep up with updated international recommendations, their guidance and best practices.

Please describe what specific changes are being considered.

Please provide a summary of the key features of the regulatory framework (general) and line of sight to the law with regards to the responsibility of the operating organization.

A–1.2.4. Response

Canada is using a risk informed approach to identify the changes and maintaining the flexibility in regulatory approach that considers the potential radiological consequences and health impacts. See the answer from Canada to Question 27 for details.

The key feature of the Canadian Nuclear Safety Commission (CNSC) framework is that it is operates within a modern and robust legislative and the regulatory framework is technology neutral which is adequate for the licensing of projects involving SMRs. CNSC REGDOC-3.5.3, 'Regulatory Fundamentals' [A–20] describes the CNSC's regulatory framework and the responsibilities of an operating organization (i.e. license holders). The responsibilities of an operating organization are defined in the NSCA [A–13] and its regulations and also in the Licence Condition Handbook (LCH) of an operating organization.

The CNSC's regulatory framework balances prescriptive and performance-based requirements based on a risk-informed approach to the regulated nuclear activity. As shown in Fig. A–1, CNSC's regulatory framework consists of laws passed by Parliament that govern Canada's nuclear industry, including the NSCA [A–13], the Nuclear Liability and Compensation Act, federal environmental legislation. As well, the CNSC has the authority to establish regulations (subject to Order in Council approval), issue licences and certificates, create regulatory documents (REGDOCs), as well accepting the use of national and international nuclear standards used to oversee all nuclear facilities and activities in Canada. The framework takes into account Government of Canada regulatory policy guidance, as well as the views of stakeholders, Indigenous peoples and the public.

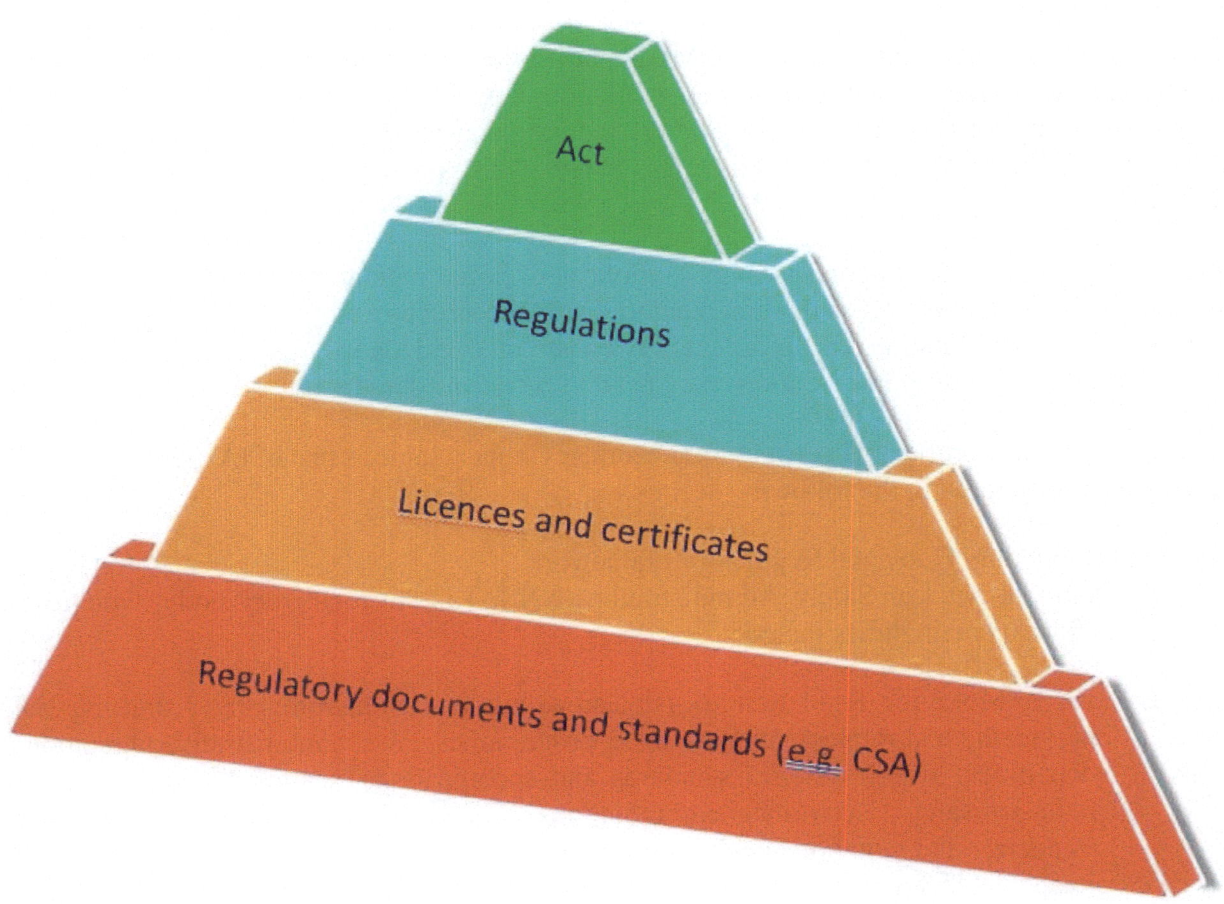

FIG. A–1: CNSC's regulatory framework

The CNSC has developed REGDOCs[2] that provide elaboration on the requirements in the NSCA and the regulations made under it. The REGDOCs are organized into three key categories: regulated facilities and activities, safety and control areas, and other regulatory areas.

REGDOCs take into account international regulatory best practices and modern codes and standards and align with the IAEA's Safety Fundamentals and Safety Requirements. Industry or international standards may be referenced in REGDOCs. Modern codes and standards include, but are not limited to, standards created by independent, third-party standard-setting organizations.

A–1.3. CHINA–NNSA

A–1.3.1 Question

Was there a need to change the legal framework (including the Authorization process) in your country as a result of proposed projects involving SMRs?

---

[2] A list of CSNC's REGDOCs:

http://www.nuclearsafety.gc.ca/eng/acts-and-regulations/regulatory-documents/index.cfm .

A–1.3.2. Response

According to the experience of HTR-PM demonstration plant and land-based SMR safety supervision and management, the legal framework does not need to be adjusted. For specific SMR design, especially for HTGR-SMR, some safety requirements need to be formulated or modified, which may relate to some Department rules, Safety Guides Technical Documents, and even Regulations.

For floating reactor, the situation is more complicated. Some regulatory adjustments are under discussion. As the foundation of floating reactor, that is the structure similar to ship, is unfamiliar to NNSA. Moreover, floating reactor is generally constructed and loaded in factories and movable, so the setting of regulatory interventions may need to be reconsidered.

The regulatory framework and licensing framework of China is same as that of IAEA's suggestion, which will be described in the following sections.

(a) China Regulatory Body and Regulatory Framework
In China, National Nuclear Safety Administration (NNSA) is the regulatory body, which was founded in 1984. Its main duties include:

(i)   Responsible for regulation of nuclear safety and radiation safety, drafting out, organizing and implementing policies, programmes, laws, administrative regulations, department rules, systems, standards and specifications relating to nuclear safety, radiation safety, electromagnetic radiation, radiation environment protection as well as nuclear and radiation accident emergency;
(ii)   Responsible for unified regulation of nuclear safety, radiation safety and radiation environment protection for nuclear facilities;
(iii)   Responsible for regulation of licensing, design, manufacture, installation and non-destructive testing (NDT) activities for nuclear safety equipment and the safety inspection of imported nuclear safety equipment;
(iv)   Responsible for control of nuclear materials and regulatory inspection and management of physical protection;
(v)   Responsible for regulation of radiation safety and radiation environment protection of nuclear technology application projects, uranium (thorium) mines and associated radioactive mines, and taking charge of radiation protection;
(vi)   Responsible for regulation of safety and radiation environment protection of treatment and disposal of radioactive waste, and for supervisory inspection of radioactive contamination prevention and control;
(vii)   Responsible for regulation of safety transport of radioactive materials;
(viii) Responsible for nuclear and radiation emergency response, investigation and treatment of the Ministry of Ecology and Environment (MME/NNSA) and participation in prevention and handling nuclear and radiation terrorist event.
(ix)   Responsible for qualification management of reactor operators, special process personnel of nuclear equipment, etc.;
(x)   Organizing and developing radiation environment monitoring and regulatory monitoring of nuclear equipment and key radiation sources;
(xi)   Responsible for taking the lead in the nuclear safety coordination mechanism;
(xii)   Responsible for domestic implementation of international conventions relating to nuclear and radiation safety;
(xiii) Directing relevant professional work in regional offices of nuclear and radiation safety inspection.

(b)   China Nuclear and Radiation Safety Regulation System

The nuclear and radiation safety regulation system of China includes the national laws, administrative regulations, ministry rules, guidance documents and other documents of regulatory requirements. The system is shown in Fig. A–2.

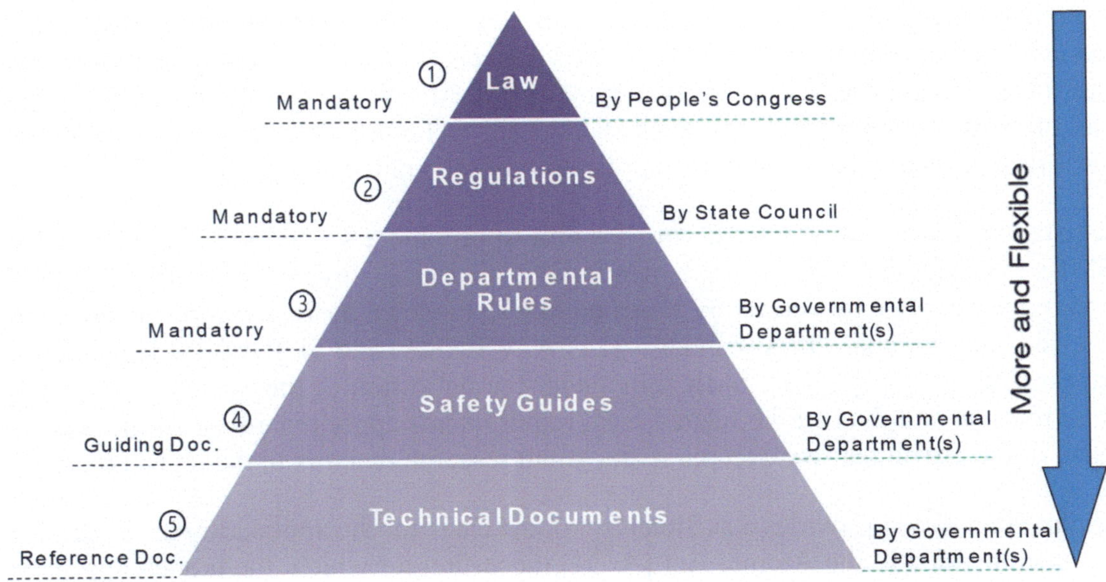

FIG. A–2: Nuclear and Radiation Safety Regulation System

A–1.3.3. Follow-up Questions

The response provides the general duties and responsibilities of the Regulator and a general hierarchy of nuclear and radiation safety regulation considerations. A description of the key considerations of the legal framework was expected, with reference to the laws. Please provide a summary of the key features of the regulatory framework (general) and line of sight to the law with regards to the responsibility of the operating organisation.

A–1.3.4. Response

The NNSA has issued the SMR Safety Review Principles, providing guidance on some review issues related to the safety characteristics of the SMR. The specific regulation adjustment is still on discussion.

At present, the PSAR review of the floating reactor is being carried out, and some specific issues encountered in the review process will be made a special discuss. The problems and experiences in the SMR review will be collected and summarized to support the further review process, and relevant regulations, guidelines and technical documents will be modified accordingly. The documents under consideration include the safety regulatory approach, emergency management, security requirements and so on.

A–1.4.   CZECH REPUBLIC–SÚJB

A–1.4.1. Question

Was there a need to change the legal framework (including the Authorization process) in your country as a result of proposed projects involving SMRs?

A–1.4.2. Response

Not applicable – there was no need to change the legal framework due to SMR deployment (currently there are no project involving SMRs proposed to be licensed in the Czech Republic).

A new system of nuclear law in the Czech Republic has recently been adopted. It entered into force in 2017, comprising the Atomic Act (No. 263/2016 Coll.) [A–21] and more than 20 implementing decrees. In general, the legislation in this field does not, as such, preclude the deployment of SMRs nor does it constitute a major obstacle to this deployment. The Atomic Act [A–21] provides a good legal basis as the legal requirements contained therein are technologically neutral and provide sufficient flexibility, in particular through the application of the graded approach.

While the legislation has not been drafted with the aim to provide a special set of requirements reflecting SMRs specificities and the implementing legislation is directed primarily to cover standard sized light water reactors currently operated in CZ, the legal basis is comprehensive and reflects the newest international safety standards (IAEA, WENRA) and EURATOM legislation). According to the Atomic Act [A–21], SMR falls under the definition of the nuclear installation (with a nuclear reactor) and thus all the related legal requirements apply regardless of the type of SMR or its size (there are only a few exceptions for small research reactors in the legislation).

The Czech atomic legislation provides a sufficiently robust basis for the application of the graded approach that is embedded in the Atomic Act [A–21] (according to which, the graded approach shall be, inter alia, commensurate to the type of the nuclear installation).

Taking into account the fact that requirements contained in secondary legislation (implementing decrees to the Atomic Act [A–21]) are specific and prescriptive and reflect currently used technology in Czech Republic (i.e. PWR), the legislative framework on the level of secondary legislation is, strictly speaking, not in its entirety technologically neutral. Hypothetically speaking, the national legislative framework would therefore need to be adjusted to reflect distinct technologies and facilitate their deployment (in particular on the level of secondary legislation). Simplifications of the licensing framework and other procedural requirements might also be considered to reflect particular SMR technology specificities and deployment model. However, there is a persistent lack of clear and reliable data of the particular technological solution which could be envisaged to be deployed. It may even occur that the legislative framework is modified but not simplified to strengthen particular requirements to reflect the specificities of a particular design and deployment model that would be considered as a viable option.

A–1.4.3. Follow-up Question

Please provide the specific areas of the secondary level legislation that considered to be very technology-specific and would likely need changing/ amending.

Please provide a summary of the key features of the regulatory framework (general) and line of sight to the law with regards to the responsibility of the operating organisation.

A–1.4.4. Response

One of the main competencies of the national regulatory authority is to issue licenses (for performing activities regulated by the Atomic Act [A–21]). The licence is issued after regulatory authority verifies that the applicant has fulfilled all the requirements established in the Atomic Act and its implementing legislation (through assessment of documentation for a licenced activity that is attached to a license application). A licence from the Office is required for: siting of a nuclear installation, construction of a nuclear installation, first physical start-up of a nuclear installation with a nuclear reactor, first power-generation start-up of a nuclear installation with a nuclear reactor, commissioning of a nuclear installation without a nuclear reactor, operation of a nuclear installation, individual phases of decommissioning of a nuclear installation and carrying out of modifications affecting nuclear safety, technical safety and physical protection of a nuclear installation.

A license is granted only if the applicant demonstrates fulfilment of all the requirements stipulated for different types of licenses by the Atomic Act and its implementing legislation. An applicant for a license is obliged to demonstrate compliance with the requirements laid down in the legislation through a documentation that needs to be submitted together with a license application.

The documentation to be provided for review and assessment together with a particular licence application is prescribed in the Appendix 1 to the Atomic Act. The documentation submitted with the application is evaluated by the SÚJB within the licensing administrative proceedings framework (some of these documents are subject to separate approval). This documentation is binding for the licensee and without proving their full compliance with legal requirements the license would not be issued.

The Atomic Act stipulates general and specific obligations that must be fulfilled by the holders of a licence for an activity related to the use of nuclear energy. Notwithstanding authorization, anyone who uses nuclear energy shall ensure, inter alia, as a matter of priority nuclear safety, safety of nuclear items and radiation protection, while respecting the present level of science and technology and good practice. Legislation clearly provides that the obligation to ensure nuclear safety, radiation protection and safety of nuclear material or other nuclear item (which is important for ensuring the non-proliferation of nuclear weapons), cannot be transferred to another person. This and other principles of the peaceful use of nuclear energy and ionizing radiation are to be found in the Article 5 of the Atomic Act.

As regards the technology-specific legislative provisions, both the Decree No. 329/2017 [A–22], on the requirements for nuclear installation design, and the Decree No. 358/2016 [A–23], on requirements for assurance of quality and technical safety and assessment and verification of conformity of selected equipment, contain relatively specific provisions that reflect the PWR technology. These documents refer to pressurized primary and secondary circuit, fuel cladding and assemblies (including specific pressure values and volumes). Both decrees, in particular the Decree No. 329/2017 [A–22], have been drafted and developed to be primarily applied to 'standard (large) size NPPs' (VVER-440 and VVER-1000 are being operated in CZ) and as such do not reflect various specificities of other different SMR designs.

As an example of a specific area of the secondary legislation that could be viewed as technology-specific is the Article 12 of the Decree No. 358/2016 [A–23], that contains a list of a specific equipment (including maximum operating pressure and temperature, and nominal diameter), for example:

> "(...) pumps, piping, and armatures providing nuclear reactor cooling, volume compensation, hermetic area cooling, emergency filling, primary circuit after-cooling, and cleaning of the pressure circuit process media, working with radioactive substances with a greatest operating pressure in excess of 0.05 MPa and nominal diameter greater than DN 70(...)"

Another example – if an SMR of a more distinct design is to be deployed in Czech Republic – may be the legal provisions pertaining to the DiD concept. For nuclear installations with a nuclear reactor, the function of physical safety barriers shall, according to the Decree No. 329/2017 [A–22], be ensured by independent systems, structures and components – namely fuel element cladding, the pressure boundary of the primary circuit of the nuclear reactor cooling and the containment system. This and other provisions are a reflection of the currently used technology in Czech Republic (i.e. PWR) but could present a challenge in case an SMR of a very distinct design is deployed.

A–1.5. FRANCE–ASN

A–1.5.1. Question

Was there a need to change the legal framework (including the Authorization process) in your country as a result of proposed projects involving SMRs?

A–1.5.2. Response

The article R.593-1 of the Environmental Code [A–24] defines a nuclear reactor as a device able to produce and control a self-sustaining nuclear reaction.

France's legal framework for nuclear installations is applicable for any nuclear reactor technology and power. Indeed, the article L593-2 from the Environmental Code stipulates that every type of and each nuclear reactor is a basic nuclear installation that has to comply with the associated requirements (defined in the Environmental Code, see book 5 – title 9 - chapter 3 and 6). These requirements are applicable to potential SMR projects and deal with project creation, licensing, operation, dismantling, and controls and sanctions from the regulatory body.

Hence, there is no need to change France's legal framework as a result of potential upcoming project involving SMRs.

A–1.6. JAPAN–NRA

A–1.6.1. Question

Was there a need to change the legal framework (including the Authorization process) in your country as a result of proposed projects involving SMRs?

A–1.6.2. Response

Legal framework was not changed for the review of High Temperature engineering Test Reactor (HTTR) since it was reviewed as a research and test reactor.

In addition, during the regulatory decision process for the HTTR, the NRA invited public comment from the scientific and technical view on the licence application. Public consultation is typically done for NPP licence applications, and this was the first instance in which public comment was invited for a research facility.

A–1.6.3. Follow-up Question

Please provide a summary of the key features of the regulatory framework (general) and line of sight to the law with regards to the responsibility of the operating organization.

Please describe why the framework was considered appropriate for the HTTR.

A–1.6.4. Response

In order to install and operate a new reactor within the Japanese legal framework, a reactor installation (establishment) permit issued by the NRA is required prior to the commencement the main phases of the lifecycle of the installation, namely: design, construction and operation.

As for reactors including HTTR which already received such permit and approval, based on the back-fitting system which is introduced by the amendment of the Reactor Regulation Act in 2012 [A–25], the conformity review process is applied. In this process, the entity is required to obtain the permission for the alteration for approved installation permit, and the approval for construction plan and operational safety program accordingly.

The Act [A–25] explicitly states the legal responsibilities of licensees that they "(…) shall be responsible for installing equipment or apparatus contributing to the improvement of the safety of nuclear facilities, enhancing education on operational safety, or taking any other necessary measures for preventing disasters resulting from nuclear source material, nuclear fuel material, and reactors, while taking into account the latest knowledge on safety at nuclear facilities."

Such Legal framework for regulation is applied not only to HTTR but also to all reactors.

A–1.7. RUSSIAN FEDERATION–ROSTECHNADZOR

A–1.7.1. Question

Was there a need to change the legal framework (including the Authorization process) in your country as a result of proposed projects involving SMRs?

A–1.7.2. Response

Safety regulation of floating SMRs such as the FNPP Akademik Lomonosov is implemented on the basis of the standards and requirements of two international organizations: the IAEA and the International Maritime Organization (IMO). The regulatory framework for the safety of ships and other vessels with nuclear installations, including nuclear icebreakers and the FNPP Akademik Lomonosov, comprises the following main international conventions:

- Convention on Nuclear Safety, 1994 (except for the provisions of Article 17 (requirements for site selection (location) of a nuclear installation);
- Convention on Early Notification of a Nuclear Accident, 1986;
- Convention on Assistance in the Case of a Nuclear Accident or Radiological Emergency, 1986;
- Convention on the Physical Protection of Nuclear Material, 1980, and Amendments to the Convention on the Physical Protection of Nuclear Material, 2005;
- Joint Convention on the Safety of Spent Fuel Management and on the Safety of Radioactive Waste Management, 1999;
- International Convention for the Suppression of Acts of Nuclear Terrorism, 2005;
- International Convention for the Safety of Life at Sea, 1974, and the Protocol to the International Convention for the Safety of Life at Sea, 1988.

The basic document for the entire regulatory framework for the use of atomic energy in the Russian Federation is Federal Law No. 170-FZ of 21.11.1995 'On the Use of Atomic Energy' [A–26]. The law establishes the basic principles and objectives of legal regulation of activities in the field of atomic energy uses, defines the types of activities and nuclear facilities. The Federal Law also describes, in sufficient detail, the functions and objectives of the regulatory bodies and operating organizations.

Taking into account the wide experience of the Russian Federation in carrying out various activities in the field of the use of atomic energy, Federal Law No. 170-FZ [A–26] establishes specific provisions for the organization of activities at certain nuclear facilities, in particular, nuclear installations, storage facilities, and radiation sources. Due to the fact that the Russian Federation has a significant nuclear icebreaker fleet, Federal Law No. 170-FZ [A–26] also introduced a dedicated Chapter VIII, which establishes special conditions for the construction and operation of ships and other floating facilities with nuclear reactors. The provisions of the abovementioned Chapter also apply to the FNPP (Floating Nuclear Power Plant) Akademik Lomonosov as its features are, to a large extent, similar to those of the ships with nuclear power installations.

Chapter VIII of Federal Law No. 170-FZ [A–26] sets out the main requirements for vessels, requirements for the safety of vessels during port calls, as well as the main provisions for limiting radiation exposure during the operation of vessels, taking into account the features of their operation. This Chapter of the Federal Law specifically emphasizes the non-stationary nature of the operation of vessels with nuclear power installations and the possible risks associated with such operation.

Provisions of Requirement 1 of GSR Part 1 (Rev. 1) [A–27] are considered in Article 2 of Federal Law No 170-FZ [A–26].

Provisions of Requirement 7 of GSR Part 1 (Rev. 1) are considered in Article 24 of Federal Law No 170-FZ.

Provisions of Requirement 10 of GSR Part 1 (Rev. 1) are considered in Article 33 of Federal Law No 170-FZ.

Provisions of Requirement 20 of GSR Part 1 (Rev. 1) are considered in Article 37.1 of Federal Law No 170-FZ.

Provisions of Requirements 16, 18, 21, and 22 of GSR Part 1 (Rev. 1) are considered in the Regulations on the regulatory bodies adopted by the Government of the Russian Federation.

In addition to the legal requirements in the field of the use of atomic energy, Federal Law No. 170-FZ establishes that ships and other vessels with nuclear installations are also subject to the requirements of the Russian Maritime Register of Shipping that covers the legal requirements in the field of commercial navigation safety. The main document in the field of commercial navigation safety is the 'Merchant Shipping Code of the Russian Federation,' No. 81-FZ of April 30, 1999 [A–28]. In accordance with the requirements of [A–28], the Russian Maritime Register of Shipping is an organization that is authorized to classify and survey vessels. The documents of the Russian Maritime Register of Shipping implement the provisions of the international agreements of the Russian Federation in the field of navigation safety, in particular the International Convention for the Safety of Life at Sea, as well as other IMO documents.

Consequently, the provisions of the existing version [A–26] have already taken into account the features of SMRs such as the FNPP Akademik Lomonosov. To this end, no changes were introduced to the Russian Federation's regulatory framework when the FNPP was undergoing the licensing process.

A–1.7.3. Follow-up Questions

The response is limited to floating NPP. Please describe considerations for non-floating SMRs. Can information be provided on the regulatory framework / changes to support the licensing of BREST-300?

Please provide a summary of the key features of the regulatory framework (general) and line of sight to the law with regards to the responsibility of the operating organisation.

A–1.7.4. Response

According to the regulatory approaches adopted by the Russian Federation, BREST-300 facility does not belong to the SMR class and hence is subject to the licensing process established for the conventional NPPs.

The licensing procedure for the construction of a power unit with BREST-300 is not over yet. By its design, BREST-300 is innovative, but it has few features inherent in other designs of small modular reactors that are being considered around the world. When licensing its construction, the existing regulatory and legal framework is used, taking into account the specifics of the fuel and coolant to be used.

In accordance with Federal Law No. 170-FZ [A–26] the operating organization bears full responsibility for the safety of the power unit.

A–1.8.   SOUTH AFRICA–NNR

A–1.8.1. Question

Was there a need to change the legal framework (including the Authorization process) in your country as a result of proposed projects involving SMRs?

A–1.8.2. Response

Since before the NNR received a nuclear installation licence application in July 2000 from Eskom, for the prospective siting, construction, operation, decontamination, and decommissioning of a demonstration unit of a 11 MW(e) Class PBMR (equivalent to an output of 268 MW(th) electricity generating power station, the NNR developed Basic Licensing Requirements for the PBMR – LG - 1037 [A–29] and elaborated the processes that must be undertaken to demonstrate compliance with the requirements.

The electrical utility in South Africa (Eskom)'s plan to construct a first of a kind Pebble Bed Modular Reactor (PBMR) led to the need to ensure that a credible and effective licensing process be developed and implemented for this technology. The scope of the regulatory assessment for the licensing of the PBMR was based on the licensing requirements and criteria defined by the NNR in regulatory documents that expanded on the current legislative requirements at the start of the PBMR project. In addition, guidance was provided on selected issues in regulatory guidance documents and position papers.

The requirements comprised, besides the general requirements to respect good engineering practice and the ALARA and defence-in-depth principle, specific risk criteria and radiation dose limits. These are categorized for normal operation and operational occurrences as well as for design basis events and beyond design basis events for workers and the public. Additional requirements and recommendations were stipulated by the NNR on safety important areas like quality and safety management, qualification of the nuclear fuel and the core structures, core design, verification and validation of computer codes, source term analysis and others. Selected NNR Position Papers have been developed to elaborate and provide further clarification on NNR requirements. For preparation of the PBMR safety case, 25 so-called 'Key Licensing Issues' were defined and agreed with the applicant. Discussions relating to these Key Licensing Issues allowed important nuclear safety aspects identified for the PBMR demonstration plant to be clarified in advance of the safety case submittal.

The principal nuclear and radiation safety requirements formulated in the Regulations in terms of Section 36, read with Section 47 of the National Nuclear Regulator Act [A–30], on Safety Standards and Regulatory Practices formed the basis for the stipulation of the Licensing Requirements for the PBMR which were elaborated further in a number of regulatory documents as listed in the table in the NNR response to the question number 02.

In view of and acknowledging the complexity and the developmental nature of the PBMR, a multi-staged licensing process was adopted by the NNR. The multi-staged licensing process provided a logical link between the various steps of the design process, the safety assessment and the development of operational support programmes. The approach adopted was that, following a satisfactory regulatory review of the safety case by the NNR (which in terms of the legislation include public participation), an initial Nuclear Installation Licence (NIL) would be granted (or refused) to the applicant for the first stage of the process. A variation to this NIL would be requested by the applicant and issued by the NNR following its satisfactory regulatory review, at each of the subsequent licensing stages.

A programme of staged licensing submissions would coincide with the application for a NIL variation to proceed to the next phase, which would need to be supported by a comprehensive safety case to demonstrate compliance with the NNR regulatory safety requirements at a level appropriate for the stage of licensing at hand.

Each stage of the licensing process would indicate the NNR Hold & Witness Points that would form the prerequisites to proceed to the next licensing stage. The applicants quality assurance (QA) Programme was to ensure traceability and credibility of results of the previous licensing stage, before the NNR issued the next stage licence variation.

The Licensing Programme included the following major licensing stages:

Stages 1&2: Site preparation, construction and manufacturing phase – typical activities would include:

- Site access, excavation, etc.
- Component manufacturing;
- Civil works;
- Installation of auxiliary systems;
- Installation of main power system;

- Cold commissioning testing up to and including non-nuclear integrated test.

(For these stages no nuclear fuel on-site).

Stage 3: Nuclear Fuel on-Site/ Commissioning and Start-up - Typical activities would include:

- Nuclear fuel on-site;
- Nuclear fuel load;
- Initial criticality;
- Low power testing;
- Full power testing.

Stage 4: Plant operation (commercialization)

Stage 5: Eventual decommissioning

A–1.8.3. Follow-up Questions

Please provide a summary of the key features of the regulatory framework (general).

Is there any need for additional authorizations to other organizations (other than the operating organization, such as manufacturers) taking part of the deployment of the SMR? Is there a need to modify the nuclear law? Please describe any changes.

A–1.8.4. Response

The NNR Safety Standards are premised on international standards such as the IAEA Safety Standards and the WENRA reference levels, as well as considering the UK ONR Safety Principles. The safety standards provides the principal safety criteria relating to risk criteria, and dose limits for normal operating conditions, applicable to members of the public and workers.

The safety standards further lay down principal radiation and nuclear safety requirements which are applied to all nuclear installations and other regulated actions, and include the following:

- Defence-in-depth;
- ALARA;
- Good engineering practice;
- Quality management;
- Accident management and emergency preparedness;
- Safety culture;
- Graded approach.

The radiological dose and risk limits for the public and workers relate directly to the objectives of nuclear and radiation safety and are therefore considered the most fundamental yardsticks against which to assess nuclear safety, contributing towards a more consistent and transparent basis for regulatory decision making. The dose limits are consistent with the IAEA Basic Safety Standards. Basic principles underlying the risk criteria are as follows:

- The risks presented by a nuclear plant must not increase significantly the total population risk;

- The nuclear risks must compare favourably with those associated with other major industrial enterprises;
- Allowance must be made for a possible increase in the standards of safety demanded by society over the period – usually several decades – represented by the working life of the plant.

The principles that must be met to ensure safety in any nuclear installation are presented in the Regulations on Safety Standards and Regulatory Practices published as Regulation No. R388 dated 28 April 2006 (SSRP) [A–31]. Regulatory requirements for any nuclear installation, as presented in NNR requirements documents[3], are based on and are established to fulfil the SSRP [A–31] principles for any nuclear installation.

Whilst design of a nuclear installation or the manufacturing of components are not expressly prohibited in terms of Section 20 of the NNR Act [A–30] in its current form, it is clear from Section 5 of the Act [A–30] that any person wishing to make use of a design of a nuclear installation or of components so manufactured in an eventual authorised nuclear installation must have performed these activities in terms of a nuclear authorisation.

Taking full cognizance of its mandate, and depending on the application, the NNR could issue the following types of authorisations for nuclear installations:

(a) Nuclear Installation Licence (NIL) to site, construct and/or operate or decommission or decontaminate the installation; or

(b) Nuclear Installation Site Licence (NISL) for new nuclear installations;

(c) Authorisation to Design a nuclear installation; or

(d) Authorisation to Manufacture components.

As may be seen from the rest of the information below, provision is made in the NNR regulatory framework in case additional authorizations are needed for other organizations (other than the operating organization), but only for the case of a designer of nuclear installations. For the case of manufacturing, authorisation to manufacture components must proceed through the applicant for the nuclear licence for the said facility obtaining an authorisation to manufacture from the NNR. More on this further down below.

For the case of a designer of nuclear installations, the following scope statement from Section 3 of the NNR position paper PP-0008, 'Design Authorisation Framework' [A–32] indicates that an additional authorization may be permitted for an organization other than the operating organization:

"This document is applicable to applicants for a nuclear installation licence to design a nuclear installation, and to regulatory assessments of the design of a proposed nuclear installation. This process is applicable where an application for a nuclear installation licence to design a nuclear installation is submitted to the NNR separate from and not

---

[3] For more information, the overview of the regulatory framework of the NNR is provided at http://www.nnr.co.za/acts-regulations/overview-of-regulatory-framework/ .

concurrent with other applications for Nuclear Installation Licenses such as for the construction or operation of the same nuclear installation."

NNR experiences from the licensing process for the PBMR (an SMR) led to the development and approval of many NNR documents relevant to both the operating organization and other organizations such as designers and manufacturers. Such NNR documents are applicable not only to licence applicants for SMRs but also to licence applicants for new nuclear build programmes involving also other nuclear designs/technologies.

Some examples of approved NNR documents[4] relevant also to other organizations such as designers and manufacturers are:

- RD-0034, "Quality and Safety Management Requirements for Nuclear Installations" [A–33];
- RG-0005, 'Guidance on Testing, Qualification and Commissioning of the PBMR DPP' [A–34];
- RG-0016, 'Guidance on the Verification and Validation of Evaluation and Calculation Models used in Safety and Design Analyses' [A–35];
- PP-0008, 'Design Authorisation Framework' [A–32];
- PP-0009, 'Authorisations for Nuclear Installations' [A–36];
- PP-0012, 'Manufacturing of Components for Nuclear Installations' [A–37];
- PP-0015, 'Emergency Technical Basis for New Nuclear Installations' [A–38];
- PP-0016, 'Conformity Assessment of Pressure Equipment in Nuclear Service' [A–39].

Based on, amongst others, NNR experiences from the licensing process for the PBMR but also with a view to possible licence applicants for new nuclear build programmes involving also other nuclear designs/technologies, the NNR also developed draft General Nuclear Safety Regulations and draft Specific Nuclear Safety Regulations: Nuclear Facilities. These regulations can only be issued once NNR Act [A–30] amendments, which were also drafted, are promulgated.

For the case of manufacturing, authorisation to manufacture components must proceed through the applicant for the nuclear licence for the said facility obtaining an authorisation to manufacture from the NNR as may be seen from the following statements in Section 6.2 of NNR position paper PP-0012 [A–37]:

"An application for construction of the nuclear installation has to be in place and being processed by the NNR as a pre-condition for issuance of an authorization to manufacture components."

A–1.9. UNITED KINGDOM–ONR

A–1.9.1. Question

Was there a need to change the legal framework (including the Authorization process) in your country as a result of proposed projects involving SMRs?

A–1.9.2. Response

---

[4] Available at http://www.nnr.co.za/acts-regulations/regulatory-documents .

There have been no changes to the legal framework as a result of considering the implications of deploying Small Modular Reactors (SMRs) in a UK context. In line with the Energy Act 2013 [A–40] and the Health and Safety at Work Act (HSWA) 1974 [A–41], ONR's regulatory philosophy is goal-setting (non-prescriptive), putting the onus on dutyholders to demonstrate that they have reduced the level of risk to As Low As Reasonably Practicable (ALARP) – this is the 'goal' in 'goal-setting'. The duty is discharged when the dutyholder demonstrates that there is gross disproportion between the cost associated with further measures to reduce risk and the benefit that would be achieved in terms of risk reduction. The specific means by which the goal is achieved is rarely defined in law, and therefore is open to dutyholder choices.

In line with the goal-setting legal framework, ONR applies risk-informed decision in the regulation of nuclear installations [A–42], and this supported by ONR's numeric targets. These consist of 9 numeric targets defining what risks ONR considers to be broadly acceptable and those which would be unacceptably high or acceptable only if they are demonstrably ALARP. ONR also has a set of specific guidance for ONR inspectors on what they should expect of a nuclear licensee or dutyholder in meeting its legal requirement to reduce risks to ALARP (NS-TAST-GD-005 - Guidance on the demonstration of ALARP (As Low as Reasonably Practicable) [A–43]). In most circumstances, demonstrating ALARP does not require explicit comparison of costs and benefits, but by applying Relevant Good Practice (RGP).

RGP is those standards for controlling the risk judged and recognized by ONR as satisfying the law, when applied appropriately. As documented in Annex 3 of NS-TAST-GD-005 [A–43], alternative approaches and standards are acceptable: it is possible that same standards of safety can be achieved through different means and therefore this complies with the law. If RGP is not clearly established in particular cases e.g. areas of innovation in the context of Advanced Nuclear Technologies (ANTs), SMRs, etc. the overarching requirement in UK law still stands and therefore dutyholders are required to establish explicitly the significance of the risks to determine the action needs to be taken.

ONR expects inspectors to use the Safety Assessment Principles (SAPs) [A–44] and Security Assessment Principles (SyAPs) [A–45] together with the associated TAGs [A–46] to judge claims of RGP and acceptability of risks are justified. The SAPs and SyAPs are considered technology-neutral and are regularly benchmarked against international guidance and explicitly recognise, and are aligned with, sources of RGP such as IAEA standards and the WENRA Reference Levels. It is considered that the technology-neutral and goal-setting regulatory framework enables innovation, by setting a goal and recognising that there may be multiple, alternative approaches to demonstrate safety and security of the design across its lifecycle.

No changes at the legal framework level are currently envisaged due to projects involving SMRs.

A–1.9.3. Follow-up Question

A reference to the IRRS mission report/conclusion may be valuable.

Please provide a summary of the key features of the regulatory framework (general) and line of sight to the law with regards to the responsibility of the operating organisation.

A–1.9.4. Response:

The IRRS mission report 'Integrated Regulatory Review Service (IRRS) to the United Kingdom of Great Britain and Norther Ireland', IAEA-NS-IRRS-2019/06 - Rev. 1, 2019 [A–47], has been published and further information in the context of licensing (and the report conclusion) follows.

A summary to the key features of the regulatory framework (goal-setting, risk-based) and line of sight to the law from a general standpoint was provided in the initial questionnaire response, including references to Acts, Regulations and Guidance. The line of sight to the law with regards to the responsibility of the operating organisation was mainly documented in Question 3. The line of sight is through the Energy Act [A–40] and the Nuclear Installations Act (NIA) 1965 [A–48] defines the concept of a Nuclear Licensed Site in section 1, prohibiting the use of a site for the purpose of installing or operating any nuclear reactor (other than a nuclear reactor comprised in a means of transport), or any other installation of a prescribed kind unless a licence has been granted. The Nuclear Installations Regulations (NIR) 1971 [A–49] provides descriptions for those installations that are prescribed by the Act.

The NIA [A–48] provides for ONR to grant a licence to a corporate body for the use a defined site for the prescribed activities, and for ONR to attach conditions to the licence (according to section 4). ONR has a standard set of 36 nuclear site licence conditions, which are explained in ONR's Licence Condition Handbook [A–50] covering from expectations on the marking of the site boundary, the consignment of nuclear matter, training, emergency arrangements, safety documentation, operating rules, decommissioning and organisational capability (the list is not meant to be exhaustive). ONR's approach and expectations to the licensing of nuclear installations are documented in Licensing Nuclear Installations [A–51]. The Licence Conditions (LCs) imposed through section 4 of NIA [A–48] include powers to control specific activities proposed by the licensee for the licensed site using Primary Powers: Directions, Approvals, Notifications, Specifications, Agreements, and Consents. The LCs provide the principal legal basis for regulation of nuclear safety on licensed sites. Indeed, as part of the licensing process, ONR considers the adequacy of the arrangements in place to meet the requirements of the LCs and can approve all or parts of these arrangements, ensuring that licensees cannot deviate from agreed programs without ONR's written approval. ONR may apply regulatory control by specifying hold-points (i.e. before a NPP enters a care and maintenance stage or between decommissioning stages). This is referred to in the IRRS report.

With regards to suggestions and recommendations to ONR regarding the licensing process, the IRRS raised a recommendation around establishing provisions for the public to be consulted in its process for making significant regulatory decisions, establishing regulatory guidance or when updating licence conditions.

A–1.10. UNITED STATES OF AMERICA–NRC

A–1.10.1. Question

Was there a need to change the legal framework (including the Authorization process) in your country as a result of proposed projects involving SMRs?

A–1.10.2. Response

No, the regulatory framework for reviewing and approving SMRs are provided for under both 10 CFR [A–52] Part 50 and Part 52; and defines small modular reactors (SMRs) as light water reactor (LWR) designs generating 300 MW(e) or less. However, recent laws enacted by the U.S. Congress have provided additional support for the licensing of SMRs, such as The Nuclear Energy Innovation and Modernization Act (NEIMA) [A–53] which provides the public with greater clarity into the process by which the NRC develops its budget and recovers its costs through fees. NEIMA [A–53] requires the NRC to develop a regulatory framework for America's innovators, who seek to deploy new and advanced nuclear technologies and directs the NRC to: (1) expedite and establish stages in the licensing process for commercial advanced nuclear reactors; and (2) increase, where appropriate, the use of risk-informed and performance-based evaluation techniques and regulatory guidance in licensing commercial advanced nuclear reactors within the existing regulatory framework. The NRC staff presented its proposed plan for a change to its regulations in NRC document, SECY-20-0032, Rulemaking Plan On Risk-Informed, Technology-Inclusive Regulatory Framework for Advanced Reactors (RIN-3150-AK31; NRC-2019-0062), dated April 13, 2020 [A–54].

A–2. REGULATIONS AND GUIDANCE

This Annex presents the responses provided by the following Member State regulatory bodies to the Question 2: "Was there a need to change regulations and guidance in your country as a result of proposed projects involving SMRs?"

A–2.1. ARGENTINA–ARN

A–2.1.1. Question

Was there a need to change regulations and guidance in your country as a result of proposed projects involving SMRs?

A–2.1.2. Response

No, there was no need to change the regulations for the CAREM 25 reactor. However, in the years after the start of the construction project, ARN began a process of reviewing the Argentine Regulatory Standards.

Law No. 24,804/97 [A–1] empowers the Regulatory Body to issue and establish the standards, which regulate and control nuclear activities, of compulsory application, along with the whole national territory.

The first Regulatory Standards related to nuclear power plant licensing were initially produced more than thirty years ago and over time, a normative system was established comprising subjects such as radiological and nuclear safety, safeguards of nuclear materials, and physical protection. The system, known as 'AR Standards' (AR Standards for Regulatory Body), has at present 65 regulatory standards of which 31 are related to NPPs.

ARN is in an on-going process of harmonization between the Argentinean Regulatory Standards and the IAEA Safety Standards. Nevertheless, Argentine Regulatory Standards are already consistent with IAEA's corresponding standards in general terms, considering that ARN has adopted a performance or goal-oriented approach.

Moreover, Argentina participates actively in the IAEA standards committee's activities and particularly in the international efforts to take account of the lessons learned from the Fukushima Daiichi NPP accident, in order to strengthen the nuclear safety in achieving the objectives of the IAEA Action Plan and the Nuclear Safety Convention, as well as to maximize the benefit of the mentioned lessons learned.

The Regulatory Body agreed with the Vienna Declaration on Nuclear Safety [A–55] and adopted it to prevent accidents with radiological consequences and to mitigate such consequences should they occur. In this sense, ARN decided to carry out a normative framework integral review that includes addressing the Vienna Declaration [A–55] in national standards.

The goals of the normative framework review are the following:

- Overall review of Argentina normative framework based on ARN regulatory experience as well as the international knowledge and Vienna Declaration [A–55]. This review would include, if necessary, the modification of the existing standards and the development of new ones;

- Update the harmonization process of ARN regulatory standards in line with IAEA's standards, according to the Convention on Nuclear Safety and the Joint Convention on the Safety of Spent Fuel Management and on the Safety of Radioactive Waste Management;
- Facilitate the presentation and exchange of information on Argentine's standards, as part of the preparation for the next Integrated Regulatory Review Service (IRRS) that will be carried out in.

In some cases where there was a lack of national requirements or guidelines, international standards were used as a guide to expand the scope of RA standards.

A–2.1.3. Follow-up Question

Please describe the standards / areas pending harmonisation with IAEA guidance.

In the response to the questionnaire, ARN mentioned that "In some cases where there was a lack of national requirements or guidelines, international standards were used as a guide to expand the scope of RA standards". Please could you give examples?

A–2.1.4. Response

Some examples of the use of international guidelines and standards are:

- Guidance on the format and content of the Safety Report (US NRC RG 1.206) [A–56];
- Classification of Structures, Systems and Components in Nuclear Power Plants(IAEA SSG-30) [A–57];
- Standards and reference guides related to specific areas of application, for example: ageing management (IAEA SSG-48 Ageing Management and Development of a Programme for Long Term Operation of Nuclear Power Plants [A–58]), specific civil engineering requirements, V&V of system codes (CNSC G-149 - Computer programs used in the design and safety analysis of nuclear power plants and research reactors [A–59]), etc.

A–2.2. CANADA–CNSC

A–2.2.1. Question

Was there a need to change regulations and guidance in your country as a result of proposed projects involving SMRs?

A–2.2.2. Response

In Canada, the regulations under the NSCA generally give licensees flexibility in how to comply with legislative requirements. The regulations typically do not specify detailed criteria used in assessing licence applications or judging compliance. The CNSC updates regulations and guidance documents periodically to take into account lessons learned, best practices and new knowledge.

All reactor facilities, including SMRs, are classified as Class IA nuclear facilities under the Class I Nuclear Facilities Regulations (CINFR) [A–60]. These regulations also encompass research reactors (Class IA) and fuel-fabrication facilities (Class IB). The CINFR [A–60] were amended in 2017 to address lessons learned from Fukushima Daiichi NPP.

Given several stakeholders expressed interest in the possible construction of new SMRs, the following related CNSC regulatory documents have been published:

- RD-367, Design of Small Reactor Facilities [A–61] (to be superseded / integrated into by a revision to REGDOC-2.5.2, Design of Reactor Facilities: Nuclear Power Plants in the near future) [A–62];
- REGDOC-2.4.1, Deterministic Safety Analysis [A–63];
- REGDOC-1.1.1, Site Evaluation and Site Preparation for New Reactor Facilities [A–64];
- REGDOC-1.1.5, Supplemental Information for Small Modular Reactor Proponents [A–65].

Much of this work was informed by the publishing and feedback received regarding a CNSC SMR discussion paper, DIS-16-04 [A–14].

It is worth noting that the CNSC has moved towards systematically integrating information related to SMRs within the broader topical documents rather than providing this specific information separately.

The Canadian regulatory approach to licensing SMRs is built on a long-established foundation of risk-informed regulation. Regulatory tools and decision-making processes are structured to enable a licence applicant for a reactor facility to propose alternative ways to meet regulatory objectives. Overall, the regulatory framework is fit-for-purpose, with novel proposals needing further discussions (potentially through pre-licensing engagement). Proposals must demonstrate, with suitable information, that they are equivalent to or exceed regulatory requirements.

Requirements and guidance for reactor facilities are generally articulated to be technology-neutral and, where possible, permit the use of the graded approach. The graded approach enables applicants to propose the stringency of design measures, safety analyses and provisions for conduct of their activities commensurate with the level of risk posed by the reactor facility. The use of a graded approach is not a relaxation of requirements, but rather the application of requirements in a manner commensurate with the risks and characteristics of a facility or activity. The use of a graded approach is articulated in section 5.4 of REGDOC-3.5.3 [A–20]. Documentation or evidence submitted to support a graded approach must have the following characteristics.

Facts and data are derived from validated and quality-assured (i.e. traceable and repeatable) scientific and engineering processes, such as:

- Computer modelling;
- Experimental or field-derived data;
- Operating experience;
- Uncertainties are characterized and accounted for;
- Information is demonstrated to be relevant to the specific proposal.

CNSC expectations are drawn from Canadian as well as modern global practices such as those in the IAEA safety framework. Requirements and guidance rarely prescribe specific methodologies but rather express safety objectives to be met.

Where the licence application relies on the use of documents not traditionally used in the Canadian nuclear industry, the applicant should submit an accompanying assessment to facilitate a timely review of the submission. This assessment may be a gap analysis between the documents referenced in the application versus Canadian industry-equivalent documents, or an independent assessment of the design against equivalent documents commonly used in Canada.

One area that has been raised by Canadian industry is regarding civil and structural design. Industry has suggested that SMR designs may not employ a traditional containment. CNSC staff believes the existing regulatory framework is sufficient to address this concern. In particular, under the existing system, applicants can propose to use confinement structures provided they demonstrate through their safety case that the dose limits for members of the public are met.

Industry is also expected to demonstrate the use of Proven Engineering Practices (described in Section 4 of REGDOC 2.5.2 [A–62]) in their safety proposals. This includes the use of applicable industry standards but also addresses how practices such as analytical models, prototypical experiments are deemed sufficient to support safety claims commensurate with importance to safety.

Upcoming REGDOC revisions will address topics such as security, safeguards for reactors using a liquid fuel or pebbles (pebble bed high-temperature gas reactor) and overarching aspects (revision to REGDOC-1.1.2, Licence Application Guide: Licence to Construct a Nuclear Power Plant [A–66] and REGDOC-1.1.3, Licence Application Guide: Licence to Operate a Nuclear Power Plant [A–67]).

REGDOC 1.1.5 [A–65] proposes a graded approach to the application of the above regulatory documents specifically to the SMRs.

With respect to the Nuclear Security Regulations in Canada, they were already in the process of being updated. With the arrival of SMR's to be captured into regulatory regimes, this has had an impact on the completion of the Regulations due to the many new considerations that must be considered with respect to Nuclear Security and SMRs. Through this process it has certainly identified that a need was present to update the regulatory requirements for SMRs. For further information, please see the response from Canada to Question 1.

A–2.3. CHINA–NNSA

A–2.3.1. Question

Was there a need to change regulations and guidance in your country as a result of proposed projects involving SMRs?

A–2.3.2. Response

As mentioned in the response from China to the question 01, some Department rules, Safety Guides and Technical Documents, even Regulations need to be formulated or modified.

For example, Safety Requirements on the Nuclear Power Plant Design (HAF102) [A–68] described in 1.1 Objectives: "It gives the nuclear safety principles for land-based thermal neutron reactors". So, for Floating reactors, the base regulation HAF102 and the accompanying Guides (HAD [A–69]) are needed to be improved or supplemented.

The following three documents for SMR have been issued by NNSA:

a) The safety review principles for small PWR nuclear power plants HAF (Trial) (referred to as the 'principles') have been issued in 2016 [A–70]. In the principle, NNSA stated his attitude to

SMR about Safety goal, Defence in Depth concept, General Design Basis, External events protection, Accident Source Terms, Emergency Planning, PRA application and Safety related software V&V. The main content of the Principles are as follows:

i)   Part 2: Safety Goal

The basic safety goal of SMR is to provide higher protection level without off-site protect actions to public than that of large light water reactor nuclear power plant with off-site protect actions. Meanwhile, according to the long-term goal of nuclear safety in the year 2020, eliminating the possibility of releasing large radioactive material is added as a part of the safety goal and PSA goal was deleted.

ii)   Part 3: Defence in Depth

Keep the five levels of Defence in Depth with the first three levels, at most the first four levels as the emphasis, so as to fulfil the goal that technically the need for off the site protect actions could be limited and even eliminated.

iii)   Part 4: General Design Basis

Content BDBA Analysis: The important event sequences of BDBA shall be considered in the design by probability, deterministic theory and engineering judgment. Take into consideration the safety system, non-safety system and additional temporary system to draw up accident management procedure.

Acceptance criteria of postulated accident radioactive consequence: Individual effective dose on-site boundary shall be controlled below 5 mSv or 10 mSv during whole accident (generally 30 days) and the equivalent dose of thyroid below 50 mSv and 100 mSv when an infrequent faults or a limiting faults occurs. Individual effective dose on-site boundary in BDBAs shall be controlled below 10 mSv during the whole accident (generally 30 days).

iv)   Part 5: External events protection

SMR must be protected from external natural disasters based on design basis with proper safety margin. As to external man induced events, protection measures shall meet the demands of existing laws and standards in China.

v)   Part 6: Source term

Important events sequences of DBA and BDBA for NPP with SMR shall be analysed to determine radioactive substances, from which conservative and enveloping ones can be chosen as source term for site selection and emergency plan.

vi)   Part 7: Emergency plan

For the important events sequences of all DBAs and BDBAs, off-site individual effective dose and the equivalent dose of thyroid respectively should be below intervention level of shelter and iodine protection. Therefore, technically it should lay the foundation for off-site emergency simplification.

b)   External events selected in the design of floating nuclear power plant (Trial) issued by the National Nuclear Safety Administration in 2018 [A–71].

This document described the external events that should be considered in Floating NPPs design and their design basic determination methods. External events conclude external natural events: such as extreme waves, extreme currents, extreme sea ice, extreme water temperature, extreme wind and snow, tropical cyclones, tornadoes, hail, salt fog, sunlight radiation, earthquake and tsunami, harmful gases; and external human events: such as ship collision, aircraft crash, explosion, missiles, sinking, reef grounding, etc.

c) The safety review principles for HTR-PM issued in 2008. The associated requirements for Safety goal, Defence-in-Depth concept, General Design Basis, Containment, Accident Source Terms, Emergency Planning, PRA application and Safety related software V&V are included. The main content of the Principles are as follows:

i) Part 2: Safety Goal

- General nuclear safety goal: to establish and maintain effective defences against radioactive hazards in HTR-PM and keep individuals, society and environment from hazards.

- Radiation protection goal: to ensure that, under all operation states, radiation exposure due to HTR-PM itself or any planned discharge is kept below prescribed limits as low as reasonably achievable. The radioactive consequences of any accident should be mitigated.

- Technical safety goal: to take all feasible measures against accidents and mitigate consequences of potential accidents. As for all the potential accidents considered in HTR-PM design, including those with low probability, to ensure in high confidence that any radioactive consequence is as low as possible below prescribed limits; to ensure that accidents with severe radioactive consequence is eliminated practically.

- Probabilistic safety goal: Based on probabilistic safety analysis, the accumulative frequency of all the Beyond Design Basic Accident sequences, which result in off-site (site boundary included) individual effective dose over 50mSv, should be less than 10^{-6} per reactor year.

ii) Part 3: Defence in Depth

HTR-PM still preserve the above five levels of defence-in-depth (DID). However, considering the safety characteristics of HTR-PM, the design consideration of each level could be different from conventional PWR and BWR NPPs. For example, Integrity of coated fuel particles, the first radioactive containment barrier, would act as a more important role. In addition, the long tolerant time of HTR-PM could be considered as another effective feature in DID. The rationality of DID levels should be verified by integrated safety assessments.

iii) Part 4: Plant States Definitions

- Anticipated Operational Occurrences (AOO)
  The frequency of AOO is equal to or greater than 10^{-2} per reactor year. For public individuals (adults), the effective dose threshold for radioactive release is 0.25 mSv per plant year. Typical examples are as follows:
    o Primary loop helium blower unplanned acceleration;
    o Loss of offsite power and so on.

- Design Basis Accidents

    Infrequent Accidents: Frequencies range from 10^{-2} to 10^{-4} per reactor year. The effective dose of public individuals (adults) should be limited below 5 mSv while thyroid dose equivalent below 50 mSv for each accident. Typical examples are as follows:

    - o Small breaks in feed water pipeline;
    - o One steam generator tube double ended break;
    - o Turbine loss of external load and so on.

    Limiting accidents: Frequencies range from 10^{-4} to 10^{-6} per reactor year. The effective dose of public individuals (adults) should be limited below 10 mSv while thyroid dose equivalent below 100 mSv. Typical examples are as follows:

    - o Main stream pipe break;
    - o Anticipated transients without scram and so on.

- Beyond Design Basis Accidents (BDBA)

    Probabilistic and deterministic safety analysis combined with engineering judgment are used to identify the BDBA sequences to be considered in HTR-PM. Realistic or best estimated assumption, methodology and acceptance criteria are expected in BDBA analysis.

iv) Part 5: Containment structure

Retain of radioactive material in HTR-PM mainly relies on the coated particle fuel elements with high reliability. They are insensitive to the loss of the coolant. Even for the most severe accident, the amount of radioactive release is limited, with enough time for taking accident management measures. As a result, it is feasible to utilize the Vented Low-Pressure Containment (VLPC) different from the traditional PWR or BWR containment. Meanwhile, VLPC must be demonstrated by the thorough safety assessment and show its safety and rationality. That is, the design must fulfil the HTR-PM safety goal and not weaken the whole defence levels including external events defence, etc.

A–2.3.3. Follow-up Question

The response states that for specific SMR designs, especially for HTGR-SMR or floating reactors, some safety requirements need to be formulated or modified, which ones?

Regarding the HAF102 [A–68] described in 1.1 Objectives it is stated that "It gives the nuclear safety principles for land-based thermal neutron reactors" and that for Floating reactors, the base regulation HAF102 [A–68] and the accompanying Guides (HAD [A–69]) will need to be improved or supplemented. Please describe the changes under consideration.

If a different SMR (to an HTGR) is to go through licensing, would NNSA need to generate safety principles specific to that technology (noting that it had to be done in 2008 for HTR-PM)

The response reads as though the VLPC is a design solution accepted by the NNSA based on reliance on TRISO fuel for a set of scenarios, and that it addresses the demands for the external events postulated. Is there a set of postulated events that support the use of VLPC? Please provide.

A–2.3.4. Response

(a)   At present, neither HAF [A–68] nor HAD [A–69] has been modified for Floating Reactors accordingly. Relevant research has been carried out, and it is proposed to be modified in the following aspects.
- Construction;
- Internal and external hazard;
- Safety systems and DEC conditions for multi-unit nuclear power plants;
- Reactor core control;
- Design for loss of off-site power;
- Air conditioning and ventilation systems, etc.

(b)   Since all of nuclear power plants previously built in China are PWR, the regulatory requirements formulated by NNSA are mainly apply for PWRs. If a small reactor technology uses PWR technology, such as ACP100, most of the regulatory requirements are applicable.

(c)   Postulated events in PSAR of HTR-PM as follows.
Typical AOOs and DBAs:
- One control rod spurious withdrawn during power operation (100% and 50%);
- Main helium blower spurious speed-up;
- Loss of Offsite power supply;
- Loss of normal feed water flow;
- Double-end rupture of the large primary piping;
- Double-end rupture of the DN10mm instrument piping;
- Double-end rupture of the steam generator tube.

BDBA (DEC):
- ATWS during the case of loss of offsite power supply;
- ATWS during the case of loss of normal feed water flow;
- ATWS during the case of one control rod spurious withdrawn;
- ATWS during the case of operational basis earthquake;
- Loss of feed water flow overlapped with the failure of blower isolation valve;
- SGTR overlapped with the failure of SG discharge system;
- Rupture of SG tube plate;
- Air ingress accident (chimney effect);
- Loss of passive heat removal system;
- The consideration of external events is no different from that of traditional NPPs.

A–2.4. CZECH REPUBLIC–SÚJB

A–2.4.1. Question

Was there a need to change regulations and guidance in your country as a result of proposed projects involving SMRs?

A–2.4.2. Response

Not applicable — currently there are no project involving SMRs proposed to be licensed in the Czech Republic so as of today no need to change regulations and guidance has been identified.

The secondary legislation – mostly decrees of the State Office for Nuclear Safety, that are implementing legislation to the Atomic Act [A–21] are inherent part of the Czech nuclear legislative framework. Therefore, the abovementioned (see answer from Czech Republic to Question 1) applies regardless of the type of the legislative document.

Taking into account the fact that requirements contained in secondary legislation (implementing decrees to the Atomic Act [A–21]) are relatively specific and prescriptive and reflect currently used technology in Czech Republic (i.e. PWR), the legislative framework on the level of secondary legislation is not in all cases, strictly speaking, technologically neutral. Hypothetically, the national legislative framework would therefore need to be adjusted to reflect distinct technologies and facilitate their deployment (especially on the level of secondary legislation). However, there is a persistent lack of clear and reliable data of the particular technological solution which could be envisaged to be deployed. It may even occur that the legislative framework is modified but not simplified to strengthen particular requirements to reflect the specificities of a particular design and deployment model that would be considered as a viable option.

As the guidance documents develop to a greater level of detail the requirements contained in the secondary legislation (in a similar way to how the secondary legislation is implementing to the Atomic Act [A–21]), the same applies in their case. Since the guidance documents are non-legally binding and serve as manuals describing one of the possible ways how to comply with the legal requirements, they are not indispensable to demonstrating the compliance with the legal requirements. For this reason, it might not be reasonable from a regulatory perspective to amend the existing ones or draft a new set of guides reflecting particular SMR technology specificities for all type of technologies. Hypothetically speaking, guidance documents would have to be modified to reflect a very new and different SMR technology and its deployment model. Changes could be more significant as the guidance documents are very detailed (at the other end of the spectrum is the Atomic Act [A–21] whose provisions are more general in nature).

A–2.4.3. Follow-up Question

What areas of the existing regulations are technology-specific and would need to be revisited or require new regulations?

A–2.4.4. Response

Some of the requirements contained in the secondary legislation (implementing decrees to the Atomic Act [A–21]) are relatively technology-specific and prescriptive and reflect currently used technology in Czech Republic (i.e. PWR). Both the Decree No. 329/2017 [A–22], on the requirements for nuclear installation design, and the Decree No. 358/2016 [A–23], on requirements for assurance of quality and technical safety and assessment and verification of conformity of selected equipment, contain several such provisions. That is because both decrees have been drafted to be primarily applied to standard (large) size NPPs and their text therefore reflect the design specificities of this type of design.

If an SMR of a very distinct design was deployed in the Czech Republic a more detailed analysis of these decrees in particular would be required. However, since there are no plans for deployment of an SMR (of a particular design) in Czech Republic, no specific challenges could be identified. Given the lack of detailed information about various SMR designs and the uncertainty over the type of SMR that might be hypothetically deployed, no specific information can be provided (for a specific examples see the response from Czech Republic to Question 1).

A–2.5. FRANCE–ASN

A–2.5.1. Question

Was there a need to change regulations and guidance in your country as a result of proposed projects involving SMRs?

A–2.5.2. Response

Regulations and guidance precise the higher-level requirements defined in the Environmental Code [A–24].

Regarding regulation, France has not identified any need to change it as a result of a future project involving SMRs yet.

Indeed, the main regulatory framework under which ASN undertake regulation of nuclear safety is the ministerial order of the 7th February 2012 [A–72]. This order sets the general rules relative to basic nuclear installations and is applicable to every nuclear installation, including potential upcoming SMR projects. It defines the rules related to the design, construction, operation, final shutdown, dismantling, management and surveillance of these installations. The order states that these rules rely on a proportionate approach, regarding the risks of the installation.

Moreover, the Decree nr 2019-190 of the 14th March 2019 [A–73] codifies the applicable requirements regarding nuclear installations, radioactive material transportation and nuclear transparency. In particular, it provides the possibility to a prospective licensee to ask ASN's opinion about its design's safety options. This decree is applicable to potential upcoming SMR projects.

However, to deal with specific issues, the Ministry in charge of nuclear safety (or ASN for individual resolutions) can enact resolutions to enforce new requirements for licensees. For example, resolution n° 2014-DC-0462 of the 7th October 2014 [A–74] establish requirements regarding criticity risk issues and is applicable to every nuclear installation concerned by this risk. These resolutions can also target specific designs (for example, resolution n° 2016-DC-0578 [A–75] is only applicable to pressurized water reactors), or target specific installation (for example, ASN resolution n° 2013-DC-0347 [A–76] provides requirements for the EPR of Flamanville). If an SMR project is proposed in France, additional resolutions to deal with its specificities could be enacted.

Regarding guidance, ASN produces guides for licensees which sets out recommendations with the aim of:

- Explaining the regulations and the rights and obligations of the persons concerned by the regulations;
- Explaining the regulatory objectives and, as applicable, describing the practices considered by ASN to be satisfactory;
- Giving practical tips and information concerning nuclear safety and radiation protection.

ASN has published guides applicable to any kind of nuclear power plant - including SMRs, and others focused on specific designs and technologies (for example, Guide n° 22 published in 2017 [A–77] provides recommendations for the design of pressurized water reactors). Considering that there is no SMR project submitted to ASN yet, ASN has not identified any need regarding the creation of new guidance specific to SMRs.

A–2.6. JAPAN–NRA

A–2.6.1. Question

Was there a need to change regulations and guidance in your country as a result of proposed projects involving SMRs?

A–2.6.2. Response

Regulations and guidance were not changed for the review of HTTR.

As a reference, enhanced points of the new regulatory requirements are shown below, which introduces lessons learnt from the Fukushima Daiichi NPP accident.

In the new regulatory requirements, earthquake resistance, tsunami resistance and fire resistance are enhanced in common for research reactor facilities and it is newly required to introduce countermeasures for internal flooding, volcano and forest fire and so on. In addition, basic design policy for gas cooled test reactor facilities is newly shown, which introduces enhancement/newly addition of countermeasures for loss of external power, radiation monitoring facility, measures for mitigating consequences of accidents which may release large amount of radioactive materials.

A–2.6.3. Follow-up Question

What specific regulations apply in this case? Also please explain the specific changes e.g. the new earthquake design criteria, tsunami, fire etc. The response states that additional countermeasures for a number of events were introduced – are these specified in the regulations/ requirements or are they goals to be met? What are they?

What were the challenges associated with introducing these requirements to the HTTR and how were they resolved?

A–2.6.4. Response

The requirements for the permission for establishment of research and test reactor are stipulated in the 'NRA Order Prescribing Standards for the Location, Structure, and Equipment of Research and Test Reactors and their Auxiliary Facilities' [A–78] and its interpretation guides. Specifically:

- Article 4 Prevention of damage from earthquake;
- Article 5 Prevention of damage from tsunami;
- Article 6 Prevention of damage from external impact;
    - This article requires damage prevention for safety features against following exemplified hazards - natural hazards such as: flood, wind(typhoon), tornado, freezing, precipitation, snow accumulation, lightning, landslip, volcanic activity, biological effect, forest fire;
    - Possible external man induced events (except intentional case) such as: flying object (e.g. aircraft crash), dam failure, explosion, fire at neighbouring factory, toxic gas, ship collision, electro-magnetic interference.
- Article 8 Prevention of damage from fire;

- Article 9 Prevention of damage from flood;
- Article 53 Prevention and mitigation of accident which release large amount of radioactive material.

On reviewing the conformity to these requirements, we also refer to evaluation criteria or evaluation guides for commercial reactors:

- Review standard for the fire protection of commercial reactors and their auxiliary facilities [A–79];
- Guide for Evaluation on Volcanic Hazards [A–80];
- Guide for Evaluation on Tornado Hazards [A–81];
- Guide for Evaluation on External Fires [A–82];
- Guide for Evaluation on Internal Flooding [A–83];
- Guide for Evaluation on Internal Fires [A–84];
- Guide for Evaluation on Surveys for geological features and structure of Site and its vicinity [A–85];
- Guide for Review on Design-Basis Earthquake and Seismic-resistance Design [A–86];
- Guide for Review on Design-Basis Tsunami and Tsunami-resistance Design [A–87];
- Guide for Review on Foundation Grounds and Slope Stability Assessment [A–88].

On reviewing HTTR, following challenges and measures are discussed:

- The (original) design had not have enough protection of emergency power supply unit against volcanic ash effect and tornado;
- Reviewed and confirmed required safety level through the alternative measures taken by the entity such as equipping mobile power supplies.

In regard to measures against internal fires, the designs for fire protection are not necessarily typical measures following requirement stipulated in the Standards and Guides:

- Self-extinguishing feature of fire-retardant-cable;
- To equip different types of fire detection system;
- To activate stationary fire extinguishing equipment from central control room;
- Separation of each fire areas.

Reviewed and confirmed required safety level in consideration of the characteristic and peculiarity of HTTR.

A–2.7. RUSSIAN FEDERATION–ROSTECHNADZOR

A–2.7.1. Question

Was there a need to change regulations and guidance in your country as a result of proposed projects involving SMRs?

A–2.7.2. Response

The provisions of the international conventions under the aegis of the IAEA are implemented in the legislation of the Russian Federation in the federal rules and regulations. The provisions of the international conventions under the IMO are implemented in the legislation of the Russian

Federation in the federal rules and regulations (related to nuclear powered ships and other vessels with nuclear reactors) and in the regulatory documents of the Russian Maritime Register of Shipping.

In the Russian Federation, the issues of ensuring the safety of ships and other vessels with nuclear reactors at all stages of their life cycle are regulated by the Federal Law 'On the Use of Atomic Energy' of 21.11.1995, No. 170-FZ [A–26]. In accordance with Article 6, federal rules and regulations in the field of atomic energy use are developed.

List of main federal rules and regulations for ship-based nuclear installations and their nuclear service vessels:

- General safety provisions for nuclear power installations of ships and other vessels (NP-022-17) [A–89];
- The basic rules of nuclear material accounting and control (NP-030-12) [A–90];
- Requirements for organization of material balance areas (NP-081-07) [A–91];
- Ensuring safety during decommissioning of nuclear facilities. General provisions (NP-091-14) [A–92];
- Nuclear safety rules for nuclear power installations of ships and other vessels (NP-029-17) [A–93];
- Safety rules for storage and transportation of nuclear fuel at nuclear facilities (NP-061-05) [A–94];
- Safety regulations for transport of radioactive material (NP-053-16) [A–95];
- Strength calculations standards for equipment components and pipelines of ship nuclear steam generating installations with VVER-type reactors (NP-054-04) [A–96];
- Requirements to quality assurance programs of nuclear facilities (NP-090-11) [A–97];
- Safety of radioactive waste management. General provisions (NP-058-14) [A–98];
- Basic rules of accounting and control of radioactive substances and radioactive waste in organization (NP-067-16) [A–99];
- Rules of reclassification of nuclear materials into radioactive substances or radioactive waste (NP-072-13) [A–100];
- Rules of physical protection of radioactive substances and radioactive sources during transportation (NP-073-11) [A–101];
- Requirements for justification of the possibility to extend the design service life for nuclear facilities (NP-024-2000) [A–102];
- Rules on safety ensuring during decommissioning of ships and other vessels with nuclear installations and radiation sources (NP-037-11) [A–103];
- Requirements for planning of actions and protection of employees (personnel) during radiation accident at a nuclear installation of a ship and (or) other watercraft (NP-079-18) [A–104];
- Regulations on order of investigation and accounting of operational violations of ships with nuclear installations and radiation sources (NP-088-11) [A–105];
- Requirements for safety analysis report of nuclear power installations of ships (NP-023-2000) [A–106].

The features of the FNPP Akademik Lomonosov were taken into account for the development of the abovementioned federal rules and regulations, and the issues of the usage of nuclear installations on vessels to produce electricity and thermal power were considered.

A–2.7.3. Follow-up Question

The response only covers Floating NPPs (whilst responding that they were developed with FNPP Akademik Lomonosov in mind) — so they appear technology specific. What other regulations would apply to a land-based SMR, e.g. BREST-300.

A–2.7.4. Response

When assessing the safety case for a power unit with BREST-300, the existing regulatory and legal framework is used. With regard to the other land-based SMRs, Rostechnadzor is planning to carry out research to adapt the existing requirements and assess the need to develop new regulatory requirements for the case of serial production of modular reactors.

At the moment, one of the promising designs of a land-based SMR is the RITM-200 reactor. The preliminary analysis carried out in general showed the possibility of using the regulatory legal framework of the NPPs to regulate its safety, as well as the need for a number of changes to the existing regulatory legal framework. It should be noted that specific proposals for revision are at an early stage of development.

A–2.8. SOUTH AFRICA–NNR

A–2.8.1. Question

Was there a need to change regulations and guidance in your country as a result of proposed projects involving SMRs?

A–2.8.2. Response

The first challenge faced by the NNR was to develop licensing requirements for this 'new' type of reactor taking cognizance of reactor operating experience, developments in international safety standards and application of these in the design of new generation of reactors. Recognizing these factors, the NNR developed and published the first revision of LG-1037 [A–29]. These requirements are essentially 'technology neutral' safety requirements according to which DiD is enforced in a graded approach based on the scale of the hazard in terms of the risk and dose criteria referred to above.

This was followed by the progressive development of many specific regulatory requirements and guidance documents in support of LG-1037 [A–29] that would have formed the basis for the NNR review of the safety case as presented by the applicant.

The next challenge faced by the NNR was to provide guidance to the applicant and the designer on the processes that will need to be undertaken to demonstrate compliance with these requirements.

During the years, LG-1037 [A–29] was followed by a number of NNR requirements and guidance documents which were developed when a need was identified based on a request from industry or where the NNR was of the opinion that further intervention is needed.

The regulatory standards should ideally be in place in advance of a licence application and should inform the licensing process and the development of the safety case.

Most of the documents listed in Table A–1 were developed subsequent to the initiation of the licensing process.

TABLE A–1: NNR LICENSING DOCUMENTS APPLICABLE TO PBMR

Licensing document number	Type	Title
RD 0018 [A–107]	Requirements	Basic licensing requirements for Pebble Bed Modular Reactor (RD-0018, Rev 0 Superseded LG-1037 [A–29])
RD-0024 [A–108]	Requirements	Requirements on the risk assessment and compliance with principal safety criteria for nuclear installations
RD-0034 [A–33]	Requirements	Quality and Safety Management Requirements for Nuclear Installations (Superseded LD-1094 issued in 2001)
RD-0016 [A–109]	Requirements	Guidance on the Verification and Validation of Evaluation and Calculation Models used in Safety and Design Analyses (Superseded LG-1038 [A–110] that was issued in 2001)
RD-0019	Requirements	Requirements for the Core Design of the Pebble Bed Modular Reactor
LD-1096 [A–111]	Requirements	Fuel qualification requirements for PBMR
LD-1097 [A–112]	Requirements	Qualification Requirements for the Core Structure Ceramics of the Pebble Bed Modular Reactor (LD-1097 was revised in 2010 and a final draft of the revised document was finalised as RD-0036 but not submitted for approval.)
RD-0014 [A–113]	Requirements	Emergency Preparedness and response requirements for nuclear installations
LG-1041 [A–114]	Guidance	Licensing guide on safety assessments for nuclear power plants
LG-1045 [A–115]	Guidance	Guidance for licensing submissions involving computer software and evaluation models for safety calculations
RG-0005 [A–34]	Guidance	Guidance on Testing, Qualification and Commissioning of the PBMR Demonstrations Power Plant

A–2.9.   UNITED KINGDOM–ONR

A–2.9.1. Question

Was there a need to change regulations and guidance in your country as a result of proposed projects involving SMRs?

A–2.9.2. Response

ONR has not been engaged in Generic Design Assessment (GDA), licensing or regulatory permissioning work associated with the deployment of an SMR. ONR has, however, proactively undertaken a review of extant guidance including the SAPs [A–44] and TAGs [A–46] for compatibility with Advanced Nuclear Technologies (ANTs). The review of the SAPs was informed by regulatory capability activities on ANTs from 2017 and the provision advice to the UK Department for Business Energy and Industrial Strategy (BEIS) in the Advanced Modular Reactors (AMR) Feasibility and Development (F&D) project. The advice to Government was provided in 2019 in line with section 89 of the Energy Act [A–40], and covered the security, safety and transport aspects associated with 7 fission reactor submissions to the Regulators. These reactor designs involved a sodium fast reactor, two lead fast reactors, three high temperature gas reactors and a molten salt reactor proposal. The environmental protection aspects were considered by the Environment Agency (EA).

The key outcome of the high-level review of the SAPs undertaken as part of ANT project was that they continue to provide a sound framework and basis for ONR inspectors to make consistent regulatory judgements on the safety of activities relating to AMR technologies. The review nevertheless highlighted that additional guidance or information may be helpful to clarify how some of the SAPs are applied in the context of potential faults and hazards specific to AMR technologies, changes in operating philosophy, novel characteristics of the fuel or the coolants amongst other features. ANT guidance review activities are consequently being undertaken to implement this recommendation and the outcomes of the work so far is documented in the relevant sections of this questionnaire.

A–2.10. UNITED STATES OF AMERICA–NRC

A–2.10.1. Question

Was there a need to change regulations and guidance in your country as a result of proposed projects involving SMRs?

A–2.10.2. Response

Yes, while the Nuclear Regulatory Commission (NRC) licenses small modular reactors (SMRs) in accordance with the framework established by the 10 CFR [A–52], Part 50 and Part 52 regulations, the associated guidance for SMRs has evolved and has been routinely updated. The principal guidance for reviewing and licensing new reactor applications is the NRC document entitled, NUREG-0800, 'Standard Review Plan for the Review of Safety Analysis Reports for Nuclear Power Plants: LWR Edition.' [A–116]

The following lists summarizes the changes to the U.S. Code of Federal Regulations, and associated NRC guidance documents, for the licensing of SMRs:

(a) The NRC staff provided a document for NRC Commission consideration entitled, SECY-20-0045, 'Population Related Siting Considerations for Advanced Reactors' [A–117]. In this document, the NRC staff provided options and a recommendation to the NRC Commission on possible changes to guidance documents to address population-related siting considerations for advanced reactors.

(b)	The NRC staff identified a possible approach for a scalable emergency planning zone for SMRs. The NRC staff proposed changes to its regulations to address EP issues for future SMRs and other new technologies, including non-LWRs and medical radioisotope facilities.

(c)	In an NRC staff document entitled, SECY-11-0178, 'Insurance and Liability Regulatory Requirements for Small Modular Reactor Facilities,' [A–118] the NRC staff identified a potential inequity between the insurance requirements for facilities of different sizes.

(d)	The NRC staff provided its Commission with a paper entitled, SECY-18-0076, 'Options for Physical Security for Light-Water Small Modular Reactors and Non-Light-Water Reactors.' [A–119]. Based on the options presented in the NRC staff paper, the NRC Commission directed the NRC staff to initiate a limited-scope revision to the regulations and guidance related to physical security for advanced reactors.

(e)	In an NRC document entitled, SECY-11-0079, 'License Structure for Multi-Module Facilities Related to Small Modular Nuclear Power Reactors,' [A–120] the NRC staff reviewed three potential licensing structure alternatives for multi module facilities and determined that the licensing of each module individually was preferred.

(f)	The NRC staff proposed an equitable assessment of annual fees for SMRs in an NRC document entitled, SECY-15-0044, 'Proposed Variable Annual Fee Structure for Small Modular Reactors,' [A–121] for NRC Commission consideration.

(g)	In Revision 3 of the NRC's Standard Review Plan (SRP) [A–116], Chapter 19, 'Probabilistic Risk Assessment and Severe Accident Evaluation for New Reactors,' the NRC documented the criteria to ensure appropriate treatment of important risk insights related to multi-module design and operation.

(h)	The NRC staff informed the NRC Commission of plans to ensure that SMR licensees provide reasonable assurance of availability of decommissioning funding in an NRC document entitled, SECY-11-0181, 'Decommissioning Funding Assurance for Small Modular Nuclear Reactors.' [A–122]. The approach allows applicants to submit a site-specific estimate of decommissioning costs with a supporting analysis and adequate justification for an exemption to the minimum funding requirements for large LWRs required in the NRC regulations.

In an NRC document entitled, SRM-SECY-15-0168, 'Recommendations on Issues Related to Implementation of Risk Management Regulatory Framework or RMRF' [A–123] the NRC Commission approved the NRC staff's recommendation that the NRC not develop a definition of and criteria for determining adequacy of defence in depth and directed the staff to expeditiously complete the revision to NRC Regulatory Guide 1.174, 'An Approach For Using Probabilistic Risk Assessment in Risk-Informed Decisions on Plant-Specific Changes to the Licensing Basis,' [A–124] on defence in depth, in order to improve the clarity of the guidance. The NRC issued Revision 3 of [A–124] in January 2018.

A–3. LICENSING PROCESS

This Annex presents the responses provided by the following Member State regulatory bodies to the Question 3: "Did you make changes to the licensing process to address SMRs? Were new processes, such as pre-licensing engagement developed?"

A–3.1. ARGENTINA–ARN

A–3.1.1. Question

Did you make changes to the licensing process to address SMRs? Were new processes, such as pre-licensing engagement developed?

A–3.1.2. Response

Yes, some changes were made to the licensing process of CAREM 25, especially at the beginning of the construction stage.

A basic aspect of the Argentine regulatory system is the approach adopted, in which the Licensee deals with the design, construction, commissioning, operation and decommissioning stages of the nuclear facility, being completely responsible for the radiological and nuclear safety of the installation as well as for the physical protection and safeguards. This responsibility goes beyond the compliance of requirements stated in the Regulatory Standards.

The Regulatory Standards establish that the construction, commissioning, operation or decommissioning of an NPP shall not be initiated without the corresponding authorization: License, which must be previously required by the Licensee and later, issued by the Regulatory Body. Despite that there is a validity period for the commissioning and operation Licenses, in all cases the validity of such Licenses is always subordinated to the compliance with the conditions stipulated in its articles of terms and conditions. There are conditions on operation issues including staff training and qualification, emergency preparedness, radiological issues on workers, emissions and waste, transport of nuclear and radioactive material, safeguards, security and communication of the Licensee towards ARN.

The non-compliance with any of the regulatory standards, conditions or requirements is enough reason for the Regulatory Body to suspend or cancel the corresponding License validity, according to the sanction regime in force.

The regulatory system considers licenses for construction, commissioning, operation, and decommissioning that establish the conditions that the Licensee must fulfil at each stage.

The Construction License is issued when regulatory standards and requirements of the siting, basic design and expected safety operation conditions have been complied with prior to start of this stage.

The applicable regulatory standards, consistent with international recommendations on the subjects, establish the safety criteria to be met in the design of the installation and define the timetable and type of mandatory documentation that must be presented together with the application for the Construction License.

The Memorandum of Understanding (MOU) was the establishment, since an early stage of the project, of the regulatory requirements and expectations in terms of licensing process and safety level that must be fulfilled by the design of the proposed plant and demonstrated through the Safety Analysis to be further submitted to ARN.

Once the Construction License is requested by the Licensee, a continuous interaction between the constructor or operator of the future installation and the Regulatory Body is initiated. It is a dynamic process, as complex as the demands involved. It should be emphasized that the Licensee's capacity to carry out its responsibilities is evaluated starting from the construction stage.

The Commissioning License establishes the conditions for the approach to criticality, operation with increasing power up to its nominal value, as well as verifications and tests of the components, equipment, and systems to determine whether they comply with the original design basis. To do so the Licensee must appoint an ad hoc Commissioning Committee of senior specialists, to evaluate the execution of the commissioning program and recommends on its continuation and adjustment (Regulatory Standards AR 3.7.1. [A–125] and AR 3.8.1. [A–126]).

The Operating License is issued when the ARN verifies that conditions, regulatory standards, and specific requirements applicable to a specific installation are fulfilled. Such conclusion is the result of analysing the submitted documentation and detailed studies, as well as the inspection results carried out during the construction and commissioning together with the ad hoc Commissioning Committee recommendations.

Bearing in mind that the CAREM 25 reactor was considered a prototype the ARN has established a license scheme applicable to construction and preliminary tests. This framework established the milestones for the beginning of construction and, in order to issue the construction authorization, the regulatory body established additional mandatory documentation requirements regarding the traditional nuclear power plant licensing scheme. This authorization license scheme was approved by the ARN in 2010 and was communicated to the responsible entity.

This Licensing scheme considering CAREM 25 as a prototype of NPP. ARN granted the construction authorization with 'license conditions' and a regulatory requirement reinforcing the authorization. The fulfilment of specific regulatory requirements conditioned the beginning of the construction of the nuclear module of the reactor about some findings of the PSAR assessment. The findings included in the regulatory requirements (RQ — 'requerimiento regulatorio') were related to engineering whose resolution has an impact on civil works, whether due to structural functions, confinement or shielding.

In December of 2014, the ARN lifted the condition to Authorization for Use of Site and Construction (AUSC) after the Responsible Entity (RE) presented the necessary information as corrective actions of the evaluation findings.

The evolution of the project and the experience gained in other projects (O.L. of Central Nuclear de Atucha II -CNA II-, LTO license of CNE, pre-licensing of an updated CANDU and of the HPR 1000) leads to an up-date of the licensing scheme. The revised licensing scheme fits completely in the licensing procedures foreseen for new NPPs in terms of mandatory documents (TOC and scope) and overall approach.

Next licensing milestones:

- Update the license scheme for the eventual issuance of the Commissioning License (completed);

- Definitions of the operating staff license and the interface between construction / testing / commissioning / operation;
- Approval of a Safety Report 'as built', complete, autonomous and with information of preliminary tests;
- Program evaluation and commissioning procedures;
- Assessment of the mandatory documentation requested to issue the Commissioning License.

The requirements of Mandatory Documentation for the application for the Commissioning License are:

- Safety Report;
- Missions and Functions Manual (Staff);
- Organization Chart and Requirements for Staff Licensing;
- Emergency Plan;
- Manual of Radiological Safety, Waste Management and Environmental Monitoring;
- Documentation of Constitution and operation of the Ad Hoc Committee for the PeM;
- Report with the Results of the Preliminary Tests;
- Plant Manuals: Maintenance, OPEX Management, In-service Inspection, Routine Tests;
- Design Questionnaire Report (Safeguards);
- Physical Protection System Design Report;
- Compliance with the requirements of the Construction Authorization.

A–3.1.3. Follow-up Question

The response does not appear to cover pre-licensing activities. Does Argentina take undertake any such activities before the formal licensing process? If so, what do they entail?

A–3.1.4. Response

The regulatory system considers licenses for construction, commissioning, operation and decommissioning that establish the conditions that the Licensee must fulfil at each stage.

The pre-licensing tasks are conducted as a regulatory practice, before the formal process of licensing and are documented in the Memorandum of Understanding (MOU).

The main objective of the MOU was the establishment, since an early stage of the project, of the regulatory requirements and expectations in terms of licensing process and safety level that must be fulfilled by the design of the proposed plant and demonstrated through the Safety Analysis to be further submitted to ARN.

Regarding the design requirements, the MOU is in line with the Vienna Declaration [A–55] as it states the mandatory fulfilment of AR standards, as well as the latest IAEA safety standards: Safety Fundamentals, General Safety Requirements (GSR Part 1) [A–22], and Specific Safety Requirements Safety of Nuclear Power Plants: Design (SSR-2/1), Revision 1 [A–127].

As part of Defence in Depth (DiD), analysis of Design Extension Conditions (DEC) shall be undertaken with the purpose of further improving the safety by:

- Enhancing the plant's capability to withstand events or conditions more challenging than those considered in the design basis;
- Minimizing radioactive releases harmful to the public and the environment as far as reasonably practicable, in such events or conditions.

For project realization, ARN states in the MOU the need for a clear rationale connecting the engineering safety requirements for systems, structures and components, as derived from the Safety Analysis, with the safety classification following the IAEA SSG-30 [A–57].

The development of the MOU for Argentine next NPP and the CAREM project are practical examples that illustrate the strong commitment that Argentina has with the Vienna Declaration [A–55].

A–3.2. CANADA–CNSC

A–3.2.1. Question

Did you make changes to the licensing process to address SMRs? Were new processes, such as pre-licensing engagement developed?

A–3.2.2. Response

The Canadian regulatory approach to licensing SMRs is built on the long-established foundation of risk-informed regulation that has been applied to traditional reactor facilities. The Canadian nuclear regulatory framework is comprehensive and in large part technology neutral, which means that that it allows for all types of technologies to be safely regulated. Regulatory tools and decision-making processes are structured to enable a licence applicant for a reactor facility to propose alternative ways to meet regulatory objectives.

Small modular reactors are not legally defined in Canada as they fall under the Class I Nuclear Facilities Regulations [A–60], which despite the name, outlines the activities that require a licence. While no changes were made to the licensing process itself, the CNSC has enhanced its optional pre-licensing engagement activities. The licensing process and pre-licensing engagement are described below.

<u>The Licensing Process</u>

Section 26 of the NSCA [A–13] prohibits any person from preparing a site, constructing, operating, decommissioning or abandoning a nuclear facility without a licence granted by the Commission. Subsection 24(4) of the NSCA states the following:

"No licence may be issued, renewed, amended or replaced – and no authorization to transfer one given – unless, in the opinion of the Commission, the applicant or, in the case of an application for an authorization to transfer the licence, the transferee

a) is qualified to carry on the activity that the licence will authorize the licensee to carry on; and

b) will, in carrying on that activity, make adequate provision for the protection of the environment, the health and safety of persons and the maintenance of national security and measures required to implement international obligations to which Canada has agreed."

The CNSC's licensing system is administered in cooperation with federal and provincial/territorial government departments and agencies in such areas as health, environment, Indigenous consultation, transportation and labour. Before the Commission issues a licence, the concerns and responsibilities of these departments and agencies are taken into account, to ensure that no conflicts exist with the provisions of the NSCA and its regulations. The Commission is obligated to comply with any federal legislation and therefore may make its licensing decisions in consultation with any department or agency government bodies at the federal level having independent but related responsibilities with the CNSC.

The Class I Nuclear Facilities Regulations [A–60] require licences for each of the five types of activities in the lifecycle of a Class IA nuclear facility:

- Licence to prepare a site;
- Licence to construct;
- Licence to operate;
- Licence to decommission;
- Licence to abandon.

If the necessary applications are filed with the required information, the Commission may, at its discretion, issue a licence that includes multiple classes of licences (e.g. a licence to prepare a site and construct, or a licence to construct and operate). A single licence may also be issued for multiple facilities, each at a different stage in their lifecycle. The Class I Nuclear Facilities Regulations [A–60] establishes a 24-month timeline for projects requiring the CNSC's regulatory review and decision on new applications for a licence to prepare a site for a Class I nuclear facility. This timeline does not include the time required by proponents to respond to information requests.

It is important to consider that timelines (based on experience from around the world) are affected by:

- The environmental review process, depending on the jurisdictions involved and the amount of time required by the applicant to prepare the necessary documentation;
- Whether the information provided with the application is comprehensive and complete so the review of the application can be carried out in an efficient and timely manner;
- Stakeholder support (communities, Indigenous and public consultations, provincial/territorial agencies);
- State of completeness of design;
- Novel features or approaches;
- State of completion of supporting R&D;
- The quality and timeliness of the applicant completing its activities at each licensing stage (prepare the site, construct and commission the nuclear facility and train and certify facility personnel);
- Outstanding safety issues at each licensing stage, which will require resolution before CNSC staff can prepare their recommendations to the Commission for the next stage.

CNSC REGDOC-3.5.1, Licensing Process for Class I Nuclear Facilities and Uranium Mines and Mills [A–128], outlines the current licensing process in the context of the NSCA. Figure A–3 below depicts the CNSC licensing process and the key activities to be carried out by the licence applicant, CNSC staff and the Commission. The Commission may choose to hold a public hearing in one or two parts. The CNSC Rules of Procedure set out the requirements for one-part and two-part public hearings.

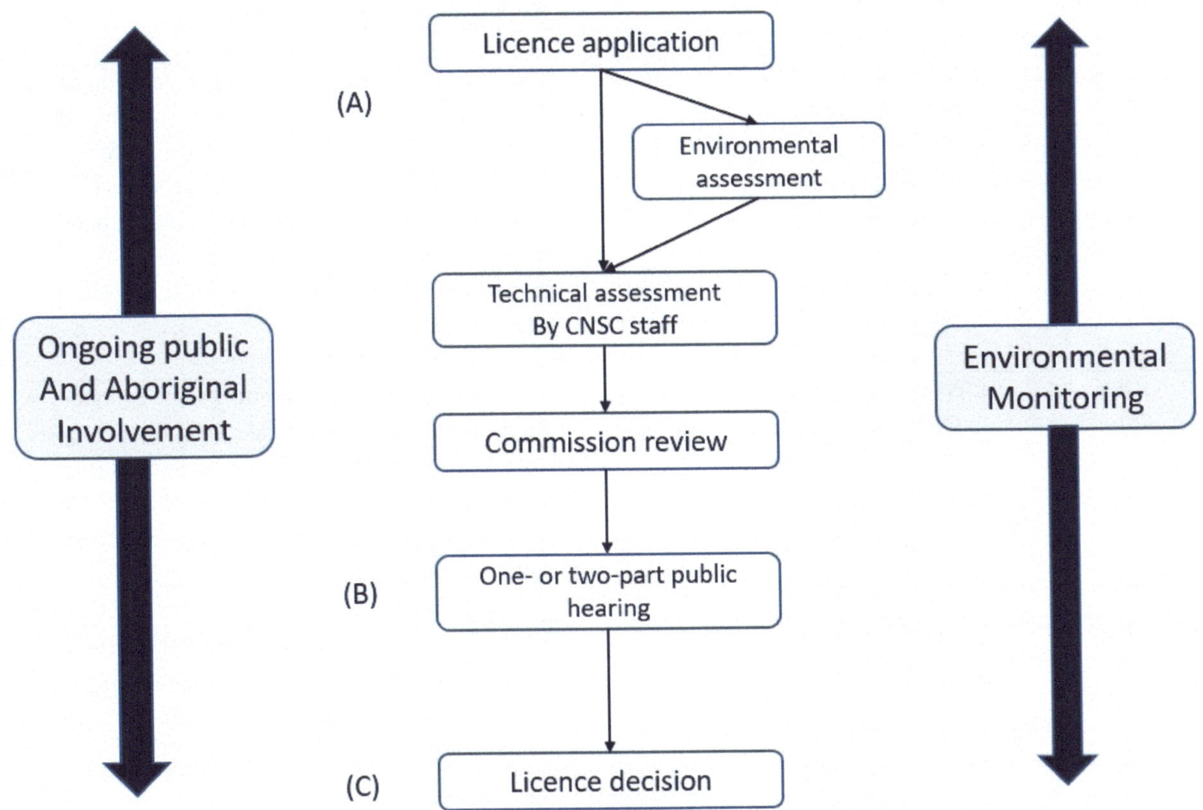

FIG. A–3: *Licensing Process under the NSCA (note: the environmental assessment is either a review carried out by the CNSC under the NSCA, or under applicable federal, provincial or territorial jurisdiction).*

The licensing process is initiated when the proponent sends an application to the CNSC. A licence application must contain sufficient information to meet regulatory requirements and to demonstrate that the applicant is qualified to conduct the licensed activity. The regulations under the NSCA provide licence applicants with general performance criteria and details about the information and programs they must prepare and submit to the CNSC as part of the application process.

To enhance clarity, the CNSC has published, or plans to publish, supporting regulatory documents for each licence type. These REGDOCs provide additional details and criteria (such as references to other CNSC regulatory documents, national codes and standards, or the IAEA safety standards) so applicants clearly understand what is necessary to satisfy the requirements of the applicable regulations under the NSCA. The Table A–2 lists published and planned CNSC REGDOCs that provide guidance on licence applications for reactor facilities.

TABLE A–2: GUIDANCE ON LICENCE APPLICATIONS FOR REACTOR FACILITIES

Document #	Title	Applicant must Demonstrate
REGDOC-1.1.1 [A–64]	Site Evaluation and Site Preparation for New Reactor Facilities	Suitability of proposed site for construction and operation of the nuclear facility considering the activities involved in preparing the site (for example, land clearing and building services requirements), and adequate consultation with stakeholders and consideration of their views (potentially affected public, Indigenous groups, etc.)
REGDOC-1.1.2 [A–66]	Licence Application Guide: Licence to Construct a Nuclear Power Plant	Proposed facility design conforms to regulatory requirements and will provide for safe operation over the proposed plant life, and responsibility for all activities pertaining to design, procurement, manufacturing, construction and commissioning.
REGDOC-1.1.3 [A–67]	Licence Application Guide: Licence to Operate a Nuclear Power Plant	Appropriate safety management systems, plans and programs have been established and resolution of outstanding issues from construction stage.

REGDOC-1.1.5 [A–65], provides information about CNSC safety and control areas as they apply to a licence application for an SMR facility. This document is intended to be used in conjunction with other licence application guides and existing regulatory documents to assist proponents in developing risk-informed proposals that take into account.

REGDOC-1.1.5 [A–65] also provides information on pre-licensing engagement activities (see below for additional information on pre-licensing):

- The VDR process;
- The process for establishing an appropriate application assessment strategy for risk-informed licensing.

In addition, REGDOC-1.1.5 [A–65] provides information that is additional to the licensing process documented in REGDOC-3.5.1 [A–128], which provides an overview of the licensing process for Class I nuclear facilities.

For new reactor facilities, information on decommissioning plans and financial guarantees is required early in the licensing process. The Class I Nuclear Facilities Regulations [A–60] require an applicant to provide information on its proposed plan for decommissioning a nuclear facility or site, while the General Nuclear Safety and Control Regulations [A–129] require information on financial guarantees to accompany a licence application. Financial guarantees are used to ensure sufficient funds are available so that the facility does not pose any unnecessary risk in the event that the licensee can no longer operate the facility. To date, these have mostly been used for decommissioning a nuclear power plant (NPP) at the end of its operating life and for the long-term management of spent nuclear fuel. Information on proposed financial guarantees should include any obligations for funding the decommissioning and long-term management of nuclear fuel waste, pursuant to the Nuclear Fuel Waste Act [A–130].

In accordance with the CNSC's current regulatory framework, the CNSC requires that the environmental effects of all nuclear facilities or activities be considered and evaluated when licensing decisions are made. All licence applications that demonstrate potential interactions with the environment are subject to an environmental review, commensurate with the scale and complexity of the environmental risks associated with the facility or activity. For the different types of environmental reviews, please refer to Question 1.

The CNSC staff assessment of an applicant's information is augmented by input from federal and provincial government departments and agencies responsible for regulating health and safety, environmental protection, emergency preparedness and the transportation of dangerous goods in relation to nuclear-related projects. The CNSC maintains memoranda of understanding with these departments and agencies. The NSCA [A–13] also requires that members of the public be invited to participate in licensing hearings of Class I facilities (NPPs, conversion facilities, research reactors) and uranium mines and mills.

CNSC staff members document the conclusions and recommendations from their reviews in Commission member documents (CMDs), submitting them to the Commission for a public hearing. For the licensing of reactor facilities, intervenors are typically allotted significant periods of time at the hearing to present their information and engage the Commission. The hearings are webcast live and the video is available online for a minimum of three months following the hearing. In addition, a verbatim transcript is prepared for these proceedings and available to the public within one week of the day of the proceedings.

During and after public hearings, the Commission deliberates upon the information provided and makes the final decision on the granting of the licence. The CNSC issues news releases to inform the public of the decisions made. The records of proceedings from the hearings, along with the reasons for the Commission's decisions, are available in both of Canada's official languages, posted on the CNSC website and sent to all participants.

Once a licence is issued, the CNSC carries out compliance activities to verify that the licensee is complying with the NSCA [A–13], associated regulations and its licence.

Optional Pre-Licensing Engagement

As mentioned, REGDOC 1.1.5 [A–65] provides information on pre-licensing activities. There are two types of pre-licensing engagement with the CNSC:

- The Vendor Design Review (VDR) process: used for reactor vendors;
- The process for establishing an appropriate application assessment strategy for risk-informed licensing: used for potential SMR applicant.

Figure A–4 illustrates these two types of pre-licensing activities at a conceptual level, including how the two processes can overlap while incorporating graded-approach considerations.

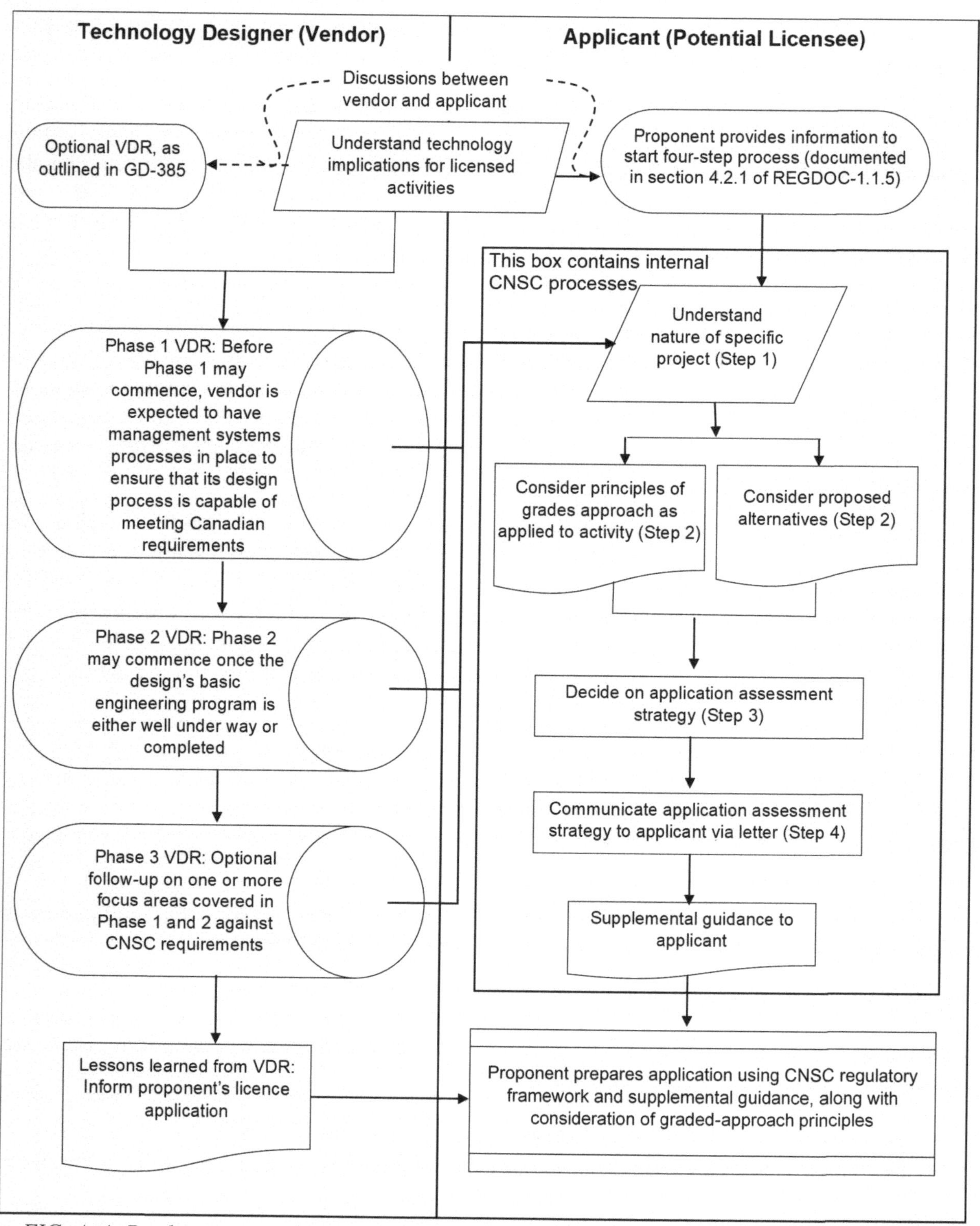

FIG. A–4: Pre-licensing engagement activities in establishing the licensing basis for an SMR facility.

Vendor Design Review

SMRs differ greatly in size, design, and operation. Each SMR design also has varying degrees of uncertainty, which SMRs of similar types may address differently. In light of this variability, a vendor may wish to consult with CNSC prior to licensing to ensure that its design meeting high-level Canadian requirements. The CNSC offers an optional VDR service in this regard.

A VDR is separate from the licensing process, and its primary purpose is to inform the vendor of the design's overall acceptability. This review provides early identification and resolution of potential regulatory or technical issues in the design process, particularly those that could result in significant changes to the design or safety case.

In a VDR, the CNSC enters into a service agreement with the vendor that is based on a fixed scope of work, under which the vendor can gain a comprehensive grasp of Canadian regulatory requirements and how its design, as it is evolving, would be capable of meeting those requirements. Similarly, this agreement helps CNSC develop a better understanding of the specific technology being presented.

The VDR process is divided into three phases, each requiring increasingly detailed technical information, and is fully described in CNSC regulatory document REGDOC-3.5.4, Pre-licensing Review of a Vendor's Reactor Design [A–131].

While the VDR process is separate from the process for determining an appropriate application assessment strategy, outputs from each VDR phase can inform the determination of such a strategy. A VDR is not a licensing process — it does not involve a potential applicant for a project, does not involve any decision making by the Commission, and does not result in any decisions that could fetter the Commission's decision making concerning a potential project. However, VDR results may be used by an applicant in the licensing process. A VDR can also take place in parallel with a licence application; for example, a vendor may decide to engage in the VDR Phase 2 or 3 in parallel with the CNSC's review of an application for a licence to prepare site.

Establishing an appropriate application assessment strategy

All licence applications are presented to the Commission for approval. In Canada, when applying for a license, the proponent should provide clearly articulated descriptions of how proposed activities would be conducted safely and would meet all applicable requirements. Clearly documented intentions facilitate fair and informed decisions. With this in mind, the purpose of establishing an application assessment strategy is to enable potential proponents to understand:

- The overall licensing process;
- The specific licensing process for the proposed activity;
- Regulatory framework tools available to support the licensing process (e.g. regulations, licence application guides and other regulatory documents) and how they are used to establish the licensing basis;
- Licensee obligations (should the licence application be approved).

The CNSC has a process for determining an appropriate application assessment strategy for an innovative activity or facility that uses technology that is new to Canada. This process ensures that a risk-informed approach is systematically and consistently applied.

While establishing an appropriate application assessment strategy is optional, it could be especially beneficial for a proponent whose application includes one or more of the following:

- New organizational models for conducting a project;
- A proposal for new types of activities, for which there is little or no past experience (e.g. potential demonstration activities to be performed in a demonstration facility);
- New ways to conduct activities (e.g. construction approaches);
- New technological approaches that require extensive interpretation of requirements.

This optional process is carried out prior to any licence application. It begins via early CNSC engagement with a potential SMR applicant to reach a common understanding of the nature of the proposed design and of the specificities of the approach to operation.

The establishment of an application assessment strategy begins with a high-level analysis of the proposed project, including applicable regulations and regulatory process. Applicable regulatory documents and practices, with recommendations on their risk-informed application, are also identified. Pre-licensing engagement and review of proposed activities may indicate that no license is required. For example, the testing of a thermal hydraulic loop (without the use of nuclear substances) is not subject to sections 24 or 26 of the NSCA [A–13] and therefore would not require a licence application to the CNSC.

The outcome of this process is an appropriate risk-informed application assessment strategy, which CNSC staff will ultimately use in developing supplemental guidance for an applicant on how to prepare a licence application for a given project. The process is expected to be iterative, with several interactions between the CNSC and an applicant before the CNSC develops this supplemental guidance.

A–3.3. CHINA–NNSA

A–3.3.1. Question

Did you make changes to the licensing process to address SMRs? Were new processes, such as pre-licensing engagement developed?

A–3.3.2. Response

According to the Nuclear Safety Law of the People's Republic of China [A–132] and Regulation of the People's Republic of China on the Supervision and Management for Civil Nuclear Installations [A–133], the State implements a licensing system for civil nuclear facilities, including nuclear power plants, research reactors. The operators of nuclear facilities may carry out corresponding construction, operation and decommissioning activities only after obtaining relevant licenses or approval documents. The types of licenses for NPPs include:

- Review Comments on Siting;
- Construction Permit;
- Operating Permit;
- Approval for Decommissioning;
- Operator license and senior operator license;
- Other documents requiring examination and approval, etc.

At present, the process of Land-based SMR, as HTR-P and ACP100, license related to nuclear safety refers to the approval process of conventional NPPs. Whether the license process of Floating Reactors needs to be adjusted is still under discussion. Also, NNSA encourages vendors to contact and communicate safety design and safety regulatory issues with us before license application, especially for innovative reactors.

A–3.3.3. Follow-up Question

Are there any changes planned to the licensing system and types of licenses to accommodate FNPP? What would the changes entail? Does NNSA consider the requirements and processes directly applicable to SMRs? Are there any pre-licensing activities / engagements / design assessment prior to the formal licensing process?

A–3.3.4. Response

1) We have adjusted the licensing process for FNPPs built and loaded in the factory. If the factory has already built similar FNPPs, it can directly apply for a Construction Permit, instead of having to conduct a site safety assessment firstly.

2) At present, the process of the license for SMR has not changed. NNSA have carried pre-licensing activities with vendors before the formal licensing process. According to the design characteristics of SMR, the reviewers will intervene in the safety review of SMR in advance and carry out corresponding technical exchanges with designers.

A–3.4. CZECH REPUBLIC–SÚJB

A–3.4.1. Question

Did you make changes to the licensing process to address SMRs? Were new processes, such as pre-licensing engagement developed?

A–3.4.2. Response

Not applicable – currently there are no project involving SMRs proposed to be licensed in the Czech Republic and therefore there was no need for changes in the licencing process.

Hypothetically speaking, simplifications of the licensing framework might also be considered to reflect particular SMR technology specificities and its deployment model and to allow for savings achieved by standardization. However, there is a persistent lack of clear and reliable data of the particular technological solution which could be envisaged to be deployed. It may even occur that the licencing framework is modified but not simplified to strengthen particular requirements to reflect the specificities of a particular design and deployment model that would be considered as a realistic possibility for construction. In general, the Atomic Act [A–21] does not, as such, preclude the deployment of SMRs nor does it constitute a major obstacle to this licencing. It provides a good legal basis as the legal requirements contained therein are more general in nature and technologically neutral and should provide sufficient flexibility, in particular through the application of the graded approach, so it does not hinder the SMR deployment as such (to put aside the question of simplified licencing procedure and savings achieved by standardization).

According to the Atomic Act [A–21] any SMR falls under the definition of the nuclear installation and as such is licenced as any other nuclear installation with nuclear reactor (siting, construction, first physical start-up, first power-generation start-up, operation). There are design requirements but the design itself is not licenced.

It is to be noted that the national licensing framework, including SMRs licensing, is only in part governed by the Atomic Act [A–21]. Other administrative procedures according to the

Construction Act [A–134] or the Environmental Impact Assessment Act [A–135] need to be also taken into account should we consider modifications to legislation related to SMRs.

Same applies for decommissioning phase (including the accumulation of the financial reserves for the decommissioning and obligatory insurance for the third-party civil liability for nuclear damage).

A–3.5. FRANCE–ASN

A–3.5.1. Question

Did you make changes to the licensing process to address SMRs? Were new processes, such as pre-licensing engagement developed?

A–3.5.2. Response

Licensing process is described in the Environmental Code [A–24] and can be divided in different steps. No need for change in this process has been identified yet to address potential future SMR licensing applications. In particular, the article R.593-15 of the same code stipulates that the set-up of multi-units can be authorized by the same approval if they are operated by the same licensee, on the same site.

Firstly, before requesting an authorization to set up a basic nuclear installation, in accordance with the article R.593-14, a prospective licensee can ask ASN's opinion about its project's safety options. This step is not mandatory.

Prior to any construction work, the licensee must submit a request file to the ministry in charge of nuclear safety to be authorized to set up a basic nuclear installation. The article R.593-16 lists the documents expected in the request file. Among these documents, we can find, for example:

- A preliminary safety report that must take into account ASN's opinion about the safety options;
- An impact assessment;
- Evidence of the technical and financial capacities of the licensee.

The ministry in charge of nuclear safety asks ASN its opinion about the request file. Moreover, different consultations are required by the legal framework (see articles R.593-20 to R.593-24). Once this authorization is granted, the construction can begin.

Finally, from article R593-30 to article R593-37, the Environmental Code [A–24] describes the commissioning procedure. To address a commissioning application for its nuclear reactor, the licensee has to provide:

- An updated safety report;
- General operating rules;
- An on-site emergency plan;
- An update of the dismantling plan;
- Elements to assess the compliance of the installation with ASN requirements (see question 2);
- An update of the impact assessment;
- An update of the risk study.

The request is addressed to ASN, and for each reactor, a specific request must be addressed.

Furthermore, ASN can enact resolutions to fix hold points. For example, ASN's resolution n° 2013-DC-0347 [A–76] stipulates that the containment challenge requires an agreement for ASN.

A–3.6. JAPAN–NRA

A–3.6.1. Question

Did you make changes to the licensing process to address SMRs? Were new processes, such as pre-licensing engagement developed?

A–3.6.2. Response

No change was done for the licensing process of HTTR.

During the regulatory decision process for the HTTR, the NRA invited public comment from the scientific and technical view on the licence application. Public consultation is typically done for NPP licence applications, and this was the first instance in which public comment was invited for a research facility.

A–3.6.3. Follow-up Question

The licensing process is not described, please can you provide a general overview of the process and the licensing organisation. This should include the key steps, expectations and requirements. Please can you also confirm whether the NRA undertakes pre-licensing activities and if so, what the pre-licensing activities entail?

A–3.6.4. Response

The NRA developed new regulatory requirements related to commercial power reactors which were significantly enhanced than the previous requirements in the light of lessons learned from the Fukushima Daiichi NPP accident and put them into force. The licensee must submit applications on compliance to the new regulatory requirements to the NRA to obtain authorization for their operation of reactors.

To install and operate a new reactor in Japan, it is necessary to obtain the permit for reactor installation (Reactor Installation Permit) and make a specific design; obtain the approval of plan for construction for a specific design ('Construction Plan') and carry out construction work; and finally obtain the approval of Operational Safety Programs prior to start of operation. For reactors on which authorization have been already obtained, the Conformity Review is to be conducted based on the back-fitting system introduced with the amendment of the Reactor Regulation Act [A–25]; amendment to Reactor Installation Permit is to be granted; and approvals of Construction Plan and Operational Safety Programs based on the amended permit are also to be obtained.

The NRA implements the Conformity Review by holding the Conformity Review Meeting where Commissioners participate. The Conformity Review Meeting is made open to the public by allowing their attendance and webcasting, along with materials for the examination disclosed in principle, thus maintain transparency of the review.

In addition to the Conformity Review Meeting where Commissioners participate, meetings and hearings with licensees are occasionally held as appropriate by the NRA staff for the purpose of regulatory activities such as confirmation of facts related to matters included in applications. Summaries of those proceedings are made open along with related materials.

A–3.7. RUSSIAN FEDERATION–ROSTECHNADZOR

A–3.7.1. Question

Did you make changes to the licensing process to address SMRs? Were new processes, such as pre-licensing engagement developed?

A–3.7.2. Response

The procedure for licensing of activities in the field of the use of atomic energy is set out in the 'Administrative Procedures for the Public Service of Licensing Activities in the Field of Atomic Energy Use to be Provided by the Federal Environmental, Industrial and Nuclear Supervision Service' approved by Order of the Federal Environmental, Industrial and Nuclear Supervision Service No. 453 of October 8, 2014 [A–136].

In the licensing procedure for ships and other floating facilities with nuclear reactors, the type of activity in the field of the use of atomic energy entitled 'siting of nuclear facilities' is not used with regard to 'ships and other floating facilities with nuclear reactors'.

In addition, according to Paragraph 10 of Annex 2 to the Administrative Procedures, for ships and other vessels with nuclear reactors, other transport and transportable means with nuclear reactors, building is a type of activity to be licensed (the license is granted to the shipbuilding organization for each nuclear installation design), in contrast to other nuclear facilities, for which the type of activity to be licensed is construction (the license is granted to the operating organization). After the construction stage, the operating license is granted to the operating organization as it is done for other nuclear facilities.

The specifics of the licensing process for ships and other floating facilities with nuclear reactors are defined in Article 40 of Federal Law No. 170-FZ [A–26]. In accordance with Article 40, the responsibility for safety of these nuclear facilities at the construction and commissioning stages rests with the leading design organization and the ship-building organization, and after operational acceptance, it rests with the operating organizations. At the same time, commissioning of ships and other vessels with nuclear facilities is allowed when the operating organization has the appropriate permits (licenses). The responsibility for the safety of vessels can be handed over when the acceptance documents for the vessel with a nuclear installation is signed. Thus, the building and commissioning permit for ships and other vessels with nuclear installations is received by the ship-building organizations, while the operating and decommissioning licenses for such facilities are received by the operating organizations.

A–3.7.3. Follow-up Question

The response does not provide information on licensing processes or the licensing organisation. Please provide a brief description of the process set out in the 'Administrative Procedures for the Public Service of Licensing Activities in the Field of Atomic Energy Use to be Provided by the Federal Environmental, Industrial and Nuclear Supervision Service' and the licensing body.

There is no information on licensing of land-based SMR installations. Please describe the approach for land-based SMR installations/ and any specific considerations for land-based SMRs, if different from the above.

The response notes the possible roles of two licensing organisations for floating NPP– how does the regulatory body assure the change in licensee?

A–3.7.4. Response

There is no concept of 'pre-licensing' in the Russian legislation.

The federal executive body providing the public service is Federal Environmental, Industrial and Nuclear Supervision Service. The results of the provision of the public services for licensing the activities in the field of the use of atomic energy are:

(a)   Granting a license to the applicant with the validity period and conditions;
(b)   Refusal to grant a license;
(c)   Amendments to the license conditions;
(d)   Refusal to amend the license conditions;
(e)   Termination of a license;
(f)   Renewal of a license;
(g)   Refusal to renew a license;
(h)   Renewal of a license;
(i)   Refusal to renew a license;
(j)   Providing the licensee with a duplicate of a license.

The process for the provision of public services for licensing with the issuance of a license with its validity period and conditions includes:

(a)   Consideration of an application for a license, including a preliminary check of the list of documents attached to the application;
(b)   Based on the results of the preliminary check, a decision is made to review the documents submitted for obtaining a license or to refuse to review the documents;
(c)   Review of the documents submitted for obtaining a license, which includes checking the accuracy of the information contained in these documents by means of:
  o Arranging of a safety review of an atomic energy facility or a type of activity;
  o Inspections of the licensee and the facility at which or in relation to which the activity is planned.
(d)   Making a decision on the issue or refusal to issue a license on the basis of the review conclusions and the receipt of an inspection report;
(e)   Granting a license, which includes a list of validity conditions.

In accordance with the legislation of the Russian Federation, the licensing body in the field of the use of atomic energy is Federal Environmental, Industrial and Nuclear Supervision Service.

This process is common for all nuclear facilities, including BREST-300 and other land-based SMRs (if an application for a corresponding license is submitted).

For the floating NPP, the licensing specifics are as follows:

At the design and construction stages, licenses must be held by the designer (design license) and the shipbuilder (construction license).

At the operation and decommissioning stages (operation and decommissioning licenses), the operating organization must have the license.

The end of the construction stage and the beginning of the next stage of operation is considered to be the end of construction with the closure of construction certificates, the conduct of comprehensive tests to confirm the declared characteristics, and the signing of a state acceptance report of the facility.

In accordance with the legislation of the Russian Federation in the field of the use of atomic energy, commissioning of a vessel with a nuclear reactor/reactors is possible only if the operating organization has an appropriate license (license to operate a nuclear installation – a vessel with a nuclear reactor/reactors).

The operating organization submits a set of documents for obtaining a license to Rostechnadzor to operate a nuclear installation — a vessel with a nuclear reactor/reactors. The set of documents includes, in particular, a safety analysis report developed by the designers of the reactor and vessel and adjusted with due regard for the changes in the initial design of the vessel made as a result of its construction. Thus, safety review of the nuclear facility is carried out for the operating organization, but on the basis of the documents of the parent design organization and the reactor designer. The safety review must be completed before the signing of the state acceptance report of the facility. In the event that the safety review reveals significant inconsistencies with the requirements of federal regulations and rules in the field of the use of atomic energy, which leads to the refusal to grant the operating organization a license to operate the vessel, the state acceptance report is not signed, the commissioning stage is not considered completed, and the operating organization does not become responsible for the safety of the atomic energy facility.

Russia has no experience in licensing land-based SMRs. However, the existing licensing process for the NPPs can be basically applied to the land-based SMRs.

A–3.8.   SOUTH AFRICA–NNR

A–3.8.1. Question

Did you make changes to the licensing process to address SMRs? Were new processes, such as pre-licensing engagement developed?

A–3.8.2. Response

Did you make changes to the licensing process to address SMRs?

Yes, broadly speaking, the NNR licensing process requires the applicant to present a safety case to the NNR which is a structured presentation of documented information, analyses and intellectual arguments to demonstrate that the proposed design can and will comply with the NNR licensing requirements.

The regulatory approach applied to the licensing of the Koeberg Nuclear Power Station (KNPS) (twin 930 MW(e) PWRs), which was a turnkey project compared to a developmental project, presented challenges to the NNR in terms of its applicability to the PBMR. One of the major aspects of the PBMR licensing process, which must be thoroughly considered as an integral part of the development (by the applicant) and review (by the regulator) of the safety case, is the credibility of the PBMR licensing basis. Unlike Light Water Reactors (LWRs), for which well-researched and documented design criteria and rules are readily available, broad international
144

consensus has not been developed on general design criteria and design rules for high temperature gas reactors. Although these type of reactors have been licensed and operated elsewhere in the world, no international 'off the shelf' package was available for defining the design basis and the safety case for the PBMR. As part of the safety case the establishment, documentation and assessment of the PBMR design basis is therefore an important step in the licensing process and received significant attention by the designers, applicant and the NNR.

In order to demonstrate that the PBMR design will meet the above licensing requirements the applicant has, in consultation with the NNR, developed and implemented a structured process to develop the PBMR safety case which takes account of the absence of well-researched and documented design criteria and rules. This process also provides a logical link between the various steps of the design process, the safety assessment and the development of operational support programmes. The main components for the development and review of the PBMR safety case are:

(a) The PBMR Safety Case Philosophy (SCP)

The SCP provides the intellectual and philosophical arguments of how PBMR safety will be demonstrated to meet the safety requirements set by the NNR. These refer to the broad safety objectives of the PBMR.

The process for developing the SCP also involved the systematic identification of Key Licensing Issues, applicable to this type of reactor technology, which will need to be addressed as part of the demonstration of the PBMR safety objectives in the Safety Analysis Report.

(b) The PBMR Safety Analysis Report (SAR)

The Safety Analysis Report (SAR) for the PBMR, and other supporting documents are to provide a detailed justification of how the safety arguments/objectives presented in the safety case philosophy are or will be demonstrated.

(c) The General Operating Rules (GOR)

The General Operating Rules (GOR) refers collectively to safety related practices or programmes that are applicable during the operational phase of the plant and may also be applicable during interim licensing stages.

Were new processes, such as pre-licensing engagement developed?

Yes, one of the lessons relating to the licensing process and design assessment relates to the engagement process with the designer. The engagement framework should allow for direct engagement with the designer/architect engineer with involvement of a potential client and eventual operator. The NNR captured this process in the NNR PP-0008 [A–32]. This process requires:

(a) The identification and agreement on key safety issues and the proposed technical resolution;
(b) An adequately developed and stable design before development of the safety case and licensing engagements.

Application of sound system engineering principles and past experience with the aim of demonstrating through robust research, test and qualification that the design will survive all postulated transient and accident conditions.

145

A–3.9.1. Question

Did you make changes to the licensing process to address SMRs? Were new processes, such as pre-licensing engagement developed?

A–3.9.2. Response

The Nuclear Installations Act (NIA) 1965 [A–48] defines the concept of a Nuclear Licensed Site in section 1, prohibiting the use of a site for the purpose of installing or operating any nuclear reactor (other than a nuclear reactor comprised in a means of transport), or any other installation of a prescribed kind unless a licence has been granted. The Nuclear Installations Regulations (NIR) [A–49] provide descriptions for those installations that are prescribed by the Act.

The NIA [A–48] provides for ONR to grant a licence to a corporate body for the use a defined site for the prescribed activities, and for ONR to attach conditions to the licence (according to section 4). ONR has a standard set of 36 nuclear site licence conditions, which are explained in ONR's [A–50] covering from expectations on the marking of the site boundary, the consignment of nuclear matter, training, emergency arrangements, safety documentation, operating rules, decommissioning and organisational capability (the list is not meant to be exhaustive). ONR's approach and expectations to the licensing of nuclear installations are documented in [A–51].

So far ONR has not received applications to grant nuclear site licences relating to the deployment of SMRs in the UK, or indeed engaged in formal 'pre-licensing' activities such as Generic Design Assessment (GDA) relating to an SMR.

However, the questionnaire queries whether a 'pre-licensing' process was introduced as a result of consideration of SMRs, or changes introduced. Certainly, in the context of ANTs and with SMRs in mind, ONR carried out a review and modernised the Generic Design Assessment (GDA) process in 2018-19.

GDA is the first of 3 phases in the regulation of nuclear New Build in the UK and relates to establishing the acceptability of the generic design and generic site. It is then followed by licensing and permissioning activities associated with construction and commissioning. GDA is not a regulatory requirement prior to licensing but given the benefits of identifying and resolving key issues on safety, security and environmental protection before licensing and construction commences, it remains a UK Government expectation.

As a result of the review, the GDA process was modernised to increase flexibility and better adapt to the differing levels of maturity and development of SMR vendors and their technologies, whilst remaining consistent with previous GDAs. The improvements were also aimed at capturing important lessons learnt from previous and on-going GDAs.

As part of the modernised GDA, ONR is to conduct its assessment in three steps instead of the current four steps, and has changed the emphasis of each step:

- The aims of Step 1 (estimated to take around 12 months) are to agree the GDA scope, define the basis for the generic safety, security and environmental cases, identify the gaps to meeting UK regulatory expectations, and agree resolution plans for how & when these will be resolved. As part of this step the arrangements to undertake GDA and the programme for subsequent Steps are defined;

- During Step 2 (estimated to take around 12 months), the focus of the assessment is towards the adequacy and suitability of the fundamental aspects of the design, the safety, security and environmental claims, the methodologies and approaches;
- During step 3 (with an indicative timescale of 24 months), the focus is an in-depth assessment of the safety, security and environmental case evidence which underpins the design, to come to a conclusion on the acceptability of the design for construction in Great Britain (GB).

The estimated timelines are indicative and ultimately will depend on the timely submission by the GDA Requesting Party of high quality documentation.

The modernised GDA process provides opportunities for the Requesting Party to make better and more effective use of existing submissions (e.g. to other Regulators) — supplemented to meet UK expectations. The GDA process will also place greater emphasis on earlier engagement and agreement of scope / submissions throughout process. It also provides flexibility in the assessment activities (with robust internal governance to agree scope of assessment that can be accepted).

A key feature that provides flexibility for SMRs is the introduction of additional outputs (GDA Statements) (as well as the Design Acceptance Confirmation [DAC] and Statement of Design Acceptance [SoDA] as previously) to show that the design (or a meaningful assessment scope from the design) has shown alignment with UK regulatory expectations. The acceptability of partial scopes and new GDA statement outputs is summarised in the Table A–3. A new suite of GDA guidance for Requesting Parties ONR-GDA-GD-006 [A–137] including detailed Technical Guidance ONR-GDA-GD-007 [A–139] with key expectations and lessons learnt were published in 2019.

TABLE A–3. ACCEPTABILITY OF PARTIAL SCOPES AND NEW GDA STATEMENT OUTPUTS

Example	Technical Assessment Topics	Steps	Output
Full plant (well developed) design	All	1,2 and 3	DAC, interim Design Acceptance Confirmation (iDAC) or no DAC
Major portions of a well-developed plant design (for example: one complete reactor module of a multi-module design where the interactions between modules are potentially safety significant but are declared out of scope by the RP)	All	1,2 and 3	Statement
Major portions of the plant design (of limited design maturity)	Most	1 and 2	Statement

Example	Technical Assessment Topics	Steps	Output
Conceptual full plant design	Most	1 and 2	Statement
Partial plant design (for example, a design where the deployment model relies on out of scope supporting SSCs)	Assessment not meaningful – no GDA undertaken		
Distinguishing safety system (for example, the control and instrumentation technology and architecture)	Assessment not meaningful – no GDA undertaken		

A–3.10. UNITED STATES OF AMERICA–NRC

A–3.10.1. Question

Did you make changes to the licensing process to address SMRs? Were new processes, such as pre-licensing engagement developed?

A–3.10.2. Response

Yes, the licensing process that addresses SMRs has evolved and changed. The NRC's licensing process requires that the applicant present a safety case to the NRC which is a structured presentation of documented information, analyses and intellectual arguments to demonstrate that the proposed design can and will comply with the NRC licensing requirements. More specifically, the regulatory approach applied to accepting a design certification application for an SMR, have evolved as new and differed licensing challenges and design difference have been presented to the NRC. Unlike large light water-cooled reactors, for which well-researched and documented design criteria and rules are readily available, broad consensus have not been developed on general design criteria and design rules for SMRs and there is not a design criterion for SMRs in the United States at this time. As part of the safety case the establishment, documentation and assessment of the SMR design basis is therefore an important step in the licensing process and received significant attention by the designers, applicant and the NRC. Throughout the process, early pre-application engagement, public meetings, and routine and structured discussion on key technical issues provide a clear and logical link between the various phases of the design review process, the issuance of request of additional information, and the ability of the staff to develop its safety assessment.

A–4. REGULATORY APPROACH

This Annex presents the responses provided by the following Member State regulatory bodies to the Question 4: "Were there changes in the regulatory approach to the review & assessment process done in support of authorization? Describe information on the prior practice and the changes in approach made."

A–4.1. ARGENTINA–ARN

A–4.1.1. Question

Were there changes in the regulatory approach to the review & assessment process done in support of authorization? Describe information on the prior practice and the changes in approach made.

A–4.1.2. Response

No, it was not necessary to change the regulatory approach for licensing CAREM 25 prototype reactor.

The Argentine Regulatory Standards are based on a set of fundamental concepts, which are part of the Performance Approach philosophy (goal-oriented approach), sustained by the regulatory system, concerning radiological and nuclear safety, safeguards, and physical protection.

ARN understands that a performance-based regulatory approach does not imply limiting the requirements to qualitative issues. Moreover, it is perfectly compatible with specific deterministic requirements and even numerical criteria:

- Defence in depth concept;
- The single failure criterion;
- Conservative approach for the demonstration of safety cases.

As mentioned earlier, Regulatory Standards are not prescriptive but of compliance with safety objectives. The compliance of these objectives must be demonstrated by the licensee by sound procedures within mandatory documents than can be objectively assessed by the Regulatory Body. The role of the latter is to be sceptical and critical, without imposing 'how', which implies an interaction between professionals of the Regulatory Body and the Licensee, in order to ensure a common understanding of the overall safety approach. This includes the statement of safety goals, the engineering solutions adopted, the analytical tools for proving safety, and the methodology for deriving safety requirements.

ARN approach is consistent with IAEA approach to the establishment of safety (engineering) requirements, derived from the safety classification of SSCs, which in turn is based on the Safety Analysis demonstrating the functional safety of a design.

Regarding the adoption of a performance-based regulatory approach, some advantages, learned by the verified application experience are the following:

- The nature of the interaction between the Regulatory Body and the Licensee contributes to early detection of possible non-compliances or deficient compliance with regulatory requirements (in early design stages), avoiding the increase in time and efforts in fulfilling such requirements in later phases of a project (fabrication or construction);

- The design solutions to comply with regulatory requirements come, in general, from the supplier (nuclear vendor) through the Licensee, that know in detail the installation and the system involved in;
- The establishment of safety objectives keeping an openness to different design solutions helps to manage projects from different vendors, i.e. nuclear reactors with different safety approaches while keeping coherence on the need of objective (factual) demonstration of the compliance with regulatory requirements.

A–4.2. CANADA–CNSC

A–4.2.1. Question

Were there changes in the regulatory approach to the review & assessment process done in support of authorization? Describe information on the prior practice and the changes in approach made.

A–4.2.2. Response

The CNSC regulates to prevent unreasonable risk to the environment, the health and safety of persons, and national security. To this end, the CNSC has established a licensing and compliance system to ensure that all persons who use or possess nuclear substances and radiation devices do so in accordance with a licence, and that regulated parties have safety and security provisions in place that ensure compliance with regulatory requirements.

The CNSC's regulatory philosophy is based on the following:

- Licensees are directly responsible for managing regulated activities in a manner than protects health, safety, security and the environment, and that conforms with Canada's domestic and international obligations on the peaceful use of nuclear energy.
- The CNSC is accountable to Parliament and to Canadians for assuring that these responsibilities are properly discharged.

The CNSC therefore ensures that regulated parties are informed about requirements and provided with guidance on how to meet them, and then verifies that all regulatory requirements are and continue to be met.

The regulations made under the Nuclear Safety and Control Act (NSCA) [A–13] provide further legislative authority with respect to topic-specific considerations, using a combination of prescriptive and performance-based approaches. Prescriptive approaches tell licensees exactly what they need to do to meet requirements, whereas performance-based approaches set specific performance measures that licensees must meet with respect to particular aspects of their licensed activities.

While the CNSC sets requirements and provides guidance on how to meet requirements, an applicant or licensee may put forward a case to demonstrate that the intent of a requirement is addressed by other means. Such a case must be demonstrated with supportable evidence. CNSC staff consider guidance when evaluating the adequacy of any case submitted. This does not mean that the requirement is waived; rather, it is an indication that the regulatory framework provides flexibility for licensees to propose alternative means of achieving the intent of the requirement. The Commission is always the final authority as to whether the requirement has been met; the purpose of the NSCA (Clause 3(a)) [A–13] is to limit risks to a reasonable level. Understanding and mitigating risks is a key part of the decision-making process.

CNSC staff members perform detailed assessments of safety in relation to reactor facility licence applications. The answer from Canada for question 3 describes the general CNSC licensing process for new-build projects. The CNSC's assessment of safety for a licence application is conducted against the application requirements set out in the General Nuclear Safety and Control Regulations [A–129], the Class I Nuclear Facilities Regulations [A–60], and other relevant regulations.

Licence application guides have been written to supplement the regulations and provide further details and clarify the CNSC's expectations of content and format. They are written in the context of the 14 CNSC safety and control areas (management system, human performance management, operating performance, safety analysis, physical design, fitness for service, radiation protection, conventional health and safety, environmental protection, emergency management and fire protection, waste management, security, safeguards and non-proliferation, and packaging and transport) as well as the other matters of regulatory interest (other matters of regulatory interest include an environmental or impact assessment, indigenous consultation, cost recovery, financial guarantees, improvement plan and significant future activities, the Applicant's public information and disclosure program and any other relevant information.), and are supported by the REGDOCs in part 2 and 3 of the regulatory framework which articulate the details of CNSC's expectations. CNSC staff members use assessment plans, along with staff work instructions, to coordinate the assessment of licence applications. The assessment plan provides the logistical structure for carrying out the integrated review while the work instructions capture knowledge and experience from senior staff.

One notable recent change has been enhancement of the review reports that document the results from carrying out the individual work instructions. This provides the reviewers with a single document that contains or references practically all information relevant to their review. It also provides a systematic structure for soliciting their contributions. Sections include:

- Background & introduction: providing relevant context specific to the project at hand, reference to previous or related assessments, key internal and external project documents and the objective of the assessment.

- Scope & depth: provides the criteria set against which the assessment is planned to be conducted.

- Variance from planned scope & depth: documents the approval of any variance from the plan (e.g. depending on the type of reactor some technology-specific criteria may not be applicable).

- Summary of information presented in the application: solicits a high level overview which outlines the relevant information provided by the applicant.

- Staff assessment: documents findings and provides any information requests made of the applicant and the associated disposition of the applicant's response(s), as well as soliciting an overview outlining the assessment activity completed by staff. The assessment may also document observations regarding construction, commissioning or operations activities including CNSC monitoring and inspections.

- Conclusions: to provide a concluding statement as to whether the applicant has met the requirements and expectations.

- Recommendations for licensing follow-up: to provide recommendations for licence conditions or hold points (e.g. commissioning).

- Critical elements of the licensing basis: for reviewers to identify the documents and sections of documents that, if changed, deserve additional regulatory scrutiny.

- Recommendations for additional information or licensing / compliance follow-up activities: to document where follow-up is required based on the content of the licence application, to verify fulfilment of commitments made in the licence application, or recommendations for compliance activities should a licence be granted.

- Lessons learned and recommendations for updates: to capture lessons learnt or recommendations for changes to CNSC documents used during the assessment.

- References.

Within the background & introduction section, results from an applicable pre-licensing Vendor Design Review (VDR) can be provided. That being said, the purpose of a VDR is to provide the early identification and resolution of potential regulatory or technical issues in the design process, particularly those that could result in significant changes to the design or safety analysis. The licence application would need to clearly identify how and where the submitted VDR documents have changed since they were submitted for review. This allows the CNSC to carry out a targeted review of the changes instead of the entire submission.

CNSC recognizes that any new reactor technology and any new licensee organization (or combination thereof) requires the regulator to address factors such as:

- Increased need to engage with proponents early to promote an understanding of the requirements and what is expected to demonstrate requirements have been met;
- Capacity and capability to assess applications that contain novel approaches;
- Identifying and responding in a timely manner to new regulatory considerations introduced by novel technologies, technology deployment methods, applicant licensing strategies and applicant organizational models.

To address these factors, CNSC published [A–15]. The strategy is a living regulatory program which requires program management that considers timely funding, training of staff, and coordination of activities with the existing fleet of Canadian facilities. The strategy includes three key elements that ensure readiness to respond to regulatory challenges through its three pillars for regulatory readiness:

- A robust and flexible regulatory framework;
- Risk-informed processes;
- A knowledgeable and capable workforce with sufficient capacity and technical expertise.

In 2019, the CNSC published REGDOC-1.1.5 [A–65]. This document provides additional guidance on the information to be provided in support of an application to prepare site, construct or operate in view of different SMR technologies.

Turning to the logistics of carrying out the regulatory assessment of an application for a licence to construct, CNSC staff have created a schedule which allows the applicant to submit information in logical and sequential topical packages. This focuses the review and prevents wasted efforts on down-steam reviews that would be impacted by issues discovered in earlier related topics.

Finally, given recent international and national operating experience, staff are working to enhance the collective understanding of what the CNSC expectations are regarding management system and of an 'intelligent customer' — particularly in regard to the capabilities of an applicant scrutinizing a technology vendor or the applicant's contractors and sub-contractors. These expectations have been documented in a draft revision to REGDOC-1.1.2 [A–66], which should be published as a draft for feedback in 2020.

A–4.3. CHINA–NNSA

A–4.3.1. Question

Were there changes in the regulatory approach to the review & assessment process done in support of authorization? Describe information on the prior practice and the changes in approach made.

A–4.3.2. Response

In the safety review of HTR-PM, we mainly refer to the conventional NPPs approach, as the format and content of Safety Analysis Report, the application of Deterministic Safety Analysis Method and Probabilistic Analysis Methods, etc.

The procedure of license application and issuance of could be illustrated as the following, which is shown in Fig. A–5.

(a) First the applicant company will submit application and related document to NNSA. After receiving the application for the safety license, NNSA will carry out the format review to decide whether the application is accepted or not within 1 month. The review and assessment will begin after the acceptance of the application.

(b) Then NNSA will authorizes the technical supporting organizations (TSO) to carry out the technical review and assessment, which is responsible to submit the review and assessment report.

(c) During the process of review and assessment, the safety licenses applicant shall reply, interpret the questions raised by the NNSA or make demonstration supplements or modifications of the relevant information without delay. Generally, it will last for about 1 year, and may run through several bouts of question-answer. At last, the review centre will provide the review and assessment report to NNSA.

(d) The NNSA will transmit the review and assessment report to the Nuclear Safety Advisory Committee for review, the Committee has the responsibility of giving the advisory opinions to the NNSA. Based on this and other department approval of the government, NNSA can decide whether to issue the license or not.

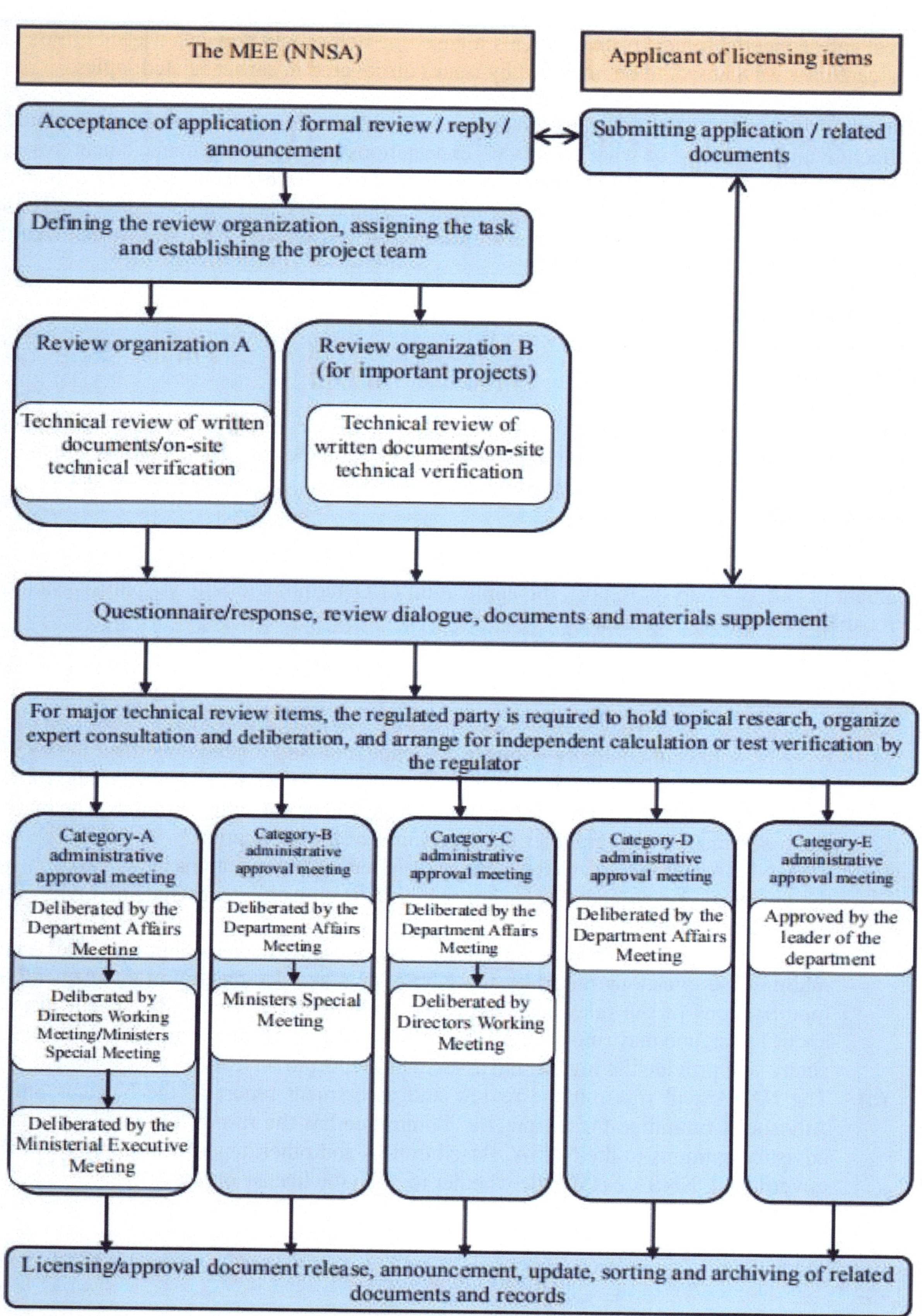

FIG. A–5. Procedure of License Application and Issuance.

A–4.3.3. Follow-up Question

Based on the information provided, the technical review by the TSO appears to be a Q&A process over approximately 1 year. The response does not provide the nature of that review process and why it is considered appropriate for SMRs. What guidance /requirements are available to the TSO to follow a consistent proportionate approach during the review process? Are these the same as for a HTR as a PWR?

Please describe any differences for the regulatory approach for floating and land-based SMRs.

A–4.3.4. Response

The guidelines for TSO review include HAF (rules) [A–68] and HAD Guides [A–69] formulated by NNSA. The technical opinions and review principles for special design, which were issued by NNSA, also be used as basis for review.

At present, the HTR was reviewed according to these guidelines too. And there is no special regulatory approach for floating reactor except for the situation mentioned in the response from China to Question 3.

A–4.4. CZECH REPUBLIC–SÚJB

A–4.4.1. Question

Were there changes in the regulatory approach to the review & assessment process done in support of authorization? Describe information on the prior practice and the changes in approach made.

A–4.4.2. Response

Not applicable – there is not an expectation of a review & assessment process involving SMRs and therefore there are no changes planned to standard regulatory approach and prior practice. Currently, no changes are foreseen for SMRs.

According to the Atomic Act [A–21] any SMR falls under the definition of the nuclear installation and as such is licenced as any other nuclear installation with a nuclear reactor. The licence application shall contain the documentation for a licenced activity as stipulated by the Annex I to the Atomic Act [A–21] (detail requirements for the content of the documentation are established by the implementing legislation). This documentation is the basis for the review and assessment of the safety case prior to issuing a licence.

The formal scope and content of the documentation for licenced activity is the same as for any other nuclear installation with a nuclear reactor and the applicant would therefore have to provide to the national regulatory authority (State Office for Nuclear Safety, SÚJB) with the same documents and to the same extent as the applicant for a licence for 'standard size NPP'. At the same time, these requirements are general enough to allow for different types of technologies (especially on the level of primary legislation). Taking into account the fact that requirements contained in secondary legislation (implementing decrees to the Atomic Act [A–21]) are more specific and prescriptive and reflect currently used technology in Czech Republic (i.e. PWR), the legislative framework in this regard is not in its entirety, strictly speaking, technologically neutral. This is the reason why the deployment of the SMR that are similar to PWR might be less complicated compared to very distinct technologies. There are 4 VVER-440 and 2 VVER-1000 under operation currently in Czech Republic so that well-documented design criteria, extensive research and operating experience are available. This results in well-established and reasonably clear and stable legislative and regulatory framework for PWR type SMRs. Hypothetically speaking, the national legislative framework would therefore need to be adjusted to reflect distinct technologies and facilitate their deployment (especially on the level of secondary legislation). This is also true for documentation for a licenced activity, its scope and content and general design criteria and design rules for assessing safety cases.

A–4.5. FRANCE–ASN

A–4.5.1. Question

Were there changes in the regulatory approach to the review & assessment process done in support of authorization? Describe information on the prior practice and the changes in approach made.

A–4.5.2. Response

There hasn't been any SMR project submitted to ASN for review and assessment yet, but the regulatory approach to the review and assessment process done in support of SMRs authorization would be similar to what has been done recently for existing pressurized water reactors.

The safety options assessment is a first step that enables ASN to control if the preliminary design is adapted to regulatory safety objectives. Also, it enables ASN to identify potential areas that will require a particular attention or improvements if the licensee decides to submit a request file for authorization to set up a basic nuclear installation. As an example, in 2019 ASN published its opinion about EPR 2's safety options.

Then, between the set-up request and the commissioning authorization (see question 3), ASN reviews and assesses the submitted file through instructions with the support of IRSN, a technical support organisation. Instructions are led on every aspect of the safety demonstration following a graded approach. For example, regarding the EPR of Flamanville, instructions were led about:

- Various safety systems;
- Qualification of SSCs;
- General operating rules;
- Commissioning tests;
- Non-nuclear risks.

Moreover, for major topics, ASN can ask the advisory committee for nuclear reactors (GPR) its opinion. For example, regarding the EPR of Flamanville, ASN asked GPR's opinion about:

- Human and organisational factors;
- Safeguard systems;
- Internal and external events;
- Safety demonstration;
- Rules for accident's studies;
- I&C;
- Safety classification of SSC.

Furthermore, ASN leads inspection to control the compliance of activities related to the protection of security, safety, public health and sanitation, nature and environment with regulatory requirements. In particular, ASN controls construction activities through on-site inspections, design and licensee's integrated management system through central services inspections, and manufacturing through vendor inspections.

A–4.6. JAPAN–NRA

A–4.6.1. Question

Were there changes in the regulatory approach to the review & assessment process done in support of authorization? Describe information on the prior practice and the changes in approach made.

A–4.6.2. Response

No change was done in the review and assessment process.

As described in the answer from Japan to Question 02, a review was performed after the introduction of new regulatory requirements following the Fukushima Daiichi NPP accident.

In addition, during the regulatory decision process for the HTTR, the NRA invited public comment from the scientific and technical view on the licence application. Public consultation is typically done for NPP licence applications, and this was the first instance in which public comment was invited for a research facility.

A–4.6.3. Follow-up Question

We are aware that there have been changes in requirements following Fukushima Daichi NPP, but these are not presented and the regulatory review process is not described in detail. Please provide this information.

Please explain what is meant by research reactor in this context of the questionnaire, and the differences with SMRs generally?

A–4.6.4. Response

The review process by NRA is described in the answer from Japan to Question 3. Please note that the definition for research and test reactor is 'reactor that is not categorized as power reactor' in Japanese legislation.

Since HTTR is not categorized as power reactor, it is categorized and regulated as research and test reactor accordingly.

Therefore, within the Japanese current regulatory framework, SMRs in the questionnaire will be:

- Regulated as power reactor if it is for providing electric power generation purpose;
- Regulated as research and test reactor which is not for power generation.

A–4.7. RUSSIAN FEDERATION–ROSTECHNADZOR

A–4.7.1. Question

Were there changes in the regulatory approach to the review & assessment process done in support of authorization? Describe information on the prior practice and the changes in approach made.

A–4.7.2. Response

The licensing processes for all nuclear facilities, including floating power plants, are similar. Part 4 of Annex 3 to the Administrative procedures [A–136] contains requirements to the contents of the set of documents justifying nuclear and radiation safety of nuclear facilities — ships and other floating, transport and transportable facilities with nuclear reactors.

On paragraph 4.1 of Annex 3 to the Administrative procedures [A–136], are stated the "Requirements to the content of the set of documents justifying nuclear and radiation safety of construction of nuclear installations — vessels and other floating, transport and transportable nuclear reactor facilities; nuclear service vessels for storage and transportation of nuclear material and radiation sources — nuclear reactor vessels transitioned to radiation sources; nuclear service vessels for storage and transportation of radioactive substances and radioactive waste." The following documents and information are required:

- "Safety analysis report."
- "Reference on design, engineering, operational and process documents for construction and commissioning."
- "Description of the range of works on operation and storage of radiation sources used in the construction process."
- "Description of the selection, training, re-training and authorization of the applicant's workers who are engaged in construction and commissioning of the nuclear installation (radiation source) and ensuring of nuclear and radiation safety."
- "Description of the range of works on radioactive waste management at each stage of construction and commissioning."
- "Reference on documents that establish general and specific requirements to nuclear and radiation safety of productions (technologies) and items."
- "Description of the structure and composition of nuclear and radiation safety services."
- "Analysis of design-basis and beyond-design-basis accidents in the conditions of construction and commissioning of the nuclear facility; measures to prevent such accidents."
- "Instruction on prevention and mitigation of nuclear and radiation accidents and fires."
- "Information on permissible rates of radionuclide emissions and discharges into the environment."
- "Information on commissioning of nuclear material storage facilities and radioactive waste storages."
- "Information on training of managerial staff; information on training, composition, qualification and appraisal of personnel responsible for the technological process, for institutional control over nuclear and radiation safety and for accounting and control and physical protection of nuclear materials and radioactive waste."
- "Reference on nuclear hazardous operations and administrative and technical measures towards their fulfilment."
- "Reference on certificates for equipment, items and technologies of nuclear installations, radiation sources and storing facilities."
- "Reference on regulatory and administrative documents for ensuring nuclear and radiation safety of operations at various stages of construction (building) and for safe management of nuclear material and radioactive waste."
- "Copy of the existing instructions for accounting and control of nuclear material and radioactive waste."

- "Reference on the organizations that engage in works and provide services during construction and commissioning."
- "Action plan for mitigation of nuclear and/or radiation accidents, protection of the personnel and population in the event of nuclear and radiation accidents."
- "Information on individuals responsible for nuclear and radiation safety at the stages of construction and commissioning."
- "Certificate of assurance of physical protection of nuclear facilities (as per Sections 2 and/or 4 of Appendix 4)." [A–136]
- "Certificate of the accounting and control of nuclear material and radioactive waste (as per Sections 1 and/or 3 of Appendix 4)." [A–136]
- "Instruction on ensuring nuclear and radiation safety."
- "Description of the comprehensive system for product quality management."
- "Reference on the structure of laboratories that conduct quality analysis of coolant in circuits and high-pressure gas, as well as tools to assess the radiological situation at the applicant's site and surrounding areas."
- "Description of the system for management of design, engineering, maintenance and operational documents."

In addition, the paragraph 4.2 of Annex 3 to the Administrative procedures [A–136], states the "Requirements to the content of the set of documents justifying nuclear and radiation safety of operation of nuclear installations on vessels and other floating, transport and transportable nuclear reactor facilities; nuclear service vessels for storage and transportation of nuclear material and radiation sources — nuclear reactor vessels transitioned to radiation sources; nuclear service vessels for storage and transportation of radioactive substances and radioactive waste, and on-land test facilities with experimental transport nuclear reactors." The following documents and information are required:

- "Safety analysis report."
- "Reference on classification and certification of the vessel by Russian Maritime Register of Shipping acting pursuant to the Charter approved by the Ministry of Transport of the Russian Federation, No. MS-16-r of March 3, 2014 (registered by the Ministry of Justice of the Russian Federation on August 25, 2014, registration No. 33 791; Bulletin of Regulatory Acts of Federal Executive Power, 2014, No. 47)."
- "Description of the structure and composition of nuclear and radiation safety services."
- "Description of key technical solutions, systems and means that ensure nuclear and radiation safety."
- "Instruction on prevention and mitigation of nuclear and radiation accidents and fires."
- "Reference on documents that establish general and specific requirements to nuclear and radiation safety during operation, performance of work and provision of services."
- "Action plan for compensating for deviations from the requirements of regulatory and technical nuclear and/or radiation safety documents, with analysis of their necessity and adequacy."
- "Reference on registration of equipment and pipelines of a nuclear installation."
- "Information on permissible rates of radionuclide releases and discharges into the environment."
- "Instruction on radioactive waste management."
- "Reference on the administrative and regulatory documents for ensuring nuclear and radiation safety."

- "Guidance on beyond-design-basis accident management during operation of nuclear installation."
- "Reference on technical and legal regulations that cover arrangements for physical protection of nuclear facilities."
- "Reference on the organizations engaged in works and providing services during operation of nuclear installations and radiation sources, indicating the type of works (services) and information on availability of relevant Rostechnadzor licenses."
- "List of nuclear and radiation-hazardous facilities where the declared activity is to be conducted."
- "Instruction on the provision of nuclear and radiation safety."
- "Reference on technical and legal regulations that cover arrangements for storage, accounting and control of nuclear material, radioactive substances and radioactive waste."
- "Information on training of managers; information on training, composition, qualification and appraisal of personnel responsible for the process, for institutional control over nuclear and radiation safety and for accounting and control and physical protection of nuclear materials, radioactive substances and radioactive waste."
- "Information on availability of conditions for storage and processing of liquid and solid radioactive waste."
- "Analysis of conformity of a nuclear installation and radiation source with the requirements of current regulatory and technical nuclear and/or radiation safety documents, and the list of deviations from the requirements of the current regulatory and technical documents."
- "Reference on the results of recent engineering certification of equipment and pipelines of the nuclear installation."
- "Reference on nuclear and radiation hazardous operations and administrative and technical measures towards their fulfilment."
- "Acceptance certificate (or its copy) of the nuclear installation and radiation source."
- "Reference on accounting and control of nuclear material and radioactive waste (as per Sections 1 and/or 3 of Appendix 4)." [A–136]
- "Reference on the latest (for the reporting period) physical inventory of nuclear materials (if any)."
- "Copies of instructions for accounting and control of nuclear material (if any)."
- "Structural chart of units that conduct radiochemical, radiometric or physical-chemical measurements, and make descriptions of technical means of monitoring over the radiological situation at workplaces and in compartments of the vessel, test facility and on the site of the license applicant."
- "Analysis of conformity of nuclear fuel storages with the requirements of nuclear and radiation safety rules and regulations."
- "Mitigation plan in the event of nuclear and/or radiation accident; plan for protection of personnel and the population."
- "Quality assurance program for operation."
- "Description of the system of selection, training, re-training and authorization of the applicant's workers who engage in operation of nuclear facility or radiation source and are responsible for nuclear and radiation safety."
- "Reference on ensuring physical protection of nuclear facilities (as per Sections 2 and/or 4 of Appendix 4)." [A–136]
- "Reference on the fulfilment of the terms of license over the previous period."

- "Reference on availability of personnel authorizations to engage in operations in the field of the use of atomic energy."
- "List of individuals responsible for nuclear and radiation safety of operation of the nuclear installation; for accounting and control of nuclear material, radioactive substances, radioactive waste and radiation sources; for physical protection of nuclear facilities."

A–4.7.3. Follow-up Question

The response is a list of topics for floating plant, and the differences from conventional large NPPs or other SMRs (land-based) are not described, nor the considerations for SMRs. Please describe the regulatory approach to the review & assessment processes carried out in support of authorization? Please describe any differences for SMRs (floating and land-based).

A–4.7.4. Response

The regulatory approach to the review and assessment process for the floating SMRs did not require any changes. The safety assessment of the floating SMRs was carried out on the basis of the current legislation in the field of the use of atomic energy. The federal regulations and rules in the field of the use of atomic energy have been amended to take into account the specifics of new nuclear facilities (e.g. NP-022-17 [A–89], NP-029-17 [A–93], NP-079-18 [A–104]).

The main difference between the floating SMRs and land-based ones is the absence of a separate stage of the life cycle 'siting.' As a result, there is no need to obtain a siting license for an atomic energy facility (during its operation, a vessel may change several sites; the list of sea ports of the Russian Federation where vessels and others floating crafts with nuclear power installations are allowed to call at is regulated by the decree of the Government of the Russian Federation). Accordingly, an assessment of the possibility of operating a vessel at a specific site can be carried out both at the stage of its construction and at the stage of operation and performed in each specific case. At the same time, according to NP-022-17 [A–89], the design of a floating SMR must formulate the requirements that the site must satisfy.

In addition, the set of documents to demonstrate the nuclear and radiation safety during the operation of nuclear installations of vessels and other floating equipment must contain a certificate of classification and examination of the vessel in the Russian Maritime Register of Shipping. Also, when licensing floating plants, unlike stationary stations, the set of demonstrating documents does not include results of observation of buildings and structures.

The review processes are determined by the 'Regulation on the procedure for safety review (safety analysis) of nuclear facilities and/or activities in the field of the use of atomic energy,' which was approved by the order of Federal Environmental, Industrial and Nuclear Supervision Service dated April 21, 2014, No. 160 [A–138]. The processes are the same for all nuclear facilities, including nuclear power plants, research nuclear reactors, and the facilities that, according to the IAEA terminology, can be classified as small modular reactors.

A–4.8. SOUTH AFRICA–NNR

A–4.8.1. Question

Were there changes in the regulatory approach to the review & assessment process done in support of authorization? Describe information on the prior practice and the changes in approach made.

A–4.8.2. Response

In addition to all the information related to the regulatory approach in the NNR responses to the other questions of this questionnaire, it was recognized that there is a need to accompany the licensing process for the PBMR with independent safety assessments to ensure that the safety case submitted by the applicant complies with the licensing requirements of the NNR.

The traditional safety assessment process was adapted to take into account the developmental nature of the project without reducing the margin of safety required for a new design. The designer, applicant and the regulator must ensure by performing independent safety assessments that the design as proposed for construction and as-built meets the safety requirements defined by the regulatory framework. The scope of regulatory assessment for licensing of the PBMR is based on the licensing requirements and criteria defined by the NNR in appropriate license documents and guides.

The dual nature of the NNR safety criteria implies that the safety analyses for demonstration of compliance of the Safety Case with the licensing criteria for the PMBR have to comprise both deterministic and probabilistic analyses. The regulatory assessment of both types of analysis follows a graded approach.

The analysis of events was not necessarily restricted to those events selected and submitted by the license applicant and neither did it necessarily cover all of those events. The Regulator defines for the assessment of the safety case a set of representative events that may combine together specific analyses of the applicant and may also contain additional assumptions, if needed. In that way, particular attention is given to events that are unique to the PBMR design and ensure that the Regulator understands the amount of margin that there may be to any safety limits.

The regulatory assessment of the PBMR Safety Case (SC) requires a number of techniques. During the NNR review of PBMR licensing documents and recognising other HTR experience and international research programmes, all areas of importance to the SC have been identified and listed in a detailed matrix and suitable techniques for assessment, computer codes (CC) and/or models identified. An example is the use of CC both to verify calculations performed by PBMR (Pty) Ltd and provide additional calculations deemed necessary by the assessors.

Dependent on the particular importance of the analysis in terms of the NNR Basic Licensing Requirements (BLR) [A–107], on the degree of uniqueness of both design solution and analysis method, and on the availability of established alternative analysis methods, the regulatory assessment methods range from independent analyses with newly developed or improved computer code supported methods, across spot check calculations with existing codes, to plausibility checks e.g. based on hand calculations. In this, two TSOs, TUEV Rheinland of Germany and AMEC NNC of the UK, support the NNR.

The most important part of the safety analyses is the accident analysis to be submitted by the applicant in the relevant chapters of the Safety Case. Since not all postulated event sequences that can challenge the BLR [A–107] can be investigated and analysed, the overall premise of safety analysis is that representative bounding cases in terms of the BLR are selected, to ensure that the most severe consequences are covered by the analyses.

On the other hand, the scope of analysis and applied methods must be sufficient to cover the required scope that starts with the initiating events and ends with the external source terms and public exposures — deterministically as well as probabilistically. In this sense, the analyses on reactivity control and heat removal must be understood as prerequisites of source term analysis.

This is of specific importance for HTRs since — different to the LWR — there are potentially no plant damage states that can be defined as indicators for potential subsequent external releases (e.g. core melt at the LWR). Thus for the PBMR, all analyses must have the objective to quantify the possible releases and the respective doses and risks.

Because the primary NNR regulatory requirements are related to limits in releases of radioactive material, particular attention has been paid to ensuring that an adequate calculation route is in place for the potential release paths. The PBMR specific parts of the release path relate primarily to the core and the helium pressure boundary (HPB). In these areas, computer code and PBMR specific model development has been required. Outside the HPB there is relatively little that is PBMR specific, apart from the identification of the radioactive species that could be released. These involve complex chemical and physical processes that must be understood, but this does not generally require substantial code or model development, although it is design specific.

In addition to the regulatory reviews of technical submissions from the applicant, the methods that can be applied to assess the various areas of Accident Analysis can vary from area to area. Typical options are:

- Independent analysis using newly developed or existing computer codes significantly enhanced for PBMR application;
- Spot check calculations with commercial computer codes using independent calculation models;
- Enveloping conservative analysis using spreadsheets or hand calculations;
- Plausibility checks.

Whatever method is applied to a specific area, attention must be paid to use methods as independent from the applicant's methods as possible, e.g. use of different computer codes. Another general aspect of all methods is that they have to demonstrate the required degree of conservatism for the specific type of analysis.

The efforts and depth of independent analysis and the choice of methods are determined by the following criteria:

- Importance of the particular area for the SC and the BLR [A–107], either directly and/or as a support of other important areas (e.g. steady-state core design as a basis for Accident Analysis);
- Uniqueness of PBMR specific phenomena and/or design features;
- Availability of alternative assessment methods;
- Capability to perform reliable and non-trivial plausibility assessments.

In order to optimise the regulatory efforts, the preferred approaches are those, which can demonstrate with minimum effort that a sufficient margin to the defined limits and ALARA targets exists. It must, however, be recognised that several of the areas addressed below require complex analytical methods since simplistic approaches fail to be accurate enough because of the nature of the phenomena to be considered.

A–4.9. UNITED KINGDOM–ONR

A–4.9.1. Question

Were there changes in the regulatory approach to the review & assessment process done in support of authorization? Describe information on the prior practice and the changes in approach made.

A–4.9.2. Response

The Nuclear Installations Act (NIA) [A–48] defines the concept of a Nuclear Licensed Site in section 1, prohibiting the use of a site for the purpose of installing or operating any nuclear reactor (other than a nuclear reactor comprised in a means of transport), or any other installation of a prescribed kind unless a licence has been granted. The Nuclear Installations Regulations (NIR) [A–49] provide descriptions for those installations that are prescribed by the Act.

The NIA [A–48] provides for ONR to grant a licence to a corporate body for the use a defined site for the prescribed activities, and for ONR to attach conditions to the licence (according to section 4). ONR has a standard set of 36 nuclear site licence conditions, which are explained in ONR's [A–50] covering from expectations on the marking of the site boundary, the consignment of nuclear matter, training, emergency arrangements, safety documentation, operating rules, decommissioning and organisational capability (the list is not meant to be exhaustive). ONR's approach and expectations to the licensing of nuclear installations are documented in [A–51].

So far ONR has not received applications to grant nuclear site licences relating to the deployment of SMRs in the UK, or indeed engaged in formal 'pre-licensing' activities such as Generic Design Assessment (GDA) relating to an SMR.

However, the questionnaire queries whether a 'pre-licensing' process was introduced as a result of consideration of SMRs or changes introduced. Certainly, in the context of ANTs and with SMRs in mind, ONR carried out a review and modernised the Generic Design Assessment (GDA) process in 2018-19.

GDA is the first of 3 phases in the regulation of nuclear New Build in the UK and relates to establishing the acceptability of the generic design and generic site. It is then followed by licensing and permissioning activities associated with construction and commissioning. GDA is not a regulatory requirement prior to licensing but given the benefits of identifying and resolving key issues on safety, security and environmental protection before licensing and construction commences, it remains a UK Government expectation.

As a result of the review, the GDA process was modernised to increase flexibility and better adapt to the differing levels of maturity and development of SMR vendors and their technologies, whilst remaining consistent with previous GDAs. The improvements were also aimed at capturing important lessons learnt from previous and on-going GDAs.

As part of the modernised GDA, ONR is to conduct its assessment in three steps instead of the current four steps, and has changed the emphasis of each step:

- The aims of Step 1 (estimated to take around 12 months) are to agree the GDA scope, define the basis for the generic safety, security and environmental cases, identify the gaps to meeting UK regulatory expectations, and agree resolution plans for how & when these will be resolved. As part of this step the arrangements to undertake GDA and the programme for subsequent Steps are defined.

- During Step 2 (estimated to take around 12 months), the focus of the assessment is towards the adequacy and suitability of the fundamental aspects of the design, the safety, security and environmental claims, the methodologies and approaches.

- During step 3 (with an indicative timescale of 24 months), the focus is an in-depth assessment of the safety, security and environmental case evidence which underpins the design, to come to a conclusion on the acceptability of the design for construction in GB.

The estimated timelines are indicative and ultimately will depend on the timely submission by the GDA Requesting Party of high-quality documentation.

The modernised GDA process provides opportunities for the Requesting Party to make better and more effective use of existing submissions (e.g. to other Regulators) – supplemented to meet UK expectations. The GDA process will also place greater emphasis on earlier engagement and agreement of scope / submissions throughout process. It also provides flexibility in the assessment activities (with robust internal governance to agree scope of assessment that can be accepted).

A key feature that provides flexibility for SMRs is the introduction of additional outputs (GDA Statements) (as well as the Design Acceptance Confirmation (DAC) and Statement of Design Acceptance (SoDA) as previously) to show that the design (or a meaningful assessment scope from the design) has shown alignment with UK regulatory expectations. The acceptability of partial scopes and new GDA statement outputs is summarised in the Table A–4. ONR-GDA-GD-006 [A–137] including detailed Technical Guidance [A–139] with key expectations and lessons learnt were published in 2019.

TABLE A–4. ACCEPTABILITY OF PARTIAL SCOPES AND NEW GDA STATEMENT OUTPUTS

Example	Technical Assessment Topics	Steps	Output
Full plant (well developed) design	All	1,2 and 3	DAC, interim Design Acceptance Confirmation (iDAC) or no DAC SoDA, iSoDA or noSoDA (Environment Agency scope only)
Major portions of a well-developed plant design (for example: one complete reactor module of a multi-module design where the interactions between modules are potentially safety significant but are declared out of scope by the RP)	All	1,2 and 3	Statement

Example	Technical Assessment Topics	Steps	Output
Major portions of the plant design (of limited design maturity)	Most	1 and 2	Statement
Conceptual full plant design	Most	1 and 2	Statement
Partial plant design (for example, a design where the deployment model relies on out of scope supporting SSCs)	Assessment not meaningful – no GDA undertaken		
Distinguishing safety system (for example, the control and instrumentation technology and architecture)	Assessment not meaningful – no GDA undertaken		

A–4.9.3. Follow-up Question

ONR explained that there are no changes to the regulatory approach other than the modernisation of the process for additional flexibility in outputs (GDA statements) and recognising that submissions developed for other regulators can be used and, so far, any gaps with UK expectations are identified and plans are putting forward to address those gaps. A question was raised on whether ONR would expect additional training, confirmatory analysis, checks, etc., for new innovative designs?

A–4.9.4. Response

Regulatory capability (training and knowledge management), research activities and proactive review of guidance (SAPs [A–44] and TAGs [A–46]) are some of the activities ONR embarked upon early on to ensure that it is ready to regulate innovative designs and technology. We covered these in separate questions so further information can be brought in from those during the next review meeting.

However, regarding the additional question on whether 'additional' analysis, checks are expected etc. The overarching regulatory requirement is for the dutyholder to demonstrate that the risks have been reduced to be ALARP. The dutyholder's demonstration that the risks have been reduced to ALARP is expected to consider engineering, operations and the management of safety, that is 'relevant good practice, RGP'.

ONR considers relevant good practice as those standards for controlling risk which have been judged and recognised by ONR as satisfying the law, when applied to a particular relevant case in an appropriate manner. When the design or technology is innovative, those standards and relevant good practice are less evident or not fully applicable, or where the consequence of the hazard is high, the onus is on the dutyholder to implement measures to the point where the costs of any additional measures (in terms of money, time or trouble – the sacrifice) would be grossly disproportionate to the further risk reduction that would be achieved (the safety benefit). This is essentially demonstrating ALARP which is expected in any other technology/case. Further guidance on RGP and ALARP is available in NS-TAST-GD-005 [A–43].

ONR applies risk-informed regulatory decision making, as described in [A–42] which specifically addresses this situation:

"Where there are significant hazards and/or the operation is complex, ONR expect dutyholders to produce adequate risk analyses and/or PSAs that are related to and underpinned by engineering substantiation and operational measures. It is important to note that we expect such analyses to highlight potential weaknesses in the engineering and operation of the facility and not solely to compare against numerical risk targets. Where it is not possible to demonstrate ALARP by good practice features and risk estimates alone, dutyholders need to explicitly compare the benefits of other risk reducing measures with the costs of their implementation, demonstrating that the costs relating to implementation would be 'grossly disproportionate' to the benefit provided (risk reduction). We expect dutyholders will implement the safest option that is reasonably practicable taking appropriate consideration of the impact of all risks to all those affected in making its balanced decision."

ONR will assess the dutyholders safety case and ALARP demonstration and come to a judgement on its adequacy. In order to do so, ONR may choose to scrutinise the analyses presented in more detail as the technology may not benefit from extensive research and/or operating experience and may seek confirmatory analysis in specific areas.

A–4.10. UNITED STATES OF AMERICA–NRC

A–4.10.1. Question

Were there changes in the regulatory approach to the review and assessment process done in support of authorization? Describe information on the prior practice and the changes in approach made.

A–4.10.2. Response

No, while guidance for SMRs has evolved, the NRC's overall existing regulatory approach to the review and assessment processes for nuclear power plants are focused on large light water-cooled reactors has remained unchanged with the introduction of SMRs. Should there be an application from an applicant to construct an SMR facility in the United States, there is flexibility in the requirement (10 CFR [A–52] Parts 50.59 and 50.9) that provide applicants and license holder the means to change or depart from a process or method of evaluation.

A–5. ON-SITE INSPECTIONS

This Annex presents the responses provided by the following Member State regulatory bodies to the Question 5: "Were there changes in the regulatory approach to on-site inspections of the applicant carried out to confirm compliance with regulatory requirements and with any conditions specified in the authorization? Describe information on the prior practice and the changes in approach made."

A–5.1. ARGENTINA–ARN

A–5.1.1. Question

Were there changes in the regulatory approach to on-site inspections of the applicant carried out to confirm compliance with regulatory requirements and with any conditions specified in the authorization? Describe information on the prior practice and the changes in approach made.

A–5.1.2. Response

No, it was not necessary to change the regulatory approach to manage the on-site inspections.

National Law No. 24,804 [A–1] on Nuclear Activity and its Decree No. 1,390/1998 [A–7] establish the Regulatory Functions for the ARN, and one of them is Regulatory Control and Licensing.

The regulatory objective, in the Licensing and Control processes of a nuclear reactor, is to ensure that the Responsible Entity achieves and maintains the Fundamental Safety Objective by complying with safety requirements (according to IAEA Safety Fundamentals SF-1).

From the beginning of nuclear activity in Argentina, the Regulatory Body has performed, as core functions, review and assessments as well as multiple and different regulatory inspections and audits as frequently as considered necessary, with the purpose of verifying that nuclear installations satisfy the regulatory standards, Licenses, and requirements in force. All these activities are performed according to written procedures.

Act No. 24,804 [A–1], entitles the ARN to carry out with such inspections and regulatory review and assessments, performed by its personnel such as:

- Routine planned inspections are carried out by resident inspectors and other ARN personnel;
- Their purpose is to verify that the Licensee complies with limits and conditions of operation established in the mandatory documentation;
- Special inspections including reactive inspections are carried out by ARN specialists (dosimetry, instrumentation and control, thermo hydraulics, etc.) in coordination with resident inspectors.

These inspections are performed under special circumstances or due to the occurrence of abnormal events in the installation.

As mentioned in GSR Part 1 (Rev.1) [A–27] the Regulatory Authority has a duty to supervise all safety-related activities in all phases of the life of a nuclear installation. Characterization of the site, Design, Construction, Commissioning, Commercial operation, and Decommissioning.

168

In CAREM 25 prototype reactor project during the manufacturing, construction, assembly, and commissioning stages, regulatory inspections are carried out by ad-hoc inspectors under an Inspection Program schedule. In the future, during the Operation stage, the inspections will be carried out by Resident Inspectors like in the other operational NPPs. Regulatory inspections are associated exclusively to ensure compliance with the safety requirements specified in SSCs and processes. Regulatory inspections should not be considered as reinforcement, complement, or replacement of the functions of the agents or inspectors of the Responsible Entity (CNEA) that are associated with compliance with the requirements of the project in general.

In the CAREM 25 construction authorization, the ARN established an inspection plan for components / regulatory control during the construction stage. Main SSCs are:

- Reactor pressure vessel;
- Steam generators;
- Hydraulic CRDs;
- RHRS and SIS (main line safety systems);
- Fuel elements;
- Protection systems (I&C);
- Ventilation systems;
- Containment.

For SSCs with the highest nuclear safety rating, compliance with the safety requirements must be verified in a complete (exhaustive) manner at the time of commissioning the installation. As the safety requirements are essentially functional, the inspection of the functional tests in the safety systems is not carried out by sampling, but in a complete way, in all the tests, which in turn must have full scope.

For Safety-Related SSCs, regulatory inspections are performed on a sample of inspections or verifications of the Responsible Entity, verifying that they are carried out in accordance with the planned procedures. Sampling should be programmed with some statistical parameters and for relevance. For example, if a tightness test in a system turns out poorly, the project will have to correct what is necessary and repeat the test until it goes well, so the inspection of its execution would not be a priority. On the other hand, the tests that declare the 'release' of a system for the Start-up stage are a milestone that is not repeated, so they are an inspection sampling focus.

To meet established regulatory requirements, ARN only conducts inspections and audits on the Responsible Entity, not vendors. The Responsible Entity oversees assuring that the supplier quality program is implemented, and the defined design criteria are accomplished. ARN reviews the mandatory documentation presented, to define and plan the inspection tasks.

Regulatory Audits are planned and carried out by ARN personnel to analyse the organization, operation, and process aspects related to radiological and nuclear safety to examine the degree of compliance with the provisions in the mandatory documentation.

As an important and recent antecedent in construction inspections and commissioning, the ARN has the experience of having participated in the projects of the completion of construction of the NPP Atucha II and the refurbishment of CNE, which were completed in the past years. These important projects gave ARN the chance to train their personnel and improve their capabilities in different areas of knowledge. Also, we can add the current experience gain of the construction of RA-10, a world-class type research reactor.

Regarding the challenges for the construction, assembly, and commissioning inspections of CAREM 25, one of them is that it is a prototype reactor and first of its kind, so the Regulatory Authority must adapt to the changes that the project proposes. In addition, another important challenge is related to managing the knowledge acquired in the other projects, so that experience and knowledge are not lost over time and the possible generational turnover of personnel.

A–5.2. CHINA–NNSA

A–5.2.1. Question

Were there *changes* in the regulatory approach to on-site inspections of the applicant carried out to confirm compliance with regulatory requirements and with any conditions specified in the authorization? Describe information on the prior practice and the changes in approach made.

A–5.2.2. Response

Scope and tasks of nuclear safety inspection:

(a) Whether the submitted information conforms to reality;
(b) Checking whether it is built according to the approved design;
(c) Checking whether it is managed according to the approved quality assurance program;
(d) Checking whether the construction and operation of nuclear facilities meet the requirements of relevant regulations and permits;
(e) Checking the ability of operators to operate safety and implement emergency plan;
(f) Checking whether the decommissioning meets the requirements of nuclear safety and environmental protection regulations.

Ways and methods of nuclear safety inspection:

Nuclear safety inspection is divided into daily, routine and non-routine (special) inspection. Non-routine inspection should be done with or without prior notice. Inspection methods mainly include: document inspection, on-site observation, interview, measurement or test.

A–5.2.3. Follow-up Question

Regarding inspections, please elaborate on the encountered challenges, if any.

Please describe any differences on-site inspections for SMRs (floating and land-based).

A–5.2.4. Response

At present, NNSA has not issued the construction license for the on-site inspection of SMR or floating reactors, and the specific on-site inspection procedures is still on development.

NNSA has developed special inspection programs and procedures for HTGR on-site inspection, mainly considering the system and components are different to PWR.

A–5.3. CANADA–CNSC

A–5.3.1. Question

Were there changes in the regulatory approach to on-site inspections of the applicant carried out to confirm compliance with regulatory requirements and with any conditions specified in the authorization? Describe information on the prior practice and the changes in approach made.

A–5.3.2. Response

The last nuclear power plant (NPP) constructed in Canada was the Darlington NPP (a 4 unit station). A construction licence was issued in June 1981. The first unit was declared in-service in October 1990 and the final unit in June 1993. Darlington 'A' was built by Ontario Hydro Construction, a fully owned subsidiary of Ontario Hydro.

Ontario Hydro Engineering and Construction hired and trained its own construction workers minimizing the need for contractors and sub-contractors. Since the plant was being built by the company that would later operate it, there was an emphasis on high construction standards. Although third parties can, and many do, place equal emphasis on high construction standards, recent experience with nuclear construction projects utilizing third parties has demonstrated that they require significant oversight by the licensee and, in some cases, by the regulator.

Prior to Darlington 'A', much of the design and engineering work of a new NPP followed the 'just in time' approach. Design work was completed just in time to allow procurement and installation of equipment to meet the construction schedule. Construction licence applications were based primarily on conceptual studies and preliminary design work with a sprinkling of detailed design. As a result, many design and safety issues arose during construction of the plant.

This situation was not acceptable from a regulatory perspective, and so for Darlington 'A' the CNSC (then AECB) required the design and safety analysis to be much more advanced prior to the issuance of the construction licence. Ontario Hydro agreed to completion of a more complete design at the start of construction. The intent was to give more certainty in construction schedules and more confidence that construction will not be disrupted by 'last minute' design revelations.

New reactors are currently under construction in several countries. Most are constructed by contractors and sub-contractors with the licensee performing primarily oversight activities. Their design and construction has presented a number of issues to their respective regulators. Examples include:

- Inadequate completion of design and engineering work prior to start of construction;
- Poor quality design and other vendor information;
- Design information that does not adequately demonstrate compliance with regulatory requirements;
- Shortage of experienced staff in design, construction and manufacturing;
- Contractors and sub-contractors compromising on quality to achieve schedules due to economic pressures and lack of safety culture;
- Worldwide shortage of qualified equipment manufacturers resulting in sub-standard components and counterfeit parts.

REGDOC-1.1.2 [A–66], clearly articulates expectations for completeness of the design prior to the issuance of the construction licence. Given recent international experience, the CNSC is revising this REGDOC [A–66] to incorporate recent lessons learned. In particular, the section on management system has been enhanced.

The applicant shall ensure that, as a contractual obligation:

- The applicant and the CNSC will have right of access to the premises of any supplier to the construction program (including off-site testing);
- All sub-suppliers will provide right of access to their premises by those clients who are suppliers to the construction program (including off-site testing).

The CNSC is also conducting a multi-disciplinary review of the lessons-learned from the Boeing 737 MAX accidents, including Boeing's interactions with the regulator. The CNSC is actively considering how this operational experience applies to SMRs and will make any necessary changes to the regulatory framework.

Activities Under a Licence to Construct

Once a licence to construct has been granted, the licensee will undertake various activities. The following are examples that indicate the scope of activities that would be performed:

- Cement pours (for example, for the nuclear island);
- On-site physical construction of structures and systems;
- Off-site manufacturing of components (such as vessels, pipes, pumps and valves);
- Detailed engineering and design including field-driven design changes and supporting analyses;
- Non-nuclear (or cold) commissioning of systems, structures and components;
- Fuel storage on-site (if requested by the applicant, and it has made appropriate provisions).

In addition, the licensee will require programs such as for environmental monitoring, security, radiation protection, criticality, safeguards, emergency planning and response, etc.

The extent of regulatory oversight for each of these activities and programs will be set in accordance with an overall oversight strategy with:

- A baseline compliance plan [5];
- Supporting processes and project management tools;
- An enforcement strategy [6].

---

[5] The baseline compliance plan is the minimum number of inspections and other oversight activities needed to give the CNSC reasonable assurance that the facility is being constructed and manufactured in accordance with the design specifications. A detailed baseline compliance plan will be produced once the technology has been chosen, the principal engineering, procurement and construction company has been identified and other decisions regarding the site have been made.

[6] An enforcement strategy will identify in broad terms the escalation of CNSC activities if CNSC finds significant non-compliances.

In addition, in parallel with the activities under the licence to construct, the licensee will carry out a number of long-lead activities in order to prepare its organisation for operation, including:

- Developing the organisation in readiness for operation;
- Obtaining a simulator, performing training and certification of authorized nuclear operators.

Types of Regulatory Oversight

To ensure regulatory oversight, various types of regulatory verification activities such as on-site inspections and technical assessments by head office staff are needed. The scope and depth of the activities will be based on the principle of 'trust but verify', recognising that the licensee has the primary responsibility for safety. Regulatory activities will include:

- Regular inspections of general on-site activities including witnessing licensee's inspections;
- Targeted on-site inspections of specific 'safety critical' items at time of construction or installation (for example, concrete pour for containment), including conducting or overseeing:
  - As-built inspections;
  - Construction testing;
  - Operations testing;
  - Welding qualification criteria;
  - Seismic walkdowns;
  - Construction completion assurance;
  - Management system inspections;
  - Verification of compliance with the licensing basis for topics such as fire protection, engineering, security, emergency planning and radiation protection.
- Targeted off-site inspections, including the vendor's facilities and manufacturing inspections and selected manufacturers of safety critical items such as pressure vessels or safety system components;
- Technical assessments of engineering documentation;
- Technical assessments of safety significant in-field deviations from design specifications by head office staff;
- Technical assessments of performance indicators;
- Inspections of licensee's oversight processes;
- Dispositioning, where necessary, the findings of inspections or engineering assessments performed by the authorised inspection agency (an approved agency that performs inspections and design reviews of boilers and pressure vessels);
- Technical assessments of third-party inspections;
- Technical assessments of other regulator's inspection reports (for example through the Multinational Design Evaluation Programme), where these are available.

As the construction phase of the project progresses, the compliance plan will be reviewed and modified on an as needed basis as a result of the feedback from previous inspections and other compliance activities.

Compliance Program Principles

The primary responsibility for safety lies with the licensee. Compliance activities performed by CNSC staff will not complement or replace those of the licensees; the CNSC is not part of the licensee's processes.

All compliance activities will be aligned with the CNSC Safety and Control Area Framework. All activities will consider the four categories of activities:

- Category 1: licensee commitments made in the application for Licence to Construct;
- Category 2: licensee operational readiness;
- Category 3: events;
- Category 4: construction and commissioning activities (including long-lead items).

Compliance activities will be risk-informed and performance based. CNSC will use a risk-informed approach to select areas for regulatory scrutiny. The approach will be based on past experience in Canada and lessons learned in other nations currently conducting new build activities. Compliance activities will measure licensee performance against performance objectives.

'Trust but verify'. CNSC have the authority to independently verify, where necessary, licensee compliance with the conditions of their licence. This may be done using:

- Inspections;
- Desk-top reviews;
- Field presence and walkdowns.

The compliance program will be planned and executed using CNSC core process 'Manage Compliance'. Existing processes in the licensing and compliance process maps are more than sufficient to plan and execute all compliance activities under the 'Licence to Construct'. Graduated enforcement tools will be used to ensure that a licensee returns to a safe and compliant state where circumstances justify. Promotion will be one of the enforcement tools used.

A single point of contact (SPOC) will be used for compliance activities. A new build construction project is highly fluid with conditions changing from day-to-day. This contrasts with operating plants which represent a more steady state from a compliance perspective. All activities being performed by the licensee are managed through a master construction schedule. CNSC staff will conduct its activities in respect of the schedule while at the same time reinforcing that the schedule is secondary to safety. Staff need to be able to adapt to the changing schedule and react, if necessary to issues it encounters in a timely manner, managed through the SPOC.

The above practices remain applicable to SMR projects and would be applied in a graded manner to adapt to the specifics of the project, related for instance related to:

- Shorter construction time;
- Commissioning of novel design features;
- Factory inspection of modules, both nationally and internationally;
- Extensive use of 3rd party contractors and licensee oversight.

A–5.4. CZECH REPUBLIC–SÚJB

A–5.4.1. Question

Were there changes in the regulatory approach to on-site inspections of the applicant carried out to confirm compliance with regulatory requirements and with any conditions specified in the authorization? Describe information on the prior practice and the changes in approach made.

A–5.4.2. Response

Not applicable at present, no SMR are planned to be deployed in the Czech Republic and therefore there have been no changes in the regulatory approach to on-site inspections carried out to confirm compliance with regulatory requirements and with any conditions specified in the authorization. Unless modified, the regulatory approach would be the same.

SÚJB performs inspections of compliance with the requirements of the Atomic Act [A–21], implementing legislation and compliance with the specific requirements contained in decisions issued on the basis thereof. SÚJB is authorized to perform both scheduled and unscheduled inspections. These specialized inspections verify that the selected area is in accordance with the approved inspection plan or on an unplanned basis, based on specific events.

Unscheduled inspections are usually performed in case of events that are preliminary classified as more serious, in case of unclear findings of the investigation done by licensee, in case of PSA safety-relevant events or events with root causes linked to more serious system failures in higher management, or in case of IAEA inspections. The scheduled (routine) inspections are performed in line with Plan of Inspection Activities which is drafted on annual basis according to the relevant internal regulations of SÚJB to reflect current priorities in the field of supervision.

The assessment commission is established for evaluation of inspection activities. Inter alia, it discusses and analyses the results of inspections, verifies the correctness of classification of the findings, updates a Plan of Inspection Activities, decides on proposals to initiate administrative enforcement procedures (to impose penalties or corrective measures), decides upon continued evaluation.

The inspection activities are comprehensive and include, inter alia, design aspects (during operation design modifications), technical aspects (in particular of selected equipment), management system aspects, documentation, activities during outages, fuel loading, nuclear material, radiation protection, security and compliance with the conditions of the decisions issued by SÚJB in general.

A–5.4.3. Follow-up Question

Collectively, for the Czech Republic, what challenges does the regulatory body anticipate if SMRs are proposed, based on which SMRs are being considered?

A–5.4.4. Response

Unfortunately, as no SMR is planned to be deployed in Czech Republic, only a general description of the legislative requirements can be provided.

Changes in the regulatory approach to on-site inspections to reflect particular SMR technology specificities and deployment model cannot be excluded. However, there is a persistent lack of clear and reliable data of the particular technological solutions and no SMR (of a particular design) is envisaged to be deployed. It cannot be excluded that the regulatory approach to on-site inspections would be different but not simplified to reflect the specificities of a particular design and deployment model.

Given the lack of detailed information about various SMR designs and the uncertainty over whether and which type of SMR could be hypothetically deployed, no further information can be provided.

A–5.5. FRANCE–ASN

A–5.5.1. Question

Were there changes in the regulatory approach to on-site inspections of the applicant carried out to confirm compliance with regulatory requirements and with any conditions specified in the authorization? Describe information on the prior practice and the changes in approach made.

A–5.5.2. Response

There hasn't been any construction of SMR in France. However, ASN intends to adapt its control to the risks of the installation. French's legal framework gives ASN the freedom to define its inspection program.

Graded approach is central in the construction of ASN's inspection program for nuclear reactors. Every year, ASN assesses its priorities regarding control of nuclear reactors. Risk analysis, return of experience and project's or installation's upcoming milestones are used to define these priorities. For example, ASN has adapted its inspection program following the different EPR's construction phases.

Different types of inspections are led by ASN: regular inspections, reactive inspections (conducted after an event) and in-depth inspections which are longer than regular inspections. These inspections can be conducted on NPPs, but also in central services or vendors' facilities.

These inspections enable ASN to confirm compliance with regulatory requirements and conditions specified in the authorization.

A–5.6. JAPAN–NRA

A–5.6.1. Question

Were there changes in the regulatory approach to on-site inspections of the applicant carried out to confirm compliance with regulatory requirements and with any conditions specified in the authorization? Describe information on the prior practice and the changes in approach made.

A–5.6.2. Response

No change was done in the on-site inspections.

Note: At the beginning of April 2020, the new revised Oversight Program, which is designed based on the US Reactor Oversight Process, started to be fully implemented.

A–5.6.3. Follow-up Question

No information on the prior practice was given.

Collectively, for Japan, what challenges do the regulatory body anticipate if SMRs are proposed, based on which SMRs are being considered?

A–5.6.4. Response

There is no specific information on new SMRs.

A–5.7.   RUSSIAN FEDERATION–ROSTECHNADZOR

A–5.7.1. Question

Were there changes in the regulatory approach to on-site inspections of the applicant carried out to confirm compliance with regulatory requirements and with any conditions specified in the authorization? Describe information on the prior practice and the changes in approach made.

A–5.7.2. Response

The regulatory approach to on-site inspections is defined in the Article 24.1 of the Federal Law No 170-FZ [A–26], 'Oversight in the Field of the Use of Atomic Energy':

> "Federal state oversight in the field of the use of atomic energy entails activity on the part of the empowered Federal authority which is intended to prevent, identify and stop violations by juridical persons operating in the field of the use of atomic energy, their managers and other officers (hereinafter – juridical persons) of the requirements laid down by the international agreements entered into by the Russian Federation, this Federal law, other Federal laws and other enactments of the Russian Federation in the field of the use of atomic energy (hereinafter – mandatory requirements), by organising and conducting checks (or inspections) on these individuals, taking the measures prescribed by the laws of the Russian Federation to stop violations which have been identified, and through the work of the aforementioned Federal authority to systematically monitor the fulfilment of mandatory requirements and analyse and forecast performance in terms of fulfilling these requirements when juridical persons pursue their activities."

Detailed description of checking (or inspections) procedures, is given according to the Federal Law No 170-FZ [A–26]:

> "The purpose of checks (or inspections) shall be to ascertain whether a juridical person is complying, while pursuing activity in the field of the use of atomic energy, with mandatory requirements and the terms of permits (or licences) necessary to maintain safety in the field of the use of atomic energy, and also to assess the compliance of facilities which use atomic energy, their components and systems with the aforementioned requirements."

The inspections are classified, according the Federal Law No 170-FZ [A–26] as scheduled or unscheduled checks (or inspections), as follows:

> "Scheduled checks (or inspections) shall be included in the annual plan of scheduled checks (or inspections) where one year has passed since the date on which:"

- "the juridical person was granted a permit (or licence) to pursue activity in the field of the use of atomic energy and the juridical person was registered in accordance with article 36.1 of this Federal Law;"
- "a decision was taken, in accordance with the procedure established by the Government of the Russian Federation, to commission facilities which use atomic energy after they have been built or undergone technical upgrading, reconstruction or major repairs, including those used during the operation of facilities which use atomic energy, their components and systems, including buildings, premises, installations, hardware, equipment and materials;"
- "the last scheduled check (or inspection) was completed."

Unscheduled checks (or inspections) shall be carried out where:

- "the state safety regulatory authority receives:"
    - "enquiries and submissions from citizens, including individual entrepreneurs and juridical persons and information from state authorities (or officials from the state safety regulatory authority), local authorities and the mass media about violations of nuclear and radiation safety requirements in the use of atomic energy, including the terms of permits (or licences) which are necessary to maintain safety in the field of the use of atomic energy, requirements concerning physical protection, state recording and monitoring of nuclear materials, radioactive substances and radioactive waste, about the performance of works and pursuit of activity which have an impact on the safety of a facility which uses atomic energy and fall outside the scope of permits (or licences) which have been issued, about the pursuit of activity without the relevant permits (or licences), about breaches of mandatory requirements when constructing, operating and decommissioning facilities which use atomic energy, their components and systems, and also when handling nuclear materials, radioactive substances and radioactive waste, if such breaches pose a threat of harm to human life or health, harm to animals, plants, the environment, state security, the property of individuals and juridical persons, state or municipal property or a danger of man-made emergencies or lead to such harm and man-made emergencies;"
    - "official data obtained through state monitoring of the radiation situation within the Russian Federation which indicate that it has changed due to the operation of facilities which use atomic energy;"
    - "a request from a juridical person to grant a permit (or licence) to pursue activity in the field of the use of atomic energy, to reissue a licence or make changes to the terms of a permit (or licence), to terminate a permit (or licence), to be registered in accordance with article 36.1 of this Federal Law or to commence works which pose a nuclear and/or radiation hazard in accordance with the regulations and rules 15 concerning the use of atomic energy;"
    - "an order to conduct an unscheduled check (or inspection) has been issued by the head (or deputy head) of the state safety regulatory authority pursuant to an instruction from the President of the Russian Federation or the Government of the Russian Federation or on the basis of a demand from a prosecutor to conduct an unscheduled check as part of oversight in

relation to law enforcement on the basis of materials and enquiries received by prosecuting authorities."

- "the time-limit for a juridical person to comply with an order to rectify an identified infringement of mandatory requirements issued by the state safety regulatory authority has passed".

A–5.7.3. Follow-up Question

Regarding inspections, please elaborate on the encountered challenges, if any.

Please describe any differences on on-site inspections for SMRs (floating and land-based).

A–5.7.4. Response

The inspection processes are identical for all nuclear facilities. A distinction is made between inspection programs (which include specific topics to be considered by inspectors) and the personnel of an inspection.

There was no need to make further changes to the on-site inspection process.

A–5.8.   SOUTH AFRICA–NNR

A–5.8.1. Question

Were there changes in the regulatory approach to on-site inspections of the applicant carried out to confirm compliance with regulatory requirements and with any conditions specified in the authorization? Describe information on the prior practice and the changes in approach made.

A–5.8.2. Response

The PBMR project did not reach a sufficiently mature stage for this topic to be developed with respect to on-site inspections.

A series of audits were conducted by the Regulator on the applicant focussing on the implementation of an Integrated Management System to be in place for the stage of the project which involved the assessment of the design, development of the safety case and oversight of components being manufactured. The NNR standards requires that all organisations delivering service or product important to nuclear safety has to have a Management System and associated process commensurate with the safety classification of the service or product.

A–5.8.3. Follow-up Question

Collectively, for South Africa, what challenges does the regulatory body anticipate, based on the SMR technology being considered?

A–5.8.4. Response

The NNR Act [A–30] requires that the regulator can perform inspections wherever products are being produced for use in nuclear installations in South Africa. As such it is expected that the contractual arrangements between the applicant and its vendor as well as the respective procurement processes make provision for regulatory oversight activities.

In terms of inspections by the applicant, RD-0034 [A–33] includes requirements giving effect to the principle of intelligent customer capability. Applicants are therefore, as the organisation ultimately responsible for nuclear safety, required to implement procurement processes that involves audits and inspections on their suppliers and sub suppliers where important to safety activities are being performed.

SMRs do not necessary introduce new challenges as the Regulatory standards are clear and transparent in this regard. It is acknowledged that modular manufacturing approaches may require more oversight, but it does not introduce a new challenge.

A–5.9. UNITED KINGDOM–ONR

A–5.9.1. Question:

Were there changes in the regulatory approach to on-site inspections of the applicant carried out to confirm compliance with regulatory requirements and with any conditions specified in the authorization? Describe information on the prior practice and the changes in approach made.

A–5.9.2. Response:

So far ONR has not been requested to license a nuclear site for deployment of an SMR, engaged in formal 'pre-licensing' activities such as Generic Design Assessment (GDA) or indeed carried out inspections associated with an SMR. There is therefore no practical experience in this area.

ONR has a suite of Technical Inspections Guides (TIGs) [A–140] covering the 36 Licence Conditions (LCs) and associated inspection topics. The LCs are considered technology-neutral and the TIGs reflect this consideration. However, similarly to the programme of guidance review activities for compatibility with ANTs, which has so far focused on the TAGs [A–46], ONR undertakes periodic revisions of the TIGs [A–140] and it is considered that this will in the future be informed by ANTs. Specifically, Inspectability of Systems, Structures and Components (SSCs) including physical space available for access as well as worker dose considerations have been identified as areas of interest in the context of the AMR F&D assessments.

A–5.10. UNITED STATES OF AMERICA–NRC

A–5.10.1. Question

Were there changes in the regulatory approach to on-site inspections of the applicant carried out to confirm compliance with regulatory requirements and with any conditions specified in the authorization? Describe information on the prior practice and the changes in approach made.

A–5.10.2. Response

No, currently there is not an application from an applicant to construct an SMR facility in the United States. When an SMR is constructed, the NRC's regulatory approach for onsite inspections will remain unchanged. Onsite inspections are conducted in accordance with NRC Inspection Manual Chapters and associated inspection procedures. The objectives of onsite inspections are as follows:

- To provide assurance that the application for a DC meets requirements specified in Subpart B to 10 CFR [A–52] Part 52;

- To verify that quality processes used in the development of the DC application are adequately described, and that technical, quality, and administrative requirements important to public health and safety are effectively implemented during the design and procurement phases of DC activities;
- To verify effective implementation of the quality assurance (QA) program, as described in the application for a DC, satisfy the appropriate provisions of Appendix B to 10 CFR [A–52] Part 50;
- To verify whether the qualification testing activities supporting the application are conducted in accordance with the requirements of Appendix B to 10 CFR Part 50.

To verify that information protection systems effectively protect Safeguards Information (SGI), as defined in 10 CFR [A–52] 73.21, and 10 CFR 73.22, and prevents unauthorized disclosure.

A–6. ASKING INFORMATION AND INSPECTIONS OF SUPPLIERS

This Annex presents the responses provided by the following Member State regulatory bodies to the Question 6: "Were there changes in the regulatory approach for asking information and inspections of suppliers?"

A–6.1. ARGENTINA–ARN

A–6.1.1. Question

Were there changes in the regulatory approach for asking information and inspections of suppliers?

A–6.1.2. Response

No, no changes were made in the regulatory approach regarding inspections of suppliers. However, due to the normative framework review, it is expected that in the future some changes may arise in order to adapt to the regulations references.

ARN only conducts inspections and audits to the Responsible Entity, not to vendors, in order to meet established regulatory requirements. The Responsible Entity oversees assuring that the supplier management program is implemented, and the defined design criteria are accomplished. ARN reviews the mandatory documentation presented, to define and plan the inspection tasks.

Regarding the ARN normative framework integral review, one of the first standards updated was the AR 10.6.1 Management system for safety in facilities and practices [A–141] in accordance with the concepts of GSR Part 2 [A–142]. The ARN assessment of the 'Management Manual' of CAREM 25 project is an ongoing task (related to each revision of responsible entity document) and it was doing against this updated regulatory standard (AR 10.6.1.) taking into account the requirements for the RE related to suppliers.

As an example of supplier's activities, Industrias Metalúrgicas Pescarmona S.A. in the province of Mendoza, Argentina, is carrying out the manufacturing of Reactor Pressure Vessel (RPV). The RE establish the Technical Specification for the RPV and the supplier had to develop the detailed engineering of the component according to the requirements of ASME Code, Section III, Division 1. ARN made inspections to the manufacturing site, by the hand of Responsible Entity, in the current stage of cladding and welding. Also, ARN performed a conceptual review of the Technical Specification and the Surveillance Program and agreed with the responsible entity the incorporation of findings. Currently, ARN is beginning with the review of the preliminary 'In-Service Inspection' report.

A–6.2. CANADA–CNSC

A–6.2.1. Question

Were there changes in the regulatory approach for asking information and inspections of suppliers?

A–6.2.2. Response

Like all businesses, safety-critical and complex organisations engage in fundamental strategic Make or Buy decisions, core tasks and processes are usually undertaken in-house by the organisation (the 'make' element). 'Non-core' activities are then outsourced to efficient third parties (the 'buy' element). However, the CNSC is seeing applicants and licensees increasing use of contractors to carry out core life cycle activities — such as engineering studies, safety analysis, periodic safety reviews, maintenance, etc. — for routine operations, refurbishment projects and new builds. This is also prevalent with SMR vendors outsourcing significant design and safety analysis work.

The impact of ineffective management of contractors has been highlighted in a number of adverse events across various high-risk industries. This highlights the need for organisations to retain the ability to understand, specify, oversee and accept contractors' technical and physical work undertaken on its behalf.

The IAEA describes an 'Intelligent Customer' in Safety Guide No. GS-G-3.5, The Management System for Nuclear Installations [A–143]. Two key questions to ask an applicant or licensee is 'could you do the work yourself if the contractor was to disappear?' along with 'when was the last time you actually did such work yourself?' To demonstrate that an organization is meeting Intelligent Customer criteria, they should:

- Be able to make informed decisions relating to risk safety issues;

- Remain in control of all work carried out by contractors that can impact on high-risk safety issues;

- Maintain oversight of work with risk safety implications, whether that work is conducted on or off the site;

- Have a clear policy for choosing between in-house activities or the use of contractors;

- Have considered whether the use of contractors could create organisational vulnerabilities arising from a dependence on contractors in relation to risk safety issues, and have in place contingency and succession arrangements;

- Be able to demonstrate that its core capability has suitable and sufficient competence, resources and arrangements to understand where and when work is needed; specify requirements to carry out that work; understand and set suitable standards; oversee and control the work; and be able to review, evaluate and accept the work carried out on its behalf.

There are some broad principles, which underpin CNSC's expectations of an applicant/licensee's arrangements for the use of contractors and for retaining control of nuclear safety:

- Maintain a core in-house staff to ensure effective control and management for nuclear safety;

- Retain overall responsibility for, and control and oversight of, the nuclear and radiological safety and security of all of its business, including work carried out on its behalf by contractors;

- Ensure choices between sourcing work in-house or from contractors is informed by a company policy that takes into account the nuclear safety implications of those choices;

- Maintain an 'intelligent customer' capability for all work carried out on its behalf by contractors that may impact upon nuclear safety;

- Ensure only contracts for work with nuclear safety significance are let to contractors with suitable competence, safety standards, management systems, culture and resources;

- Ensure all contractor staff is familiar with the nuclear safety implications of their work and interact in a well-coordinated manner with its own staff;

- Ensure contractors' work is carried out to the required level of safety and quality in practice.

CNSC Regulatory Documents include requirements and guidance that address the need for applicants and licensees to be an intelligent customer, even if the terminology isn't explicit.

REGDOC-1.1.1, Site Evaluation and Site Preparation for New Reactor Facilities [A–64], section 4.3 'Management System' addresses contractor accountability, activities, authority and oversight. It also addresses the need for the applicant to:

"(…) assess the technical and safety assessment capabilities in the context of the reactor technology organization being an intelligent user of consortium members and subcontractors (…)".

REGDOC-1.1.2 [A–66], clearly articulates expectations for completeness of the design prior to the issuance of the construction licence. The CNSC is revising this REGDOC to incorporate recent lessons learned. The current version of REGDOC-1.1.2 [A–66] Section 8.3.1 on procurement programs specifies that the overall approach to procurement and manufacturing will be systematically controlled in all respects. Considerations that should be taken into account include:

- Ensuring that, as a contractual obligation, the applicant and the regulatory body will have right of access to the premises of any suppliers to the construction program;

- Ensuring that, as a contractual obligation, all sub-suppliers will provide right of access to their premises by their clients who are suppliers to the construction program.

Intelligent customer roles and principles are also highlighted in sections 7 on safety analyses and 5 on general design aspects and support programs of REGDOC-1.1.2. [A–66]

REGDOC-2.3.1, Conduct of Licensed Activities: Construction and Commissioning Programs [A–144], section 3.1 on role of the licensee highlights the need for 'capability to understand the nuclear safety significance of purchased expertise or equipment' and the procurement and oversight of the purchased goods and/or services. It also notes that:

"Contractors at all levels in the supply chain should expect to be audited on a regular basis as part of contractual arrangements. Contractors could also be visited by the CNSC as part of regulatory oversight, particularly if the equipment they are manufacturing has high nuclear safety significance."

REGDOC-2.5.2 [A–62] section 5.1 on design authority requires that if design work is contracted, there is formal documentation of the arrangement and that overall responsibility remains with the design authority. The applicant or licensee also need to confirm that the design authority has achieved appropriate oversight of the responsible designers and other suppliers and maintained the necessary engineering and scientific skills.

184

REGDOC-2.6.2, Maintenance Programs for Nuclear Power Plants [A–145] addresses contractor work practices and contractor quality assurance in section 3.2.4 on contract workers.

There are also a number of standards published by the CSA Group which touch on the principles of an Intelligent customer, particularly CSA N286-12, Management system requirements for nuclear facilities [A–146] and CSA N286.7-16, Quality assurance of analytical, scientific, and design computer programs [A–147].

A–6.3. CHINA–NNSA

A–6.3.1. Question

Were there changes in the regulatory approach for asking information and inspections of suppliers?

A–6.3.2. Response

There is no need to change the regulatory approach for asking information and inspections of suppliers, according to the experience of HTR-PM project. For transportable SMRs, such as floating reactors, which may build and load the fuel in factory. Regulatory approaches to building factories may need to be reconsidered.

In this aspect, there are many related items in nuclear regulations and guidance. For example, in Implementation Rules of Regulations on the Safety Supervision and Management for Civilian Nuclear Installations (HAF001/01) [A–148] 'Part one: Application and Issuing of Safety License for Nuclear Installation: the Nuclear Installation Applicant contact with NNSA representing all suppliers and contractors, who must timely provide all material that NNSA and its review centre required.' And '(…) during the siting, construction, commissioning, operation, and etc. period NNSA could assign regional inspection official to equipment supplier and nuclear installation site, whose main responsibility is to check whether previous providing material is consistent with the reality.'

A–6.3.3. Follow-up Question

Please explain if there are changes in the current regulatory approach considering the floating reactors are built in a factory.

Please elaborate on the changes the regulatory body is considering for the floating reactors that may be built in a factory.

A–6.3.4. Response

At present, the regulatory approaches have not been changed.

The licensing process of FNPPs has been modified as described by China in the answer to Question 3.

A–6.4. CZECH REPUBLIC–SÚJB

A–6.4.1. Question

Were there changes in the regulatory approach for asking information and inspections of suppliers?

A–6.4.2. Response

Not applicable — as no SMR is currently planned to be deployed in Czech Republic, no changes to the regulatory approach towards suppliers have been considered and the process of ensuring quality, technical safety and the conformity assessment and the verification thereof would follow the standard procedures as prescribed in the Atomic Act [A–21] and the Decree No. 358/2016 [A–23] Coll., on requirements for assurance of quality and technical safety and assessment and verification of conformity of selected equipment.

Conformity of the selected equipment with technical requirements shall be assessed prior to its use only by a person entitled to perform an assessment (graded approach has been applied by the legislator and the assessment is performed either by authorized person, accredited person or the manufacturer of such equipment). Only equipment that has passed the conformity assessment might be used.

If the selected equipment is imported and it complies with:

- Technical regulations for nuclear installations that are binding for the manufacture of this selected equipment;

- Technical standards or rules of good practice intended for nuclear installations, which are issued by a national standardisation Authority or a body of equivalent status;

- International technical standards for nuclear installations legitimately applied in that State; or

- Manufacturing procedures which are applied for nuclear installations in accordance with the legislation of the State in which it was manufactured and for which sufficiently detailed technical documentation exists.

These requirements guarantee the same level of protection as national requirements, the imported equipment is exempted from conformity assessment as such (since its conformity must have been assessed already).

SÚJB is authorized to inspect licence holders as well as manufacturers and authorized and accredited persons to verify that they comply with the Atomic Act [A–21] and its implementing legislation.

A–6.5. FRANCE–ASN

A–6.5.1. Question

Were there changes in the regulatory approach for asking information and inspections of suppliers?

A–6.5.2. Response

The article L.593-33 of the Environmental Code [A–24] enables ASN to enact requirements to the licensee about safety related activities, even if they are realised off the site by other stakeholders.

Moreover, according to the article L.596-14 of the Environmental Code [A–24], these activities are controlled by ASN under the same conditions as basic nuclear installations.

The licensee keeps the responsibility for these activities, except for nuclear pressure equipment, for which the responsibility is carried by the manufacturer during the design and manufacturing processes. Licensee's oversight is also assessed during inspections on suppliers' activities.

A–6.6. JAPAN–NRA

A–6.6.1. Question

Were there changes in the regulatory approach for asking information and inspections of suppliers?

A–6.6.2. Response

No change was done in the regulatory approach for asking information and inspections of suppliers.

A–6.6.3. Follow-up Question

Could Japan please provide info or references to the current approach, so that can be looked at by Member States?

A–6.6.4. Response

The result of inspection has been quarterly reported to the NRA commission meeting which is open to the public, and the result for each facility has also been disclosed on NRA website.

A–6.7. RUSSIAN FEDERATION–ROSTECHNADZOR

A–6.7.1. Question

Were there changes in the regulatory approach for asking information and inspections of suppliers?

A–6.7.2. Response

Federal Law No. 170-FZ [A–26], Article 37:

> "Organisations which undertake scientific investigation and surveying, perform design activity, build and operate nuclear facilities, radiation sources or storage facilities, design and manufacture equipment for them, perform other work and provide other services in the field of the use of atomic energy shall ensure that their work is performed and their services are provided in a quantity and to a level of quality which comply with the regulations and rules concerning the use of atomic energy, and shall be responsible for the quality of the work performed and the services provided throughout the design life of the nuclear facility, radiation source or storage facility or the manufacturing of equipment for it." The nuclear management authority may choose, in accordance with the established procedure, leading scientific organizations, leading engineering organizations, and leading design organizations out of the organizations specified in the first part of this article.

> Equipment, items and technologies for nuclear facilities, radiation sources or storage facilities must undergo conformity assessment in accordance with the laws of the Russian Federation."

When the organizations specified in the first and the second part of the Article, carrying out works and providing services in the field of the use of atomic energy for the operating organizations, cease their operations, the responsibility for all types of activity of these organizations is imposed on another organization, which is recognized by the respective nuclear management authority.

To support the implementation of the provisions of the Article, there are federal rules and regulations in the field of the use of atomic energy use 'Rules of conformity assessment of equipment, component materials and semifinished products supplied to nuclear facilities' that establish major requirements to the products supplied to nuclear facilities.

However, Rules for assessment of compliance of products, for which requirements related to safety in the field of the use of atomic energy, as well as requirements for processes of their design (including surveys), production, construction, installation, adjustment, operation, storage, shipment, sale, recovery, and final disposal - NP-071-18 [A–149])] do not apply to ships and other vessels with nuclear installations (Part 3 of NP-071-18: "These rules establish the requirements to the assessment of conformity of products used at nuclear power plants at the stages of their life cycle, facilities and complexes with research nuclear installations, critical and subcritical nuclear facilities. Ships and other vessels with nuclear installations do not belong to the specified nuclear facilities").

When the floating unit Akademik Lomonosov was being built, Rostechnadzor and Rosatom developed Decision No. 00-03-10/641 of April 25, 2014 "On the temporary procedure for assessment of conformity of equipment, components, materials and semi-finished products supplied to the nuclear-powered ships and floating structures and support facilities, with mandatory requirements" [A–150] that defines a temporary procedure for evaluation of conformity of equipment, components, materials and semi-finished products supplied to the nuclear-powered ships and floating structures and support facilities, with mandatory requirements.

According to this Decision [A–150], the activity on assessment of conformity of the items with the mandatory requirements to be used on nuclear powered ships and floating structures, as well as on the maintenance vessels is carried out by the Russian Maritime Register of Shipping in accordance with the applicable Rules and Guidelines of the Register (Paragraph 1 of the Decision).

Rostechnadzor assesses the compliance of the items supplied to the nuclear powered ships and floating structures and their maintenance facilities with the mandatory requirements established in the regulatory legal acts in the field of the use of atomic energy in the form of federal state control (supervision) in accordance with its powers under the Decree of the Government of the Russian Federation, No. 401 of July 30, 2004 'On Federal Environmental, Industrial and Nuclear Supervision Service' [A–151] (Paragraph 2 of the Decision).

Thus, in accordance with Decision No. 00-03-10/641 [A–150], the State Atomic Energy Corporation Rosatom, the Department for Shipbuilding Industry and Marine Equipment of the Ministry of Industry and Trade of the Russian Federation, and Rostechnadzor must consider Paragraphs 1 and 2 of the Decision when they develop and revise regulatory legal acts in the field of the use of atomic energy that define the rules of assessment of conformity of the products, for which the requirements related to safety in the field of the use of atomic energy are established, and also the processes of their design (including surveys), production, construction, installation, commissioning, operation, storage, transportation, sale, recovery, and disposal, in relation to nuclear powered ships, floating structures, and nuclear maintenance vessels (Paragraph 3 of the

Decision). Decision No. 00-03-10/641 became invalid when the regulatory legal acts specified in Paragraph 3 of the Decision came into force.

A–6.8.   SOUTH AFRICA–NNR

A–6.8.1. Question

Were there changes in the regulatory approach for asking information and inspections of suppliers?

A–6.8.2. Response

The regulatory process for the PBMR required that the licence applicant establish processes to ensure compliance with several requirements on Quality and Safety Management before design, manufacturing, testing and commissioning of safety important components can be initiated. As a result, suppliers of important to safety products had to comply with the relevant requirements.

To this effect, the NNR performed joint monitoring activities with the applicant and its designer and was involved as part of its assessment process in the following quality assurance (QA) related activities:

- Qualification of PBMR (Pty) Ltd as designer focusing on design and analysis methods, application of codes and standards, configuration management, etc.;
- Qualification of the applicant (Eskom Client Office) as operating organization;
- Qualification of PBMR suppliers.

Many NNR requirements for suppliers are stipulated in RD-034 [A–33]. For example:

- Activities delegated by the applicant / licensee to a designee and / or suppliers must be classified with respect to the importance of their products / services to nuclear safety to allow for the identification of the applicable requirements of this RD.

- For important to safety activities outsourced by the applicant / licensee or his designee or suppliers to other suppliers / sub-suppliers, the delegating organisation must implement adequate oversight measures to retain intelligent customer capability.

- The applicant/ licensee must submit to the NNR for acceptance a detailed description of the relevant organisational structure, the applied Management Systems including processes, supporting functions and related human resources for the respective stages the lifecycle of the nuclear installation. This description must include the designee(s) and the suppliers.

- Where there is collaboration between different organisations and contractors involved in the performance of design, manufacturing and/or construction, each organisation must define its responsibilities and tasks. The interfaces between the organisations must be clearly specified and described.

- The applicant / licensee, his designee and all suppliers of products important to nuclear safety must implement a QMS considering the requirements specified in ISO 9000:2000 Series including ISO 9001:2000 or an equivalent QM standard and the additional QM requirements specified in this RD and referred to as 'level 2' requirements. The QA measures in terms of the products form part of the organisation's QMS.

- If an organisation intends to introduce different standards for QM, a clear structure or framework must be provided in the QM manual to indicate the intended use of the standards as well as their compliance with the requirements of this RD.

- The applicant / licensee, the designee and the suppliers of safety important products must provide a set of documents describing their management system. This set must include a management system manual supported by additional documents describing the management policy, priorities and objectives.

- The structure of the organization and the internal and external interfaces must be described.

- The functional responsibilities, levels of authority and interactions of departments and persons responsible for managing, performing and assessing work must be adequately reflected in the documents.

- A description of the processes and supporting information to reflect how work is prepared, reviewed, carried out, recorded, assessed and improved must be provided.

- The documents must reflect the characteristics of the organization and its activities. The complexities of processes and the interaction of processes must be clear.

- The documents must define how process improvements will be achieved, giving special attention to personnel training and qualification.

The NNR must be informed of important organisational and/or process changes before implementation.

A–6.9. UNITED KINGDOM–ONR

A–6.9.1. Question

Were there changes in the regulatory approach for asking information and inspections of suppliers?

A–6.9.2. Response

There have been no specific changes in ONR's regulatory approach to asking for information from, and undertaking inspections of, suppliers to the nuclear industry, arising from ONR's fore-knowledge of SMR business models.

In line with UK law, ONR can ask for information from suppliers to the nuclear industry and powers are therefore given to ONR Inspectors to ask for this information for the purposes of nuclear safety. The law applies as much to suppliers to SMRs as to suppliers of the existing nuclear facilities, and suppliers to any new facilities currently being built. These arrangements include inspection of suppliers, irrespective of whether they supply existing facilities, new build or (potentially) SMRs.

ONR's regulatory approach to suppliers in the nuclear industry is detailed in TAG NS-TAST-GD-077 (Rev 5) Supply Chain Management Arrangements for the Procurement of Nuclear Safety Related Items or Services [A–152]. This Guide references the particular UK law that applies, specifically, the Energy Act [A–40] and the Health and Safety (Enforcing Authority) Regulations 1998 [A–153]. The latter identifies ONR as the enforcing authority for subsections 1, 2, 4 and 5 of section 6 of the Health and Safety at Work Act 1974 [A–40], but only in so far as those requirements relate to:

- Articles for use at work which are designed, manufactured, imported or supplied; or

- Substances which are manufactured, imported or supplied, where the articles or substances are to be used exclusively or primarily in the installation, operation or decommissioning of a GB nuclear site or authorised defence site.

Appendix A4 to the above TAG [A–152] explains how ONR will undertake this enforcement responsibility through inspection of suppliers to the GB nuclear industry.

NS-TAST-GD-077 [A–152] was revised in July 2019, in an exercise that involved a supply chain specialist inspector who was also active in the IAEA SMR Regulators' Forum and its MCCO working group. Whilst the content was wholly reviewed in knowledge of SMRs, the wording was changed to acknowledge SMRs in only one place:

> Paragraph 5.5.3: "Where the Licensee organisation is not in place or is not fully developed, some or all of the Licensee's responsibilities may be undertaken by the purchaser. This may be the case for procurements in support of new civil reactor build, where the reactor vendor may choose to place orders for long lead items, or for small modular reactors, in advance of an order from the future reactor operator/future Licensee. Where this is the case the ONR preferred approach is for the purchaser to organise the issue of a 'Licensee Certificate'."

The TAG remained the same because ONR judged that the regulatory approach to all new build activities is also applicable to SMRs. This is in line with the conclusion of the SMR Regulators' Forum MCCO working group, which said in its Interim Report [A–154] Executive Summary:

> "The common regulatory position is that the existing arrangements to regulate activities involving large nuclear power plants are also suitable arrangements to regulate activities involving SMRs, with some adjustments and balancing to take into account novel deployment approaches under the SMR business model."

The Interim Report discusses aspects of the SMR business model, such as modularity, and identifies aspects that all regulators should consider as SMRs move from concept to reality. At this stage, ONR envisages that its strategic regulatory approach (as set out in TAG 77 [A–152]) will remain in place, although our practical stance on some areas of detail may need to be adjusted in light of the SMR business model as it evolves. For example, ONR may need to increase its inspections at suppliers' factories and reduce its activities at SMR construction sites, when compared to current ONR inspection activity at a conventional new build nuclear power plant.

A–6.10. UNITED STATES OF AMERICA–NRC

A–6.10.1. Question

Were there changes in the regulatory approach for asking information and inspections of suppliers?

A–6.10.2. Response

Yes, the current DC application differs from those received in the past and special circumstances resulted in the use of vendors or suppliers with specialized capabilities, some of whom may not be well known to the regulator. While regulations allow applicants to use these alternate vendors or suppliers, the NRC must assess the vendors' or suppliers' technical and financial qualifications. To determine such qualifications the NRC has dispatched multi-disciplinary teams of technical specialist to conduct audits and find answers to questions:

- Has the applicant adequately assessed the ability of vendor or supplier (including their subcontractors) to provide the information that must be reconstituted?

- Is there reasonable assurance that the process employed by the applicant was adequate to identify all information that must be reconstituted?

- Do we have a reasonable assurance that the vendors or suppliers and their subcontractors will be able to assume the duties assigned? Do they have the expertise and technical competence to manage and control design changes and support the licensing process?

- Do the vendors or suppliers have adequate financial resources to provide the services required for the duration of the project?

This Annex presents the responses provided by the following Member State regulatory bodies to the Question 7: "Were there changes to safety objectives and numerical targets (e.g. dose limits, environmental release limits, PSA goals) introduced or needed to address proposed SMRs? For example, introducing multi-unit facilities. Describe implication of multi-unit facilities, multiple facilities or multiple licensees on the same site with respect to the interpretation of the numerical targets."

A–7.1. ARGENTINA–ARN

A–7.1.1. Question

Were there changes to safety objectives and numerical targets (e.g. dose limits, environmental release limits, PSA goals) introduced or needed to address proposed SMRs? For example, introducing multi-unit facilities. Describe implication of multi-unit facilities, multiple facilities or multiple licensees on the same site with respect to the interpretation of the numerical targets.

A–7.1.2. Response

Not applicable.

A–7.1.3. Follow-up Question

In the Atucha site there are two NPPs in operation and the CAREM prototype in construction. Are there any safety consideration about these multi-unit facilities?

A–7.1.4. Response

In Argentina, a NPP construction must not be initiated without a previous Construction License issued by the Regulatory Body, upon request from the Licensee.

In line with this approach, at the time of applying for the Construction License, the Licensee must submit to the Regulatory Body all the documentation required to evaluate the radiological and nuclear safety of the installation to be built, including the site characteristics in relation to:

- Natural and man-induced external events that could affect the installation safety;
- Dispersion of radionuclides to the environment, both in normal and accidental conditions.

ARN issues the Construction license once the Licensee has demonstrated that the design of the NPP to be built complies with the regulatory standards and other specific regulatory requirements for the selected site, taking into account the NPP-site interaction.

Besides, the Regulatory Standard AR 10.10.1. 'Site Evaluation for Nuclear Power Plants' [A–155], take into account the lessons learned from the Fukushima Daiichi NPP accident and the corresponding IAEA standards.

The CAREM 25 prototype reactor is under construction in the Atucha site, next to CNA I and CNA II. The main requirements for the site selection are related to the protection of the public and the environment from the radiological consequences due to accidents and their mitigation in case they should occur.

The site selected includes the necessary infrastructure and effective security measures with an established response force which has proven its competence to handle emergencies effectively during annual drills.

The site is suitable for building the CAREM as demonstrated during the many studies carried out during the CNA I Project. These studies undertaken by NA-SA (Nucleoeléctrica Argentina S.A.), duly extrapolated and updated were made available to CNEA responsible for the design, construction and operation of the CAREM prototype reactor. These studies include, inter alia, external hazards, population density and distribution, and NPP lay out.

These studies were complemented by others related to the CAREM 25 Project and its specific location, such as the geological studies and the impact of CAREM 25 on CNA I and CNA II and vice versa.

Agreements have been signed to ensure co-operation and feedback of experience between CNEA and NA-SA in connection with radiological and nuclear safety, physical protection, security, exchange of technical information including environmental monitoring data and the balance of plant design.

Regarding the authorized limits of the environmental releases of the Argentinian NPPs under operation, that were set by the ARN for relevant radionuclides. The radiological protection criteria used by ARN to control the dose received by workers are consistent with the latest ICRP recommendations.

For the design purposes of every facility, the Regulatory Body has established a constraint of 0.3 mSv for the annual effective dose of the representative person, due to the release of liquid and gaseous radioactive effluents (AR 3.1.2 [A–156], AR 4.1.2 [A–157] and AR 6.1.2 [A–158] standards).

In addition, since June 2013, the ARN has established that in the case of the design of a nuclear power reactor, a research reactor or a Type I radioactive facility within a site with multiple facilities, enough retention against the release of radioactive effluents should be considered, so that the annual dose value in the representative person does not exceeds 0.5 mSv, taking into account the release of radioactive effluents of all facilities included in the site (ARN 191/2013 Resolution [A–159]). This is expressed, particularly for nuclear power reactors, in AR 10.10.1 [A–155].

It refers only to restriction in the public, in accordance with requirement 31 of the Radiation Protection and Safety of Radiation Sources: International Basic Safety Standards, GSR Part 3 [A–160].

The provisions of the aforementioned Resolution [A–159] will be incorporated into the regulations on radioactive discharges to the environment in the next review carried out.

About the regulatory control used by the ARN on this subject, the quarterly releases presented by the facilities are evaluated (in compliance with their respective Operating Licenses) and the determinations of the downloaded activities carried out by each facility are verified (procedures, methodology of measurement, equipment calibration, etc.).

194

A–7.2. CANADA–CNSC

A–7.2.1. Question

Were there changes to safety objectives and numerical targets (e.g. dose limits, environmental release limits, PSA goals) introduced or needed to address proposed SMRs? For example, introducing multi-unit facilities. Describe implication of multi-unit facilities, multiple facilities or multiple licensees on the same site with respect to the interpretation of the numerical targets.

A–7.2.2. Response

The regulatory experience for existing multiple-unit facilities in Canada should be applicable to multi-module SMR deployments.

A–7.2.2.1. Introduction / How do we regulate multiple-unit facilities in Canada

The Canadian Nuclear Safety Commission (CNSC) has decades of regulatory experience with multiple-unit facilities. This includes the facilities listed in Table A–5.

TABLE A–5. MULTIPLE-UNIT FACILITIES - CANADA

NPP	Description (# of Units, power)	In-service since
Pickering NGS A	4 units - 515 MW(e)	1971–1973
Pickering NGS B	4 units - 516 MW(e)	1983–1986
Bruce NGS A	4 units - 750 MW(e)	1977–1979
Bruce NGS B	4 units – 817 MW(e)	1985–1987
Darlington NGS	4 units - 881 MW(e)	1992–1993

There are architectural features that are common to all these stations. For example, these stations feature a vacuum building for common containment and a common services building which includes a central fuel handling area for on-line refuelling and used fuel pools. See Fig. A–6.

The Canadian licensing model for these facilities is one facility, one licence. A licence is issued for all activities concerning a reactor facility regardless of the number of units. If differences exist between units, they are reflected in the licensee's licensing basis documents, such as design manuals, operating manuals, etc. The CNSC can impose operating restrictions on individual units.

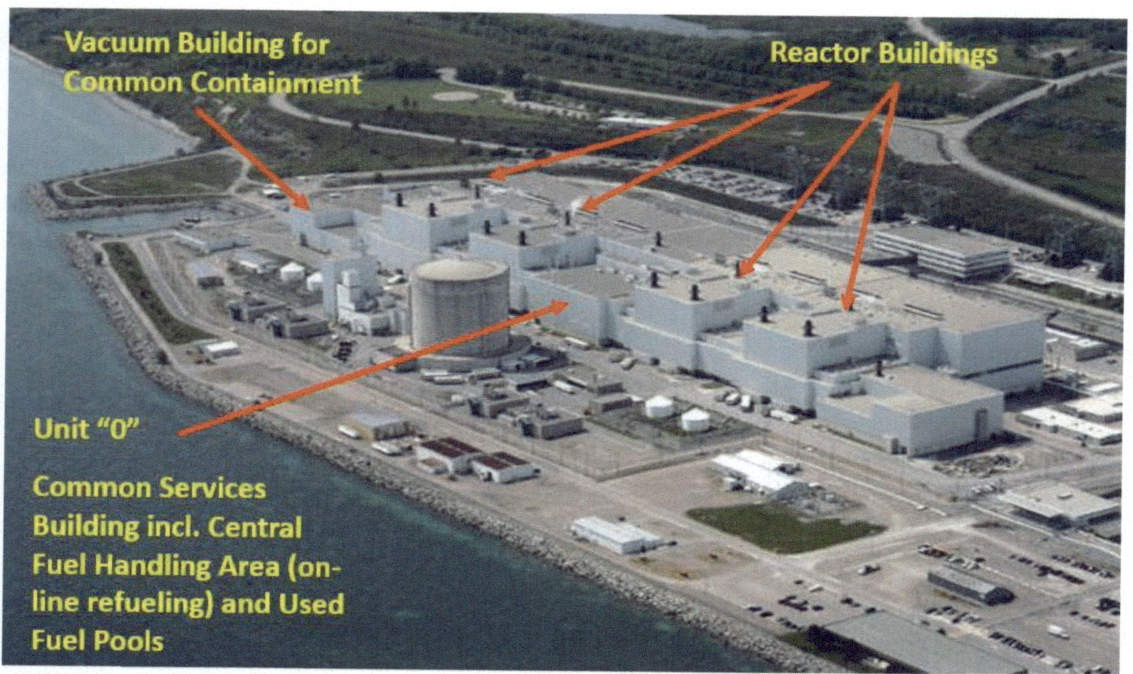

FIG. A–6. Image of Darlington Nuclear Generation Station labelled with architecture features common to multiple-unit facilities in Canada

A–7.2.2.2. Shared SSCs in multiple-unit facilities

Shared system, structure and component (SSC) features are designed to supplement unit specific DiD. Some of these features include:

- One main control room (MCR) with dedicated space allocated for each operating unit panels, including Unit 0 and fuel handling;
- Common containment, including one vacuum building;
- Common emergency coolant injection system (ECI) functions;
- Emergency power system to supplement unit-specific electrical supply architecture;
- Emergency service water to supplement unit specific water cooling systems.

A–7.2.2.3. Role of unit 0

In multiple unit facilities, Unit 0 SSCs play various roles:

- Supply common station needs (power, water, lighting, compressed air);
- Infrastructure for station-wide fuel handling systems;
- Ensure common station safety mitigation outside the dedicated unit systems, for example:
    - Used fuel pools (bays) environmental support systems;
    - Maintain station-wide confinement/containment envelope;
    - Fire protection and response.
- Provide supplemental support to each unit during all plant states including transients and accidents;
    - Executes and maintains containment 'button-up';
    - Central pressure relief and dousing on a unit pressure excursion event;
    - Emergency coolant injection supply function.

- For example, station containment system (see Fig. A–7) serves multiple functions, such as:
  - o Allows fuelling machines to move between units and Unit 0 Common Services Area (new fuel rooms and irradiated fuel bays);
  - o Very large containment volume reduces effects of pressure excursions;
  - o Common emergency coolant recovery.

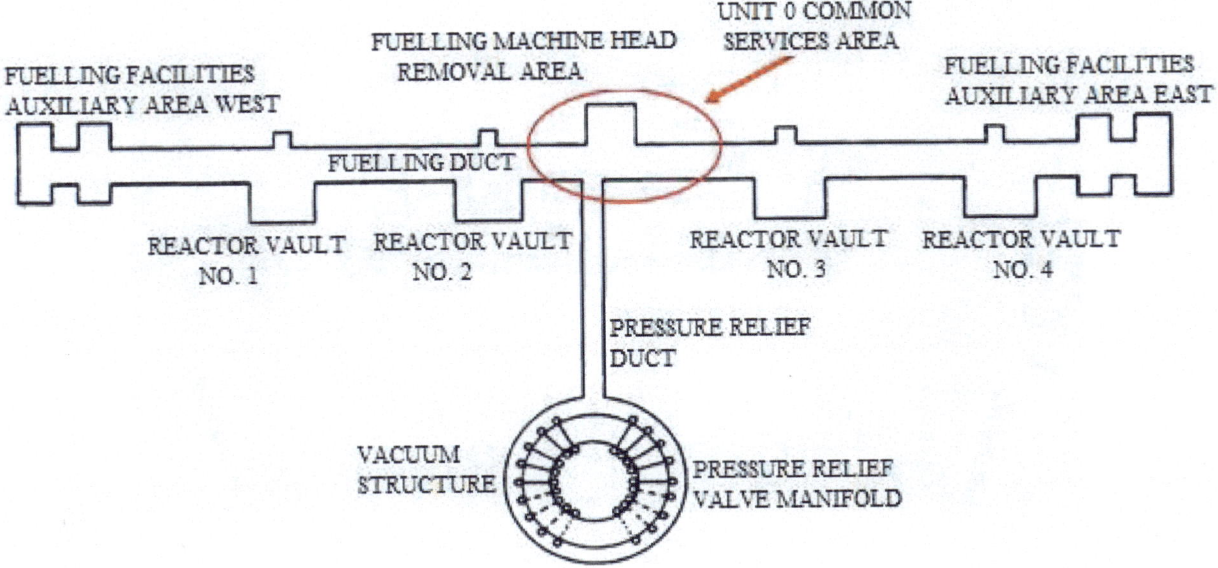

FIG. A–7: Diagram of Darlington Nuclear Generating Station containment system.

A–7.2.2.4. Unit 0 control room description

The Unit 0 control room fulfils many functions (see Fig. A–8), such as:

- Oversight of common station system operations, including:
  - o Station-wide containment systems (including access control);
  - o Common power supply systems and emergency power transfer;
  - o Station heavy water inventory quality assurance (upgrading, cleaning, etc.).
- Coordination of station wide emergency plan execution (e.g. fire response, providing emergency power):
  - o Support each unit during transients and accidents with Unit 0 systems.

There is significant use of (digital) automation in the MCR. Here, very little equipment is operated manually. The MCR is designed to give panel operators time to focus on big-picture situational awareness to always understand where they are inside the operating envelope. Operators are notified when automatic actions occur or do not occur within prescribed operating limits. Any manual operation or actions are driven by procedures based on diagnostic information such as alarm manuals and abnormal incident manuals (AIM). Operator responses to changing conditions are rehearsed and inter-unit communication is addressed in procedures and training.

For the conduct of control room operation, clear rules are in place to ensure safety in multiple configurations. Consideration must be taken that each unit can be in a different operating state, for example:

- Operating normally at full power;
- Maneuvering from one state to another;
- Shut down in guaranteed shutdown state (GSS) – outage;
- Unit transient or event.

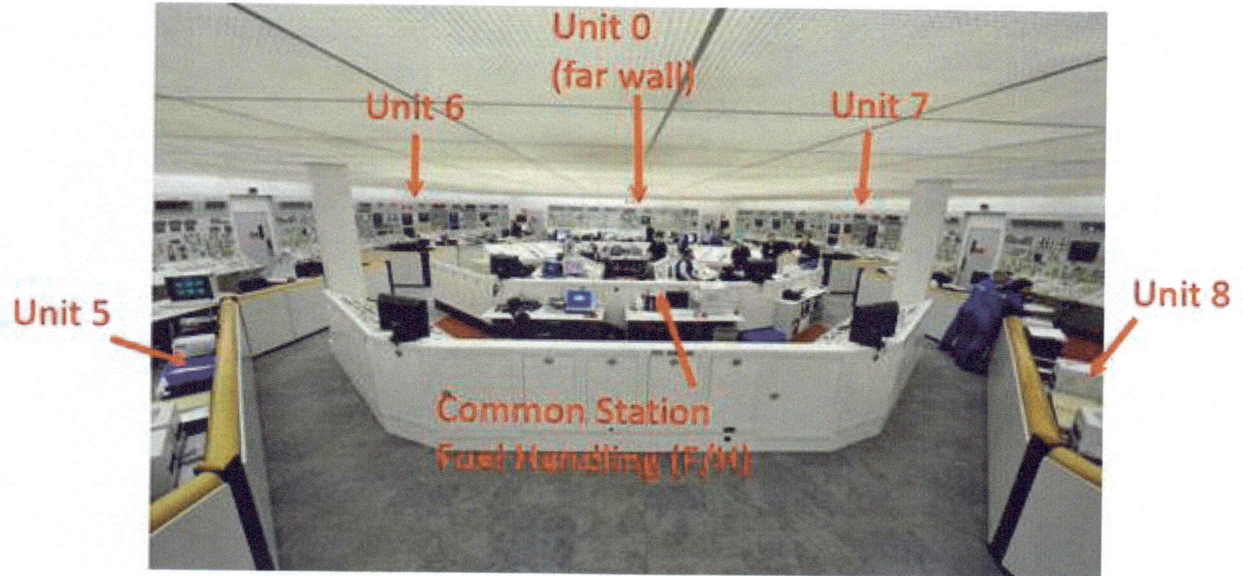

FIG. A–8: Labelled image of Bruce B MCR.

A–7.2.2.5. Minimum staff complement in multiple-unit facilities

Minimum staff complement applies to the entire facility. This topic is further discussed in REGDOC-2.2.5, Minimum Staff Complement [A–161]. Overall:

- CNSC expects a licensee to maintain a systematic analysis to determine the basis of the minimum staff complement while considering:
  - The most resource-intensive initiating events and credible failures considered in the Safety Analysis and the PSA;
  - Required actions;
  - Operating strategies;
  - Required interactions among personnel;
  - Staffing demands associated to the required tasks;
  - Staffing strategies under all operating conditions including normal operation, AOO, DBA and emergency conditions.
- Validation to show safe operation and response to the most resource-intensive conditions (including events that affect more than unit) under all operating states including normal operations, AOO, DBA and emergency conditions.

With a shared MCR, minimum shift complement is typically achieved by the following:

- Each unit has:
  - Dedicated ANO(s) and panel operators in MCR;
  - Field operators are also assigned (dedicated) to each unit;
  - Switching staff between units can be done but is not a regular practice (day-to-day configurations between units may be different).

- Under transient conditions, additional ANO and an ANO from a stable adjacent unit can provide support but dedicated ANO leads;
- Unit 0 has two Certified Control Room Operators (CRO) plus dedicated Field Operators;
- Fuel Handling has dedicated Panel and Field Operators;
- Control Room Shift Supervisor (CRSS) and Shift Manager (SM) are also licensed staff:
  - CRSS Coordinates MCR activities and is available to assist on incident unit upon request by the Unit ANO;
  - SM is responsible to maintain global oversight of station activities and leads execution of the emergency plan.
- Provisions, with shifting responsibilities and priorities, are also made for multiple unit transients:
  - Done through procedures and regular training exercises.

A–7.2.2.6. Probabilistic safety assessment for multiple-unit facilities

In multi-unit facilities, PSA has to reflect the station design, not just the unit design. This includes the following considerations:

- How do the units interact with each other in different station states?
- How are shared/common systems divided up in the PSA to reflect individual unit safety?
- Site based PSA versus unit based PSA?
  - External events (human-induced or natural);
  - Common-cause failures;
  - What is the modelled release size and inventory?

A–7.2.2.7. Overall considerations

The following lessons Learned were observed in multi-unit facilities in Canada over the years:

- Multiple unit facilities require a particularly strong configuration management program, since every unit has some differences from others and the impacts of this must be understood for the whole facility;
- Be aware of the unit you are in when performing operations or maintenance! 'Wrong unit' work represents significant operational risk.
  - Units are colour-coded and equipment is tagged by unit;
  - Field/Control room communication always confirms correct unit during evolutions;
  - Procedure-use-and-adherence requires a check that the correct procedure is being used for the right unit (configuration can vary between units);
  - Where does human error fit into Safety Analysis?

A–7.2.3. Follow-up Question

Need to state there were no changes to objectives and targets, RD-367 [A–61] states objectives and goals apply to the facility?

A–7.2.4. Response

There were no changes to the safety objectives and numerical targets to address proposes SMRs. RD-367 [A–61] sets out the requirements for the design of SMRs that is defined as "(…) a reactor facility containing a reactor with a power level of less than approximately 200 megawatts thermal

(MWt) that is used for research, isotope production, steam generation, electricity production or other applications".

This document is technology neutral and includes:

- Safety goals and objectives for the design;
- Safety concepts in the design;
- Safety management principles;
- Design of structures, systems and components;
- Safety, security and engineering aspects of the reactor facility features and layout;
- Integration of safety assessments and the design process.

When an applicant proposes to construct more than one reactor on a site, the design of the multi-reactor site shall meet the safety objectives in this regulatory document. The design of each reactor facility shall also satisfy the safety and design requirements in this document. In addition, the applicant shall ensure that the impact on the safety of all reactors on the site due to interactions between reactors; common-cause failure events; and any sharing of structures, systems and components (SSCs) between reactors is assessed for normal operation, anticipated operational occurrences (AOOs) and accident conditions.

A–7.3. CHINA–NNSA

A–7.3.1. Question

Were there changes to safety objectives and numerical targets (e.g. dose limits, environmental release limits, PSA goals) introduced or needed to address proposed SMRs? For example, introducing multi-unit facilities. Describe implication of multi-unit facilities, multiple facilities or multiple licensees on the same site with respect to the interpretation of the numerical targets.

A–7.3.2. Response

The regulation HAF102 [A–68] stated safety objective of nuclear facilities in section 2.1 "The fundamental safety objectives is to establish and maintain an effective defence against radioactive hazards, so as to protect people, society and the environment from hazards", this is suitable for all nuclear facilities.

The Nuclear Safety Review Principles for Small PWR Nuclear Power Plants [A–70] also announced "The basic safety goal of SMR is to provide higher protection level without off-site protect actions to public than that of large LWR nuclear power plant with off-site protect actions."

In additional facilities also have PSA goals (i.e. core damage frequency and large early release frequency) which are currently set as per reactor. Site level goals covering multi-unit facilities haven't been formally agreed throughout the country and are still under discussion.

Detailed numerical acceptance criteria have been established by NNSA in the case of planned exposures, including mandatory dose limits for occupational and public exposures. The dose limits and environmental release limits under normal operation condition are the requirements for NPP site, and the total release amount of multi-unit in one site should meet these requirements.

In the condition of accidents, the acceptance criteria for radiation dose is more strict for SMRs than current PWRs. For example, in Section 4 of The Nuclear Safety Review Principles for Small PWR Nuclear Power Plants [A–70], it stated "Acceptance criteria of postulated accident
200

radioactive consequence: individual effective dose on-site boundary shall be controlled below 5 mSv or 10 mSv during whole accident (generally 30 days) and the equivalent dose of thyroid below 50 mSv and 100 mSv when an infrequent faults or a limiting faults occurs. Individual effective dose on-site boundary in BDBA shall be controlled below 10 mSv during the whole accident (generally 30 days)".

A–7.4. CZECH REPUBLIC–SÚJB

A–7.4.1. Question

Were there changes to safety objectives and numerical targets (e.g. dose limits, environmental release limits, PSA goals) introduced or needed to address proposed SMRs? For example, introducing multi-unit facilities. Describe implication of multi-unit facilities, multiple facilities or multiple licensees on the same site with respect to the interpretation of the numerical targets.

A–7.4.2. Response

SÚJB has experience with existing multiple unit facilities on the same site (Dukovany power plant has four VVER-440 and Temelín power plant two VVER-1000). Each unit is licenced as a separate nuclear installation in line with the definition of nuclear installation (facility or plant comprising a nuclear reactor). If an SMR multi-unit facility would be deployed, it would be done in the same way unless the legislation is modified.

In accordance with the IAEA safety standards, each unit is expected to have its own safety systems and the legislation does not provide for sharing of such systems. At the same time, when ensuring compliance with the principles for the safe use of nuclear energy (basic safety functions) for external design events and scenarios which, due to their frequency of incidence and severity, fall within the scope of design extension conditions, nuclear installation design shall, inter alia, take into account the impact of the external design event on failures of multiple nuclear installations on the same site and ensure sufficient capacity and means for managing accident conditions and radiation accidents caused by external design events on the site with multiple nuclear installations expected to share support equipment and services.

According to the Atomic Act [A–21] and the Decree 162/2017 [A–162], on the requirements for safety assessment, a PSA model for each nuclear installation shall be created and PSA assessment shall be performed (on a regular and ad-hoc basis). Results are to be compared to probabilistic acceptance criteria. The interdependencies between multiple nuclear installations that are located on the same site shall be reflected in probabilistic safety assessment model.

When radiation extraordinary event analysis and assessment are prepared (to determine the emergency planning zone), the possibility of simultaneous occurrence of a radiation extraordinary event on two and more nuclear reactors located on the nuclear installation grounds shall be taken into account (see the answer from Czech Republic for Question 24).

A–7.5. FRANCE–ASN

A–7.5.1. Question

Were there changes to safety objectives and numerical targets (e.g. dose limits, environmental release limits, PSA goals) introduced or needed to address proposed SMRs? For example, introducing multi-unit facilities. Describe implication of multi-unit facilities, multiple facilities or multiple licensees on the same site with respect to the interpretation of the numerical targets.

A–7.5.2. Response

The Environmental Code [A–24], especially the articles L.593-4, L.593-6 and following, in addition to the Order [A–72], require that the basic nuclear installations are designed, located, built, commissioned, operated and dismantled with the objective to prevent accidents, and, if an accident occurs, to limit its consequences. These requirements are applicable to any basic nuclear installation, including SMRs.

Even though there is no SMR in France, ASN has regulatory experience for existing multiple-unit facilities. Indeed, for example, Gravelines nuclear power plant is composed of six nuclear reactors. These reactors can share some SSCs (see the answer from France to Question 16 for more information).

For pressurized water reactor designs, ASN published a guide that provides recommendations about the main safety objectives. These recommendations consider WENRA's safety objectives for new power plants, IAEA's SSR-2/1 [A–127] and the Directive 2009/71/EURATOM [A–163] establishing a Community framework for the nuclear safety of nuclear installations and its amendment, Directive 2014/87/Euratom [A–164].

For example, one objective shall be to limit radioactive or dangerous discharges, or dangerous effects for people or environment induced by incidents and accidents at a level as low as reasonably achievable in economically acceptable conditions. Also, another objective shall be to prevent nuclear incidents and accidents and to limit consequences of those which could occur despite the preventive provisions.

In particular:

- Core meltdown accidents likely to lead to massive radioactive discharges with a kinetic that would not enable the implementation of the necessary measures to protect the population must be physically impossible, or failing that, extremely unlikely with a high confidence level;
- The necessary measures to protect populations in case of core meltdown accidents must be limited in terms of scope and duration (no permanent rehousing, no emergency evacuation far from the installation, no long-term restriction regarding food consumption at the site neighbourhood's exterior).

These safety objectives shall be applicable for every pressurized water reactor, regardless its power. These safety objectives don't incorporate numerical exposure limits. Through the safety demonstration, the licensee must define criteria to comply with these objectives and to demonstrate that they are met. They are subject to continuous improvement.

Considering that France has mainly experienced pressurized water reactors, the need for a guide regarding the design of this type of reactors has been identified. However, the other nuclear installations are in general unique in France. ASN provides its opinion about safety options and objectives before the creation authorization. If necessary, as explained by France in the answer to Question 2, additional requirements can be enacted for specific designs.

Regarding public and worker exposure, Public Health Code [A–165] and Labour Code [A–166] establish principles and numerical dose limits.

The article R.1333-8 of the Public Health Code [A–165] limits the annual exposure for the population from nuclear activities. An exceeding exposure shows an unacceptable situation.

The article R.4451-6-8 of the Labour Code [A–166] fix a one-year rolling exposure limits for workers. Exceptional and limited derogations can be provided for justified activities.

These objectives shall be applicable to SMRs.

A–7.6. JAPAN–NRA

A–7.6.1. Question

Were there changes to safety objectives and numerical targets (e.g. dose limits, environmental release limits, PSA goals) introduced or needed to address proposed SMRs? For example, introducing multi-unit facilities. Describe implication of multi-unit facilities, multiple facilities or multiple licensees on the same site with respect to the interpretation of the numerical targets.

A–7.6.2. Response

Dose target value for research reactor facilities is the same with it for commercial reactors, and the target value is 50 μSV/year under normal operation as a total of nuclear facilities at outside of the environmental monitoring.

In response to an accident, all necessary measures are taken at each nuclear facility, and the effective dose value to public for a nuclear facility at the accident is set not to exceed 5 mSv.

In Japan, probabilistic risk assessment (PRA) is not required because the PRA methodology for research reactor facilities is not considered sufficiently mature.

A–7.6.3. Follow-up Question

Clarify if it means for all reactors on a site, or per unit.

Please provide references of the regulatory standards and / or guidelines.

A–7.6.4. Response

In the regulatory standards, both the radioactivity concentration in the air at outside of environmental monitoring area and in the water at the boundary of the said area, are required to be sufficiently reduced.

Dose objective is set to 50 μSv/year as a reference of 'Guide for Dose Target Value for commercial light water reactors' [A–167].

On reviewing HTTR, it is confirmed that the total value does not exceed 50 μSV/year under the condition of all research reactor facilities located in Oarai-Site (JMTR, HTTR, Joyo and DCA) are in normal operation.

On the other hand, target value of 5 mSv at the accident is for one reactor facility (5 mSv per unit) — as in the previous response.

In the previous response, we mainly provided countermeasures for external hazard — which enhanced in the new regulatory standards - as the examples of common cause failure.

In the actual review, other PIEs within facility are taken into account — such as:

- Random failure of equipment which compose safety function;
- Internal fires;
- Internal floods.

Basic approach on reviewing the design basis event (DBE) which caused by postulated initiating event (PIE) are stipulated in the 'Guide for the safety review of water cooled research and test reactors (July 18th, 1991, specified by the Nuclear Safety Commission)' [A–168].

HTTR which is gas cooled also applies this concept.

PIEs assume the events mainly failure, damage of equipment or system in prevention system (PS), or mis-operation of relevant equipment by human errors of the operator.

On the analysis of DBEs, other than analysis for PIEs, it is required to assume single failure of equipment that cause most severe impact to each system which mainly compose reactor shut down function, core cooling function or confinement function.

Additionally, on the review for safety design, it is required that the design for the system that has significant importance on safety function should fulfil its functionality under the condition of single failure of component in the system within external power failure condition.

A–7.7. RUSSIAN FEDERATION–ROSTECHNADZOR

A–7.7.1. Question

Were there changes to safety objectives and numerical targets (e.g. dose limits, environmental release limits, PSA goals) introduced or needed to address proposed SMRs? For example, introducing multi-unit facilities. Describe implication of multi-unit facilities, multiple facilities or multiple licensees on the same site with respect to the interpretation of the numerical targets.

A–7.7.2. Response

The floating NPP (Akademik Lomonosov), currently in operation, comprises two KLT-40C reactors.

Paragraph 22 of NP-022-17 [A–89] states that safety targets for the nuclear power installations of the ship are:

- The cumulative severe accident probability for each reactor is not more than 10^{-5} within the period of one year;
- The cumulative large-scale emergency release probability for each reactor is not more than 10^{-7} within the period of one year.

Paragraph 98 of NP-022-17 [A–89]: 'Localizing safety systems must be provided for each reactor and perform the assigned functions in case of design-basis accidents as well as beyond-design-basis accidents. Combined usage of individual components of the localizing safety systems for several reactors is permitted in general provided that prevention of any impact of accidents at one reactor on the other reactor is substantiated in the ship design.'

Paragraph 109 of NP-022-17 [A–89]: 'When several reactors are installed on the ship, the power supply systems must be designed and calculated with due regard for the possibility to arrange backup and emergency power supply for each reactor.'

In the Russian Federation, the main regulatory documents regulating radiation safety of personnel and the public are NRB-99/2009 'Radiation Safety Standards' [A–169] and OSPORB-99/2010 'Basic Rules for Provision of Radiological Safety' [A–170]. These documents are generally based on the IAEA GSR Part 3 [A–160]. According to these documents, radiation exposure to the public and personnel should be limited to the same levels as for the units of large-capacity nuclear power plants or other nuclear installations. In particular, due to the fact that two reactors are located in close proximity, the limit values of emissions and discharges will be set for two reactors simultaneously.

A–7.8. SOUTH AFRICA–NNR

A–7.8.1. Question

Were there changes to safety objectives and numerical targets (e.g. dose limits, environmental release limits, PSA goals) introduced or needed to address proposed SMRs? For example, introducing multi-unit facilities. Describe implication of multi-unit facilities, multiple facilities or multiple licensees on the same site with respect to the interpretation of the numerical targets.

A–7.8.2. Response

The PBMR project did not reach the stage where this topic was developed as all attention was focused on the envisaged first PBMR demonstration power plant.

Notwithstanding, the NNR fundamental safety criteria of dose and risk considers all facilities that may contribute to the dose and risk to the public. These are applicable to any activity or facility posing a potential nuclear risk.

A–7.8.3. Follow-up Question

Could it be stated there were no changes to objectives and targets?

A–7.8.4. Response

The principles that must be met to ensure safety in any nuclear installation are presented in the 'Regulations on Safety Standards and Regulatory Practices (SSRP)' published as Regulation R388 [A–31]. For the PBMR, the NNR requirements document RD-0018 [A–107] 'Basic Licensing Requirements for the Pebble Bed Modular Reactor', was developed and is based on and expands the SSRP principles and requirements.

Section 6 of RD-0018 sets down the principal radiation protection and nuclear safety requirements as formulated in Section 3 of the SSRP, for their application to the PBMR.

Section 7 of RD-0018 defines the Basic Licensing Requirements (BLR) for the PBMR (based on the principal safety requirements) that inter alia include the Dose and Risk limits applicable to the PBMR.

Section 8 of RD-0018 specifies the processes which the applicant/licensee and the constructor must undertake to demonstrate compliance with the BLR.

Although all the relevant requirements formulated in the SSRP [A–31] are applicable to the design, construction, operation and subsequent decommissioning of the PBMR, some specific Principal Safety Requirements of Section 3 of the SSRP are elaborated further in RD-0018 [A–107] in terms of their particular application to the PBMR. However, there were no changes to safety objectives and numerical targets.

And, as far as implications of multi-unit facilities, multiple facilities or multiple licensees on the same site with respect to the interpretation of the numerical targets are concerned, the NNR requirements make provision for such circumstances by statements such as:

In Section 2 of Annex 2 of the SSRP:

"The annual effective dose limit for visitors to the sites and those not deemed to be occupationally exposed is 1 mSv." [A–31]

In Section 2 of Annex 2 of the SSRP:

"The annual effective dose limit for members of the public from all authorised actions is 1 mSv." [A–31]

In Annex 3 of the SSRP [A–31], the probabilistic risk limits for the public and the workers are specified as criteria per site.

Therefore, in the event that there are multi-unit facilities, multiple facilities or multiple licensees on the same site, their dose and risk contributions must be added up where applicable when comparing to the dose and risk criteria, which are independent of the number of units or facilities on the site.

A–7.9. UNITED KINGDOM–ONR

A–7.9.1. Question

Were there changes to safety objectives and numerical targets (e.g. dose limits, environmental release limits, PSA goals) introduced or needed to address proposed SMRs? For example, introducing multi-unit facilities. Describe implication of multi-unit facilities, multiple facilities or multiple licensees on the same site with respect to the interpretation of the numerical targets.

A–7.9.2. Response

No changes are proposed to safety goals or numerical targets to specifically address SMRs. ONR Safety objectives and Numeric Targets are documented in the SAPs [A–44]. SAP FA.7, for example expects that analysis should demonstrate, so far as is reasonably practicable, that the correct performance of the claimed passive and active safety systems ensures that: (a) none of the physical barriers to prevent the escape or relocation of a significant quantity of radioactive material is breached or, if any are, then at least one barrier remains intact and without a threat to its integrity; (b) there is no release of radioactivity; and (c) no person receives a significant dose of radiation if the criteria (a) to (c) cannot be fully met within the design, SAP FA.7 nevertheless seeks minimal consequences. This is reflected in Numerical Target 4 which defines the Basic Safety Objectives and the Basic Safety Level for the mitigated radiological consequences of design basis fault sequences — see Table A–6.

TABLE A–6. DESIGN BASIS FAULT SEQUENCES (FROM SAP FA.7 – TARGET 4, UK) [A–44]

Design basis fault sequences – any person	**Target 4**
The targets for the effective dose received by any person arising from a design basis fault sequence are:	
On-site:	
BSL: 20 mSv for initiating fault frequencies exceeding 1×10^{-3} pa	
200 mSv for initiating fault frequencies between 1×10^{-3} and 1×10^{-4} pa	
500 mSv for initiating fault frequencies between 1×10^{-4} and 1×10^{-5} pa	
BSO: 0.1 mSv	
Off-site:	
BSL: 1 mSv for initiating fault frequencies exceeding 1×10^{-3} pa	
10 mSv for initiating fault frequencies between 1×10^{-3} and 1×10^{-4} pa	
100 mSv for initiating fault frequencies between 1×10^{-4} and 1×10^{-5} pa	
BSO: 0.01 mSv	

Three examples of numerical risk targets are presented in Tables A–7, A–8 and A–9, and then discussed in the context of multi-unit facilities.

TABLE A–7. INDIVIDUAL RISK TO PEOPLE OFF-SITE FROM ACCIDENTS (FROM SAP FA.7 – TARGET 7, UK) [A–44]

Individual risk to people off the site from accidents	**Target 7**
The targets for the individual risk of death to person off the site, from accidents at the site resulting in exposure to ionising radiation, are:	
BSL: 1×10^{-4} pa	
BSO: 1×10^{-6} pa	

TABLE A–8. TOTAL RISK OF 100 OR MORE FATALITIES (FROM SAP FA.7 – TARGET 9, UK) [A–44]

Total risk of 100 or more fatalities	Target 9
The targets for the total risk of 100 or more fatalities, either immediate or eventual, from accidents at the site resulting in exposure to ionising radiation, are:	
BSL: 1×10^{-5} pa	
BSO: 1×10^{-7} pa	

TABLE A–9. FREQUENCY DOSE TARGETS FOR ACCIDENTS ON AN INDIVIDUAL FACILITY – ANY PERSON OFF THE SITE (FROM SAP FA.7 – TARGET 8, UK) [A–44]

Frequency dose targets for accidents on an individual facility – any person off the site	Target 8

The targets for the total predicted frequencies of accidents on an individual facility, which could give doses toa person off the site are:

Effective dose, mSv	Total predicted frequency per annum	
	BSL	BSO
0.1-1	1	1×10^{-2}
1-10	1×10^{-1}	1×10^{-3}
10-100	1×10^{-2}	1×10^{-4}
100-1000	1×10^{-3}	1×10^{-5}
>1000	1×10^{-4}	1×10^{-6}

As can be seen from the above, the scope of the example targets relates to either the site or the facility. Therefore, multiple reactors on the same site or within the same facility are considered within the extant numerical targets.

Assessment of a new build nuclear power plant in the UK against the numerical targets should consider the results of a Level 1, 2 and 3 Probabilistic Safety Analysis (PSA). Traditionally, PSA in the UK has been conducted on a single unit basis, with a factor applied to scale the results to the number of units on a site if required. For SMRs, the PSA may need to explicitly consider multiple units or modules. The precise content and scope of the PSA are likely to be dependent on specific design details, such as interactions or dependencies between units or modules.

Consideration of multi-unit PSA has been an on-going topic of research for ONR. ONR has contributed to the OECD/NEA WGRISK task on-site level PSA [A–171] and the development of an IAEA Safety Report on multi-unit PSA [A–172]. ONR has also conducted research on the effects of multiple releases from multiple units on the same site on Level 3 PSA consequences [A–173]. These activities have been considered in the recent update to the PSA TAG [A–174] providing guidance to ONR inspectors when conducting regulatory assessment of PSA. Further guidance may be developed as specific SMR designs are considered under GDA by ONR.

A–7.10. UNITED STATES OF AMERICA–NRC

A–7.10.1. Question

Were there changes to safety objectives and numerical targets (e.g. dose limits, environmental release limits, PSA goals) introduced or needed to address proposed SMRs? For example, introducing multi-unit facilities. Describe implication of multi-unit facilities, multiple facilities or multiple licensees on the same site with respect to the interpretation of the numerical targets.

A–7.10.2. Response

No, the same regulations that apply to large light water reactors (LLWRs) will apply to SMRs. The introduction of smaller multi-unit facilities requires the NRC to evaluate the safety objectives that were previously used. A small modular reactor module can generate over 50 MW of electricity using a smaller, scalable version of pressurized water reactor technology. A plant can house multiple SMRs for a total output in excess of 700 megawatts. The NRC licenses and regulates the Nation's civilian use of radioactive materials to provide reasonable assurance of adequate protection of public health and safety and to promote the common defence and security and to protect the environment and that hasn't changed with the introduction of SMRs.

The dose limits for the plant workers and for public exposure are the same as for any single unit large light water reactor (LLWR). It is not assumed that more than one module would have a DBA at the same time and, therefore, the radioactive releases and associated doses for the site are based on a single module DBA. As with existing LLWRs, radiation exposure to occupational workers or the general public resulting from a DBA must be maintained within the limits of 10 CFR [A–52] Part 20, Standards for Protection Against Radiation, and other regulatory requirements applicable to LLWRs. Likewise, the same regulations that apply to LLWRs regarding radioactive waste and discharges apply to this SMR.

A–8. REGULATORY ASSESSMENT OF THE COMPLETENESS AND CLASSIFICATION OF INITIATING EVENTS.

This Annex presents the responses provided by the following Member State regulatory bodies to the Question 8: "Describe challenges associated with regulatory assessment of the completeness and classification of initiating events."

A–8.1. ARGENTINA–ARN

A–8.1.1. Question

Describe challenges associated with regulatory assessment of the completeness and classification of initiating events.

A–8.1.2. Response

The Construction Authorization of CAREM 25 prototype reactor, was granted by ARN against a safety demonstration based on comprehensive deterministic and probabilistic safety analysis.

The deterministic safety assessment (DSA) includes more than 40 sequences, including those with the failure of the First Shutdown System (Anticipated Transient Without Scram, ATWS). The analysis for most of the sequences was extended to 36 hours (plant grace period).

In order to simulate the plant response in case of design basis events, a plant model was developed using the plant code RELAPSIM. For severe accidents, reactor and containment models were developed using MELCOR code. The review and evaluation of safety demonstrations were carried out using ARN's own resources and external TSOs.

Acceptance criteria set by ARN include compliance with the design criteria, regulatory standards, current regulatory requirements, and good design practices. In all cases, the acceptance criteria are fulfilled with large margins.

In addition, ARN could verify an adequate derivation of safety requirements on engineering. For ARN, the engineering was consistent with the safety demonstration and consolidated enough to begin the construction.

Regarding the scope of the event postulation for Safety Analysis (deterministic and probabilistic), internal initiating events were considered for full-power operating conditions of the CAREM 25 reactor, analysing the reactor core and a fuel element in manoeuvre in the pool of irradiated fuel elements as a radioactive source. This scope was agreed between the Regulatory Authority and the RE for the issuance of the Construction Authorization as a preliminary stage.

For the preliminary stage mentioned, the identification, selection, and application of initiating events was carried out based on the RE's internal operating procedures, according to IAEA SSG-3 [A–175]. Based on the definition of the scope, event identification methods were used, which are mentioned below:

- Initiating events postulated in other reactors;
- Lists of events prepared by other institutions;
- Engineering judgment or technical study of the plant;
- Operational experience.

In selection tasks, the initiating events excluded by the reactor design itself with their justification were established.

In the advanced stage of the safety analysis, in addition to the identification methods used in the preliminary stage, other methods of selecting initiating events were included, such as Master Logic Diagram (MLD), Failure Mode and Effects Analysis (FMEA) and other sources of information such as operation manuals, HAZOP (Hazard and operability studies) carried out for plant systems, which could describe situations that can lead to initiating events not previously identified.

A–8.2. CANADA–CNSC

A–8.2.1. Question

Describe challenges associated with regulatory assessment of the completeness and classification of initiating events.

A–8.2.2. Response

The CNSC assess the PIE identification and classification approaches at a high level during pre-licensing vendor design reviews. CNSC Staff have applied this regulatory framework model to SMR vendor design technical reviews on multiple occasions. However, the CNSC requires detailed information from an applicant in order to support a licence application.

Vendors or licence applicants may identify PIEs using engineering judgment, deterministic and probabilistic assessments. To ensure completeness of the identified PIEs, an applicant or licensee is required to use a systematic process (such as MLD, HAZOP, FMEA and expert judgment) to identify a comprehensive set of PIEs so that all foreseeable events that have the potential for serious consequences are anticipated and considered. The list of PIE includes credible failures or malfunctions of SSCs, operator errors, common-cause internal hazards, and external hazards.

The list of PIEs should be iteratively reviewed for accuracy and completeness as the plant design and safety analyses proceed. As per Section 8.2.1 of REGDOC 2.4.1 [A–63] Part II DSA for Small Reactor Facilities, the list of identified events shall be reviewed for completeness during the design and DSA process and verified for the 'as-built' state after construction. Reviews should also be periodically conducted throughout the NPP lifecycle, to account for new information and requirements.

The regulatory challenge relating to the completeness of the PIE list is with regard to a first of a kind SMR where the availability of information to support PIE identification and classification may be limited. In this particular case, detailed information is required from an applicant in support of a licence application to address uncertainties arising from gaps in operating experience. Section 7.4 of REGDOC 2.5.2 [A–62] requires licensees to provide detailed information if analytical tools (e.g. hazard and operability analysis (HAZOP), failure modes and effects analysis (FMEA), and master logic diagrams (MLD)), and/or expert judgment are used.

The regulatory challenge relating to event classification is with regard to the uncertainty in the quantification of the PIE frequency, especially for a FOAK SMR where OPEX to support such quantification may be limited.

An acceptable approach is the use of consensus models to quantify PIE frequencies (e.g. Bayesian update process or equivalent statistical process combining evidence from generic and plant-specific information, methods for pipe rupture frequency evaluation, fault tree development based upon plant-specific design features, expert judgment), followed by the characterization of PIE uncertainties.

Section 8.2.3 of REGDOC 2.4.1 [A–63] in Part II specifically states that an event with a predicted frequency that is on the threshold between two classes of events, or with substantial uncertainty in the predicted event frequency, shall be classified into the higher frequency class.

A–8.3. CHINA–NNSA

A–8.3.1. Question

Describe challenges associated with regulatory assessment of the completeness and classification of initiating events.

A–8.3.2. Response

Generally, the selection of SMR PIEs could take the experience of large-scale NPPs as reference. Besides, engineering deduction and fault tree method could be adopted as supplementary methods to make additional evaluations.

Possible challenges associated with the regulatory assessment of IE completeness and classification include: Initiating Events (IE) affecting multiple units are roughly identified and might need a systematic and comprehensive study. And some IEs occurring in one unit might have a non-immediate cascading effect on other units during the accident evolution.

A–8.3.3. Follow-up Question

The response indicates two challenges identified (initiating event completeness and classification). Please can you describe these challenges in more detail and how they were resolved in the context of licensing the HTR-PM.

Were there other challenges associated with the approach followed i.e. identification of PIEs from LWRs expectations as the basis for HTRs? How does the regulator verify the approach to identification for PIEs? How does the operator demonstrate completeness?

A–8.3.4. Response

a) The review process of PIEs and classification of HTR is complicated. The general principle is to refer to the existing guidelines which are based on PWR, and analysis case by case for HTR design. For example, PSA is used to re-evaluate the condition classification of the initiating event, and the risk-inform method is used to review the safety classification of systems and equipment.

b) A review of the integrity of initiating events is generally conducted in the following ways:

(i)   Review of the list of previous initiating events;
(ii)   Identify specific initiating events resulting from the failure or mis-operation of the frontier system or supporting system;
(iii)   Reference to the list of initiating events for nuclear power plants of the same type;
(iv)   Use the main logic graph method for deductive analysis.

A–8.4.   CZECH REPUBLIC–SÚJB

A–8.4.1. Question

Describe challenges associated with regulatory assessment of the completeness and classification of initiating events.

A–8.4.2. Response

Not applicable — as there was no application for a licence for SMR, no challenges associated with regulatory assessment of the completeness and classification of initiating events have been identified.

The identification of the initiating events and its review (assessment of the completeness and their classification) would follow then standard procedure — nuclear installation design shall set a list of design basis internal postulated initiating events using conservative approach, which shall be established on the basis of an engineering judgement using deterministic and probabilistic methods of analysis or a combination thereof. The list of design basis internal postulated initiating events shall comprise events that may randomly arise during the operation of the nuclear installation in accordance with the nuclear installation design and may have a significant relevance on nuclear safety of the installation or be caused by events triggered by site characteristics or human activity (in particular random single failure of SSC, incorrect intervention by an operator or a combination of thereof with such failures or events triggered by site characteristics). The nuclear installation design shall also set a list of postulated initiating events and scenarios for design extension conditions.

The internal postulated initiating events and scenarios for design extension conditions shall be categorized and DiD in nuclear installation design shall be based on this categorization. This categorization shall be determined with a respect to their anticipated frequency of occurrence and severity of the possible radiation extraordinary event.

A–8.5.   FRANCE–ASN

A–8.5.1. Question

Describe challenges associated with regulatory assessment of the completeness and classification of initiating events.

A–8.5.2. Response

According to ASN's guide for designing PWRs, to determine the events that need to be analysed in the safety case, every event that may affect installation's safety during normal operation (including shutdown states) has to be identified by the licensee. These triggering events include:

- Unique initiators events;
- Internal hazards;
- External hazards.

The potential accumulation of these events with a failure of the foreseen provisions to deal with these events must be taken into account in the safety demonstration.

In the safety demonstration, the triggering events must be 'excluded' or 'treated'. A triggering event can be 'excluded' if the licensee demonstrates that it is physically impossible or extremely unlikely to occur with a high confidence level regarding the safety objectives. Sufficient design and construction provisions completed by operating provisions shall be implemented to justify this exclusion.

If a triggering event can't be excluded, it has to be treated, which means that its occurrence is postulated, and its consequences assessed. Provisions shall then be implemented to prevent its occurrence and reduce its consequences in order to meet the safety objectives.

Moreover, the article 3.2 of the Order [A–72] stipulates that the safety demonstration shall include potential accumulation of triggering events.

ASN's guide for designing pressurized water reactors provides a non-exhaustive list of events that should be taken into account. Also, it defines the different categories of accidents and the associated acceptance criteria.

ASN, with the technical support of IRSN, assesses the completeness and the classification of the events presented in the safety case. This assessment requires a lot of discussions between the licensee, ASN and IRSN.

Indeed, for example, regarding the Flamanville EPR, discussions about classification of events started from the beginning of the project with the incorporation of a preliminary list of events in the technical guidelines in 2004, 3 years before the authorization to set up the reactor. Then, in 2015, ASN, with the support of IRSN, examined the completeness of the events and the associated acceptance criteria. In 2016 and 2018, ASN asked the advisory committee for nuclear reactors (GPR) its opinion about the accident's studies presented in the Flamanville EPR's safety report. Furthermore, when necessary, ASN, with the support of IRSN, examined specific events in dedicated instructions.

These principles and methodology remain fully applicable to SMRs.

A–8.6. JAPAN–NRA

A–8.6.1. Question

Describe challenges associated with regulatory assessment of the completeness and classification of initiating events.

A–8.6.2. Response

Countermeasures against natural hazards and human induced phenomena that may cause common cause failures are enhanced, as well as countermeasures against the other common cause failure events such as fire and internal flooding, in the new regulatory requirements.

It is required to prevent damage to facilities resulting from an earthquake or tsunami, and also to maintain safety functions against flooding, typhoon, tornado, freezing, rainfall, snow, lightning strike, land slide, volcano, biological phenomena, forest fire, or combination of these phenomena.

As countermeasures for human induced phenomena (excluding intentional ones), it is required to maintain safety functions against flying object (aircraft fall), dam collapse, explosion, fire at

214

neighbouring factories, toxic gas, ship collision or electromagnetic interference and so on at the site or site vicinity.

Also, it is required to assume Design Extension Condition (DEC) which may cause excessive radiation exposure to public at site vicinity, although the frequency of occurrence is low, and required to take necessary measures to mitigate accidents. To assume DEC, external events which may be a common cause like natural hazards, and multiple failures caused by internal events based on the characteristics of reactor facilities, should be taken into account.

A–8.6.3. Follow-up Question

The response identifies a number of external hazards for which countermeasures are expected. Does NRA have expectations for other PIEs (equipment malfunction, internal hazards etc.) — please describe and provide a reference to relevant NRA guidance, requirements or regulations. Have any additional considerations been made in the context of SMRs and, specifically, the HTTR? Does the NRA have specific expectations on the types of failure to be postulated and PIE identification techniques? How does NRA judge that the list of PIEs is complete?

A–8.6.4. Response

In the previous response, we mainly provided countermeasures for external hazards — which are enhanced in the new regulatory standards — as the examples of common cause failure.

In the actual review, other PIEs within facility are taken into account, such as:

- Random failure of equipment which plays a role in the safety function;
- Internal fires;
- Internal floods.

Basic approach on reviewing the Design Basis Event (DBE) which caused by postulated initiating event (PIE) are stipulated in Ref. [A–168].

HTTR which is gas cooled also applies this concept.

PIEs assume the events are mainly failure, damage of equipment or system in Prevention System (PS), or mis-operation of relevant equipment by human errors of the operator.

On the analysis of DBEs, other than analysis for PIEs, it is required to assume that single failure of equipment causing the most severe impact to each system are mainly those related to reactor shutdown function, core cooling function or confinement function.

Additionally, on the review for safety design, it is required that the design for the system that has significant importance on safety function should fulfil its functionality under the condition of single failure of component in the system within external power failure condition.

The following shows the concept of HTTR-specific measures that are different from those of power reactor.

Emergency power supply equipment (generators and storage batteries) had ensured multiplicity or diversity and independence since the beginning of construction. For that reason, in case of single failure or design basis accident, HTTR is designed to be able to supply sufficient power to ensure its safety function.

For external hazards whose requirements have been strengthened by the new regulatory standards, basically, the above-mentioned conventional safety design can be used, and no additional design support is required.

In the case of volcanoes or tornadoes, if there is loss of external power supply due to the event (due to blockage of the intake port of the emergency generator flying objects and volcanic ash, if the emergency generator cannot be expected, take the following new measures.

These measures for HTTR are different from a LWR for power generation. If it is assumed that reactor is stopped, HTTR as gas reactor can allow the decay heat from reactor core to be removed by natural convection and radiation without expecting design basis safety functions that requires an AC power source. This is due to the inherent safety design of HTTR.

Furthermore, the following measures have been taken.

- If it is determined that the influence of a volcano or tornado will affect the facility, the reactor will be shut down;
- The power supply required to monitor the cooling status after the reactor shutdown is designed to be supplied from a storage battery (power supply is possible for 60 minutes);
- After the storage battery is exhausted (after 60 minutes), power will be supplied from the portable generator, and necessary monitoring of temperature, pressure, etc. will be continued until the commercial power is restored.

For BDBA that exceeds the design basis accident, JAEA (the licensee of HTTR) will take measures assuming that the external power source and emergency generator cannot be expected, and we will be applied a portable generator will be available. In addition JAEA was required to prepare measures for external hazards mitigation and control. JAEA was required to develop operating procedures to prevent the expansion of BDBA; these include establishing a reserved shutdown system different from the normal control rods and deploy a portable generator to operate it.

A–8.7. RUSSIAN FEDERATION–ROSTECHNADZOR

A–8.7.1. Question

Describe challenges associated with regulatory assessment of the completeness and classification of initiating events.

A–8.7.2. Response

In developing designs of ships and other vessels with nuclear reactors, as well as of designs of small nuclear power plants with SMRs, the requirements of the current regulatory legal acts of the Russian Federation (Paragraph 14, 20 of NP-022-17) [A–89] provide for analysis of all possible initial events of design basis accidents. In particular, the list of the initial events includes:

- Increase of heat removal from the primary circuit;
- Decrease of heat removal from the primary circuit;
- Decrease of the primary coolant flow;
- Unauthorized change in reactivity;
- Increase of the mass of primary coolant;
- Loss of integrity of the primary circuit;

- Release of radioactive media from the systems and equipment;
- Occurrences in the power supply system;
- Occurrences in nuclear fuel management;
- Occurrences of habitability conditions in the MCR and premises of the nuclear installation.

A–8.7.3. Follow-up Question

Does the list of initial events provided apply generally for land-based reactors including SMRs as well? Are there any further specific tools and reference documents recommended to derive initiating events specific to the design in question that the regulator expects designers or licensees to follow? How does the regulator ensure that the list of PIEs is complete?

A–8.7.4. Response

A preliminary list of initial events that should be considered in the safety case is available in the regulatory documents that determine the requirements for the content of the safety analysis report. There are several documents of this kind, including:

- Requirements for the content of the safety analysis report for a nuclear power plant unit with a VVER reactor (NP-006-16) [A–176];
- Requirements for the content of the safety analysis report for nuclear power plants with fast reactors (NP-018-05) [A–177];
- Requirements for the content of the safety analysis report for nuclear research facilities (NP-049-17) [A–178].

The above list of initial events applies in general to SMRs.

However, this list is considered on a case-by-case basis and can be expanded based on the results of an assessment of the analysis of design basis accidents, including those caused by possible failures of the unique equipment used in SMRs.

This analysis is developed by the licensee and evaluated as part of the safety review.

A–8.8. SOUTH AFRICA–NNR

A–8.8.1. Question

Describe challenges associated with regulatory assessment of the completeness and classification of initiating events.

A–8.8.2. Response

Overview

The NNR required for demonstration of adherence to RD-0018 [A–107], acceptance criteria, the establishment of a rigorous method for identification and analysis of Postulated Initiating Events (PIEs).

The NNR requires to be shown an agreed process and methodology for the selection, screening and analysis of the PIEs.

PIE is defined as an identified initiating event that leads to anticipated operational occurrences or accident conditions. PIEs are the enveloping IEs (covering one or several IEs) and/or combinations of IEs but excluding mitigators. Based on justified frequencies, taking uncertainties into account, the PIEs are to be allocated to the Categories A, B and C (as defined in RD-0018 [A–107]).

The comprehensive set of PIE forms the basis for both deterministic and probabilistic safety analyses.

The list of all initiating events to be considered for the PBMR and the site must be investigated for determination of representative (bounding) PIEs to be allocated to category A, B and beyond category B. The PIEs analysed and implemented in the safety case are the Licensing Basis Events (LBE).

Class B events have a wide spectrum of initiating frequency 1.0E-2/year to 1.0E-6/year.

The wide spectrum may cause problems later in the safety analysis and licensing processes due to the uncertainties in the 1.0E-6/year region and with seismic events.

The PRA may not be sufficiently advanced or complete to support any proposed resolution.

Inconsistencies may arise between LBEs analysed deterministically and by PRA.

Failure to agree beforehand on the process for selection and categorisation of licensing basis events (LBE) will lead to rejection of the construction safety case, as well as having an impact on the design requirements.

Examples of further considerations

The applicant has to define a coherent heat removal scenario for every LBE using the full scope of systems prepared by design. Assuming that the (final) passive mode of the reactor cavity cooling cystem will always be able to cope with every event does not fit in with the requirements of the DiD principle, the components, the equipment and the maintenance strategy.

Starting with the requirements of the events the active system functions have to be coherently considered by the applicant. Based on that the applicant has to describe how the required actions will be carried out (by operator or automatically).

The derivation of LBE needs to be clarified to provide an auditable trail from the list of initiating events. Following this, the safety classification of the systems claimed will need to be re-examined.

The treatment of common mode failure, in particular for passive components, should be addressed. In addition, human factors and potential dependencies between actions claimed need to be well defined.

There are difficulties with the balance between the deterministic and probabilistic safety analysis approaches. It is to be expected that both would be applied, but there are many examples where the initial approach is probabilistic, and it would have been better to carry out a deterministic analysis as a starting point. Use of a probabilistic approach alone results in lack of detail or a more complicated analysis than would otherwise be required. The balance between probabilistic and deterministic approaches needs to be examined and the overall approach presented.

In many cases, the acceptability of a fault is justified by reference to the safety design principle that activity remains within the fuel because fuel limits are never exceeded. This must be justified by analysis for all credible initiating events. Transient analysis has to be carried out to justify acceptable consequences of faults.

The method and results of the deterministic analysis of a bounding set of PIE is to be described. By nature of the conservative, deterministic analysis rules, design basis accident scenarios will not credit any benefit from non-safety classified systems. This conservative accident analysis alone cannot demonstrate ALARA or DiD.

Because its realistic event sequences model the success or failure of a greater number of SSC, the probabilistic analysis can be used to help to demonstrate the principle of DiD.

ALARA is demonstrated with realistic, best estimate analysis such as those for worker dose due to normal operation.

A–8.9.   UNITED KINGDOM–ONR

A–8.9.1. Question

Describe challenges associated with regulatory assessment of the completeness and classification of initiating events.

A–8.9.2. Response

ONR does not prescribe a list of initiating events that need to be analysed by the licensee. Regulatory assessment of the list of initiating events postulated by the Requesting Party (RP) or Licensee is performed on a sampling basis and considers the process used for identifying faults, including the appropriateness of the individuals involved, the documentation and audit trail, in addition to the final output (i.e. the list of faults identified). This approach is thus technology neutral, and as such is not particularly challenged by application to ANTs.

ONR expects that fault analysis should identify all initiating faults having the potential to lead to any person receiving a significant dose of radiation, or to a significant quantity of radioactive material escaping from its designated place of residence or confinement. This expectation is set out in the SAPs [A–44], in particular SAP FA.2 - Identification of initiating faults. The SAPs are clear that the process for identifying faults should be systematic, auditable and comprehensive.

IAEA SSG-2 'Deterministic Safety Analysis for Nuclear Power Plants' [A–179] identifies categories of postulated initiating events typically considered for design basis accidents. This list was formulated on the basis of experience with LWRs, although it is sufficiently high-level that it could potentially be applicable to ANTs. This, however, represents the main challenge associated with the identification of initiating events for ANTs; the lack of operational experience (OPEX) and a well-established list of faults informed from decades of international experience such is the case for LWRs. ONR is currently undertaking an activity to gather transferable OPEX from past reactor operation, which may begin to address this challenge. ONR has also proposed gathering of international OPEX as a potential topic for phase 2 of the SMR Regulators' Forum. However, as the key extant ONR guidance (SAPs and TAGs [A–46]) are goal-setting and technology neutral, no imminent update is considered necessary to address identification of initiating events for ANTs.

A–8.10. UNITED STATES OF AMERICA–NRC

A–8.10.1. Question

Describe challenges associated with regulatory assessment of the completeness and classification of initiating events.

A–8.10.2. Response

For the SMR design, some challenges existed because the assumed initiating event frequencies contain large uncertainties, as plant-specific operating experience and associated data are not available to inform design-specific initiating event frequency estimates.

The NRC staff used existing NRC Commission Policy statements regulatory requirements as well as existing regulatory guidance and NRC Staff review guidance from NUREG-0800 [A–116], Section 19.0, Revision 3, 'Probabilistic Risk Assessment and Severe Accident Evaluation for New Reactors' issued December 2015.

Applicants use a structured, systematic process, which accounts for design-specific features, to identify initiating events. The applicant used a failure modes and effects analysis and a master logic diagram to identify design-specific system and support system faults that could lead to an initiating event or adversely affect the module's ability to respond to an upset condition. These approaches supplemented the review of potential initiating events from industry operating experience data sources.

The design, in conjunction with the use of simplifying assumptions, allows the potential accident sequences to be reasonably represented by designed initiators. This was possible because the design uses fail-safe features, passive core cooling, and heat removal capabilities, thereby relying less on active systems than a traditional large PWR. The assumed frequency estimates are reasonably estimated based on comparisons with industry databases. The NRC found that the applicant provided initiating event analysis sufficiently consistent with SRP [A–116] Section 19.0 and DC/COL-ISG-028 [A–180] and the appropriate regulations, regulatory guidance and NRC Commission Policy.

A–9. EXTERNAL EVENTS FROM OTHER FACILITIES COUPLED TO THE SMR PROPOSED

This Annex presents the responses provided by the following Member State regulatory bodies to the Question 9: "Describe any challenges associated with regulatory assessment of external events from other facilities (e.g. chemical plant, steam plant) coupled to the SMR proposed."

A–9.1. ARGENTINA–ARN

A–9.1.1. Question

Describe any challenges associated with regulatory assessment of external events from other facilities (e.g. chemical plant, steam plant) coupled to the SMR proposed.

A–9.1.2. Response

The objective of the siting studies is to select a suitable site for a NPP, including appropriate assessment and definition of the related design bases, taking into account that a NPPs design implies the consideration of site dependant factors which may affect, directly or indirectly, the plant safety. For instance, the capability and reliability of the ultimate heat sink and power supply networks, the potential occurrence of natural and/or man-induced events, and the characteristics of communication routes and accesses.

Therefore, the siting studies aim at determining the effects of external events occurring in the region of the site, to evaluate the potential radiological impact on the environment due to the plant operation and the feasibility of the emergency plans.

In Argentina, the studies (for selecting the location of a NPP) are part of the requirements that the Licensee shall comply at the time they request a Construction License, or included in the Periodic Safety Review (PSR), document necessary to require the Operating License renewal. A previous and independent licensing of a site is not explicitly required.

The results of siting studies of the NPPs were used in determining parameters required for the application of models describing radionuclide dispersion to the environment. These models enabled the evaluation of dose exposure due to radioactive effluents released during normal operation.

Moreover, the information supplied by siting studies enabled us to foresee the implementation of actions required to protect the public from accidental situations. These actions were taken into account in the elaboration of the corresponding Emergency Plans.

In Argentina, a NPP construction must not be initiated without a previous Construction License issued by the Regulatory Body, upon request from the Licensee.

In line with this approach, at the time of applying for the Construction License, the Licensee must submit to the Regulatory Body all the documentation required to evaluate the radiological and nuclear safety of the installation to be built, including the site characteristics in relation to:

- Natural and man-induced external events that could affect the installation safety;
- Dispersion of radionuclides to the environment, both in normal and accidental conditions.

The Regulatory Body issues the license once the Licensee has demonstrated that the design of the NPP to be built complies with the regulatory standards and other specific regulatory requirements for the selected site, taking into account the NPP-site interaction. Besides, the Regulatory Standard AR 10.10.1 [A–155] had been developed and put into force, taking into account the lessons learned from the Fukushima Daiichi NPP accident and the corresponding IAEA standards.

The CAREM 25 prototype reactor is under construction in the Atucha Site, next to CNA Unit I and II, therefore it is postulated as acceptable for the evaluation of external events to consider that said events will have similar impacts to the resulting ones for said facilities. The main requirements for the site selection are related to the protection of the public and the environment from the radiological consequences due to accidents and their mitigation in case they should occur. The site selected includes the necessary infrastructure and effective security measures with an established response force which has proven its competence to handle emergencies effectively during annual drills. Also, facilitates communication with the community and a co-operation program to assist its needs, contributing to its social and economic development.

The site is suitable for building the CAREM 25 as was demonstrated during the many studies carried out during the CNA I Project. These studies undertaken by the Responsible Entity of Atucha NPPs (NA-SA, Nucleoeléctrica Argentina S.A.), duly extrapolated and updated were made available to, responsible for the design, construction and operation of the CAREM 25 prototype reactor. As a consequence of the Fukushima Daiichi NPP accident and in order to apply the corresponding lessons learned using methodologies and databases according to the state of the art. The Regulatory Body requested perform a stress test to each Argentinian NPP consisting in a reassessment of the NPPs safety margins assuming the occurrence of a sequential loss of the lines of DiD caused by extreme initiating events and, among others safety related issues.

These studies were complemented by others related to the CAREM 25 Project and its specific location, such as the geological studies and the impact of CAREM 25 on CNA I and CNA II, and vice versa. The sources of risk due to external man-induced events for the installation site were identified following the guidelines of the IAEA Guide NS-G-3.1 [A–181] and it was verified for each of them that they do not represent a potential risk for the installation, since either because they are at a distance greater than the safety distance (SDV: screening distance value) or because their annual probability of occurrence is very low.

A–9.2. CANADA–CNSC

A–9.2.1. Question

Describe any challenges associated with regulatory assessment of external events from other facilities (e.g. chemical plant, steam plant) coupled to the SMR proposed.

A–9.2.2. Response

CNSC expects an applicant to develop, document and implement a systematic approach for identifying all external events that may be linked with significant radiological risk. The possible interaction of external and internal events should be considered by the designer submissions.

While SMRs may be subject to unique external events as a result of hazards associated with cogeneration facilities and other site-specific features, the approach for identifying external events described in CNSC's current regulatory framework remains applicable. CNSC verify if for a site with multiple units, the design should take account of the potential for specific hazards (including external events) simultaneously impacting several units on the site.

For some design submissions, the site-specific information and assessment of all external events are not available yet. The CNSC assess external event identification methodologies at a high level during a pre-licensing vendor design review and have applied the regulatory framework to SMRs without challenge so far. However, more detailed information is required from an applicant in support of a licence application.

CNSC examines interactions between the plant, the environment and other site-specific information, such as population in the surrounding area, meteorology, hydrology, geology and seismology which an applicant is expected to identify during licensing and account for in the design basis. Also, CNSC verifies if the vendor submission considers the concept of potential cliff-edge effects when analysing external hazards, where a small change of conditions may lead to a catastrophic increase in the severity of consequences.

CNSC expectations are further described in REGDOC-1.1.1 [A–64], REGDOC-1.1.2 [A–66], REGDOC 2.4.1 [A–63], REGDOC 2.4.2 [A–182] and REGDOC 2.5.2 [A–62].

A–9.3.   CHINA–NNSA

A–9.3.1. Question

Describe any challenges associated with regulatory assessment of external events from other facilities (e.g. chemical plant, steam plant) coupled to the SMR proposed.

A–9.3.2. Response

In the Part 5 of The Nuclear Safety Review Principles for Small PWR Nuclear Power Plants [A–70], it is stated that:

> "Small PWR nuclear power plants must provide reliable protection against external natural disasters and human-caused events, which can be achieved through the design of containment or the combination of nuclear island design. For external natural disasters, external natural disasters of design basis should be protected and appropriate safety margin should be reserved. The protection of external human-caused events should meet the requirements of current nuclear safety regulations and standards in China, and properly refer to the latest international practices and relevant regulations and standards."

Floating reactor could refer to relevant requirements of nuclear power plant on land. NNSA issued External events selected in the design of floating nuclear power plant in 2018. It described in detail the external events that should be considered in floating reactor design, as well as the design basis. For example.

> "External events mainly consider external natural events, such as extreme waves, extreme currents, extreme sea ice, extreme water temperature, extreme wind and snow, tropical cyclones, tornadoes, hail, salt fog, sunlight radiation, earthquake and tsunami, harmful gases; and external human events, such as ship collision, aircraft crash, explosion, missiles, sinking, reef grounding, etc."

A–9.3.3. Follow-up Question

The response is limited to the high level expectation and considerations for floating reactors. Were there any challenges associated with the HTR-PM or other SMRs under consideration in China? Please describe those challenges and how they were resolved. What considerations have been taken into account with regards to interactions between facilities?

A–9.3.4. Response

The HTR and other SMRs in China are used for power generation and civil building heat supply. These uses are no different from those of traditional NPPs, so we have no problem mentioned.

A–9.4. CZECH REPUBLIC–SÚJB

A–9.4.1. Question

Describe any challenges associated with regulatory assessment of external events from other facilities (e.g. chemical plant, steam plant) coupled to the SMR proposed.

A–9.4.2. Response

Not applicable — as there was no licence application related to SMR, no challenges associated with regulatory assessment of external events from other facilities (e.g. chemical plant, steam plant) have been identified.

According to the current legislative system, the fundamental design basis shall specify the external fundamental design basis events for the site. When determining external fundamental design basis events, all events triggered by the site characteristics included in the site assessment shall be considered — including phenomena originated in human activity as operation of an installation where readily flammable, explosive, toxic, suffocating, corrosive or radioactive material are located or are released therefrom. In this perspective, the regulatory assessment of external events from other facilities would be performed in the same way as in case of a 'standard sized NPP'.

A–9.5. FRANCE–ASN

A–9.5.1. Question

Describe any challenges associated with regulatory assessment of external events from other facilities (e.g. chemical plant, steam plant) coupled to the SMR proposed.

A–9.5.2. Response

The article 3.6 of the Order of the 7th February [A–72] provides a preliminary list of external events that must be considered in the safety demonstration. In particular, this list includes the risks induced by any external industrial activity that could affect the safety of the installation. These events must be excluded or treated (see the answer from France to Question 8 for more information). The regulatory requirement remains applicable for SMRs.

ASN published several guides and basic safety rules that deal with external events:

- ASN guide n° 22: pressurized water reactors design (2017) [A–77];
- ASN basic safety rule 2001-1: Determination of the seismic risk for the safety of surface basic nuclear installations (2001) [A–183];
- ASN basic safety rule I.2.D: Risks related to industrial environment and transportation arteries (1982) [A–184];
- ASN basic safety rule I.2.B: Risks induced by projectile emission following the burst of a turbine-generator unit (1980) [A–185];
- ASN basic safety rule I.2.A: Risks induced by an aircraft crash (1980) [A–186].

ASN intends to update its basic safety rules. These guides may not be fully applicable to SMRs.

The potential effects of an accident or a hazard on the other units of the same NPP shall be considered in the safety demonstration. For example, in the design of the Flamanville EPR main control room, the potential effects of a nuclear (or non-nuclear) accident that would occur on the other units nearby has been taken into account.

However, there is an exception for Flamanville's EPR, which doesn't share its safety report with Flamanville 1 and 2. In this case, potential effects of a nuclear (or non-nuclear) accident that would occur on the other units nearby have been assessed and taken into account in the design of the main control room for example.

ASN, with the support of IRSN, assesses the completeness of the eternal events taken into account and their compliance with the safety objectives.

A–9.6. JAPAN–NRA

A–9.6.1. Question

Describe any challenges associated with regulatory assessment of external events from other facilities (e.g. chemical plant, steam plant) coupled to the SMR proposed.

A–9.6.2. Response

There is no need to take into account external events from other facilities, because there are no facilities coupled to the HTTR.

Although there are sodium handling facilities and fuel oil storage tanks in the same site as the HTTR, it was confirmed in the review that there is no temperature effect when these facilities are on fire, and that there is no need to consider the effect of toxic gas generated by the fire as a second effect in the design.

A–9.6.3. Follow-up Question

The response indicates that the hazards associated with those other facilities were analysed and deemed not to impact the plant. Please provide a reference to the relevant regulatory expectations (standards, guidance and requirements) that apply to the assessment of external events relating to other facilities.

A–9.6.4. Response

Among the Standards and Guides used in the review for HTTR, 'Guide for Evaluation on External Fires' [A–82] is related to evaluate external hazard associated with other facilities, in this Guide, evaluation is required for:

- Fire or explosion at industrial facility that is located in the vicinity of the site;
- Effect of fire caused by aircraft crash.

A–9.7.   RUSSIAN FEDERATION–ROSTECHNADZOR

A–9.7.1. Question

Describe any challenges associated with regulatory assessment of external events from other facilities (e.g. chemical plant, steam plant) coupled to the SMR proposed.

A–9.7.2. Response

Accounting for external natural and man-induced impacts for the nuclear power facilities must be done in accordance with the requirements of NP-064-17 [A–187]. However, NP-064-17 does not apply to ships and other vessels with nuclear installations. Requirements for determining the limits and conditions of safe operation of the FNPP under external natural and man-induced impacts are set in NP-022-17 [A–89]. It should be noted that the requirements for accounting for external conditions that have impact on ships with nuclear reactors and their equipment are specified in the documents of the Russian Maritime Register of Shipping.

In developing the designs for ships and other floating facilities with nuclear reactors, as well as of designs of small nuclear power plants with SMRs, the requirements of the current regulatory legal acts of the Russian Federation (Paragraph 19 of NP-022-17 [A–89]) provide for analysis of the impact of all possible external effects on the functioning of the systems and components of the nuclear power plant. In particular, the list of external effects includes:

- Ship accidents, which include stranding, collision with a ship (pier), and water ingress into the power and auxiliary compartments, capsizing, flooding in the shallows, flooding in deep water;

- Shock waves caused by explosions on board the ship, human activity while the ship is in port, fire in the MCR, power compartment, engine room, electrical compartment, reactor compartment, and rooms with the equipment of the Integrated Marine Automation Systems;

- Helicopter crash, including that on the premises of the nuclear power plant and on the hull structures of the vessel containing potentially dangerous equipment (equipment working under pressure, equipment filled with hydrogen, oxygen, aviation fuel);

- Loss of cooling water.

A–9.7.3. Follow-up Question

The response is focused on floating plant generally and is not specific NPP. Please describe approach and potential challenges for nuclear plant and also land-based SMRs. Is the list of events

226

provided meant to be exhaustive or just a selection of events extracted from a longer list since it does not cover the full range of potential external hazards e.g. winterisation, weather, etc.

A–9.7.4. Response

In accordance with the requirements of NP-064-17 [A–187], when conducting engineering surveys and studies of the area of the nuclear facility site, a list of processes, phenomena, and factors of natural and man-made origin that can affect the safety of the nuclear facility should be established. A list of possible influencing factors is shown below.

(a)　Hydrometeorological processes and phenomena:
- Flood;
- Tsunami;
- Ice phenomena at streams;
- Coastal area of water bodies;
- Seiches;
- Low and high tide;
- Water resource variations: extremely low flow; abnormal fall of the water level;
- Tornado;
- Wind, hurricane;
- Tropical cyclone (typhoon);
- Precipitation;
- Extreme snowfalls and snow cover;
- Air temperature;
- Snow avalanche;
- Glaze ice;
- Lightning strike.

(b)　Geological engineering — geological processes and phenomena:
- Seismotectonic fault displacements, seismic dislocations, seismotectonic uplifts, subsidence of crustal blocks;
- Recent differentiated movements;
- Tectonic creep;
- The latest movements of the earth crust;
- Residual seismic deformations of the earth crust;
- Earthquake (of any genesis);
- Volcanic eruption;
- Mud volcanism;
- Landslides;
- Soil collapse and slope collapses;
- Mud flow;
- Snow and rock avalanches; crushed stone and block avalanches;
- Erosion of banks, slopes, channels;
- Subsidence and sinkholes;
- Underground erosion, including karsts;
- Permafrost-geological (cryogenic) processes;
- Deformation of specific soils (karst, thermokarst, liquefaction, solifluction, suffusion processes);
- Aeolian processes (wind erosion, dune formation);

- Corrosive aggressiveness of groundwater soils;
- Water table depth;
- Climatic (solar) thermal destruction;
- Atmospheric corrosion.

(c)   Factors creating external biological phenomena.

(d)   Factors creating man-induced external impacts (anthropogenic factors):

- Crash of an aircraft or other missiles;
- Fire caused by external factors;
- Explosion at the facility;
- Radiological accident;
- Release of explosive, flammable or toxic vapours, gases or aerosols into the atmosphere; drifting cloud explosion;
- Discharge of corrosive effluents into near-surface groundwater;
- Electromagnetic interference;
- Spill of oil or petroleum products on water body coastal surface;
- Break of a natural or artificial water reservoir.

(e)   A nuclear facility design must establish and substantiate the values of the parameters of external natural impacts with an estimated probability of occurrence in the interval of one year 10^{-4} or higher and external anthropogenic impacts with an estimated probability of occurrence in the interval of one year 10^{-6} or higher.

The above list is comprehensive.

A–9.8. SOUTH AFRICA–NNR

A–9.8.1. Question

Describe any challenges associated with regulatory assessment of external events from other facilities (e.g. chemical plant, steam plant) coupled to the SMR proposed.

A–9.8.2. Response

It was claimed that the PBMR has been designed to be resistant to external events such as tornadoes, floods, and missiles including the potential for aircraft crashes. These events are based on past experience and current US NRC regulatory requirements. Given world-wide terrorist events, the adequacy of nuclear reactor designs, including the PBMR, has been called into question. These concerns may require that a well thought-out approach, which minimises the threat of external events and improves the resistance of the PBMR structural design to such threats, be included in the PBMR design, security and emergency planning programmes.

The work on external events has interfaces with work on the selection and deterministic and probabilistic analysis of events. A major part of work on external events is to develop a methodology for the analysis and design of, as an example, important-to-safety civil engineering structures subject to the external hazards (including aircraft impact) and other PIEs identified. It also entails the development of design basis envelopes for 'Category B' and 'Beyond Category B' loads. Load envelopes are developed to minimise the number of load combinations for analysis and design.

The analysis methodology developed for this must consider the means to ensure that a balance is obtained in the design for internal and external events.

228

Civil engineering design criteria must be developed for the civil structures based on the PBMR set of General Design Criteria. The civil engineering design criteria consider both 'Category B' and 'Beyond Category B' loads.

The development of a methodology for evaluating the structural performance of important to safety civil structures under external hazards is linked to work on containment with regard to the support and protection of safety classified plant against, for example, depressurisation events, which may be initiated by external events.

The analysis of external and beyond design basis aircraft impact events considers both the initial loading such as the impact from an aircraft crash into the structure, as well as consequent loads such as explosions and fires within the structure if penetration occurs, and also the possibility of an aircraft impacting other facilities or combination of facilities on the site.

For 'Category B' and 'Beyond Category B' events it must be demonstrated that no 'cliff edge' effects exist.

The external extreme events considered for the design of the PBMR Module Building include seismic loads, aircraft crash, tornado loads, straight wind, fire effects, flooding due to precipitation, high water and high tide, tsunamis or seiches, and high intensity loads generated by industrial activities in the close vicinity of the site.

Local failure of the outer structural elements to the Reactor building will be permitted, but the Citadel including the reactor cavity, the spent fuel storage facility, the rooms housing the helium purification system and other systems, which are identified as significant sources of fission products, will be designed to remain intact.

Damage to the outer walls of the building will be limited to ensure that access into the reactor building is not prevented by the 'Beyond Category B' accident and that operator intervention is still possible. 'Cliff edge' effects are considered in the analysis to ensure that failure of the Citadel, spent fuel and helium purification system facilities would not result from 'Beyond Category B' events and that any failures would not prevent access into the building for mitigating intervention.

A–9.8.3. Follow-up Question

The response outlines the claims made by the vendor and the Regulators expectations. What regulatory insights did the assessment of the safety cases provide? Were there any challenges in during that assessment or any changes to regulatory expectations or the design as a result? Is there any information on interactions between facilities?

A–9.8.4. Response

The general nuclear safety requirements are equally applicable to external events from other facilities on the site compared to external event emanating off-site. The design must consider all hazards posed by the facility in compliance with the general nuclear safety requirements.

The PBMR project had not yet reached the stage where other facilities other than power generation were considered as part of the nuclear installation. It is however expected that specific regulatory guidance should be developed to support the designers and eventual regulatory assessments. This guidance should specifically address issues such as proximity (hazard), separation, possible coupling, feedback effects, etc. of the other facility with the reactor systems.

A–9.9. UNITED KINGDOM–ONR

A–9.9.1. Question

Describe any challenges associated with regulatory assessment of external events from other facilities (e.g. chemical plant, steam plant) coupled to the SMR proposed.

A–9.9.2. Response

As set out in the SAPs [A–44], external hazards are those natural or man-made hazards to a site that originate externally to both the site and its processes i.e., the dutyholder may have very little or no control over the initiating event. Internal hazards are those hazards to the facility or its structures, systems and components that originate within the site boundary and over which the dutyholder has control in some form. ONR classes man-made or industrial hazards that occur on-site as internal hazards. ONR considers that natural external hazards can originate on-site [A–188]. However, man-made or industrial hazards may also be considered as external hazards if they arise outside of a nuclear licensed site e.g. hazards arising from nearby chemical facilities, ship collision, aircraft impact.

For events arising from other facilities, whether these facilities are on-site or off-site, ONR expects dutyholders to adopt a systematic and comprehensive approach to hazard identification and assessment, including consideration of credible combinations of hazards. As part of this hazard identification, ONR expects dutyholders to consider hazards from adjacent nuclear sites, and potential hazards arising where there may be tenants and pre-existing facilities on a site.

ONR SAPs [A–44] on Siting Considerations ST.4 to ST.6 set out ONR's expectations for suitability of a site (relating predominantly to external hazards and civil engineering issues), hazard interactions between facilities and multi-facility sites. Other relevant SAPs include EHA.1 to EHA.19 (external and internal hazards), ELO.4 (engineering principles – layout) and FA.1 to FA.25 (fault analysis). The extant National Policy Statement (NPS) for Nuclear Power Generation (EN-6) [A–189] lists sites determined by the UK Government to be potentially suitable for the deployment of new nuclear power stations in England and Wales before the end of 2025. ONR's expectations for siting and land-use planning are also available on the ONR's website.

Whilst external facilities (e.g. associated with cogeneration or desalination) may or may not be within the licensed site, they can present hazard resulting in nuclear safety consequences depending on the interactions with other facilities, and they could be concurrently affected during an external hazards event.

ONR has not yet assessed SMRs coupled with cogeneration of heat and/or power, heat generation or desalination etc. However, in those cases, analysis by dutyholders should demonstrate that threats to nuclear safety from internal and external hazards are adequately identified and characterised, and measures are put in place to prevent, control and/or mitigate the consequences to ensure risks are ALARP.

Through regulation of existing nuclear licensed sites in Great Britain, ONR has already developed its expectations for the identification and characterisation of external events from facilities both on-site and off-site. ONR's TAGs [A–46] on Internal Hazards NS-TAST-GD-014 [A–190] and External Hazards NS-TAST-GD-013 [A–191] provide such guidance to ONR inspectors. These TAGs highlight the importance of hazard combinations and interactions at multi-facility sites, how external hazards could challenge multiple facilities on a single site simultaneously and how external hazards could threaten neighbouring installations that in turn threaten the plant under consideration.

ONR's expectations for multi-facility sites are set out in ONR SAP [A–44] ST.6 and the supporting guidance (paragraphs 133-139): "Interactions between facilities, between facilities and shared services and between shared services, where events in one may adversely affect others, should be considered explicitly. This entails analyses of events that can have physical effects outside the boundaries or limits for the particular facility or service." Further information on multi-unit facilities is provided in ONR's response to Question 7.

ONR considers that the existing guidance for the regulatory assessment of external events from other facilities is relevant to ANTs and SMRs, irrespective of the size of the facility, the delineation of the nuclear site license and deployment model.

Other relevant ONR and international guidance:

- Probabilistic Safety Analysis TAG, NS-TAST-GD-030 Revision 6 [A–174];
- Western European Nuclear Regulators' Association (WENRA) (2014). Reactor Harmonisation Working Group: Safety Reference Levels for Existing Reactors [A–192];
- IAEA. Safety Guide No. NS-G-1.7. Protection against Internal Fires and Explosions in the Design of Nuclear Power Plants, (2004) [A–193];
- IAEA Safety Guide No. NS-G-1.11. Protection against Internal Hazards other than Fire and Explosions in the Design of Nuclear Power Plants, (2004) [A–194];
- IAEA (Under publication). Safety Standards Series No. SSG-64, Protection against Internal Hazards in the Design of Nuclear Power Plants [A–195];
- IAEA (2009). Specific Safety Guide No.SSG-2. Deterministic Safety Analysis for Nuclear Power Plants [A–179];
- IAEA (2016). Specific Safety Requirements No. SSR-2/1. Safety of Nuclear Power Plants: Design [A–127];
- IAEA (2017). Safety Reports Series No. 86, Safety Aspects of Nuclear Power Plants in Human Induced External Events: General Considerations [A–196];
- IAEA (2018). Safety Reports Series No. 87, Safety Aspects of Nuclear Power Plants in Human Induced External Events: Assessment of Structures [A–197];
- IAEA (2017). Safety Reports Series No. 88, Safety Aspects of Nuclear Power Plants in Human Induced External Events: Margin Assessment [A–198];
- IAEA (2002). Safety Guide No. NS-G-3.1, External Human Induced Events in Site Evaluation for Nuclear Power Plants [A–181];
- IAEA (2018). Safety Reports Series No. 92, Consideration of External Hazards In Probabilistic Safety Assessment For Single Unit And Multi-Unit Nuclear Power Plants [A–199];
- IAEA (2019). Safety Reports Series No. 96, Technical Approach for Multi-Unit Site Probabilistic Safety Assessment [A–200].

A–9.10. UNITED STATES OF AMERICA–NRC

A–9.10.1. Question

Describe any challenges associated with regulatory assessment of external events from other facilities (e.g. chemical plant, steam plant) coupled to the SMR proposed.

A–9.10.2. Response

Currently there is not an application from an applicant to construct an SMR facility in the United States. When an SMR is constructed, the NRC's regulatory approach for reviewing external events will remain unchanged.

In an application for a design certification, the description of locations and transportation routes provides information about potential external hazards or hazardous materials that are present or may reasonably be expected to be present during the projected lifetime of the proposed plant. The purpose of including a description of location and transportation routes in a design certification application is for the NRC staff to evaluate the sufficiency of information on the presence and magnitude of potential external hazards, so that the staff can perform the reviews as described in SRP [A–116] Section 2.2.3; SRP Section 3.5.1.5, 'Site Proximity Missiles (Except Aircraft)' and SRP Section 3.5.1.6, 'Aircraft Hazards'.

Any applicant referencing a design under 10 CFR [A–52] Part 52, will address the locations and distances from the plant of nearby industrial, military, and transportation facilities, and such data agree with data obtained from other sources, when available.

A–10. SAFETY CLASSIFICATION OF SSC

This Annex presents the responses provided by the following Member State regulatory bodies to the Question 10: "Describe the challenges associated with regulatory assessment of safety classification of SSC basis, methodology and results."

A–10.1. ARGENTINA–ARN

A–10.1.1. Question

Describe the challenges associated with regulatory assessment of safety classification of SSC (basis, methodology and results).

A–10.1.2. Response

Regulatory Review of Methodology of Safety Classification of SSC

One of the most challenging aspects of the licensing process of the CAREM 25 prototype reactor, carried out by the Nuclear Regulatory Authority (ARN), was assessing the safety demonstration of a new reactor design, unique in Argentina.

The regulatory approach of CAREM 25 licensing was developed under the concept of an 'integral' evaluation of mandatory documentation, essentially the PSAR. This assessment links the Demonstration of Safety and the Safety Classification of Structures Systems and Components (SSCs). From the Safety Classification of SSCs, engineering requirements are derived and the compliance of these have to be demonstrated in the chapters of the PSAR, within the description of design and engineering of SSCs. Acceptance criteria set by ARN include the adequacy of the Design Criteria and engineering requirements considering regulatory standards, current regulatory requirements, and good practices.

Consequently, engineering features are verified as consistent with the demonstration of safety and consolidated enough to allow ARN to grant the construction authorization of CAREM 25 prototype reactor.

Safety Classification Methodology

The CAREM 25 Project developed a comprehensive and detailed methodology for establishing the engineering requirements for SSCs.

Their objective was stablishing an SSC Safety Classification system in order to set the design, manufacturing, assembly, testing, operation and quality assurance (QA) requirements among others that will apply to each SSC, namely safety requirements.

The engineering requirements applicable to SSCs establish rules and acceptance criteria related to:

- Capability — to fulfil the required function (functional capability);
- Reliability — to perform the function with a low-enough failure rates;
- Robustness — to ensure that the operational loads of the demanding sequence, do not affect the performance.

The requirements on these aspects are implemented under the internalization of the Defence in Depth principle (based on the approach of WENRA).

The safety requirements coming from the post-Fukushima Daiichi NPP accident lessons were included in the design (actually before the occurrence of the accident). These include the provision of devices and systems supported by autonomous external means that allow extending the operation of the Safety Systems, covering cooling functions of the primary system and of the containment, and recovering the water level of fuel elements pools.

Guidelines for the Development of Classification Methodology

The guidelines for the development of a Classification Methodology were based on:

- Importance of Safety Functions to be met by SSCs, avoiding a prescriptive approach (as far as possible);
- A clear implementation of the Defence in Depth (DiD) concept into design;
- Deterministic assessment through functional simulations at the plant level (using codes, modelling-conservativeness approach and acceptance criteria according to the kind of PIE);
- Probabilistic assessment through well-known computer codes.

Basis for Classification Methodology Development

a) Establish criteria for assigning Categories to Specific Safety Functions according to their importance, in the framework of internalization in the design of the Defence in Depth concept (that is, according to their contribution in prevention, in the control of PIE or in mitigation);

b) Establish criteria for assigning Classes to SSC that fulfil the categorized functions (Functional Safety Groups, FSG);

c) Establish the Design Requirements at the structural level and to components, to be applied to each of defined safety Category-Class;

d) Establish a Procedure for assigning Categories to Functions and Classes to the set of SSCs that complies with it.

Detailed Classification Methodology Development

(a) Identification of CAREM 25 design features for the safety management of relevant PIEs. Preliminary design of relevant SSCs;

(b) List of Safety Functions "at the Plant Level", (FSNP, by the acronym in Spanish), revised and adapted to item 1. Four groups: 3 for the Fundamental Safety Functions, and a 4th for cross-cutting functions;

(c) Identification PIEs, grouped according deterministic and frequency criteria;

(d) Implementation of the DiD concept to control and mitigate the PIEs;

(e) Criteria to assign Safety Categories to Safety Functions;

(f) Elaboration, based on 1, 2 and 4, of a list of the 'Safety Functions at Basic Level' (FSNB, by the acronym in Spanish) to be accomplished with preventive goals;

(g) From 6, list of preventive functions FSNBs, identifying associated Safety Functions at Plant Level (FSNP, by the acronym in Spanish) and the SSCs that perform it, the Safety Category assigned by criteria of 5 and Safety Class of the SSC assigned according guidelines;

(h) List of FSNB similar to item 7, acting in case of DBA PIEs;

(i)   List of FSNB similar to item 7, acting in case of Events of Design Extension Conditions (with / without core damage);

(j)   From lists of 7 to 9, review the completeness of CAREM SSCs and its functions, and feedback on the design. Some items are treated as 'blocks' instead of individual SSCs;

(k)   Detail of actuation blocks (including monitoring, detection of IE and actuation of Safety Systems), main and diverse lines;

(l)   Condensation of information of previous items for SSCs with more than one specific safety function - FSNB.

Review of Safety Classification of SSCs.

ARN reviewed the methodology in depth. The review and assessments were carried out using ARN own resources and external TSOs. Special attention was paid to the SSCs with passive operation and to the rules defined to assign the safety class of SSCs from the safety category and engineering rationale.

Some of ARN findings were related to:

- Completeness and consistency of the list of Safety Functions at Plant Level (FSNP);
- Completeness and consistency of the list of Criteria for assigning Categories to Safety Functions;
- Nomenclature consistency;
- Classification of Monitoring Functions methodology.

A–10.2. CANADA–CNSC

A–10.2.1. Question

Describe the challenges associated with regulatory assessment of safety classification of SSC (basis, methodology and results).

A–10.2.2. Response

CNSC requires the design authority to classify SSCs using a consistent and clearly defined classification method. The SSCs shall then be designed, constructed, and maintained such that their quality and reliability is commensurate with this classification.

In addition, all SSCs shall be identified as either important or not important to safety. The criterion for determining safety importance is based on:

- Safety function(s) to be performed;
- Consequence(s) of failure;
- Probability that the SSC will be called upon to perform the safety function;
- the time following a PIE at which the SSC will be called upon to operate, and the expected duration of that operation.

SSCs important to safety shall include:

- Safety systems;
- Complementary design features;

- Safety support systems;
- Other SSCs whose failure may lead to safety concerns (e.g. process and control systems).

Appropriately designed interfaces shall be provided between SSCs of different classes in order to minimize the risk of having SSCs less important to safety adversely affecting the function or reliability of SSCs of greater importance.

Guidance on safety classification of SSCs can be found in REGDOC-2.5.2 [A–62].

To date, the CNSC has not encountered challenges with regulatory assessments of safety classification methodologies as the existing framework is flexible and aligns with the international best practices captured in IAEA SSG-30 [A–57] and the accompanying TECDOC-1787, Application of the Safety Classification of Structures, Systems and Components in Nuclear Power Plants [A–201].

Challenges are more likely to arise when assessing the adequacy of safety classification bases which rely on outputs from the safety analysis and research and development programs. For SMRs incorporating more passive and inherent safety features, the influencing phenomena will need to be properly understood before they can inform the safety classification of SSCs. Examples of relevant information include mission times and reliability targets for the shutdown means and containment isolation.

For more information on CNSC's experience with safety classification of SSCs related to fundamental safety functions: control, cooling, and containment, please refer to the answers from Canada to Questions 17, 18, and 19 respectively.

A–10.3. CHINA–NNSA

A–10.3.1. Question

Describe the challenges associated with regulatory assessment of safety classification of SSC (basis, methodology and results).

A–10.3.2. Response

Regarding the safety classification of the HTR-PM and ACP100 SSCs, methodology in the NNSA safety guide 'Safety Functions and Component Classification for BWR, PWR and PTR (HAD102/03, 1986)' [A–202] was applied. Thus, the deterministic method (or prescriptive approach) was used. For example, the reactor coolant pressure boundary was classified as Class 1. At present, a new safety guide is under development. The methodology recommended in the IAEA SSG-30 [A–57] and the risk-informed approach will be adopted in the new document.

A–10.3.3. Follow-up Question

What challenges were encountered, or do you expect to encounter with regard to safety classification, which was the reason for the need of the revision of the guide?

A–10.3.4. Response

As described in the answer from China to Question 8:

a) The review process of PIEs and classification of HTR is complicated. The general principle is to refer to the existing guidelines which are based on PWR, and analysis case by case for HTR design. For example, PSA is used to re-evaluate the condition classification of the initiating event, and the risk-inform method is used to review the safety classification of systems and equipment.

b) A review of the integrity of initiating events is generally conducted in the following ways:

- Review of the list of previous initiating events;
- Identify specific initiating events resulting from the failure or mis-operation of the frontier system or supporting system;
- Reference to the list of initiating events for nuclear power plants of the same type;
- Use the main logic graph method for deductive analysis.

A–10.4. CZECH REPUBLIC–SÚJB

A–10.4.1. Question

Describe the challenges associated with regulatory assessment of safety classification of SSC (basis, methodology and results).

A–10.4.2. Response

The requirements contained in secondary legislation (implementing decrees to the Atomic Act [A–21]) are more specific and prescriptive and reflect currently used technology in Czech Republic (i.e. PWR), and therefore the legislative framework is not in its entirety, strictly speaking, technologically neutral. This is the reason why the deployment of the SMR that are similar to PWR would be less complicated compared to very distinct technologies. Hypothetically, the national legislative framework could therefore be adjusted to reflect specificities of a different technological solution and facilitate its deployment (especially on the level of secondary legislation).

The Czech system of classification and conformity assessment is in line with IAEA standards. However, one of the more challenging aspects of SMR licencing would be the classification of the SSC as the current system of classification and conformity assessment is mainly adapted to the pressurized light water reactors and thus the most detailed requirements are very specific and focused on this technology.

Should the current legislative framework be applied, selected equipment (systems, structures, components or other parts that are relevant in ensuring nuclear safety) should be classified into safety classes 1 to 3 depending on the safety functions to the performance of which it contributes (graded approach applied in the legislation). The nuclear installation design shall, in line with the legislative requirements, specify the requirements for selected equipment in terms of the safety functions and classify selected equipment into safety classes accordingly.

The conformity assessment of the selected equipment with the technical requirements is performed either by authorized or accredited person or manufacturer (based on the safety classification and the type of equipment and function it performs – application of graded approach on the legislative level). Both the Decree No. 329/2017 [A–22], on the requirements for nuclear installation design, and the Decree No. 358/2016 [A–23], on requirements for assurance of quality and technical safety and assessment and verification of conformity of selected equipment, contain very specific provisions for this classification that reflect PWR technology – pressurized primary and secondary circuit, fuel cladding and assemblies (including specific pressure values and volumes).

Current system would therefore probably not provide (presuming no changes would be adopted) as robust basis for licencing of a very different type of SMR technology compared to PWR type SMR.

A–10.5. FRANCE–ASN

A–10.5.1. Question

Describe the challenges associated with regulatory assessment of safety classification of SSC (basis, methodology and results).

A–10.5.2. Response

According to the article 2.5.1 of the Order [A–72], the licensee must identify the elements and activities important for the protection of security, safety, public health and sanitation, nature and environment. The licensee must also identify the requirements associated to these elements and activities.

ASN's guide about PWRs provides guidance regarding the safety classification methodology.

For example, the safety classification must guarantee a manufacturing quality and in operation monitoring that is proportional to the SSC's role for nuclear safety. The safety classification must take into account SSC's role for prevention and mitigation of hazards' consequences and for the accomplishment of safety functions. The licensee can define several levels of safety classification.

Safety classification methodology relies on 3 successive steps:

- Identification and categorisation of safety functions, depending on their role for nuclear safety;
- Identification and classification of SSCs realising these functions;
- Definition of relevant requirements regarding design, manufacturing and monitoring.

Safety functions categorisation must be built on a deterministic approach, completed by probabilistic analysis and experts judgements when it's relevant.

SSC's safety classification must be consistent with safety functions they accomplish. If a SSC accomplish several safety functions, its classification must be consistent with the more demanding safety function accomplished.

The licensee shall consider the following principles when it comes to the design of important to safety SSC:

- Single failure criterion;
- Alternative electric source;
- Physical separation.

Also, the licensee must keep in mind that the following criteria must be consistent with the level of classification:

- In operating monitoring possibility;
- Qualification;
- Quality insurance;
- Hazards resistance;
- Use of codes and standards.

Moreover, interfaces between SSCs must be specified and designed in a way that guarantees that the failure of an SSC doesn't affect the functioning of another SSC with a higher classification. Also, SSCs serving important to safety SSCs must have the adequate classification.

The safety classification of SSCs relies on a methodology assessed by ASN with the support of IRSN. Indeed, for example, ASN asked the advisory committee for nuclear reactors its opinion about EPR's safety classification methodology in 2014. Through other instructions and inspections, ASN controlled the compliance of SSCs with the requirements induced by their safety class.

A–10.6. JAPAN–NRA

A–10.6.1. Question

Describe the challenges associated with regulatory assessment of safety classification of SSC (basis, methodology and results).

A–10.6.2. Response

It is required that the Structures, Systems and Components (SSC) that make up the safety facility must be secured with safety functions according to importance. Among them, for SSCs which have especially high importance, it is required to be secured with the redundancy or diversity and independence in consideration of the function, structure and operational principle, in order to function even if a single failure of the machine or equipment occurs and the external power is unavailable.

Safety facilities are categorized into an abnormality prevention system (PS) and an abnormal effect mitigation system (MS), with 3 levels in each system. Under a design basis accident, it is assumed that functions which belong to MS-1 and MS-2 are estimated to contribute to convergence the accident.

A–10.6.3. Follow-up Question

What challenges were encountered, or do you expect to encounter with regard to safety classification?

Please provide more information about, methodology and results. Please could you describe an example of the SSC safety classification methodology and the results obtained from its application?

A–10.6.4. Response

As for safety classification, it is based on:

- Article 12 of Ref. [A–78];

- 'Guide for Evaluation on Safety Design of water cooled research and test reactors' and its attachment 'Basic concept for classification on the importance of safety function of water cooled research and test reactors' [A–203].

And we also referred to them for gas cooled HTTR.

Table A–10 and Table A–11 illustrate the examples of important SSCs in HTTR.

TABLE A–10. ABNORMALITY PREVENTION SYSTEM (PS-1) – JAPAN

Definition	Function	Facility, system, equipment
Critical facility, system, equipment — its failure or damage may cause fuel damage and may result excessive release of radioactive substances outside of the site boundary.	Reactor coolant pressure boundary	Equipment piping which forms pressure boundary (except small diameter piping such as instrumentation)
	Preventing from applying excessive reactivity	Standpipe Standpipe closure
	Forming reactor core	Reactor core support steel structure(except restricting band of core restraint mechanism) and support post of reactor core support graphite structure (support function only)

TABLE A–11. ABNORMAL EFFECT MITIGATION SYSTEM (MS-1) – JAPAN

Definition	Function	Facility, system, equipment
Facility, system, equipment which prevent excessive radiation effect to the public at surrounding the site in an emergency	Emergency shutdown	Control rods system
	Maintaining subcriticality	
	Prevention of overpressure within reactor coolant pressure boundary	Safety valve for primary cooling equipment (opening function)
	Suppression of excess reactivity	Fixing device for standpipe
Other features essential for safety	Engineered safety features and signal for activating reactor shutdown system	Safety protection system (shutdown system)
	Other essential relevant safety feature	Central control room

A–10.7. RUSSIAN FEDERATION–ROSTECHNADZOR

A–10.7.1. Question

Describe the challenges associated with regulatory assessment of safety classification of SSC (basis, methodology and results).

A–10.7.2. Response

Section 3 of the federal rules and regulations in the field of the use of atomic energy 'General safety provisions for nuclear power installations of ships and other vessels' (NP-022-17 [A–89]), establishes the following requirements to the classification of ship nuclear power unit systems and components:

Paragraph 31. The systems and components of the ship nuclear installation are distinguished by:

- Their purpose;
- Safety impact.

Paragraph 32. The systems and components of the ship nuclear installation are divided into the following groups according to their purpose:

- Normal operation systems and components;
- Safety systems and components;
- Systems and components of special-purpose hardware for beyond design basis accident management.

The safety systems and components are also distinguished by the nature of their safety functions.

Paragraph 33. The systems and components of the ship NPU are divided into the following groups according to their safety impact:

- Safety-related systems and components;
- Other non-safety related systems and components.

The safety-related systems and components include:

- Safety systems and components;
- Normal operation systems and components if their failures result in exceeding the basic dose limits, permissible ionizing radiation exposure doses, permissible radioactive substances release or discharge limits or permissible radioactive contamination levels for the work premises of the ship;
- Normal operation control and monitoring systems and components included into the control and protection system as well as other components of the normal operation systems directly associated with the reactor if their single failure disturbs its normal operation or results in any failures of the systems intended for elimination of operational occurrences;
- Systems and components provided for accident management within the first three days after the occurrence of the beyond design basis accident initiating event;
- Radiological monitoring systems and components.

Paragraph 34. The safety systems and components are divided into the following groups according to the nature of their functions:

- Protective;
- Localizing;
- Supporting;
- Controlling.

Paragraph 35. Four safety classes are established in accordance with the safety impact of the nuclear installation components.

Class 1: includes fuel elements and nuclear installation components, the failures of which are initiating events for accidents resulting in damage of the fuel elements with exceeding the maximum design limit expressed in terms of volumetric coolant activity while the safety systems perform their design function.

Class 2: includes the following nuclear installation components not included into Class 1:

Components, the failures of which are initiating events resulting in FE damage without exceeding the maximum design limit while the safety systems perform their design function, with due regard for the number of failures in these systems specified for design basis accidents; safety system components, the single failures of which in case of a design basis accident result in exceeding the design limits specified for such accidents.

Class 3: includes safety-related nuclear installation components not included into Classes 1 and 2.

Class 4: includes non-safety-related normal operation nuclear installation components not included into Classes 1, 2 and 3.

The components used to manage beyond design basis accidents and not included into safety classes 1, 2 and 3 also belong to safety class 4.

Paragraph 36. When a component has, at the same time, features of different safety classes, this component must be assigned to a higher safety class.

Paragraph 37. The components sharing items of different safety classes must be assigned to a higher safety class.

Paragraph 38. The safety classes of the nuclear installation components shall be assigned by the reactor designers and ship designers in accordance with the requirements of these General Provisions.

The list of safety-related nuclear installation systems, indicating the components assigned to safety classes 1–3, must be defined by the leading design organization and presented in the ship design.

Paragraph 39. Requirements for quality of the nuclear installation components assigned to safety classes 1, 2 and 3 and assurance thereof must be defined in the regulations and other regulatory documents establishing requirements for arrangement and operation of the nuclear installation components. In this case the above-mentioned regulatory documents must set more stringent requirements for quality and quality assurance of the components assigned to higher safety classes.

Paragraph 40. Pertinence of the components to safety classes 1, 2 and 3, and applicability of the regulations and other regulatory documents to these components must be substantiated and specified in the documentation for design, development and manufacture of the nuclear installation systems and components and captured in the SAR.

Paragraph 41. Class designations reflect pertinence of the component to safety classes 1, 2, 3, 4. The class designations must be supplemented with a symbol reflecting the purpose of the component and/or the nature of safety functions performed by this component:

- N - normal operation component;
- P - protection component;
- L - localizing component;
- S - supporting component;
- C - safety system control component;
- T - component of special-purpose hardware for beyond design basis accident management.

If a component has multiple purposes, all these purposes must be included into the component designation.

A–10.7.3. Follow-up Question

Did not provide much info on how this topic was/is being approached?

What challenges were encountered, or what do you expect to encounter with regard to safety classification?

A–10.7.4. Response

The general approach to the classification of systems and components of a nuclear power plant of floating SMRs does not differ from the approach to classification of NPP SSC.

Clause 35 of NP-022-17 [A–89]: Regarding the impact of nuclear power installation components on safety, four safety classes are established.

Class 1. Includes fuel rods and components of a nuclear power installation, the failures of which can be initiating events of accidents leading to, during design basis operation of the safety systems, damage to fuel rods exceeding the maximum design limit expressed through the volumetric activity of the coolant.

Class 2. Includes components of a nuclear power installation that are not part of class 1 as follows:

- Components, the failures of which are initiating events leading to damage to fuel rods without exceeding the maximum design limit during the design basis operation of the safety systems, taking into account the number of failures in these systems established for design basis accidents;

- Components of the safety systems, single failures of which, in the event of a design basis accident, lead to exceeding the design limits established for such accidents.

Class 3. Includes safety important components of a nuclear power installation that are not part of classes 1 and 2.

Class 4. Includes components of normal operation of a nuclear power installation that do not affect safety and are not part of classes 1, 2, and 3.

The components used to manage a beyond design basis accident that are not part of safety classes 1, 2, or 3 are also part of safety class 4.

Clause 36 of NP-022-17 [A–89]: When a component simultaneously contains features of different safety classes, it should be assigned to a higher safety class.

The assessment of the impact of failures of the systems and components on safety is done on the basis of the deterministic and probabilistic safety analyses performed, the results of which are presented in the safety analysis report. In case the failure of a component may result in:

- The occurrence of a beyond design basis accident, it belongs to safety class 1;
- The occurrence of a design basis accident or exceeding the limits established for design basis accidents, it belongs to safety class 2;
- If a component is classified as a safety important one in accordance with clause 33 of NP-022-17, but does not have signs safety classes 1 or 2, it belongs to safety class 3.

Clause 38 of NP-022-17 [A–89] states that the safety classes of nuclear power installation components are established by the designers of the reactor and the vessel in accordance with the requirements of NP-022-17. The list of safety important nuclear power systems, indicating components of safety classes 1 to 3, must be determined by the parent design organization and presented in the design of a vessel.

In addition, all the safety systems and components are divided into protective, confining, support and control ones. Depending on the nature of their functions, additional requirements are imposed on such systems and components.

Moreover, Article 40 of Federal Law 170-FZ [A–26] states that the requirements of the legislation in the field of the use of atomic energy must be observed at the stages of the life cycle, in particular, the requirements to radiation safety of the population.

In accordance with the legislation in the field of radiation safety, all radiation facilities should be divided into 4 categories regarding their potential radiation hazard:

- Category I includes radiation facilities that may have, in the event of an accident, a radiation impact on the population and require measures to protect the population;
- For the facilities of category II, the radiation impact in the event of an accident is limited to the territory of the sanitary protection zone;
- Category III includes facilities that may have, in the event of an accident, a radiation impact which is limited to the territory of a facility;
- Category IV includes facilities that may have, in the event of an accident, a radiation impact which is limited to the premises where radiation sources are handled.

Establishing a category of a radiation facility is based on assessment of the consequences of accidents, the occurrence of which is not associated with transportation of radiation sources outside the territory of a facility and a hypothetical external impact (explosions as a result of a missile hit, an aircraft crash, or a terrorist act). Radiation facilities receive their categories at the design stage. For the operating radiation facilities, the categories are established by the administration in agreement with the bodies exercising federal state sanitary and epidemiological supervision.

Similarly, to the requirements of federal regulations and rules in the field of the use of atomic energy, depending on the established category of potential radiation hazard of a facility, the corresponding requirements apply to them. For example:

- Radiation facilities of categories I and II should be sited with regard for the wind rose, mainly on the leeward side in relation to the residential area, treatment and preventive institutions, children's institutions, recreation sites, and sports facilities;

- Siting of a radiation facility must be coordinated with the bodies exercising federal state sanitary and epidemiological supervision, taking into account the development prospects of both the facility and the area where it is sited;

- A sanitary protection zone is established around the radiation facilities of categories I to III, and a surveillance zone is established around the radiation facilities of category I. For the radiation facilities of category III, the sanitary protection zone is limited to the territory of a facility object. For the radiation facilities of category IV, no zones are established;

- The dimensions of the sanitary protection zone and the observation zone around the radiation facility are established taking into account the levels of external exposure, as well as the values and areas of possible spread of radioactive emissions and discharges. When a complex of radiation facilities is located on the same site, the sanitary protection zone and the observation zone are established taking into account the total impact of the facilities. The inner border of the surveillance zone always coincides with the outer border of the sanitary protection zone;

- Radiation exposure to the population living in the observation area of a radiation facility of category I or located in the zone of impact of several facilities should be limited by permissible exposure levels for each radiation facility, ensuring that the average annual dose limit for the population is not exceeded;

- The sanitary protection zones and the observation zones around vessels and other floating crafts with nuclear installations are established at the locations of their commissioning, in the moorage ports, and at the locations of decommissioning;

- The boundaries of the sanitary protection zone and the observation zone of a radiation facility at the design stage must be agreed with the bodies exercising federal state sanitary and epidemiological supervision;

- In the sanitary protection zone of a radiation facility, permanent or temporary residence, placement of children's institutions, as well as medical institutions, catering establishments, industrial facilities, ancillary and other structures, and facilities that are not related to the operation of the radiation facility is prohibited. The territory of the sanitary protection zone should be developed and landscaped;

- In the sanitary protection zone, a regime of restriction on economic activity is introduced in accordance with the legislation of the Russian Federation. The use of the lands of the sanitary protection zone for agricultural purposes is only possible with the permission of the authorities exercising federal state sanitary and epidemiological supervision. In this case, all manufactured products are subject to radiation control;

- In the observation zone, in the event of an emergency release of radioactive substances, the administration of the territory must provide for a set of protective measures in accordance with the requirements of section IV of NRB-99/2009 [A–169] and OSPORB-99/2010 [A–170].

Verification of the compliance with the requirements of the legislation in the field of radiation safety is carried out by the bodies exercising sanitary and epidemiological supervision. The results of the verification are taken into account by Rostechnadzor when licensing an atomic energy facility.

A–10.8. SOUTH AFRICA–NNR

A–10.8.1. Question

Describe the challenges associated with regulatory assessment of safety classification of SSC (basis, methodology and results).

A–10.8.2. Response

The safety classification of SSCs shall establish the basis for designing, manufacturing and operating the SSCs of the PBMR according to their safety relevance and thereby contribute to achieving the safety objectives by ensuring the necessary quality and reliability of the SSCs.

All SSCs that are important to safety shall be identified and then classified based on their function and significance with regard to nuclear safety. They shall be designed, constructed and maintained such that their quality and reliability is commensurate with this classification.

The safety classification determines the reliability and integrity requirements for each SSC commensurate with the SSC's significance to nuclear safety.

Increased reliability is achieved by applying suitable control during the design, manufacturing, installation, commissioning and operational phases of the demonstration power plant.

Safety classified equipment is required to comply, inter alia, with defined minimum requirements in the following areas:

- Design and manufacturing codes that define methods for design calculation, procurement, handling, storage, construction and layout, such that the equipment's quality and reliability are commensurate with this classification;
- Application of quality assurance;
- Application of conservative design criteria;
- Performance of pre- and in-service inspection tests;
- Capability to withstand seismic events;
- Environmental qualification for all plant conditions;
- Appropriately designed interfaces between SSCs of different classes.

There was no applicable international standard that establishes the nuclear safety criteria and functional design requirements of SSCs of the PBMR (as does ANSI/ANS 51.1 [A–204] for the PWR).

Therefore, there was a need to develop the methodology to identify the nuclear safety criteria for the design of the PBMR.

By identifying the PBMR safety functions that support the Fundamental Safety Functions and subsequently the challenges to them, SSCs can be allocated to the prevention and mitigation of such challenges. The specific design, manufacturing, construction, commissioning, qualification and operational requirements (functional and reliability/integrity) can then be allocated to the SSC.

246

The safety classification process also considers the importance of SSCs in terms of control of worker radiation exposure.

The PBMR took advantage of passive safety characteristics or attributes. Components and systems are categorised in the design, as defined in IAEA TECDOC 626 [A–205], to indicate their level of passivity.

Whether active or inherent/passive SSC characteristics are used to perform a Safety Function, the safety classification is still dependent on its significance. In the single failure analysis, it may not be necessary to assume the failure of a passive component designed, manufactured, inspected and maintained in service to an extremely high quality, provided that it remains unaffected by the PIE.

However, when it is assumed in analysis that a passive component does not fail, that approach shall be justified, taking account of loads, environmental conditions, and the total period of time after the initiating event for which the component is necessary.

A–10.8.3. Follow-up Question

Please could you describe an example of the SSC safety classification basis, methodology and the results obtained from its application?

A–10.8.4. Response

In addition to the previous NNR response to this question, the following information provides indications of the status of the regulatory assessment of safety classification of SSCs (basis, methodology and results) reached towards the time the PBMR programme was terminated.

The major requirements to be considered in this field are given in RD-0034 [A–33].

The logical flow describing the safety classification methodology is depicted in the following diagram (Fig. A–9) in which:

- DFNS – Design for Nuclear Safety Standard;

- FSF – Fundamental Safety Functions;

- Dependability – described in the IAEA Safety Glossary [A–206] as 'a measure of the overall trust that may be placed on system safety performance' is an all embracing term to include the Reliability, Availability, Capability and Robustness of Design solutions.

In addition, the licence applicant also developed the, 'PBMR Functional System Allocation List', as well as 'PBMR IE & Safety function list'. These two documents allowed a better understanding of the overall safety classification process.

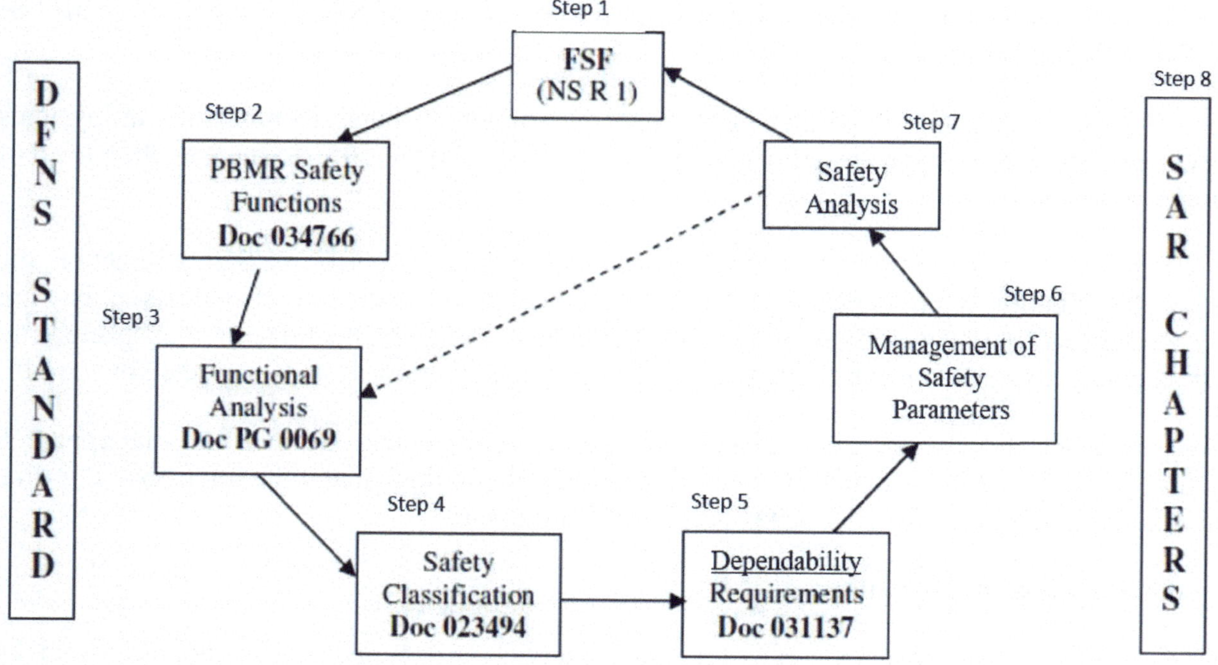

FIG. A–9. Safety classification methodology logical flow – South Africa.

Synopses of the documents mentioned in the above diagram (with shortened titles) are as follows:

(a)   PBMR Safety Functions

The purpose of this step is to describe and record the process used to determine the PBMR high level safety functions derived from the fundamental safety functions (FSF) and used in turn as the basis for PBMR System Safety Functions. It does this by taking the safety functions identified in IAEA documents NS-R-1 [A–207] and TECDOC 1366 [A–208] and reviewing them in the light of the equivalent PBMR safety objectives. The combination of IAEA Safety Functions and the PBMR challenges (safety objectives) are combined to form a set of PBMR Safety Functions. The document also identifies a number of issues which need consideration during subsequent functional analysis processes.

(b)   Functional Design Process, with Specific Reference to Nuclear Safety

This step is intended to describe the process for Functional Design for Nuclear Safety. It is used to identify system safety requirements for the SSC that are needed to prevent or mitigate the undesired consequence of challenges posed by the Postulated Initiating Events. The system safety requirements are allocated to DiD levels and classify the system safety requirements in terms of their nuclear safety significance. Amongst others, it explains how active and passive heat removal is considered.

(c)   Principles and High Level Process for PBMR Safety Classification

This step defines the principles upon which the development of the safety classification methodology for the purpose of applying a graded approach to nuclear safety in the design, manufacture, installation, commissioning, operation and support of PBMR SSCs is based.

(d)   Application of the PBMR Safety Classification Principles

The step describes how the principles of safety classification will be applied and how the graded approach to safety, derived from the functional design process, is to be implemented within PBMR by defining the requirements associated with three different safety classes. The scope of the document includes the safety, seismic, environmental, and quality classifications that form part of the overall classification methodology. The special treatment arising from availability requirements, investment protection and first of a kind equipment is also discussed in this document.

The licence applicant proposed three categories of safety class — high (SC-H), medium (SC-M) and low safety significance (SC-L). The remainder of SSCs that are not assigned with a Safety Function will, for nuclear safety purposes, be designated as Non Class (NC):

- SSCs that are designed to perform Safety Functions that prevent or mitigate specified DBA, and are required for that purpose in order to maintain dose within the category B limits imposed in RD 0018 [A–107], must be classified as Safety Class High (SC-H) [7];

- SSCs that are designed to prevent or mitigate AOO and are required to maintain dose within the category A limits imposed in RD-0018 [A–107] must be classified as Safety Class Medium (SC-M) [8];

- SSCs that are required to prevent deviations from normal operation, or are used to achieve ALARA targets, are classified as Safety Class Low (SC-L) [9].

A–10.9. UNITED KINGDOM–ONR

A–10.9.1. Question

Describe the challenges associated with regulatory assessment of safety classification of SSC (basis, methodology and results).

A–10.9.2. Response

ONR expectations on the Categorisation of Safety Functions and Classification of structures, systems and components (SSCs) are documented extensively in NS-TAST-GD-094 Rev. 1 [A–209]. In line with this TAG [A–46], ONR expect nuclear facilities to be designed and operated with layers of DiD, the purpose of which should be to prevent faults arising, to provide protection in the event that prevention fails and to provide mitigation should an accident occur, (see SAP [A–44] EKP.3 at paragraph 5.2.1.2).

---

[7] Related design rules are employed to achieve the necessary dependability (Reliability, Availability, Capability and Robustness i.e. the ability to perform without failure under a wide range of conditions).

[8] Related design rules are employed to achieve the necessary dependability. Some SC-M SSC may have the 'capability' to mitigate DBA, but are not required to have the overall dependability that would enable them to be credited in the deterministic analysis for DBA e.g. perhaps being less robust.

[9] Related design rules are employed to achieve the necessary dependability and ensure SC-L SSCs have a low failure probability.

The identification and categorisation of safety functions and the identification and classification of SSCs are key activities that are required to support reasonable and balanced implementation of DiD. The TAG [A–46] includes examples of SSC classification schemes, discipline-specific classification guidance, and examples to illustrate the classification process. As with the rest of ONR TAGs, the aim of the TAG is providing advice to inspectors. The TAG is formulated on a technology-neutral basis and broadly aligns with international guidance including IAEA guidance.

It is important that SSCs are classified on the basis of their safety significance as determined by fault analysis. For designs under development, safety classification may be an iterative process, with preliminary assignments of the safety class that are later confirmed by fault analysis. For structural integrity, it is important that all structures, systems and components are designed, manufactured, installed and then subsequently commissioned, operated and maintained to a level of quality commensurate with their classification.

Where a metal SSC forms the primary means of ensuring the safe operation of a plant, and there are no practical means to mitigate its failure, then it may be necessary to identify a special category of SSC whose reliability needs to be shown to be so high that its gross failure can be discounted from the deterministic safety analyses. It is analogous to the Practical Elimination of an initiating event in IAEA terminology. An example of this would be the reactor vessel in conventional light water reactors. These SSCs are termed 'highest reliability components' in the UK, and SAPs [A–44] EMC.1 to EMC.3 are applicable. It is probable that SMR designs may have SSCs that fall into this category, and it will be necessary to identify such components, as additional controls on the design, manufacture and through life management may be necessary.

Where an SMR reactor technology does not feature metal SSC 'highest reliability components', this should be on the clear basis that there is adequate protection to limit both the direct and indirect consequences of gross failure of a component within tolerable limits. Importantly, leak-before-break arguments have not been accepted in the UK as an alternative means by which a new-build safety case can justify discounting the consequences of gross failure.

ONR considers that the expectations on SSC classification outlined in ONR guidance are applicable to SMRs. As part of the ONR ANT project, ONR has recently reviewed IAEA's Specific Safety Guides SSG-30 [A–57] and SSG-34 [A–210], IAEA TECDOC 1787 [A–201], Specific safety requirements SSR 2/1 [A–127], WNA CORDEL Reports [A–211] and [A–212], and the IEC document 61226 [A–213] in the context of ONR SAPs [A–44] ECS1 (Safety Categorisation) and ECS2. Safety Classification of Structure Systems and Components, as well as ONR's experience of classification of electrical power supply in AMR facilities.

ONR's experience of SMRs is that they are often claimed to be designed with passive safety systems. However, Class 1 electrical power supplies may still be required to initiate safety systems and to maintain the plant in a controlled state for a period of time following a design basis accident. Due to the short time required to initiate the systems (and limited ongoing power requirements of safety systems with passive characteristics) these electrical power supplies could be provided on-site battery systems.

The safety systems that are required to establish a safe state (e.g. cooling, ventilation, I&C systems etc.) and to support the maintenance of this safe state (e.g. I&C based plant monitoring systems) can have longer initiation times so may be assigned a lower safety classification. The power supplies to meet this requirement should generally be Class 2 and could be supplied from AC power supply sources.

250

There is lack of consistency in the guidance and standards applicable to safety classification, for example between the IAEA and IEC guidance compared to NRC and IEEE. Individual countries apply variations to the international guidance and there is not a consistent international application of terminology between the terms safety, safety related, non-safety, non-safety related, not important to safety and non-classified. Also, there is no consistency in the definition of shutdown between the controlled state and safe state defined in IAEA guidance and safe shutdown defined in IEEE standards. The ONR SAPs [A–44] are consistent with IEC and IAEA guidance. Depending on the approach and descriptor used, the importance to nuclear safety of the above class 2 power supplies may not be recognised.

Based on the above considerations, it would be useful to develop guidance to define the requirements for the classification of electrical power supply systems as part of the plant safety assessment of SMRs. This guidance should be based on a safety classification system based on ONR SAPs, IEC 61226 [A–213] and IAEA SSG-30 [A–57]. It also concluded that the guidance should define the terminology to be used based on the IAEA Safety Glossary [A–206] definitions.

A–10.10. UNITED STATES OF AMERICA–NRC

A–10.10.1. Question

Describe the challenges associated with regulatory assessment of safety classification of SSC (basis, methodology and results).

A–10.10.2. Response

This area of review did not present extra challenges as the NRC staff used the same guidance and regulations applicable for LLWRs. The NRC performs the safety classification for SSCs review, in accordance with Standard Review Plan (SRP) [A–116] Section 3.2.2, 'System Quality Group Classification', which references Regulatory Guide (RG) 1.26 'Quality Group Classifications and Standards for Water, Steam, and Radioactive Waste Containing Components of Nuclear Power Plants' [A–214]. RG 1.26 is the principal document used by the staff to identify, on a functional basis, the pressure retaining components of those systems important to safety as NRC QG A, B, C, or D. In GDC 1, the NRC requires, in part, that nuclear power plant SSCs important to safety be designed, fabricated, erected, and tested to quality standards commensurate with the importance of the safety functions they perform. Where generally recognized codes and standards are used, they shall be identified and evaluated to determine their applicability and adequacy and modified as necessary to assure a quality product in keeping with the required safety function. As stated in SRP Section 3.2.2, these SSCs will be relied upon for the following functions:

- Prevent or mitigate the consequences of accidents and malfunctions originating within the reactor coolant pressure boundary (RCPB);
- Permit the shutdown of the reactor and maintain it in a safe shutdown condition;
- Ensure the integrity of the RCPB.

In accordance with 10 CFR [A–52] 50.55a(c)(1), components classified as QB A that are part of the RCPB must meet the requirements for Class 1 components in ASME Boiler and Pressure Vessel Code (BPV Code), Section III, except as provided in 10 CFR 50.55a(c)(2) through (4). In accordance with 10 CFR 50.55a(d)(1), components classified as QG B must meet the requirements for Class 2 components in ASME BPV Code, Section III. In accordance with 10 CFR 50.55a(e)(1), QG C components must meet the requirements for Class 3 components in ASME BPV Code [A–215] Section III.

A–11. INTERPRETATION AND IMPLEMENTATION OF DEFENCE-IN-DEPTH

This Annex presents the responses provided by the following Member State regulatory bodies to the Question 11: "Describe the challenges associated with interpretation and implementation of Defence-in-Depth (DiD). Provide information on how multiple provision and/or measures to fault progression are achieved, with adequate independence."

A–11.1. ARGENTINA–ARN

A–11.1.1. Question

Describe the challenges associated with interpretation and implementation of DiD. Provide information on how multiple provision and/or measures to fault progression are achieved, with adequate independence.

A–11.1.2. Response

CAREM 25 design features have an enhanced implementation of the DiD concept, and can, therefore be considered to be an example of how the basic objective in the Vienna Declaration [A–55] could be implemented in future projects.

A summary of basic design aspects of the CAREM 25 Reactor in relation to DiD concept is presented below:

- Level 1 of DiD eliminates some initiating events with the potential to threaten the reactor's integrity. The integrated primary, featuring natural circulation and self-pressurizing, implies eliminating events as large LOCAs, LOFA, and control rod ejection;

- Level 2 of DiD identifies the specific systems that prevent the demand of Safety Systems and in general, that reduce the occurrence of fault sequences, namely Risk Reduction Systems;

- Level 3 of DiD prevents initiating events from escalating to a severe accident, and it is unfolded in:
    - Sub-level 3A, with the goal of controlling PIEs plus single failure events within the Design.

    Basis scenarios account for both the short and the long term.

    - The Controlled State, namely grace period, is achieved by means of Safety Systems featuring passive driving forces (require no Power Supply) and is extended up to 36 hours without the requirement of operator intervention;
    - In the second step, a Safe State is kept as long as necessary, by means of active systems actuated manually with no urgency, at any moment within the grace period.

    - Sub-level 3B, with the goal of controlling multiple failures or extremely rare events, accounts for two conditions in which the additional failures can take place:
    - For failures of the Safety Systems in Sub-level 3 A during step 1, the goal is Controlled State by means of diverse Safety Systems, also passive;

— For failures in the Safe State (Sub-level 3A during step 2), the goal is to extend the grace period beyond 36 hours, by means of Safety-Related Systems. It allows the operator intervention to recover the availability of the Safe State Systems.

- Level 4 of DiD mitigates conditions of core damage by the preservation of the confinement function, preventing releases to the environment. Design features dealing with preventing high-pressure failure of the RPV, hydrogen deflagrations and detonations, corium-concrete interaction, and Containment failure in the long term (pressure increase is prevented by sprinklers and a Suppression Pool cooling system).

In reference to the development of the licensing activities, ARN follows a proactive, rather than retrospective, approach accompanying the project realization. As in other licensing projects review and assessment, inspections and audits are performed following a safety-oriented graded approach. In the particular case of CAREM, the regulatory activities observed an enlargement in its scope for the purpose of analysing the inclusion of the design aspects destined to comply with the safety functions for events occurring in sub-level 3B of DiD. As mentioned earlier, the objective of sublevel 3B is the control of multiple failure events (Design Extension Conditions), with a very low probability of occurrence, which defines a series of SSCs with particular engineering requirements designed to deal with these events. Among other conditions, the ARN was able to verify the correct implementation of requirements associated with the DiD principle, internalized in the design.

A–11.1.3. Follow-up Question

What are the challenges?

What future regulatory activities have been planned to address the challenges?

A–11.1.4. Response

Like we mentioned in the response from Argentina to Question 11, the design features of CAREM 25 have an improved implementation of the DiD concept and, therefore, it can be considered as an example of how the basic objective of the Vienna Declaration [A–55] could be implemented in future projects.

The regulatory activities of CAREM 25 observed an enlargement in its scope for the purpose of analyse the inclusion of the design aspects destined to comply with the safety functions for events occurring in sub-level 3B of DiD. The objective of sublevel 3B is the control of multiple failure events (Design Extension Conditions), with a very low probability of occurrence, which defines a series of SSCs with particular engineering requirements designed to deal with these events.

To address these challenges of licensing activities of CAREM 25, the regulatory authority reinforced licensing activities which include review & assessment, inspections, audits activities and enforcement actions, designed to verify compliance with safety requirements defined in the safety report. A so-called 'integrally concept' is used by which the connection between the engineering requirements for SSCs as derived from the Safety Analysis are verified to be consistent with those identified during the safety classification process.

A–11.2. CANADA–CNSC

A–11.2.1. Question

Describe the challenges associated with interpretation and implementation of DiD. Provide information on how multiple provision and/or measures to fault progression are achieved, with adequate independence.

A–11.2.2. Response

The CNSC is aware that designers of new reactor technologies, including SMRs, may propose alternative approaches to address the levels of DiD. For example, they may propose different physical barriers to be used. The principles of DiD, which play a large role in nuclear safety, are reflected in CNSC requirements and guidance and are expected to be addressed in all activities involving nuclear reactors, regardless of facility size and technology type. The CNSC's regulatory framework details requirements and guidance with respect to implementing DiD approaches. The CNSC requires all levels of DiD to be addressed in a safety case.

CNSC requires the implementation of DiD in the design, construction and operation of nuclear facilities or the undertaking of nuclear activities. With DiD, more than one level of defence is in place for a given safety objective, so that the objective will still be achieved even if one of the protective measures fails. To achieve this, multiple independent level of defence must be put into place to the extent practicable, taking organizational, behavioural, and engineered safety and security elements into account, such that no potential human or mechanical failure relies exclusively on a single level of defence. For reactor facilities, DiD consists of different levels of equipment and procedures to maintain the effectiveness of physical barriers placed between radioactive materials and workers, the public, or the environment.

Level 1: Normal operation – to prevent deviations from normal operation, and to prevent failures of structures, systems and components (SSCs) important to safety. This is implemented through:

- Conservative design;
- High-quality materials, manufacturing and construction (e.g. appropriate design codes and materials, design procedures, SSC classification, control of component fabrication and plant construction, operational experience);
- A suitable site was chosen for the plant with consideration of all external hazards (e.g. earthquakes, aircraft crashes, blast waves, fire, flooding) in the design;
- Qualification of personnel and training to increase competence;
- Strong safety culture;
- Operation and maintenance of SSC in accordance with the safety case.

Level 2: Operational occurrences – to detect and intercept deviations from normal operation, to prevent AOOs from escalating to accident conditions and to return the plant to a state of normal operation. This is implemented through:

- Inherent and engineered design features to minimize or exclude uncontrolled transients to the extent possible;
- Monitoring systems to identify deviations from normal operation;
- Operator training to respond to reactor transients.

Level 3: Design basis accidents – to minimize the consequences of accidents and prevent escalation to beyond design basis accidents. This is implemented through:

- Inherent safety features;
- Fail-safe design;
- Engineered design features, procedures that minimize design basis accident (DBA) consequences;
- Redundancy, diversity, segregation, physical separation, safety system train/channel independence, single-point failure protection;
- Instrumentation suitable for accident conditions;
- Operator training for postulated accident response.

Level 4: Beyond design basis accidents – to ensure that radioactive releases caused by beyond design basis accidents, including severe accidents, are kept as low as practicable. This is implemented through:

- Beyond design basis accidents guidance to manage accidents and mitigate their consequences as far as practicable;
- Robust containment design with features to address containment challenges (e.g. hydrogen combustion, overpressure protection, core concrete interactions, molten core spreading and cooling);
- Complementary design features to prevent accident progression and to mitigate the consequences;
- Features to mitigate radiological releases (e.g. filtered vents).

Level 5: Mitigation of radiological consequences – to mitigate the radiological consequences of potential releases of radioactive materials that may result from accident conditions. This is implemented through:

- Emergency support facilities;
- Onsite and offsite emergency response plans and provisions;
- Plant staff training on emergency preparedness and response.

Figure A–10 depicts how those levels are integrated into the overall safety approach for a facility and this is reflected in a licensee's management systems, which include oversight over design, construction, operation, and interfaces with key external stakeholders who are part of offsite response plans.

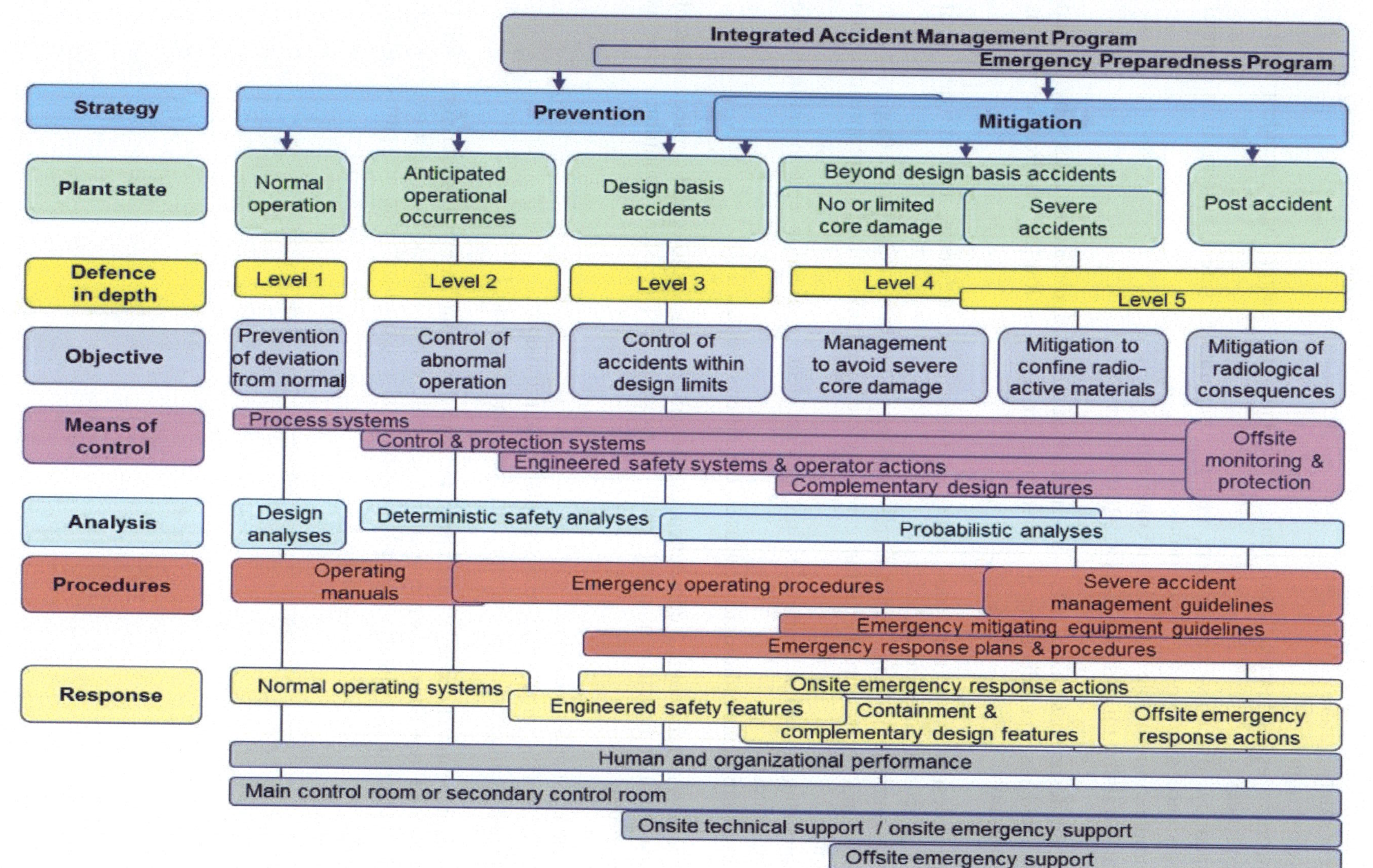

FIG. A–10. *How levels of defence in depth ensure integrated and overlapping safety provisions.*

Each reactor technology's characteristics along with where it will be located (i.e. a site) influences how the objective of each level of DiD is addressed. The overall safety approach used by a proponent must address both of these factors both on the site and with stakeholders in the surrounding regions.

The CNSC is aware that designers of new reactor technologies, including SMRs, are placing a greater emphasis on implementing engineered preventative measures to reduce the need to rely on mitigation measures. The general reason for doing this, in theory, is that stronger preventative measures should increase certainty around:

- Reducing the probabilities of entering into accident situations that would result in significant consequences;
- Ensuring that such an accident would have the smallest possible consequences if it progressed.

Some examples of preventative measures being proposed by SMR developers include:

- Smaller reactor core inventories on a per reactor basis — to improve ability to control, cool and contain fuel during and following facility events;
- New fuels with much higher robustness to withstand plant events without degrading — this would act to reduce releases;
- Alternative heat removal technologies to passively cool the fuel during and following an event;
- Alternative reactor component configurations to reduce or even eliminate high energy events associated with loss-of coolant accidents;
- Using greater amounts of automation to assist operational and maintenance staff with their oversight of the facility.

Many of these technical measures were studied decades ago, but the technology was not mature enough to permit their use in practice at the time. Newer engineering materials and improved computational tools are leading to these measures being proposed for use again. In many cases, technical claims are being made that these designs will either reduce potential beyond design basis accidents (with significant consequences) to a very low probability or eliminate them altogether.

A reactor facility design is required to meet the safety objectives of all five levels of DiD, including physical barriers to prevent uncontrolled releases of radioactive materials to the environment. The levels of defence of depth are expected to be independent to the extent practicable. In the licensing process, an applicant will be required to demonstrate how the levels of DiD are sufficient in their implementation. A demonstration — such as that from research and development activities, including results from physical experiments — will be expected to include credible evidence. The requirements for both design and safety analysis include consideration of areas including:

- External hazards that can breach multiple levels of DiD simultaneously;
- Common-cause/mode failures that cut across the concept of independence of levels;
- Proven-ness of design tools.

The Canadian approach to reactor safety evolved from the recognition that even well-designed and well-built systems may fail. However, when the DiD strategy is properly applied, no single human error or mechanical failure has the potential to compromise the health and safety of persons or the environment. Emphasis has been placed on designs that incorporate 'fail-safe' modes of operation, should a component or a system failure occur. The approach also recognizes the need for separate, independent safety systems that can be tested periodically to demonstrate their availability to perform their intended functions.

For addressing DiD in physical design, REGDOC 2.5.2 [A–62] provides the framework and basis. REGDOC-2.4.1 [A–63] and REGDOC -2.4.2, Safety Analysis: Probabilistic Safety Assessment (PSA) for Nuclear Power Plants [A–182] provide the CNSC's expectations for assessing the adequacy of DiD levels. REGDOC-2.3.2, Accident Management, version 2 [A–216] and REGDOC-2.10.1, Nuclear Emergency Preparedness and Response [A–217] provide the requirements and guidance regarding management of accidents and the application of DiD.

The following documents are also applicable:

- N293, Fire protection for nuclear power plants [A–218];
- N285.0, General requirements for pressure retaining systems and components in CANDU nuclear power plants [A–219].

Where CSA standards are CANDU-specific, the CNSC expects applicants to review the documents and, as far as is practicable, address the intent of the clauses.

A–11.2.3. Follow-up Questions

I take from the response that specific challenges during extant reviews are not yet available.

What are the challenges?

What future regulatory activities have been planned to address the challenges?

A–11.2.4. Response

What are the challenges?

The nature of our arrangements with specific vendors to conduct Vendor Design Reviews do not permit the release of specific results of our reviews.

In general, challenges are as follows:

a) Requirements for independence of levels of DiD are being challenged through the introduction of new multiple function features. That is, some developers are proposing specific designs for structures, systems and components that perform multiple safety functions and in some cases provisions are being claimed to satisfy multiple levels of DiD. Incomplete supporting information on behaviours of passive and inherent features increases uncertainties in safety claims. CNSC staff are challenging vendors to justify the independence of levels of DiD to the extent practicable with validated data.

(b) A safety 'means' does not necessarily mean a dedicated system or structure. For example: Containment was traditionally understood to mean a hardened concrete/steel structure. Now, 'means of containment' is being interpreted more broadly to include all factors that contribute to confinement of radionuclides and containment of releases. For molten salt concepts, this may include retention in fuel salts coupled with inherent temperature behaviours and physical barriers including the reactor 'pressure boundary' and surrounding civil structures working in concert.

258

(c) Sufficient and high quality supporting information to support safety claims (and performance characteristics) are not yet fully available from vendors due to incomplete R&D activities. CNSC can provide feedback in the interim but findings/statements in our VDR process are contingent on vendors providing this information at the time their designs is referenced in a licence application. In the meantime, CNSC expect the vendor to have documented an understanding of associated uncertainties and approaches that would be incorporated in design decision-making to address those uncertainties. Where information from historic records is used, the vendor is expected to justify the relevance of the data and how they will address gaps between the historic data and what is necessary to support safety claims. (e.g. quality methodologies, sufficiency of research).

Addressing uncertainties could lead to:

- The need for additional features to establish appropriate margins and confidence in predicted system performance;
- Supplemental control measures in the first of a kind (FOAK) facility ranging from operational constraints to extended commissioning testing.

(d) Provisions for failure detection (typically in I&C systems) are not always established until later in the design process. For example, aggressive chemistry/temperature regimes may challenge traditional I&C equipment and require new types of instruments to be developed to support operations or condition assessment of SSCs.

(e) Experience shows that vendors are not operators. Operating Limits and Conditions are not in a mature state unless a future applicant/representatives of operators have provided specific operating requirements to the vendor to consider in their design activities. Again, CNSC can provide feedback in the interim but findings/statements in our VDR process are contingent on vendors and applicants submitting this information at the time their designs is referenced in a licence application.

(f) Some vendors may not be conducting sufficiently systematic safety classification to arrive at an appropriate safety classification for systems important to safety. For example, CNSC staff are seeing examples of traditional systems important to control, cool and contain the reactor being proposed as not important to safety and therefore at a lower safety class. CNSC staff require the vendor to show systematically how they have derived their classifications for specific SSCs recognizing that, regardless of claims, control, cooling and containment remain essential safety functions. The use of a lower safety class will need to be justified from a confidence-in-reliability perspective.

(g) Due to the early state of many vendors' designs, operability/maintainability is at an early stage of consideration. This impacts on the confidence in performance characteristics of systems. This issue becomes more pronounced if there are security/safeguards considerations to be incorporated along with safety considerations.

What future regulatory activities have been planned to address the challenges?

CNSC's Vendor Design Review process provides not only an opportunity for early identification and resolution of potential regulatory or technical issues, but also opportunities for CNSC staff to develop knowledge and early regulatory positions for complex new technical proposals.

CNSC is engaging in significant regulatory cooperation ranging from bilateral arrangements with other regulators looking at the same/similar technologies to large issue specific fora such as NEA and IAEA. Leveraging information from other regulators and sharing information and insights to regulatory assessment as part of the cooperative activities will risk-inform CNSC's technical assessment. This promotes a better ability to interpret existing regulatory requirements and guidance especially when assessing alternative approaches/methodologies.

The licensing process provides significant flexibility to staff to regulate effectively in the face of project/technological uncertainties. For example, for First of a Kind facilities, CNSC staff will use risk-informed decision-making processes to assess a safety case with particular attention to technological novelties, complexity and potential for harm in event of failures. CNSC staff will expect the applicant to establish safety and control measures to address the uncertainties to a high degree of confidence, in particular, until sufficient operating experience has been collected.

A–11.3. CHINA–NNSA

A–11.3.1. Question

Describe the challenges associated with interpretation and implementation of DiD. Provide information on how multiple provision and/or measures to fault progression are achieved, with adequate independence.

A–11.3.2. Response

In HAF102 [A–68] of China, the concept of defence in depth for nuclear power plants is defined that all activities related to safety, including organization, personnel behaviour or design, are under the defence of overlapping measures, and five levels of defence in depth are introduced.

The Safety Review Principles for Small PWR Nuclear Power Plants [A–70] also put forward requirements for the application of defence in depth in the design of SMR. It is also mentioned that five defence in depth levels are generally maintained and the characteristics of the reactor type are considered. The specific defence in depth levels may have some differences from the traditional nuclear power plants. For example, it announced:

> "The concept of defence in depth is applied to the design of small PWR nuclear power plant to provide a series of multi-level defence (inherent characteristics, equipment and regulations) to prevent accidents or ensure appropriate protection. While the small PWR nuclear power plant should maintain five defence levels in depth as a whole. But the key point of defence level setting will be different from the traditional large light water reactor nuclear power plant when considering the characteristics of its reactor type. For example, small PWR nuclear power plants should focus on the first three levels, at most the fourth level of defence, so as to achieve the technical demand for external interventions can be limited or even exempted."

The safety review principles for HTR-PM , Part 3 stated as follows.

> "HTR-PM still preserve the above five levels of defence-in-depth (DID). However, considering the safety characteristics of HTR-PM, the design consideration of each level could be different from conventional PWR and BWR NPPs. For example, Integrity of coated fuel particles, the first radioactive containment barrier, would act as a more important role. In addition, the long tolerant time of HTR-PM could be considered as another effective feature in DID. The rationality of DID levels should be verified by integrated safety assessments."

A–11.3.3. Follow-up Questions

Please describe the challenges in DiD from the licensing of the HTR-PM. Fuel and core performance matters e.g. a defective batch, hot stops would undermine safety given the reduced provision at other levels when compared with LWRs. How is this consideration taken into account in the regulators assessment and conclusion that there is adequate DiD? Please provide further examples of differences in DiD for SMRs regulated in China.

What are the challenges?

What future regulatory activities have been planned to address the challenges?

A–11.3.4. Response

HTR-PM adopt similar defence-in-depth phenomenon.

But level 4 target to mitigate the consequence of DEC according to ALARA principle, and level 5 is technically not required. There are some changes in the definition of level 4.

For multiple barriers, the coating layer of TRISO coated particle is especially important, for normal operation, and accident condition.

In the licensing of HTR-PM, the uncertainty on the maximum fuel temperature which is related to the integrity of TRISO coated particles is investigated in detail.

A–11.4. CZECH REPUBLIC–SÚJB

A–11.4.1. Question

Describe the challenges associated with interpretation and implementation of DiD. Provide information on how multiple provision and/or measures to fault progression are achieved, with adequate independence.

A–11.4.2. Response

Not applicable — as regards the DiD concept, it is a topic related to SMR that is being discussed. However, since there are no plans for deployment of an SMR of a particular design in Czech Republic, no specific challenges have been identified. Given the lack of detailed information about the various SMRs designs and the uncertainty over whether and which type of SMR could be hypothetically deployed, no specific information can be provided (SMR designs vary significantly in terms of addressing the DiD).

The requirement for the application of DiD in the context of ensuring nuclear safety using multiple physical safety barriers and applying the safety functions to protect the integrity and functionality of these barriers at the various levels of DiD is set out in the Decree No. 329/2017 [A–22] Coll., on the requirements for nuclear installation design. The concept of DiD in Czech Atomic Act [A–21] and its implementing legislation has been developed to be primarily applied to 'standard (large) size NPPs' (VVER-440 and VVER-1000 are being operated in Czech Republic). Therefore, as mentioned elsewhere, the implementing legislation, in particular the Decree No. 329/2017 has been drafted to reflect the design specificities of the standard PWR and as such it does not reflect various specificities of other different SMR designs. Therefore, it reflects the standard approach to DiD principle as it is interpreted and applied worldwide (independent 5 level DiD).

According to current legislation, the nuclear installation design shall set out requirements ensuring the application of DiD for all activities relevant to use of nuclear energy. It shall set requirements for creation of a series of successive physical safety barriers that are placed between radioactive materials and the surrounding area of the nuclear installation, systems, structures and components and procedures for the application of the safety functions to protect the integrity and functionality of these physical safety barriers at the various levels of DiD and thus prevention of the occurrence of a radiation extraordinary event using physical safety barriers.

For nuclear installations with a nuclear reactor, the function of physical safety barriers shall, according to the Decree No. 329/2017 [A–22], be ensured by independent systems, structures and components — namely fuel element cladding, the pressure boundary of the primary circuit of the nuclear reactor cooling and the containment system. This reflects currently used technology in Czech Republic (i.e. PWR) but could create challenge should an SMR of a very distinct design be deployed.

In order to create systems of subsequent DiD levels, the nuclear installation design may only use those systems, structures and components of the systems of the preceding DiD level that has been broken which have not been compromised in the course of the development of the nuclear installation's response to an off-site or on-site initiating event or scenario and which are separable from the compromised or unusable parts of the systems of the preceding DiD level that has been broken.

Passive SSC (the function of which does not require activation, mechanical propulsion or supplies of a medium or energy from another system in order to be performed) are also addressed in the implementing legislation. If reasonably practicable, the use of passive functions of systems, structures and components to ensure safety functions is one of the prescribed means to ensure the compliance with the design requirements (principles for the safe use of nuclear energy — basic safety functions).

A–11.5. FRANCE–ASN

A–11.5.1. Question

Describe the challenges associated with interpretation and implementation of DiD. Provide information on how multiple provision and/or measures to fault progression are achieved, with adequate independence.

A–11.5.2. Response

The article 3.1 of the Order [A–72] stipulates that the licensee must apply DiD, which consists in the implementation of successive and sufficiently independent protections in order to:

- Prevent incidents;
- Detect incidents and implement actions to stop the situation aggravation and get back to normal operation, or failing that, reach and maintain a safe state;
- Control accidents that couldn't be prevented, or failing that, limit their aggravation by getting the installation under control to reach and maintain a safe state;
- Manage accidental situations that couldn't be controlled to limit their consequences for the population and the environment.

Also, according to the article 3.1, DiD implementation rely on:

- A proper site selection, taking into account natural and industrial risks;
- Identification of functions necessary to the safety demonstration;
- A prudent design approach, employing margins, and, if needed, redundancy, diversification and physical separation for important to safety SSCs in order to reach a satisfactory level of reliability;
- Quality of design, building, operating, dismantling, maintenance and oversight activities;
- Preparation of incident and accident management.

This article remains fully applicable to any kind of SMRs.

ASN controls its application through instructions and inspections. ASN gives a particular attention to the independence of levels of DiD, especially for new reactors.

A–11.6. JAPAN–NRA

A–11.6.1. Question

Describe the challenges associated with interpretation and implementation of DiD. Provide information on how multiple provision and/or measures to fault progression are achieved, with adequate independence.

A–11.6.2. Response

It is required to take necessary measures in design of research reactor facilities for DiD. In addition to conventional 3 layers, it is required to assume Design Extension Condition (DEC) which may cause excessive radiation exposure to public, although the frequency of occurrence is low, and required to take necessary measures to mitigate accidents.

For required equipment of countermeasures against the DEC, it is required to function under the condition of the DEC, not to lose function at the same time of safety functions for a design basis accident facilities and to have certain earthquake resistance and so on. Also, when using permanent equipment, high reliability is required. When using mobile equipment, it is required to comply with standards for general industrial products and arrange several equipment.

A–11.6.3. Follow-up Question

How is DiD delivered across 5 levels in the HTTR, were there any specific challenges during regulatory assessment and how were they resolved? Please describe approach to level 4, and containment as an example.

What are the challenges?

What future regulatory activities have been planned to address the challenges?

A–11.6.4. Response

Basic concept for DiD of HTTR are:

- Level 1: Prevention of abnormal operation and failures of safety significant equipment;

- Level 2: Prevention of accident by controlling events (anticipated operational occurrence);
- Level 3: Control of accidents within the design basis;
- Level 4: Control of severe conditions including prevention of accident progression and mitigation of the consequences of a severe accident;
- Level 5: Mitigation of the radiological consequences of significant external releases of radioactive materials.

On reviewing the application for alteration in Reactor Installation Permit, the subject to be reviewed was Level 4 which assumed additional new events and measures. (Review was not required for Level 1–3 because of no alteration, and Level 5 was not the subject for the NRA review).

As for the Level 4 countermeasures for HTTR is achieved by:

- Monitoring dose in the reactor building and surrounding area;
- If higher dose detected, sealing the gaps or cracks outside of the reactor building to maintain airtightness and controlling the release of fission products from higher position of reactor building.

By above mentioned measures, the effective dose that public exposed could be reduced by reducing radioactivity concentrated in the area outside of facility compared with the case released from ground level.

A–11.7. RUSSIAN FEDERATION–ROSTECHNADZOR

A–11.7.1. Question

Describe the challenges associated with interpretation and implementation of DiD. Provide information on how multiple provision and/or measures to fault progression are achieved, with adequate independence.

A–11.7.2. Response

In accordance with Paragraph 9 of NP-022-17 [A–89], the safety of a nuclear-powered vessel shall be provided through consistent implementation of the DiD philosophy which rests upon the use of a set of physical barriers on the way of ionizing radiation and radioactive substance spread in the environment and a system of technical and organisational measures meant to protect the barriers and maintain their effectiveness.

The physical barrier system for a vessel must consist of the fuel matrix, fuel cladding, reactor coolant circuit boundary, containment, safety enclosure of reactor installation, and biological shielding.

The physical barrier system of a spent nuclear fuel storage facility (if provided for in the vessel design) must incorporate fuel matrix, fuel cladding, and a leak-tight physical barrier precluding radioactive substance release in the environment (identified and justified in the vessel design proceeding from the method and conditions selected for the safe storage of spent nuclear fuel).

The system of technical and organisational measures encompasses measures intended to ensure the vessel safety, as well as the measures meant to provide the safety of vessel crew, special personnel and passengers, and also measures to ensure the safety of the public when vessel is in a mooring location or at a shipyard.

The system of technical and organisational measures must form five levels of DiD.

Level 1. Prevention of abnormal operation:

- Development of the design documentation for a vessel based on a conservative approach, with the provision of the mature inherent safety of the reactor and the measures to prevent a cliff edge effect;
- Assurance of the proper quality of vessel's systems and components important for safety, and of activities pertaining to the use of nuclear energy;
- Vessel operation in compliance with the requirements of pertinent guides and operating procedures;
- Keeping safety-significant systems and components in a good working condition, by timely identifying defects, implementing preventive measures, monitoring the system/component lifetime, arranging effective maintenance system, and documenting the work results;
- Selection of vessel crew members and special personnel with appropriate level of expertise to perform activities in the field of nuclear energy in normal operation and in abnormal conditions, including pre-accident situations and accidents;
- Development of the safety culture;
- Arrangement of basic support for vessel operation.

Level 2. Prevention of design-basis accidents by the normal operation systems:

- Timely detection and elimination of deviations from normal operation;
- Safety management in case of abnormal operation.

Level 3. Prevention of beyond-design-basis accidents by the safety systems:

- Prevention of initiating event progression into a design-basis accident, and progression of a design-basis accident into a beyond-design-basis accident, by using the safety systems;
- Should an accident occur despite the preventive measures, mitigation of its consequences by confinement of radioactive substances.

Level 4. Management of beyond-design-basis accidents:

- Bringing reactor back to a controlled state, in which the chain reaction is terminated, nuclear fuel is continuously cooled, and radioactive substances are confined within the established boundaries;
- Prevention of progression of beyond-design-basis accidents and mitigation of their consequences, in particular, by using dedicated engineered means to manage a beyond-design-basis accident, as well as any technical tools capable of fulfilling the required functions under existing conditions;
- Protection of the containment and/or the safety enclosure against destruction in beyond-design-basis accident conditions, and maintenance of their operability.

Level 5. Emergency planning:

- Preparation and implementation of action plans to protect the workers (personnel) and the public in case of vessel emergency;
- Assistance to vessel crew and/or special personnel through mobilisation of extra resources.

DiD must be provided in all stages of safety assurance activities at a nuclear installation, in the areas affected by these activities. The top-priority strategy is prevention of adverse events. Particular attention must be given to DiD Levels 1 and 2.

The vessel design must include measures to keep the DiD levels independent of each other.

A–11.7.3. Follow-up Questions

Same questions as previously. Please provide information on how independence between the levels of DiD is achieved in the context of the example provided.

What are the challenges?

What future regulatory activities have been planned to address the challenges?

A–11.7.4. Response

Clause 9 of NP-022-17 [A–89] states that the safety of a vessel must be ensured through consistent implementation of the defence in depth principle based on the application of a system of physical barriers on the way of the spread of ionizing radiation and radioactive substances into the environment and a system of engineered and organizational measures to protect the barriers and preserve their effectiveness.

The system of physical barriers for a reactor is traditional for nuclear icebreakers. A new requirement in NP-022-17 is the need to establish a system of physical barriers for a spent nuclear fuel storage facility when such a storage facility is provided for by the design of a vessel.

The established levels of DiD of a vessel with nuclear reactors are generally similar those of nuclear power plants. The implementation of the DiD concept is a method of ensuring the safety of a vessel with nuclear reactors. When it is applied, measures are taken to compensate for the negative impact on safety of potential failures and human erroneous actions, the effectiveness of the physical barriers is maintained, measures are taken to protect them, measures are taken to protect the population and the environment when the physical barriers are not fully effective. If one level of protection fails, the engineered and organizational measures provided for the next level of protection come into effect, and the process of development of the occurrence is stretched over time, which allows it to be monitored and controlled. As indicated in para. 9 of NP-022-17, the priority strategy of DiD is to prevent adverse events. This means that the most important thing for ensuring safety is to prevent the occurrence of failures and prevent them from developing into an accident, if they do occur. This requires special attention to levels 1 and 2 so that more serious operational occurrences are as rare as possible. At the same time, since it is impossible to completely exclude the occurrence of accidents, which is associated with a potential hazard to personnel, the public and the environment, the other levels of DiD are also necessary on a vessel with nuclear reactors.

The first level occupies a special (systemic) place in DiD: its aspects such as, in particular, ensuring the required quality of the systems (components) and work performed, maintaining the systems and components in good working order, ensuring the required level of qualification of the crew and special personnel, and building a safety culture are a prerequisite for the effective function of all levels of DiD.

The requirement of clause 9 of NP-022-17 [A–89] on measures to ensure the independence of the levels of DiD is one of the key factors in the effectiveness and reliability of multi-level protection and its ability to withstand common-cause failures of the levels. Among other things, this is ensured through the use of various technical means or the ones operating on different principles for different levels of DiD. In addition, in order to fulfil this requirement, the NPP design should take measures so that the impacts (internal and external) capable of disabling several levels of DiD at once are the less likely, the more levels they make ineffective. Of course, it is impossible to ensure the complete independence of the DiD levels from each other as there will always be a number of engineered means involved in several levels of protection at once. However, measures to ensure such independence must be taken, and their sufficiency justified and presented in the safety analysis report.

As an example, we can cite the implementation of DiD to protect the primary circuit from overpressure:

- DiD Level 1: the primary circuit must withstand, without destruction, static and dynamic loads and temperature effects arising in any of its parts in case of abnormal operation up to design basis accidents, including unintentional energy release into the coolant (clause 64 of NP-022-17 [A–89]). When designing the primary circuit, the maximum possible pressure values are selected that can be achieved in the event of a sequential failure of all envisaged measures to protect it from overpressure (the principle of conservative approach, clause 47 of NP-022-17). It is envisaged that the parameters of the primary circuit are monitored during operation with the provision of necessary information to the operating personnel. Based on the obtained values, the characteristics of the primary components and their materials are selected. The primary circuit is operated in accordance with the requirements of the manuals and operating instructions by the personnel authorized for independent control;

- DiD Level 2: the control systems of normal operation during operation of the reactor plant maintain the characteristics of the primary circuit within the operational limits, if they are exceeded, they contribute to the return of the parameters to the specified limits. At the same time, the operating personnel has the ability to control the parameters of the primary circuit, determine the root cause of the deviation that has arisen and influence the controls (for example, reduce power) in order to eliminate violations of the operational limits;

- DiD Level 3: the control safety system provides for the impact on the controls of the reactor in case of violation of the established limits of safe operation associated with the pressure of the primary circuit, which are designed to bring the reactor into a safe state. At the same time, the safety control system must be separated from the control system of normal operation to such an extent that the disruption or outage of any component or channel of the control systems of normal operation does not affect the ability of the safety control system to perform its functions (clause 87 of NP-022- 17). In addition, the safety control systems, like any safety systems, must comply with the principles of single failure, priority, safe failure, approbation, and irreversibility of the function (clause 43 of NP-022-17). In addition, along with the active protection systems of the primary circuit against overpressure, passive protective components based on actuation from the direct effect of the environment are provided (the principle of diversity is implemented, clause 43 of NP-022-17). If the pressure build-up could not be prevented (the low probability of such development of an accident should be justified in the design of a vessel), the rupture of the primary circuit occurs on the weakest component located at the upper point of the primary circuit (for example, the cover of the primary circuit filter). Thus, the inflow of radioactive substances into the reactor containment premises is confined, and the outflow of the primary circuit is minimized. To prevent the destruction of the reactor containment in the event of a rupture of the primary circuit, special safety systems for reducing the pressure in the containment are also provided. Moreover, an additional physical barrier against the spread of radioactive substances — a safety enclosure fence — is provided;

- DiD Level 4: violation of the limits of safe operation by pressure in the primary circuit leads to the actuation of the reactor emergency protection and transfer of the reactor to a controlled state, in which a fission chain reaction is prevented. Constant cooling of nuclear fuel is provided. At the same time, depending on the specific conditions, the cooling of nuclear fuel is possible with the use of both normal operation channels (through the steam generators with steam discharge to the condenser and the cleaning and cooling system with heat transfer along the primary circuit to the third circuit to outside water) and the safety systems (emergency cooling systems, emergency core cooling systems). The safety systems have active primary circuit cooling channels and passive ones.

A–11.8. SOUTH AFRICA–NNR

A–11.8.1. Question

Describe the challenges associated with interpretation and implementation of DiD. Provide information on how multiple provision and/or measures to fault progression are achieved, with adequate independence.

A–11.8.2. Response

NNR requirements for DiD

According to RD-0018 [A–107], the principles of DiD must be applied to the PBMR in a manner consistent with the DiD processes described in the appropriate international safety standards and related documents (e.g. Safety Reports produced by the IAEA) so that there are multiple layers of PBMR Functions provided by the structures, systems and components (SSC), and procedures, (or a combination thereof) to ensure that the fundamental safety functions (FSF) of heat removal / reactivity control / confinement of radioactivity are met. Event prevention and event mitigation are natural consequences of the DiD principle. The application of the DiD principle to the design and operation of the PBMR is elaborated further in the following Appendix C of RD-0018:

Appendix C: Explanation of the DiD Principle for the PBMR

C-1   Safety functions

The DiD approach has to be implemented in respect of the fundamental safety functions (FSF):

- Reactivity control;
- Heat removal;

- Confinement of radioactivity.

Sufficient PBMR safety functions shall be provided to ensure that the FSF are maintained and to provide the required levels of DiD.

As a result of the adoption of the DiD principle, the PBMR shall be designed so that DiD can be substantiated for the PBMR by the provision of:

- Sufficient independent reactivity control functions;
- Sufficient independent heat removal functions;
- Sufficient independent barriers for confinement of fission and activation products.

C-2   Levels of defence in depth

The DiD concept as described in the IAEA documents (Please bear in mind that the PBMR project was terminated roundabout 2010, hence, the references to IAEA documents of that period.): e.g. INSAG-12 'Basic Safety Principles for Nuclear Power Plants 75-INSAG-3 Rev 1' [A–220] and IAEA TECDOC-1366 'Considerations in the development of safety requirements for innovative reactors: Application to modular high temperature gas cooled reactors' [A–221].

The DiD principle requires that various lines of defence are provided by design and appropriate procedures to ensure the FSF.

Detailed analysis and assessment of the design of the facility and the various systems and procedures are required to ensure that the lines of defence or barriers are of satisfactory quality and independence, taking into account all the facility provisions and operating procedures.

The safety philosophy is aimed primarily at the prevention of events but also gives attention to the mitigation of the consequences of events that could give rise to radioactive releases. The aim is to reduce both the probabilities of the events and their associated radiological consequences (inside and outside the facility).

The use of the following well established principles of DiD is required:

- Prevention of deviation from normal operation;
- Detection of deviations from normal operation and provision of means to prevent such deviations leading to category B events;
- Provision of engineered safety features (active and passive to control and mitigate the category B events;
- Prevention and mitigation of beyond category B events through the consideration of events or combinations of events with an annual frequency <10-6. Emphasis shall be put on prevention of beyond cat B events. Realistic assumptions and best estimate methods may be used to analyse these conditions;
- Mitigation of radiological consequences of significant releases of radioactive materials by means of off-site emergency response.

C-3   Barriers

A second complementary aspect of the DiD principle is the concept of multiple, independent physical barriers to the uncontrolled release of radioactive material to the environment. The demonstration of the adequacy of these barriers is an important part of the safety analysis.

These barriers shall be designed on the basis of the facility's lifetime, both for steady states and transients occurring in any operational conditions and accident conditions.

The facility shall be designed so that:

- Sufficient independent barriers for confinement of fission products are provided;
- The confinement of the fission products is ensured by these barriers with sufficient margins for all category A events;
- The integrity of nuclear fuel is maintained for all category A and B events and fuel failures due to accidental conditions are minimised even for beyond category B events;
- The integrity of the primary pressure boundary (PPB) is maintained for all category A and B events except for the failure assumptions to be set for the PPB itself;
- The overall radioactivity confinement function of the civil structures forming the confinement functional design shall be ensured with sufficient margins for all category A events;
- The integrity of the civil structures forming the confinement functional design of the building shall be ensured for the category B events. Provisions shall be made to minimise the damage of the civil structures for beyond category B events;
- For beyond category B events at least one confinement function must be adequately maintained in such a way that no cliff edge effects occur.

C-4   Accident prevention

The importance of prevention of accidents as the main basis of the safety is emphasised.

The primary objective of nuclear power facility designers is to provide a sound and balanced design. The SSC of the facility shall have the appropriate characteristics, specifications and material composition and shall be combined and laid out in such a way as to meet the facility specifications. These specifications shall be consistent with the requirement to meet the safety objectives, the specified duty in terms of electrical output, availability, projected lifetime, and the operations necessary to meet system demands. In respect of the principle of DiD INSAG-12 [A–220] (46-55), and accident prevention INSAG-12 (56-62, and 159), the design shall ensure that exposures to the personnel and the public exceeding the category A dose criteria are unlikely to occur during the lifetime of the facility.

Fuel element design, fabrication and inspection, and the conditions under which the fuel is operated shall be such as to ensure a high degree of integrity.

The integrity of the reactor coolant system as well as that of the systems connected to it shall be ensured by the design with adequate margins.

The design shall aim to provide a facility that is simple to operate and maintain. At the design stage, consideration shall be given to the performance capabilities of the personnel who will operate and maintain the facility. The designer shall supply information and recommended practices for incorporation into operating procedures. The design shall aim for simplicity, adequate margins and forgiving characteristics to minimise the consequences of operator errors.

Experience feedback from nuclear operating power facilities and, as applicable, from other industrial facilities shall be extensively and systematically used in the design process. Proven components are to be preferred unless alternatives provide clear advantages in one or more specific areas (e.g. safety, cost, reliability) without significantly affecting the others.

270

Attention shall be paid to the requirements for inspections, testing, on-line monitoring and maintenance, also in their potential to prevent accidents.

The controls shall maintain the reactor within the parameters set for normal operation. The objective shall be to reduce the number of challenges to the reactor protection system.

If deviations from normal operation conditions occur which cause specific limits to be exceeded, the operational control systems shall detect such conditions and prevent them from leading to category B or beyond category B events.

C-5   Accident mitigation

Notwithstanding all preventive features to prevent radiological consequences of events, mitigative measures shall be provided to minimise the radiological consequences through the barriers.

For the design basis the confinement system of the building shall be designed to meet the radiological targets specified to meet the basic licensing requirements (BLR). The maximum allowable source terms from the confinement (including leakage rates and depressurisation) shall be defined to satisfy the BLR [A–107] for the various PIE, and the means to monitor and maintain such leak rates and releases shall be provided.

The engineered safety features providing the PBMR safety functions to control the development of accidents shall be shown to meet the BLR.

The use of inherent characteristics and the simplification of systems are seen as important design aims. Passive safety features shall be used where appropriate and of overall safety benefit. Adequate time scales are required for any operator actions. Simplification of systems design should facilitate elimination of adverse system interactions.

Measures shall be addressed to prevent fuel damage or to mitigate the consequences of event sequences that go beyond the deterministic framework of category B, using appropriate design rules. Such measures shall be implemented taking account of probabilistic safety analyses where such sequences make a significant contribution to risk.

Examples of further considerations

It is expected that licensing submissions would indicate the approach to be adopted in the SAR, or elsewhere, for each of the criteria in Section 3 of INSAG-10 [A–222] for the demonstration of the implementation of DiD. Care needs to be taken to describe the design measures at each of the levels 1 to 5 of DiD. The overall picture must be apparent.

Amongst others, the following practices are advised:

- Requirements and guidance must be established within the design process;
- General guidance references should be placed into the existing development specification documents;
- Provide training as part of the PBMR (Pty) Ltd. engineering training programme;
- Training course for all managers and design engineers regarding general DiD regulations and guidance documents;
- Emphasise the 'principles' of DiD;

- Specific requirements must be incorporated in PBMR (Pty) Ltd. design processes: e.g. the functional analysis process, reliability engineering process, design failure analysis of SSC;
- Verification (of adequate DID) must be accomplished before manufacture of the SSC commences;
- Design features and processes implemented for the purpose of DiD should be specifically recorded in PBMR data management systems;
- Specific design inputs required for DiD should be specified in the appropriate design reports and the design phase baseline defining documents (i.e., in the 'frozen design');
- Features such as conservatism in set points or design parameters may change as uncertainties in the PBMR reactor and material performance are lessened;
- Assessment of DiD adequacy should be verified at the design review for each design phase;
- Determine the necessary margin, redundancy, diversity and independence.

PBMR adopted a system engineering model, based upon IEEE 1220 [A–223] and other standards, which has been adapted to provide a systematic approach to the identification of required safety functions having assigned levels of importance to safety.

The general design criteria, which have been developed with clear links to the fundamental safety functions, address some of the aspects of DiD, including philosophies such as redundancy, diversity, and independence.

Because of the broad scope and different discipline applications of DiD provisions (such as application of margin and redundancy, diversity, and independence) it is necessary to provide guidance on these topics in separate documents to make that guidance useable for the design engineers.

The safety evaluation results for each system will be described in the SAR, which will demonstrate the adequacy of the DiD measures, safety limits, margins, analytical limits, and design limits.

DiD measures must be consistent with any related proposals in safety classification and general design criteria deliverables.

A–11.8.3. Follow-up Questions

One can take for the response that the specific challenges of application to the PBMR were not subject to regulatory assessment by the time the project concluded? Please describe the challenges in assessing the adequacy of the measures in the context to the PBMR.

What are the challenges?

What future regulatory activities have been planned to address the challenges?

A–11.8.4. Response

As part of the Safety Case Early Intervention Process, the main focus of the regulatory assessment was to reach agreement with the licence applicant on their strategy to resolve the key licensing issue of DiD.

In the 'Specification for the PBMR Safety Case', the following specification for DiD was listed:

Explanation of PBMR philosophy and methodology for the application of DiD:

- Requirements of 'Defence levels' for multiple barriers;
- Approach to common mode failure;
- Application of diversity & redundancy requirements;
- Application of conservative assumptions.

Other information regarding DiD associated with operation and control was addressed the link with OTS. It was on the basis that this and other information would be provided in the SAR that the strategy to resolve the key licensing issue (KLI) of DiD was accepted. The KLI strategy considered amongst others determination of safety margins, redundancy, diversity and Independence.

The SAR should demonstrate the adequacy of the DiD measures, safety limits, margins, analytical limits and design limits. These demonstrations would be reviewed and discussed during the SAR chapter early intervention process, and further reviewed during the subsequent SAR formal review.

It was also expected that the DiD approach must be consistent with any related proposals in the KLI strategies for safety classification and for the general design criteria and related deliverables. And, based on earlier regulatory reviews which found insufficient clarity for levels 3 and 4 DiD, the licence applicant was reminded to describe the design measures at each of the levels 1 to 5 of DiD.

A–11.9. UNITED KINGDOM–ONR

A–11.9.1. Question

Describe the challenges associated with interpretation and implementation of DiD. Provide information on how multiple provision and/or measures to fault progression are achieved, with adequate independence.

A–11.9.2. Response

The concept of DiD is incorporated into ONR guidance, particularly SAP [A–44] EKP.3 - Defence in depth: "Nuclear facilities should be designed and operated so that DiD against potentially significant faults or failures is achieved by the provision of multiple independent barriers to fault progression."

As part of the ANT project, ONR has undertaken a high level review of the SAPs and their applicability to AMRs. The review highlighted the advisability of ensuring that ONR inspectors are aware of novel features of AMRs and some key aspects that should be considered when assessing these technologies. One of the areas identified for further consideration was DiD.

As such, the ONR ANT project embarked on a review of information on DiD and its typical implementation in small modular high temperature gas-cooled reactors (HTGRs) and molten salt reactors (MSRs). ONR has gathered this experience through literature reviews, training, engagement with the AMR industry, and engagement at international working groups on HTGR technology, the review of ONR guidance and by considering the output of a dedicated workshop with a multidisciplinary group of ONR inspectors.

The key aspects of the HTGR and MSR designs and typical safety arguments have been documented, and a number of challenges to the interpretation of DiD in the context of current LWR practice have been identified. Some examples of potential challenges include:

- The sharing of safety systems between reactor modules;
- Reduction of protection provided at level 4 of DiD on the basis of enhanced passive safety features;
- Absence of severe accident analysis;
- Reduced independence between levels of DiD;
- Absence of safety classified I&C.

Increased reliance on arguments of 'practical elimination'. Based on the above, ONR is developing examples of aspects to be considered when assessing the adequacy of DiD provisions for AMR designs (and HTGRs specifically) through multi-disciplinary workshops. Key conclusions from the workshops so far undertaken have been that the concept of DiD, and related expectations as laid out in the SAPs, remain fully applicable to HTGRs. Although the implementation of DiD in HTGR designs may differ, the regulatory expectations, methodology and key considerations for assessment are expected to remain largely unchanged from assessment of a mature technology reactor. A judgement on the adequacy of DiD implementation will need to be made on a case-by-case basis, considering the particular design features and supporting substantiation.

Whilst the ONR SAPs [A–44] allow for alternative approaches, it is important to note that ONR will need to be assured that such approaches demonstrate that the risks have been reduced to ALARP. The demonstration may in this case need to be examined in greater detail to gain that assurance, as the technology may not benefit from extensive research and/or operating experience.

In the above context, ONR intends to maintain engagement with international working groups on ANTs (such as via IAEA, NEA and WENRA) to monitor developments in the consideration of the concept of DiD as applied to ANTs and consider the implications to extant guidance.

A–11.9.3. Follow-up Questions

What are the challenges?

What future regulatory activities have been planned to address the challenges?

A–11.9.4. Response

A bulleted list of potential challenges associated in the interpretation of DiD was provided from information seen by ONR including:

- The sharing of safety systems between reactor modules;
- Reduction of protection provided at level 4 of DiD on the basis of enhanced passive safety features;
- Absence of severe accident analysis;
- Reduced independence between levels of DiD;
- Absence of safety classified I&C;
- Increased reliance on arguments of 'practical elimination'.

As part of a proactive review of guidance for compatibility with advanced nuclear technology, ONR is developing examples of aspects to be considered when assessing the adequacy of DiD provisions for AMR designs (SFRs, LFRs, HTGRs and MSRs specifically) through multi-disciplinary workshops.

Key conclusions from the workshops have been that the concept of DiD, and related expectations as laid out in the SAPs, remain fully applicable. Although the implementation of DiD in some designs may differ, the regulatory expectations, methodology and key considerations for assessment are expected to remain largely unchanged from assessment of a mature technology reactor. In the short term ONR is consolidating the outcome of the multidisciplinary reports in technical notes. ONR may develop additional guidance to inspectors to provide further clarity on those expectations in the context of advanced technologies.

A–11.10. UNITED STATES OF AMERICA–NRC

A–11.10.1. Question

Describe the challenges associated with interpretation and implementation of DiD. Provide information on how multiple provision and/or measures to fault progression are achieved, with adequate independence.

A–11.10.2. Response

The NRC staff did not find any significant challenges in this area of review. The SMR design has been reviewed by the staff using existing NRC Commission Policy statements regulatory requirements as well as existing regulatory guidance and NRC Staff review guidance from NUREG-0800 [A–116], with respect to defence in depth (DiD), the NRC defines DiD as follows:

> "An approach to designing and operating nuclear facilities that prevents and mitigates accidents that release radiation or hazardous materials. The key is creating multiple independent and redundant layers of defence to compensate for potential human and mechanical failures so that no single layer, no matter how robust, is exclusively relied upon. Defence in depth includes the use of access controls, physical barriers, redundant and diverse key safety functions, and emergency response measures." [A–224]

A–12. CHALLENGES WHERE THE INDUSTRY CODES AND STANDARDS ARE ABSENT OR NOT FULLY APPLICABLE

This Annex presents the responses provided by the following Member State regulatory bodies to the Question 12: "Describe the challenges where the industry codes and standards are absent or not fully applicable. Provide information on how regulatory assessment or judgement was done in view of the situation."

A–12.1. ARGENTINA–ARN

A–12.1.1. Question

Describe the challenges where the industry codes and standards are absent or not fully applicable. Provide information on how regulatory assessment or judgement was done in view of the situation.

A–12.1.2. Response

As a framework of use, the design/construction standards are tools to 'qualify' the fulfilment of the safety (engineering) requirements (for SSCs), defined by the safety classification, in terms of offering 'guarantees' that assurance the fulfilment of the engineering requirements related to three characteristics:

- Capability - to fulfil the required function;
- Reliability - to perform the function with a low-enough failure rate;
- Robustness - to ensure that the operational loads of the demanding sequence, do not affect the performance.

In this framework, the standards are presented as tools for project management, by the Responsible Entity, and for their evaluation by the Regulatory Authority.

As mentioned earlier, the Argentine Regulatory Standards are based on a set of fundamental concepts that are part of the goal-oriented approach, sustained by the regulatory system, concerning radiological and nuclear safety, safeguards, and physical protection.

In reference to Argentina's regulatory framework, there is no prescriptive treatment that sets design/construction standards. ARN is committed to the goal-oriented approach to regulation to handle several and diverse reactor projects. This is to have guidelines covering all the aspects of a safety approach and how to do:

- A sound safety analysis proving that the safety goals are achieved;
- To develop / implement a safety classification out of the successful safety analysis;
- Define engineering requirements for every class, for every aspect / stage / activity (constructability / viability of the design);
- Define operational limits and conditions (operability);
- To establish the content of the safety analysis report ensuring it covers previous information. Idem for other mandatory documents;
- Assess all mandatory documents;
- Issue facility licenses.

The International standards are used as a guide to expand the scope of ARN standards.

The use of prescriptive industrial standards (e.g. ASME codes for mechanical devices) is only completely coherent if applied for a safety-graded process-system belonging to a specific certified design. Rules and requirements are necessary for almost everything. An evolutionary design reformulates functions/scopes of systems. An innovative design reformulates systems completely.

A–12.1.3. Follow-up Question

Were there any challenges with using prescriptive industrial standards for CAREM?

A–12.1.4. Response

It could be mentioned as a challenge for the regulatory authority, the monitoring and reviewing of the activities related to the use of prescriptive standards, for example, the manufacture of RPV under the ASME code. In this case, the ARN integrated into the training program, the development of the personnel that integrates the sectors related to the assessments and inspections of these SSCs.

A–12.2. CANADA–CNSC

A–12.2.1. Question

Describe the challenges where the industry codes and standards are absent or not fully applicable. Provide information on how regulatory assessment or judgement was done in view of the situation.

A–12.2.2. Response

In Canada, industry codes and standards such as those produced by CSA Group as well as international standards, can be used as part of a safety case to demonstrate CNSC requirements are met.

While CNSC rarely impose specific codes and standards, certain CSA standards are used by industry as benchmark practices often supplemented by a combination of international codes (i.e. ASME, IEEE, etc.) and other technical documents.

In cases where existing CSA standards are not fully applicable to SMRs, alternate standards may be used providing they meet CNSC requirements. Typically, CNSC expects a systematic gap analysis between the applicable Canadian standards and the proposed standards and the vendor's approach to addressing the identified gaps. The extent of applicability of various standards to SMR differs from one design to another, and this is assessed on a case by case basis. For example, it is expected that the standards pertaining to safety analysis (e.g. CSA N286.7 [A–147]) are largely applicable to all types of SMRs, whilst CANDU specific standards (e.g. CSA N290.5 Requirements for support power systems of CANDU NPPs [A–225]) may have a more limited relevance for SMRs.

It is noted that a major part of design codes and standards are identified based on the results of safety classification, therefore the proposed standards are assessed in an integrated manner and a relevant context.

In cases where codes and standards are absent, the case for demonstrating CNSC requirements are met should be supported by research and development activities and conservative safety margins to address risk arising from any gaps in operating experience.

A–12.2.3. Follow-up Question

Describe any experience with vendors proposing standards not traditionally used in Canada.

A–12.2.4. Response

Canada differs from many countries in that the regulator <u>rarely</u> endorses or prescribes the use of specific standards. Standards are understood to be the documentation/codification of proven industry practices. That is, standards have been developed and informed by actual industry experience.

It was observed during the Vendor Design Review projects, a vendor's selection of the industry codes and standards is often influenced by the design and practice of the country of origin, as well as a vendor's engagement with other nuclear regulators.

One area of concern for some reactor concepts is whether the country of origin standards can actually be applied to the vendor's design concept. This is particularly concerning if the country of origin regulator has not assessed the vendor's claims.

It is important to understand that Canadian industry may have limited or no experience in the development and use of standards from other countries. There are some exceptions where Canadian industry has adopted other countries' practices as their own but even this is done through a systematic industry led process. In other words, the benchmark for Canadian industry is the standards package they already work with.

In the review of a new technology proposal by a vendor, CNSC will not make any assumptions about the standards that future operators will use. The vendor is expected to provide a systematic gap analysis between the applicable Canadian standards and the alternative standards to ensure that fundamental safety objectives referenced in Canadian standards are addressed and understood. This will act to eventually enable an operator/applicant to make an informed decision on which practices to impose on the vendors and how to verify that the practices are adequate.

The vendor is expected to present approaches to address the identified gaps (e.g. through R&D activities).

A–12.3. CHINA–NNSA

A–12.3.1. Question

Describe the challenges where the industry codes and standards are absent or not fully applicable. Provide information on how regulatory assessment or judgement was done in view of the situation.

A–12.3.2. Response

There are many types of SMRs, the industry codes and standards may be not fully applicable. The Safety Review Principles for Small PWR Nuclear Power Plants [A–70] stated as follows.

> "Small PWR nuclear power plants comply with the national standards that have been issued and applied in China. If industrial standards and codes are not applicable to SMR, some internationally recognized standards and codes with good practice can be applied. These standards and codes should be approved or recognized by the nuclear safety regulatory authorities.

Special attention should be paid to the appropriateness and applicability of the reference standards, and the approval of NNSA should be obtained. When using standards and specifications of different systems, the interface consistent should be addressed."

Currently there is no separate systematic framework for the industry codes and standards for HTR. Most of current industry code and standards were used for HTR-PM, with minor revision. Some specific standards were available for HTR too, such as standards for graphite and pebble bed property.

A–12.3.3. Follow-up Question

Any specific examples of use of internationally recognized standards?

A–12.3.4. Response

For example, some equipment of HTR-PM and ACP100 are designed and manufactured according to ASME [A–215] Section III.

A–12.4. CZECH REPUBLIC–SÚJB

A–12.4.1. Question

Describe the challenges where the industry codes and standards are absent or not fully applicable. Provide information on how regulatory assessment or judgement was done in view of the situation.

A–12.4.2. Response

Not applicable — as there was not an application for a licence for SMR, no challenge related to the industry codes and standards have been identified. However, considering the lack of detailed information about the various SMRs designs and the fact that industry codes and standards might not be fully applicable or be absent, it would constitute additional complication within the regulatory context.

In any case, the selected equipment (systems, structures and component that can affect nuclear safety) shall be designed and manufactured in accordance with the nuclear installation design, technical requirements and with technical requirements in the Annex 1 of the Decree No. 358/2016 [A–23] Coll., on requirements for assurance of quality and technical safety and assessment and verification of conformity of selected equipment. The general technical requirements are specified in the aforementioned Annex 1 of the Decree No. 358/2016 for pressure equipment, control (electrical) equipment, and structural equipment. These requirements reflect currently used technology in Czech Republic (i.e. PWR).

According to this decree, technical regulations, standards or specifications that are to be used shall be specified in the selected equipment design, together with acceptance criteria etc. Therefore, in the case of selected equipment for which there are no appropriate codes or standards established, either new ones would have to be created or derived from existing codes or standards for similar equipment (reflecting the differences). Prior to the use of selected equipment, the conformity assessment by the entitled person (authorized or accredited person or manufacturer) the compliance of the equipment with all of these requirements is verified and the SÚJB is authorized by Atomic Act to perform inspections to verify this assessment is in line with the legislative requirements, including the prescribed scope and procedures for conformity assessment (see the answer from Czech Republic to Question 15).

A–12.4.3. Follow-up Questions

For technologies Czech Republic might be considering, are items for which no appropriate code or standard available.

A–12.4.4. Response

Currently, no SMRs are envisaged to be deployed in the Czech Republic. Given the lack of detailed information about various SMR designs and the uncertainty over whether and which type of SMR could be hypothetically deployed, no further information can be provided.

A–12.5. FRANCE–ASN

A–12.5.1. Question

Describe the challenges where the industry codes and standards are absent or not fully applicable. Provide information on how regulatory assessment or judgement was done in view of the situation.

A–12.5.2. Response

Industry codes and standards are not a substitute for regulation but they are industrial tools that can be used to comply with regulation's requirements and with objectives given by ASN.

Regarding industry codes and standards, ASN follows their elaboration and their evolution. Also, ASN controls how they are used during inspections. However, ASN doesn't completely review industry codes and standards. ASN encourages codes and standards elaboration that lead to a better application of the regulation.

The article 3.2.3 of ASN's resolution n° 2015-DC-0532 [A–226] relating to a basic nuclear installation's safety report stipulates that the safety report must list industry codes and standards adopted by the licensee. The safety report must precise the eventual conditions and limits of their application.

A–12.6. JAPAN–NRA

A–12.6.1. Question

Describe the challenges where the industry codes and standards are absent or not fully applicable. Provide information on how regulatory assessment or judgement was done in view of the situation.

A–12.6.2. Response

Designs are basically based on the code and standards for research reactor facilities. However, there are cases to use guides, standards or general industrial codes which are used in design of commercial reactors as references, if needed.

A–12.6.3. Follow-up Question

For technologies Japan might be considering, are items for which no appropriate code or standard available?

Please, could extend more about your experience? Could provide references of the regulatory standards and / or guidelines, examples, etc. ?

A–12.6.4. Response

In HTTR review case, we did not encounter the case that couldn't apply appropriate standards and guidelines available.

If no appropriate code or standards were available, NRA might have to confirm the safety level of the reactor design in any suitable means such as referring similar standards and guides or cases applicable.

A–12.7. RUSSIAN FEDERATION–ROSTECHNADZOR

A–12.7.1. Question

Describe the challenges where the industry codes and standards are absent or not fully applicable. Provide information on how regulatory assessment or judgement was done in view of the situation.

A–12.7.2. Response

The currently operating floating NPP (Akademik Lomonosov) comprises the KLT-40C reactors, which are an evolutionary design of the icebreaker reactors that have been manufactured for decades for the Russian icebreaker fleet.

To this end, we do not have to face any challenges regulating equipment manufacture for the floating power plants in Russia.

Paragraph 8 of NP-022-17 [A–89]: "The vessel safety is provided by a high-quality design, construction and manufacture of vessel components; vessel construction and operation in compliance with the requirements of the federal laws, federal nuclear safety regulations, and standards adopted in conformity with the Russian legislation on standardisation; development and enhancement of the safety culture; the use of operating experience and state-of-the-art in science, technology, and production."

Paragraph 12 of NP-022-17 [A–89]: "The technical and organisational solutions adopted to ensure the safety of vessel's nuclear power installation must be proven by previous experience, tests, studies, and the operating experience of prototypes. These requirements must be applied in the design, construction and operation of nuclear power installation; in the design, manufacture, maintenance, and modification of its systems and components important for safety, as well as in vessel decommissioning."

The equipment for the icebreakers, floating NPPs and other nuclear facilities is manufactured against the international standards (GOST) which are similar to the ISO documents. A broad bank has been compiled for national standards (GOST R), which set requirements for the production and manufacture of bespoke equipment. The provisions of the shipbuilding industry standards (OST) were taken into account as well. An incomplete list of the standards taken into account during the construction and commissioning of the floating nuclear co-generation plant Akademik Lomonosov is given below:

- GOST 1062-80: Main dimensions of surface ships and vessels. Terms, definitions and symbols [A–227];
- GOST 13641-80: Structural components of the metallic hull of surface ships and vessels. Terms and definitions [A–228];
- GOST 19439.2-74: Operational documentation for vessels. Logbooks [A–229];
- GOST 25056-81: Cast deck and board hawsepipes. Specification [A–230];
- GOST 26069-86: Deck machinery and hull gear. Terms and definitions [A–231];
- OST 5.0099-74: Surface ships and vessels. Calculation methodology for trim and original stability [A–232];
- OST 5.0369-83: Technological preparation of the shipyard process. Terms and definitions [A–233];
- OST 5P.0737-2001: Design documentation for vessels. Rules for development, concurrence (approval) and endorsement [A–234].

A–12.7.3. Follow-up Question

Acceptable answer for the floating plant, but please provide any/some info for other SMRs under consideration.

A–12.7.4. Response

At the moment, the most promising design for a land-based SMR is RITM-200. The preliminary analysis showed the need to develop new GOSTs based on the existing GOSTs for vessel nuclear power installations, taking into account the materials and alloys used, as well as design features and manufacturing methods of the RITM-200 reactor equipment. The newly developed GOSTs will become part of the list of standardization documents for nuclear power plants captured in the federal regulations and rules in the field of the use of atomic energy. The development of these GOSTs is at an early stage.

In accordance with Federal Law No. 317-FZ dated 01.12.2007 'On the State Atomic Energy Corporation Rosatom' [A–235], the State Atomic Energy Corporation Rosatom carries out work on standardization in the field of the use of atomic energy, including the formation, maintenance, and updating of a consolidated list of documents on standardization.

A complete list of the current standards is presented in the 'Consolidated List of Standardization Documents in the Field of the Use of Atomic Energy' of the State Corporation Rosatom.

Rostechnadzor considers it necessary to carry out work on the identified specifics of ensuring the safety of land-based SMRs that do not comply with the current federal regulations and rules, as well as on the development of separate national standards and their subsequent inclusion in the Consolidated List.

A–12.8. SOUTH AFRICA–NNR

A–12.8.1. Question

Describe the challenges where the industry codes and standards are absent or not fully applicable. Provide information on how regulatory assessment or judgement was done in view of the situation.

A–12.8.2. Response

The SSC shall be designed according to the latest or currently applicable approved standards and consistent with the plant reliability goals necessary for safety. Where an unproven design or feature is introduced a complete analysis supporting the design and the codes and standards used shall be provided. An example of a technical area that might pose challenges in this respect is High Temperature Materials, but the same approach has to be applied for all codes and standards to be selected.

The major requirements to be considered in this field are given in RD-0034 [A–33].

For example, for the Main Power System Turbo- Machinery Casing ASME Code Class Selection Justification, the licence applicant was requested to submit, amongst others, the following documents:

- Process Standard, Selecting Codes and Standards;
- Process Description, Selecting Codes and Standards;
- Codes and Standard Justification Form;
- Process Competency Requirements for Selecting Codes and Standards.

As an example of the above, the Codes and Standard Justification Form would contain many questions to be responded to under each of headings such as:

- SSC definition – List the following SSC attributes;
- Overview of codes and standards;
- Select Code(s) and/or Standard(s), taking all attributes into account. Codes and Standards will be evaluated as below:
  - Evaluate functional appropriateness;
  - Evaluate regulatory compliance;
  - Evaluate operational experience;
  - Evaluate compliance with classifications;
  - Evaluate interfaces;
  - Identify and evaluate deviations;
  - Identify and evaluate exceptions;
  - Codes or standards not from plant baseline inventory.

The PBMR design differs significantly from current LWRs, and as a consequence, the existing rules for the choice of design codes, standards, guidelines and regulations cannot simply be applied. The risk, of a blanket application of the current LWR code selection rules, is that it could result in the choice of an inappropriate code, resulting in a deficient design, or irreconcilable inconsistencies in the subsequent code choices.

Nevertheless, wherever possible, PBMR SSC important to safety shall be designed according to either the latest or currently applicable approved standards.

A–12.9. UNITED KINGDOM–ONR

A–12.9.1. Question

Describe the challenges where the industry codes and standards are absent or not fully applicable. Provide information on how regulatory assessment or judgement was done in view of the situation.

A–12.9.2. Response

ONR has not formally assessed the application of extant industry codes and standards in the context of SMR design assessment. It is however expected that definition and consideration of RGP in areas of innovation or where technology is not mature will introduce challenges for vendors, designers and regulators.

ONR's goal-setting regulatory approach means that specific design solutions, codes and standards are not prescribed and vendors can propose alternative approaches in demonstrating that the risks have been reduced to ALARP. As previously mentioned, application of standards in situations that do not fall within the scope of the circumstances under which the specific RGP has been based is unlikely to represent an adequate ALARP demonstration, and under those circumstances a deeper level of scrutiny and consideration of failure modes, consequences and risk at a fundamental level will be expected when judging that the legal duty has been discharged.

SSCs that are important to safety should be designed, manufactured, installed, examined and inspected using codes, specifications and standards commensurate with their safety classification. The starting point for design is compliance with relevant national and international codes and standards. In addition, depending on the nuclear safety significance, safety case claims for the structural integrity of SSCs may require further substantiation.

UK Regulatory expectations for metal SSC 'highest reliability components' (see the answer from United Kingdom to question 10) are particularly high. Principle EMC.1 is that the safety case should be especially robust and the corresponding assessment suitably demanding, in order that a properly informed engineering judgement can be made that: (a) the metal component or structure is as defect-free as possible; and (b) the metal component or structure is tolerant of defects. Elastic-plastic defect tolerance assessment, qualified inspections (generally volumetric ultrasonic), and material toughness testing may be needed that exceeds existing industry code requirements.

A–12.10. UNITED STATES OF AMERICA–NRC

A–12.10.1. Question

Describe the challenges where the industry codes and standards are absent or not fully applicable. Provide information on how regulatory assessment or judgement was done in view of the situation.

A–12.10.2. Response

Codes and standards applicable to systems, structures and components for an SMR are the same as those used for LLWRs. One area that required additional evaluation and challenged the staff related to the applicable code was the of small diameter ($\leq$2" [10]) nuclear components.

For this SMR area of review, challenges were present because the industry codes and standards are different from the regulations, are interpreted differently, or are not fully applicable. The issue concerns application of ASME Code requirements related to the design, construction, inspection, etc. of small structures, systems, and components of the SMR design for only those small

---

[10] 5.1 cm

components of the containment pressure boundary whose failure could result in a loss of inventory during an emergency core cooling system (ECCS) actuation (containment pressurization) event.

For an LLWR design, there would typically be a strong correlation between meeting the design, fabrication, inspection, material testing, etc. requirements of the ASME Code and meeting associated general design criteria. However, because the ASME Code takes a graded approach based on component or piping size in regard to the requirements imposed to ensure the quality and integrity of structures, systems, and components, application of the ASME Code requirements to the generally smaller components and piping of an SMR design required additional review to ensure the desired level of confidence was achieved given their potential safety significance.

Though an SMR's nuclear small bore ($\leq$2") piping meets the ASME Class 1, Class 2, and Class 3 requirements, further consideration of small-bore components that are safety related should be used based on system design parameters and not pipe size. Based on using risk-informed principles, the staff resolved issues associated with the fabrication of small diameter components identified with the NuScale SMR design certification and ensured that the design not only complied with all applicable NRC regulations (e.g. 10 CFR [A–52] Part 50, Appendix A, General Design Criteria; 10 CFR 50.55a; etc.) but also ensured that reasonable assurance of public health and safety would be met.

A–13. ADEQUACY OF COMPUTER CODES FOR SAFETY DEMONSTRATION OF SMRS.

This Annex presents the responses provided by the following Member State regulatory bodies to the Question 13: "Describe the challenges associated with the adequacy of computer codes (availability, validation and verification status) for safety demonstration of SMRs."

A–13.1. ARGENTINA–ARN

A–13.1.1. Question

Describe the challenges associated with the adequacy of computer codes (availability, validation and verification status) for safety demonstration of SMRs.

A–13.1.2. Response

ARN does not have specific standards for the system codes used for safety analysis/demonstration (TH, neutronics, fuel behaviour, etc.). Most commonly used codes have been already validated and verified within the respective correlations range (RELAP5, TRACE5, CATHENA, CATHARE, MARS, ATHLET, and others).

However, due to special design features of CAREM reactor, ARN required a comprehensive analysis of each condition and physical phenomena that could occur during PIEs transients/accidents and checking codes capability for capturing and representing them. This procedure holds mainly for thermal-hydraulics and neutronics codes.

Some specifics codes have been developed for engineering/demonstration, e.g. steady-state conditions, DNB, instabilities maps. ARN provided guidance on how to develop models/codes, its documentation, and the quality assurance process required (based on CNSC G-149 [A–59]).

An independent input deck (in RELAP5 patch 3.4) was developed for running simulating events with level 3 of DiD that includes main protection line design basis accidents and some DECs. ARN developed an independent thermohydraulic model of the CAREM 25, to perform analysis of postulated single failure or low-frequency events. The code RELAP5 mod3.3 patch 4 distributed by the US-NRC was used. The model includes:

- The primary system;
- The secondary system;
- The safety systems, SSECR, PSE, SSE, SIS;
- The logics of control for the performance are included.

Postulated initiating events simulations was:

- Blackout station with residual heat removal system availability;
- Break 2A in the steam generator (SG) power line, out of containment;
- Break 2A in the live steam line at the exit of the SG, out of containment;
- 1.5" diameter breaking in the RPR steam dome liquid zone;
- Control bar extraction at nominal speed 1 cm/s with first reactor protection system success, first shutdown system (PSE) failure and SSE success.

The simulation of the previous events may be carried out with or without the availability of the normal or emergency power supply.

A–13.1.3. Follow-up Question

What activities are done by regulator in V&V?

How is this specifically include, is it limited to safety analysis or all design codes?

A–13.1.4. Response

As mentioned in answer from Argentina to Question 13, ARN doesn´t have specific standards that contain the regulatory requirements applicable to the validation and/or verification of codes or calculation software for the evaluation of nuclear safety in nuclear reactors. In these cases, the ARN may consult or review the international documentation available (IAEA, NRC, ONR, and other agencies) to study its applicability and/or acceptance as a guide to follow.

In the case of licensing of CAREM 25, the Licensee developed some specific system codes used for safety analysis/demonstration. Regarding this, ARN developed a guidance document that is an adaptation of Guide G-149 [A–59] of the Canadian Nuclear Safety Commission (CNSC). This guide proposes a series of guidelines and principles to be taken into account regarding the proper use of the codes used in safety design and analysis. The document also included guidelines prepared by ARN personnel.

The ARN request for the verification and validation of the calculation codes used for the design of CAREM 25, consists of assessing the degree to which the mentioned specific codes comply with the content of this guide. Once the deviations were identified, the resolution of the findings was agreed between the Licensee and the ARN.

A–13.2. CANADA–CNSC

A–13.2.1. Question

Describe the challenges associated with the adequacy of computer codes (availability, validation and verification status) for safety demonstration of SMRs.

A–13.2.2. Response

In Canada, computer codes used for the nuclear industry have been developed to support the CANDU technology. The applicability of these codes to advanced reactor designs is not proven. In many cases, the same holds true for available international codes due to the differences in design and physical phenomena involved. In any case, any computer code used in safety analysis must undergo appropriate verification and validation.

CNSC's REGDOC-2.4.1 [A–63] requires computer codes used in the safety analysis to be developed, validated and used in accordance with a quality assurance program that meets the requirements and expectations of CSA N286.7 [A–147].

REGDOC-2.4.1 and N286.7 require that validation process:

    (a)   Assess adequacy and applicability of the models employed in the computer program;
    (b)   Demonstrate computer program capabilities and limits;

(c)   Determine computer code accuracy.

The validation should ideally cover the full range of values of parameters, conditions and physical processes that the computer code is intended to be used. The code accuracy obtained as a result of validation should be used as a source of uncertainties of relevant modelling parameters.

Validation is conducted by comparing computer program predictions with one or more of the following:

(a)   Applicable experimental or operational data;
(b)   Relevant solutions to standard or benchmark problems;
(c)   Relevant closed-form mathematical solutions;
(d)   Relevant results of another validated computer program.

The data used for validation should cover the full range of values of parameters, conditions and physical phenomena that the computer code is intended to model, in the specific applications for which it is to be used.

Each computer code should be adequately documented to facilitate review of validation results. The validation documentation should contain:

(a)   A statement of the application for which the computer program is being validated;
(b)   A description of the methods used;
(c)   Description of data against which validation was performed including data selection and qualification criteria;
(d)   A description of computer program inputs and output;
(e)   Assessment of validation results with respect to computer program accuracy.

While the availability of validated computer codes for safety demonstration of SMRs may be limited, vendors engaged in the CNSC's pre-licensing vendor design review process understand CNSC expectations and continue to undergo the necessary research to generate the data for code validation.

Code verification process is extensively covered in CSA N286.7-16 [A–147]. It is expected that the licensee will document and follow the process used for model and system code verification during computer code design phase. The process should cover implementation of the intended conceptual or mathematical models and the review of source coding in relation to its description in the system code documentation. The level of regulatory review depends on the importance of the code to the safety case. The assumptions, available information, uncertainties estimated margins to safety are taken into account when establishing the level of review.

A–13.2.3. Follow-up Question

What activities are done by regulator in V&V?

How is this specifically included, is it limited to safety analysis or all design codes?

A–13.2.4. Response

What activities are done by regulator in V&V?

The CNSC uses risk informed approach to determine the level of regulatory review and it is dependent on the importance of the code to the safety case. The CNSC will examine the information submitted and take into account the assumptions, available information, uncertainties estimated margins to safety when establishing the level of review.

Due to limited computer codes that are validated for safety demonstration of SMRs, CNSC expects a vendor to generate quality data for code validation via their research program. The CSA N286.7 - 16 [A–147] standard covers the code verification process that CNSC expects the vendor to follow.

How is this specifically included, is it limited to safety analysis or all design codes?

The computer codes cover safety analysis, and the information are provided in REGDOC 2.4.1 [A–63]. The computer codes used in the safety analysis are expected to be in accordance with a quality assurance program that meets or exceeds the CSA N286.7. The G-149, 'Computer Programs Used in Design and Safety Analyses of Nuclear Power Plants and Research Reactors' [A–59], provides guidance on computer code expectations.

A–13.3. CHINA–NNSA

A–13.3.1. Question

Describe the challenges associated with the adequacy of computer codes (availability, validation and verification status) for safety demonstration of SMRs.

A–13.3.2. Response

Select the HTR-PM as an example to illustrate this issue. The codes used for the design and analysis of the HTR-PM were designed by the Research Centre Jülich (FZJ), Germany. These codes were validated by some experimental data, as well as the operating data from AVR, and had been used in HTR-Module design. The information has been provided by the authorized party (INET), to explain the code validation and verification carried out in FZJ. Some experiments were also carried out by INET for further validation of the codes.

For further study and development of the HTGR technology, the most challenging problems associated with the computer codes remains the validation and verification.

(a)   Compared with the water reactor, the experimental data and operating data available for code validation of HTGR is relatively limited.
        The conceptual difference (for example, pebble bed core and prismatic core) of HTGR development between relevant countries also results that some data could not be shared.

(b)   Code to code verification is expected and should be encouraged in the future.

(c)   More experiments should be designed and carried out to study the important phenomena of the HTGR and validate the models and correlations employed in the existing codes.

A–13.3.3. Follow-up Questions

What are the regulator standards? What was the need for and the scope of the additional validation carried out by INET?

What activities are done by regulator in V&V?

How is this specifically included, is it limited to safety analysis or all design codes?

A–13.3.4. Response

HAD 102-16 'Computer Based Safety Important System Software for Nuclear Power Plants' [A–236] provides the V&V related content.

A–13.4. CZECH REPUBLIC–SÚJB

A–13.4.1. Question

Describe the challenges associated with the adequacy of computer codes (availability, validation and verification status) for safety demonstration of SMRs.

A–13.4.2. Response

Not applicable — no challenges associated with the adequacy of computer codes, their availability, validation or verification status for safety demonstration of SMRs have been identified as no SMR has been or is being currently assessed.

As such, the Czech legislation does not contain specific detailed requirements for computer codes. In this regard, the Czech legislative system is quite flexible and does not specify the means or the extent of the validation and verification of the computer codes. The general requirements for validation, verification and review are laid down in the Decree No. 408/2016 [A–237] Coll., on management system requirements. Processes and activities and their inputs and outputs shall be reviewed, verified, and validated prior to their first use (including computer codes). Details concerning the computer codes, the scope and manner of their verification are contained in the safety guides issued by the SÚJB (including assessment of uncertainties, sensitivity analyses). Any code is to be verified and validated in a documented manner to demonstrate its suitability and satisfactory accuracy for the field of use and shall correspond to the level of knowledge achieved in the relevant area. A brief description of the code, calculation model and their validation and verification is part of the safety analysis.

A–13.5. FRANCE–ASN

A–13.5.1. Question

Describe the challenges associated with the adequacy of computer codes (availability, validation and verification status) for safety demonstration of SMRs.

A–13.5.2. Response

According to the article 3.8 of the Order [A–72]:

"I. The nuclear safety case is based on:

- up-to-date and referenced data; it more specifically takes account of the available information mentioned in article 2.7.7;
- appropriate, clearly explained and validated methods, containing hypotheses and rules appropriate to the uncertainties and the extent of the knowledge of the phenomena involved;
- calculation and modelling tools qualified for the fields in which they are used.

II. The licensee specifies and substantiates its criteria for methods validation, computing and modelling tools qualification and assessment of the results of the studies performed to demonstrate nuclear safety."

ASN's guide n°28 published in 2017 [A–238] presents the ASN and IRSN recommendations for qualification of scientific computing tools used in the nuclear safety case. In particular, it deals with:

- The intended scope of utilization of the scientific computing tools in the safety case, which must be defined before the process of verification, validation and transposition;
- The process of verification and validation of the SCT, which lies at the heart of qualification;
- The process of transposition of the validation cases to the intended scope of utilization;
- The declaration of qualification;
- Several points concerning certain software (pre- and post-processing, coupling, etc. and certain uses of the SCT;
- The description of the content of the qualification file to be submitted to ASN.

An English version of this guide is available on ASN's website.

The Order's requirements [A–72] and ASN's guide n°28 [A–238] are applicable to SMR designs.

A–13.6. JAPAN–NRA

A–13.6.1. Question

Describe the challenges associated with the adequacy of computer codes (availability, validation and verification status) for safety demonstration of SMRs.

A–13.6.2. Response

Proven computer codes are used at the time of permission of instalment of the HTTR. The newly used computer code for graphite oxidation phenomenon in the review of the HTTR, was verified through comparing experimental data, and it was confirmed that the code was applicable to analysis of the phenomena in the verified range.

A–13.6.3. Follow-up Questions

What are the standards governing V&V?

What activities are done by regulator in V&V?

How is this specifically included, is it limited to safety analysis or all design codes?

A–13.6.4. Response

In HTTR review, V&V for the design codes was confirmed by reviewing the record for quality assurance activity of the entity.

A–13.7. RUSSIAN FEDERATION–ROSTECHNADZOR

A–13.7.1. Question

Describe the challenges associated with the adequacy of computer codes (availability, validation and verification status) for safety demonstration of SMRs.

A–13.7.2. Response

The Russian Federation has amassed formidable experience in the use of various software tools for comprehensive analyses of various safety aspects. Federal Law No. 170-FZ [A–26] establishes that any software tools used in safety analysis must undergo a review in an organisation providing scientific and technical support to the safety regulation authority (TSO).

The existing review process (approved by Rostechnadzor Order No. 325 of August 30, 2018 [A–239]) provides for computer testing of declared software capabilities and verification of their adequacy for conducting declared calculations. Apart from that, an expert review includes an independent appraisal of the outcome of software validation against experimental data or other software tools. The reviewers are experienced software developers and representatives of the design and technical support organisations.

The review findings are discussed at the meetings of a dedicated Board that is composed of representatives of the regulatory body, the design and operating organisations, and the TSO. The meeting conclusions provide an input to make a decision on the software eligibility for being used within the declared scope of application. After that, a code certificate is prepared, with indication of the scope of code application, the parameters calculated by the code, and the verified errors. The certificate must be renewed every 10 years, considering the code application experience and the state-of-the art in the science and technology.

The Russian Federation has a formidable nuclear icebreaker fleet. Same as with other nuclear facilities, the safety analysis for vessels equipped with nuclear power installations largely uses the software already reviewed by experts as appropriate against applicable requirements. The design of the floating SMR (FNPP Akademik Lomonosov) is in many aspects similar to the designs of existing vessels with nuclear power installations. Thus, for example, the SMR design uses as a power source the KLT-40C type nuclear reactors that are similar to the nuclear power installations used at some nuclear icebreakers. Considering this, the safety analysis for the design of the FNPP Akademik Lomonosov was performed using the computer codes that had already been tested and verified for the safety justification of nuclear-powered ships and have a relevant certificate. No new software tools have been developed and verified/validated specially for the analysis of the Akademik Lomonosov safety aspects. The safety documentation for the floating plant also includes a description of the computational models, errors and estimated potential uncertainties that have been additionally considered in the expert review of the safety justification. Thus, the following software tools are certified for the Akademik Lomonosov:

- MCU-TR;
- VIBROS 2.2;

- Gidr-3M;
- KUPOL-MT (mod 1.0);
- RiskSpectrum PSA;
- Other technology-neutral computer codes.

A–13.7.3. Follow-up Questions

What activities are done by regulator in V&V?

How is this specifically included, is it limited to safety analysis or all design codes?

A–13.7.4. Response

In accordance with clause 14 of NP-022-17 [A–89], the software used for safety demonstration must be certified. Information on the software used for safety demonstration of a nuclear facility is available in the safety analysis report. When reviewing the safety analysis report of a nuclear facility, the technical support organizations of the regulatory body checks the correctness of the use of the provided software. At the same time, the safety analysis report must provide information on the calculations performed in an amount sufficient for their verification using both the specified software tools and alternative ones.

A–13.8. SOUTH AFRICA–NNR

A–13.8.1. Question

Describe the challenges associated with the adequacy of computer codes (availability, validation and verification status) for safety demonstration of SMRs.

A–13.8.2. Response

At the time of the PBMR project (up to 2010), the NNR requirements and guidelines for computer codes were contained in RD-0016, 'Requirements for licensing submissions involving computer software and evaluation models for safety calculations' [A–109] and LG-1045, 'Guidance for licensing submissions involving computer software and evaluation models for safety calculations' [A–115].

In addition to the information contained in RD-0016 and LG-1045, the following is stated here:

Special attention needed to be paid to limitations in the QA status of 'Legacy Codes' and the status of experimental and analytical verification and validation of these codes. Compliance with the regulatory requirements regarding 'Legacy Codes', as well as for 'Codes under Development', needed to be demonstrated.

It was also required that the design of the test facilities is such that they should directly be able to support the validation of the computer codes for the phenomena modelled. The test facilities will, however, not necessarily identify phenomena that have not been included in the models, as the test facility design is often tied to the model predictions. They will not, therefore, adequately demonstrate that the models are a reasonable representation of the phenomena occurring on the PBMR plant. According to the PBMR V&V strategy, the relevant phenomena have to be identified comprehensively by a PIRT process in an early stage of evaluation model development. The licence applicant was requested to make the PIRT reports and related documents available. Once this additional information has been reviewed, the scope of the testing programme may need to be extended to address the fundamental issue of linkage between the behaviour of the proposed PBMR plant, the computer codes used to model it and the role of the tests used to validate claims made in the safety case. Specifically, the relevance of the tests proposed on the test facility to the code validation programme must be clear.

To support the resolution of the NNR concerns in this area the licence applicant was requested to submit the latest version of the code verification and validation master plan to the NNR. This update was required to include the following:

- A list of all test facilities that PBMR proposes to use in support of the code validation and plant licensing process;

- A clear description of how each of the test facilities will be utilised to support the various aspects of the code validation programme and plant licensing process. This does not need to describe the justification for the individual tests used to validate each code (this should be included in the individual code V&V plans). However, sufficient detail should be included in the master plan to allow the NNR to form a judgement as to the completeness and appropriateness of the overall programme;

- It should demonstrate how the key phenomena expected to occur on the PBMR plant are identified by PIRT or other similar process and evaluated as part of the validation programme;

- The description of the proposed tests should be based on facility rather than computer code and should indicate in general terms the nature of the tests and which computer codes they will support. For each facility the following items should be addressed, either in the V&V master plan, or later in the individual code V&V plans:

    o   Scaling reports;
    o   Outline test programme;
    o   Arrangement for test monitoring by ESKOM and the NNR as required;
    o   Data to be obtained from the tests;
    o   Assessment reports;
    o   Uncertainty analysis reports.

A–13.8.3. Follow-up Question

What activities are done by regulator in V&V?

How is this specifically included, is it limited to safety analysis or all design codes?

A–13.8.4. Response

NNR interest in V&V is naturally enhanced for the PBMR licensing process due to the first of a kind nature of many aspects associated with a demonstration project.

At the time of the PBMR project (up to 2010), the NNR requirements and guidelines for computer codes were contained in RD-0016 [A–109] and LG-1045 [A–115].

As stated at the end of Section 1 of Regulatory Guide RG-0016 , RD-0016 [A–109] and LG-1045 have subsequently been consolidated and superseded by RG-0016.

294

NNR interest in the V&V activities on the licence applicant's side is also indicated in for example the following statement in Section 6.2.1 f RG-0016:

"6) The efforts to verify and validate software products should be documented adequately. Elements of a comprehensive documentation are V&V plans, interim V&V reports where appropriate, and final V&V reports. All this documentation should be available to the NNR."

To the extent that NNR work in the regulatory independent analysis area is V&V related, the following is mentioned:

The regulatory assessment of the PBMR Safety Case requires a comprehensive set of sophisticated independent analytical tools that use a variety of techniques. Computer codes are used both to replicate calculations performed by PBMR and to provide additional calculations deemed necessary by the assessors. All areas of importance to the Safety Case that may require the use of computer codes for assessment have been identified and listed in a detailed matrix and suitable techniques for assessment, computer codes and/or models identified. This has defined a 'toolbox' of computer codes, models and methods that the Regulator may need in order to fully assess the SAR. It is also important to consider interfaces and data flow between analysis areas with a view to defining the interface requirements or a strategy in order to be ready to address the SAR on the agreed timescale.

During the PBMR project, NNR computer code development took place in the areas of neutronics and thermal hydraulics of the core as well as the whole plant model to seek to understand the potential interactions between the various areas of the plant. For the evaluation of fission product release from the fuel spheres a specific computer code was developed.

Efforts were made to develop methods to assess other areas of the safety case. Important areas relevant to the licensing criteria are:

- Source term analyses (including graphite dust);
- Civil structures (including seismic loads, external events and internal pressurisation);
- Internal fluid-dynamics of the reactor building;
- Risk analysis and probabilistic risk analysis (PRA) of all postulated initiating events (PIE);
- Structural analysis of SSC including the core structure ceramics;
- Radiation protection;
- Chemical attack.

Specific tools were needed and have been developed for source term analysis of the Primary Circuit (releases from the coated particles and the fuel spheres, code FPRC) as no appropriate codes covering the various phenomena were available. Enveloping assumptions are also not appropriate since high uncertainties have been identified and the results are directly related to the licensing criteria and the demonstration of the ALARA principle.

The following statements from RG-0016 [A–35] indicate that NNR interest in a licence applicant's V&V activities is not limited to safety analyses but also includes design analyses to the extent that design analyses have nuclear safety implications:

- According to Section 2 of RG-0016, "(…) the document consolidates all regulatory requirements and guidance in the area of verification and validation of evaluation and calculation models used in safety and/or design analyses."

- According to Section 3 of RG-0016, the " (…) document provides guidance on the verification and validation of evaluation and calculation models used in both safety and design analyses and should be used by applicants, authorisation holders, designer of nuclear facilities as well as service providers performing important to safety analyses and designs."

This guidance is not directly applicable for software used directly for plant operational control and protection. International guidance for this type of software is given for example in Verification, Validation, Reviews, and Audits for Digital Computer Software Used in Safety Systems of Nuclear Power Plants, Regulatory Guide 1.168 [A–240] and Documentation of Computer Software ANSI/ANS-10.3 [A–241].

- Item 1) in Section 5.1 states: "Information about computer software and evaluation models for safety calculations and design analyses important to nuclear safety should be comprehensive."
- Item 1) in Section 8.1 states: "In order to assess the verification and validation of software used in design and/or safety analyses the submission (validation package) should cover:(…)"
- The first sentence of Section 9.1 states: "The above guidance is applicable to all computer codes used in important to safety design and safety analyses."

A–13.9. UNITED KINGDOM–ONR

A–13.9.1. Question

Describe the challenges associated with the adequacy of computer codes (availability, validation and verification status) for safety demonstration of SMRs.

A–13.9.2. Response

ONR expectations related to assurance of validity of data and models are captured in the ONR SAPs [A–44] - AV.1 to AV.8 and further guidance for ONR Inspectors is available in NS-TAST-GD-042 Validation of Computer Codes and Calculation Methods [A–242]. ONR assesses the application of codes and methods in the context of GDA or permissioning assessments associated with new facilities against the above SAPs and TAG [A–46]. ONR has so far not undertaken formal regulatory assessment of SMR-related safety analysis.

ONR experience from early engagement and exposure to SMR vendor claims and preliminary analysis indicated that the level of maturity of computer codes and the availability of data to validate the intended application of the code may be limited in the context of some designs. In this context SAP [A–44] AV.3 covers the use of data in fault analysis and states: "The data used in the analysis of aspects of plant performance with safety significance should be shown to be valid for the circumstances by reference to established physical data, experiment or other appropriate means."

It is therefore expected that significant amount of code development, validation and verification activities as well as experimental tests to develop representative data for some of the designs under consideration will feature in design development activities by vendors and research organisations.

A–13.9.3. Follow-up Questions

What activities are done by regulator in V&V?

How is this specifically included, is it limited to safety analysis or all design codes?

A–13.9.4. Response

ONR assesses the adequacy of the dutyholder's validation and verification of codes (and does <u>not</u> undertake this validation or verification on behalf of the dutyholder). ONR has documented its expectations on validation and verification in a Technical Assessment Guide (TAG) [A–242].

The guide provides advice to inspectors making judgements in the topic area. It applies mainly to the assessment of the validation of physics, thermal and structural analysis computer codes and calculation methods in safety studies. The principles in the TAG are applicable to transient, radiological and other analyses forming part of fault analysis and also to other areas of the safety case which are underpinned by analysis and/or data, e.g. engineering substantiation.

A–13.10.   UNITED STATES OF AMERICA–NRC

A–13.10.1. Question

Describe the challenges associated with the adequacy of computer codes (availability, validation and verification status) for safety demonstration of SMRs.

A–13.10.2. Response

The same regulations that apply to existing LWRs regarding computer codes apply to SMRs, and staff did not find any significant challenges in this area of review. The applicant's design has been reviewed by the NRC staff using existing NRC Commission Policy statements regulatory requirements as well as existing regulatory guidance and NRC Staff review guidance from NUREG-0800 [A–116].

To meet the requirements of 10 CFR [A–52] Part 50, Appendix B and GDC 1, a list of computer programs used in dynamic and static analyses to determine the structural and functional integrity of seismic Category I components, ASME BPV Code [A–215] and non-Code components should be provided. For each program, as a minimum, the following information should be provided to demonstrate its applicability and validity:

- The author, program source, dated version, and facility;
- A description and the extent and limitation of its application;
- The computer program solutions to a series of test problems demonstrated to be compatible with solutions obtained from any one of sources (i) through (iv) within the acceptable margin using benchmark problems acceptable to the staff (e.g. NUREG/CR-1677, 'Piping Benchmark Problems' Volumes I and II [A–243]):
    o   Hand calculations;
    o   Analytical results published in relevant engineering literature;
    o   Acceptable experimental tests;

o   A similar computer program previously accepted by NRC or acceptable to the staff.

A summary comparison of the solution obtained from sources (i) through (iv) should be provided in either graphical or numerical form. In addition, the complete computer printout of the input and the solution should be submitted for every benchmark problem.

A–14. REGULATORY ASSESSMENT OF NOVEL/INNOVATIVE DESIGN FEATURES (SSCS) IN ABSENT OF PREVIOUS APPLICATION IN NUCLEAR INDUSTRIES

This Annex presents the responses provided by the following Member State regulatory bodies to the Question 14: "Describe the challenges associated with regulatory assessment of novel/innovative design features (SSCs) in absent of previous application in nuclear industries. Provide information on how regulatory judgement was reached in view of the situation."

A–14.1. ARGENTINA–ARN

A–14.1.1. Question

Describe the challenges associated with regulatory assessment of novel/innovative design features (SSCs) in absent of previous application in nuclear industries. Provide information on how regulatory judgement was reached in view of the situation.

A–14.1.2. Response

The concept of CAREM that belongs to the very low or low power nuclear plants was put forward from the very beginning as an advanced designed reactor, being the precursor of innovative concepts as regards safety. CAREM 25 is a light water reactor with new design solutions, which contributes to its high level of safety, being the followings its main innovative aspects:

- Integrated Primary System;
- Self-pressurization;
- Passive Safety Systems.

Even though both technical and engineering solutions associated to the NPP's technology and, the innovative design characteristics are correctly verified during the design phase, it was considered convenient to construct a reactor prototype to validate its design, manufacturing, installation and operational aspects as well as verification of SSC's reliability. Due to this fact, CNEA, proposed to the National Government, to carry out the construction of the CAREM 25 Prototype Reactor, by means of the construction of a CAREM NPP of 25 MW(e).

Regarding Safety demonstration by tests, during the 1990s, a High-Pressure Natural Circulation Rig (CAPCN) was built and run by INVAP for CNEA in Pilcaniyeu, Río Negro (near Bariloche). CAPCN was designed to reproduce intensive parameters (P, T, velocities, flow patterns, heat transfer regimes, void fraction generation (in core) and collapse) and most of the dynamic phenomena (single and two-phase natural circulation, elements of the self - pressurization of the dome, stratification) of the CAREM reactor coolant system. This installation was used for parametric studies of the dynamic response to perturbations, changing steam dome volume, hydraulic resistance, Pressure, and Temperature. CAPCN modelling, plus experimental data and similarity analyses allow CNEA to assess the codes used for CAREM modelling, and Reactor control and operating techniques were assessed.

The Research Reactor RA-8 was built to test the CAREM 25 prototype core. This was a critical facility of enriched uranium moderated in an open pool with 10 W of power. It was designed and built by INVAP for CNEA in Pilcaniyeu and operated between 1997 and 2001. This installation was used for tests of the fuel and core design of CAREM 25 into a full-scale physical model element, according to the 1990's limitations for the qualification of SoA neutronic models/codes. Currently, the installation is in the process of decommissioning.

The CAPEM facility is a high-pressure rig for testing of hydraulic CRDs, characterization, endurance tests of the control and shutdown systems. The design of the CRDs mechanisms is relatively innovative. The feasibility of the concept and technological solutions have already been verified at low-pressure conditions, plus engineering. This installation is used for qualification tests of the safety function rapid extinction in the proposed scenarios of operation, DBA and DEC, for engineering tests on functions of adjustment and control and for verification on the positioning measurement.

CAREM 25 was presented for its analysis in several international forums, for example, between 2001 and 2002 the US-DOE (Department of Energy) and the Generation IV International Forum (USA), evaluated different technological alternatives of nuclear electric generation, including the CAREM. Argentina is one of the countries that integrate the above-mentioned Forum. The CNEA is also active representing Argentina at the INPRO (International Project on Innovative Nuclear Reactors and Fuel Cycles), within the scope of the IAEA. The above-mentioned evaluation was in charge of approximately 100 experts of different countries, belonging to governmental organisms, universities, and associations.

A–14.2. CANADA–CNSC

A–14.2.1. Question

Describe the challenges associated with regulatory assessment of novel/innovative design features (SSCs) in absent of previous application in nuclear industries. Provide information on how regulatory judgement was reached in view of the situation.

A–14.2.2. Response

The Canadian regulatory approach to licensing SMRs is built on the long-established foundation of risk-informed regulation that has been applied to traditional reactor facilities. The Canadian nuclear regulatory framework consists of a comprehensive and evolving set of requirements based on more than 70 years of operating experience. The framework is intended to be technology-neutral, which means that it allows for all types of technologies to be safely regulated. Regulatory tools and decision-making processes are structured to enable an applicant to propose alternative ways to meet regulatory objectives.

Most SMR concepts, although based on technological work and operating experience from older NPPs, employ a number of novel approaches. Novel approaches, or even proven approaches used in different ways, can affect the certainty of plant performance under both normal operation and accident conditions, raising regulatory questions during the licensing process. When applying for a licence, the applicant is to address CNSC requirements in a manner that is commensurate with the novelty, complexity and potential for harm that the activity represents. Proposals must demonstrate, with suitable information, that they are equivalent to or exceed regulatory requirements.

The CNSC's requirements and guidance for reactor facilities are generally articulated to be technology-neutral and, where possible, permit the use of the graded approach. The graded approach is a method or process by which elements such as the level of analysis, the depth of documentation and the scope of actions necessary to comply with requirements[11] considers:

(a)   The relative risks to health, safety, security, the environment, and the implementation of international obligations to which Canada has agreed;
(b)   The characteristics of a facility or activity.

The graded approach enables applicants to propose the stringency of design measures, safety analyses and provisions for conduct of their activities commensurate with the level of risk posed by the reactor facility. Key factors are:

- Reactor power;
- Source term;
- Amount and enrichment of fissile and fissionable material;
- Spent fuel, high-pressure systems, heating systems and the storage of flammables, all of which may affect the safety of the reactor;
- Type of fuel elements;
- Type and the mass of moderator, reflector and coolant;
- Amount of reactivity that can be introduced (and its rate of introduction), reactivity control, and inherent and additional features;
- Quality of the confinement structure or other means of confinement;
- Utilization of the reactor;
- Siting, which includes proximity to population groups or extent of isolation from emergency responders.

The CNSC has engaged with industry stakeholders to seek their views on how a graded approach could be applied to SMRs; this consultation took the form of a workshop. The result of this consultation is documented in a Stakeholder Workshop Report [A–244]. This report provides specific examples of how the graded approach would be applied to the regulation of SMRs.

Closely related to, yet distinct from, a graded approach is the use of an alternative approach, which is identified in REGDOC-2.5.2 [A–62]. While a graded approach seeks to scale a given protective measure with the risk posed by the facility, an alternative approach looks to assess a different way of achieving an equivalent or superior level of safety. An alternative approach will also be considered where CNSC requirements may conflict with other rules or requirements or where the application of CNSC requirements would not serve the underlying purpose or is not necessary to achieve the underlying purpose.

Any alternative approach shall demonstrate that safety and security protections are maintained or improved. Where risk characteristics contain uncertainties, the amount of evidence required for the applicant to demonstrate a credible decision increases. Suitable evidence may include results of research and development, computer modelling and consideration of operating experience, and the evidence must be demonstrated to be relevant to the specific proposal. All of these types of

---

[11] It is important to note that use of a graded approach is not a relaxation of requirements. Where novelties are proposed, the CNSC expects the proponent or applicant to demonstrate how the project accounts for the uncertainties associated with the novelties. This can be through means such as sacrificial wall-thickness for uncertain corrosion rates (conservative design), more frequent or in-depth inspections (conservative maintenance), or a reduced / more restrictive operating envelope (conservative operation). The proponent or applicant must propose and justify an appropriate means.

evidence should be documented, traceable and quality-assured. A proponent that is considering a licence application for an SMR is encouraged to engage with the CNSC early on, well in advance of submitting the application, in order to understand CNSC expectations for management systems and quality assurance. This will inform research and development work, with a view to supporting a potential future licence application.

In their assessment of alternatives proposed by an applicant, CNSC staff evaluate if the alternatives:

- Meet the intent of the stated requirements;
- Meet high-level safety objectives;
- Meet fundamental safety functions of 'control, cool, contain'.

At the same time, the alternatives need to demonstrate:

- Defence in Depth (DiD);
- Safety margins in view of the uncertainties in the safety case and of specific hazards over the facility's lifecycle.

An example of an alternative approach would be where the applicant or licensee wishes to use a code or standard not recognized in Canada. In such a case, the applicant or licensee would be expected to submit a code comparison to the CNSC to make their case that the use of the code would result in an equivalent or superior level of safety. Regardless of the specific alternative being proposed or consider, the approach must demonstrate equivalence to the outcomes associated with the use of the requirements set out in the regulatory framework.

Qualification of SSCs is not a one-time discussion with a single answer of 'yes' or 'no'. Qualification of SSCs is a living discussion that is revisited as the body of knowledge grows. It starts with a discussion about whether there is sufficient information supporting the qualification claims for SSCs taking into account safety and control provisions that address uncertainties.

Considerations in deciding whether sufficient information exists to make regulatory decisions concerning 'sufficient qualification':

- Different SSCs will have differing levels of qualification data during the design process;
- The interfaces between SSCs means that these differing levels of qualification may present compounding/stacking of uncertainties;
- Does the vendor/applicant understand the long term implications of the uncertainties? What are they doing about them? (long term R&D, data gathering from the first facility etc);
- What interim safety and control measures will be proposed to address uncertainties?
- What would the regulator need to impose (licence conditions/hold points etc) to further address any potential residual uncertainties?

The Commission makes independent and objective decisions to ensure that unreasonable risks are prevented, taking into consideration regulatory requirements, best available information from regulatory or credible third-party research, and all information provided by applicants/licensees, Indigenous peoples, other stakeholders and staff. CNSC staff make recommendations to the Commission based on thorough assessments of factual evidence, measured against regulatory requirements.

302

The Commission recognizes the role of professional judgment, particularly in areas where no objective standards exist. Its independence and transparency in decision making are supported by fair, open, transparent and predictable regulatory processes.

Understanding risks, including associated uncertainties, and ensuring that these risks are mitigated plays a significant role in making regulatory recommendations and decisions. The risks and mitigation approaches need to be clearly described and well understood in order for the Commission to make an informed decision. Supporting evidence and the quality of that evidence are critical.

The Commission and CNSC staff apply a risk-informed approach to all safety and control areas in order to place an appropriate amount of regulatory scrutiny on activities, depending on the level of risk. Primary considerations for the extent and depth of application are the degree of novelty, complexity and potential harm posed by the proposed activity or facility. The degree of scrutiny, which may vary upward or downward, is further informed by:

- Technical assessments of submissions;
- Safety performance history of the licensee (if applicable);
- Relevant research;
- Information supplied through the Commission's public processes;
- National and international activities that advance knowledge in nuclear and environmental safety;
- Cooperation with other regulatory bodies.

When the Commission assesses applications that reflect a graded approach, its primary consideration is to ensure that risk is demonstrated to be at a reasonable level. This includes ensuring that:

- Regulatory requirements will be met;
- Fundamental safety functions are adequately addressed by design;
- DiD is demonstrated;
- Safety margins are appropriate and in line with specific hazards over the facility's lifecycle.

Regulatory requirements and expectations provide a starting point for regulatory review, but each case will be reviewed on its own merits. More detailed information on risk-informed techniques and other methodologies can be found in CAN/CSA-IEC/ISO 31010-10, Risk management – Risk assessment techniques [A–245], and CSA N290.19, Risk-informed decision making for nuclear power plants [A–246].

Section 2 of REGDOC-1.1.5 [A–65], provides topical guidance and information to be provided when a proponent or applicant is considering a novel or innovative design feature or approach. Guidance is provided for all the CNSC's safety and control areas as well as the other matters of regulatory interest. For example, in considering the management system safety and control area, and the relative emphasis it should be given, the following considerations should be addressed:

(a) Complexity of the facility or activity, elements of which may include:
    (i) Complexity of required managed processes;
    (ii) Complexity of the organization;
    (iii) Number and size of radioactive or nuclear sources present;

      (iv)   Number of radioactive sources being used at any one time;
      (v)   Degree of automation.
(b)   Structure of the operating organization;
(c)   The need for effectively managed processes to control identified hazards, elements of which may include:
      (i)   Change control;
      (ii)   Design control;
      (iii)   Document control;
      (iv)   Work planning and control;
      (v)   Corrective action;
      (vi)   Maintenance;
      (vii) Configuration management;
      (viii) Operations;
      (ix)   Operating experience.
(d)   Safety culture;
(e)   Extent of activities involving risk (to health, safety and the environment) and requiring managed processes and controls;
(f)   Frequency, extent and need for critical human involvement in the activities of the facility;
(g)   Remote or local operation;
(h)   Number and type of barriers to accident progression or radioactive release;
(i)   Access control to process or equipment;
(j)   The relative significance of integration points between process and programs.

The use of proven engineering practices is addressed in both REGDOC-2.5.2 [A–62] and RD-367 [A–61] (Note that RD-367 will be merged into REGDOC-2.5.2 in the next revision, which is expected to be published for stakeholder review in 2020).

When a new SSC design, feature or engineering practice is introduced, adequate safety shall be proven by a combination of supporting research and development programs and by examination of relevant experience from similar applications. An adequate qualification program shall be established to verify that the new design meets all applicable safety design requirements. New designs shall be tested before being brought into service and shall be monitored in service to verify that the expected behaviour is achieved.

A–14.3. CHINA–NNSA

A–14.3.1. Question

Describe the challenges associated with regulatory assessment of novel/innovative design features (SSCs) in absent of previous application in nuclear industries. Provide information on how regulatory judgement was reached in view of the situation.

A–14.3.2. Response

For regulatory judgment of these novel designs, NNSA stated identify requirements in HAF102 [A–68].

> "4.6.3 Where an unproven design or feature is introduced or where there is a departure from an established engineering practice, safety shall be demonstrated by means of appropriate supporting research programmes, performance tests with specific acceptance criteria or the examination of operating experience from other relevant

applications. The new design or feature or new practice shall also be adequately tested to the extent practicable before being brought into service and shall be monitored in service to verify that the behaviour of the plant is as expected."

A–14.3.3. Follow-up Questions

Please provide examples of challenges identified during the assessment of novel SMR design features e.g. in relation of HTR-PM or other SMRs. See the answer from China to the question 11 (on DiD). Provide information on how regulatory judgement was reached in view of the situation.

A–14.3.4. Response

Carry out relevant test verification for new design or new equipment. For example, SMR has carried out the test verification of new fuel assembly and safety facilities. The supervisor shall judge whether the new design or equipment meets the safety requirements according to the verification process of the test verification scheme, the verification process and verification result.

A–14.4. CZECH REPUBLIC–SÚJB

A–14.4.1. Question

Describe the challenges associated with regulatory assessment of novel/innovative design features (SSCs) in absent of previous application in nuclear industries. Provide information on how regulatory judgement was reached in view of the situation.

A–14.4.2. Response

Not applicable — challenges associated with regulatory assessment of novel/innovative design features (SSCs) in absent of previous application in nuclear industries have not been identified as no SMR is currently planned to be deployed in Czech Republic.

In general, it is the nuclear installation design that shall specify the detailed requirements for technical procedures and organisational measures for the whole life cycle of a nuclear installation. It shall set requirements for the selected equipment in terms of the safety functions and for the testing of this equipment and individual parts during and after the construction of the nuclear installation, periodically during operation and after repair of the individual equipment with the aim to verify compliance with requirements and to detect defects.

The Atomic Act [A–21] leaves relative freedom with regard to the choice of the NPP technology as long as all the legislative requirements are met. The obligation to use proven method, processes and technologies in the nuclear installation design is directly mentioned in the Atomic Act. When designing a nuclear installation, a design basis shall be established and proven methods, procedures and technology shall be used. The obligation to verify the required characteristics of systems, structures and components important to nuclear safety is subsequently set out in the Decree No. 329/2017 [A–22] Coll., on the requirements for nuclear installation design. The nuclear installation design process shall comprise an evaluation of compliance of the design with the legislative requirements. In practice, a compliance with the IAEA recommendations and WENRA documents is required for the project of a new NPP in the Czech Republic by the SÚJB.

The principle of the use of proven technologies in the design and construction of a nuclear installation was always set out in applicable Czech legislation. On the basis of legislation requirements, proven materials consistent with the relevant regulations, technical standards and technical specifications were used for the design and manufacturing of reactor cooling systems and their components including reactor pressure vessels in the Dukovany NPP and the Temelín NPP. Their sufficient rating was demonstrated by theoretical calculations and experimental verifications, and a reserve for degradation during their operation was considered. The program and methods for the detection of the state of primary circuit were also defined. Supervision of such activities is within the scope of competence of SÚJB. Proven technologies are preferred even when modifying or adding new systems, structures and components.

Preliminary evaluation of the nuclear installation design is already underway in the framework of the licence for the siting of a nuclear installation. The nuclear installation design is then evaluated in greater detail, in particular with regard to the site characteristics, within the framework of the licence for construction of a nuclear installation. Pursuant to the Atomic Act [A–21], the application for the licence for construction of a nuclear installation shall be accompanied with the comprehensive set of documentation for a licenced activity. In some cases, the documentation for a licensed activity is subject to approval by decision of the Office.

It is not possible to entirely exclude that for novel/innovative design features that cannot be considered sufficiently verified in the absence of previous similar application in nuclear industries, the experimental verification results might be used to demonstrate the reliability and effectiveness of the particular technological feature (in such case these shall be available in adequate quality — an adequate demonstration of effectiveness and reliability of an innovative/novel feature by performing sufficient analyses and tests would be necessary).

As mentioned elsewhere, the provisions of the Decree No. 329/2017 [A–22], on the requirements for nuclear installation design, are relatively specific and reflect PWR technology. One example of specific requirement that is related to regulatory assessment of novel/innovative design features of SMRs and its testing is that the characteristics of fuel assemblies in nuclear installation design shall be tested, either experimentally or in operation of another nuclear installation, with regard to the ability of fuel assemblies to perform their design function safely. Such specific requirements would therefore have to be probably adapted to different type of SMR technology, should it be deployed in Czech Republic.

A–14.5. FRANCE–ASN

A–14.5.1. Question

Describe the challenges associated with regulatory assessment of novel/innovative design features (SSCs) in absent of previous application in nuclear industries. Provide information on how regulatory judgement was reached in view of the situation.

A–14.5.2. Response

According to the article 3.8 of the Order [A–72], safety demonstration relies in particular on appropriate, clarified and validated methodologies, integrating hypothesis and rules adapted to uncertainties and knowledge limits at stake. Moreover, the article 3.2 of the same order stipulates that the safety demonstration follows a prudent and deterministic approach.

Considering that, the licensee must provide sufficient guarantees and information, including tests and mock-ups to demonstrate the safety level of novel and innovative design.

ASN is attentive to the use of safety innovative design features and assess them following a graded approach. ASN expects from the licensee to compensate the lack of operating experience by organizational provisions for example.

Even though the use of proven and reliable features is often preferred by licensees, innovative features are not excluded but might require long technical discussions. For example, ASN gave a particular attention to EPR's core catcher. ASN asked IRSN to expertise this innovative feature in 1999, 8 years before the authorization to set up the installation. For this purpose, IRSN developed its own experiments and computer codes to challenge licensee's results.

These requirements remain fully applicable to potential upcoming SMR projects.

A–14.6. JAPAN–NRA

A–14.6.1. Question

Describe the challenges associated with regulatory assessment of novel/innovative design features (SSCs) in absent of previous application in nuclear industries. Provide information on how regulatory judgement was reached in view of the situation.

A–14.6.2. Response

There is no introduction of novel/innovative design features in the review of the HTTR.

A–14.7. RUSSIAN FEDERATION–ROSTECHNADZOR

A–14.7.1. Question

Describe the challenges associated with regulatory assessment of novel/innovative design features (SSCs) in absent of previous application in nuclear industries. Provide information on how regulatory judgement was reached in view of the situation.

A–14.7.2. Response

Introduction of innovative design solutions must be preceded by appropriate studies, in particular, at experimental facilities, by prototyping, and by calculations using the software certified as appropriate. Novel safety-significant systems must be commissioned with additional testing of innovative equipment.

Paragraph 12 of NP-022-17 [A–89]: "The technical and organisational solutions adopted to ensure the safety of vessel's nuclear power installation must be proven by previous experience, tests, studies, and the operating experience of prototypes. These requirements must be applied in the design, construction and operation of vessel's nuclear power installation; in the design, manufacture, maintenance, and modification of its systems and components important for safety, as well as during the vessel decommissioning."

Paragraph 52 of NP-022-17: "The systems and components important for safety must undergo a direct and full compliance testing during commissioning, after modifications and maintenance, and periodically throughout their lifetime. If a direct and (or) full verification is not possible, an indirect and (or) partial testing must be conducted."

A–14.7.3. Follow-up Questions

Has any assessment of innovative features been made in the context of SMRs, and can you share the learning from those? Provide information on how regulatory judgement was reached in view of the situation.

A–14.7.4. Response

The introduction of innovative design features must be accompanied by preliminary research, including that on test grounds, making of prototypes, and calculations using duly certified software. When new systems important to safety are put into service, additional testing of such equipment must be carried out.

Clause 12 of NP-022-17: "Technical and organizational decisions taken to ensure the safety of a nuclear power installation of a vessel must be proven by previous experience, testing, research, and experience in operating prototypes. These requirements must be applied when designing, constructing and operating a nuclear power installation of a vessel, designing, manufacturing, repairing and upgrading its systems and components important to safety, and when decommissioning the vessel."

Clause 13 of NP-022-17: "The system of technical and organizational measures to ensure the safety of a vessel and the design basis of the systems and components important to safety must be presented in the SAR. Any non-compliance affecting the safety of a nuclear power installation with the information contained in the SAR and in the vessel design, or non-compliances between the vessel design and its implementation are not allowed."

The SAR is the main document demonstrating the safety of a nuclear facility and submitted to Rostechnadzor as part of a set of documents demonstrating the safety.

Accordingly, the assessment of the innovative characteristics of the floating plant was carried out as part of the safety review by the technical support organization of the regulatory body.

A–14.8. SOUTH AFRICA–NNR

A–14.8.1. Question

Describe the challenges associated with regulatory assessment of novel/innovative design features (SSCs) in absent of previous application in nuclear industries. Provide information on how regulatory judgement was reached in view of the situation.

A–14.8.2. Response

The test character of the PBMR DPP is apparent considering the absence of prototype facilities. Examples of new and innovative features in the DPP design are the innovative core design with a solid graphite centre column resulting in high uncertainties on pebble flow caused by friction as well as the corresponding temperature profile, the Brayton Cycle with turbines and compressors on a single shaft and some of the passive engineered safety features.

The NNR had been informed in 2008 that the reactor power and the hot gas temperature will be significantly reduced for initial operation of the PBMR DPP to address the uncertainties on fuel performance. Whilst the approach appeared sensible, the NNR still required a systematic approach on the test and verification aspects that will be covered off-site, during commissioning and operation. In addition, areas where further research and development is necessary should be clearly identified. Similar research and development programmes have been initiated in Europe (RAPHAEL project) and other HTR related programmes. The test, qualification and commissioning approach must inform the safety case for licensing by addressing the outstanding issues, the consequential measures as well as the safety implications.

RD-0034 [A–33] defines inter alia, the following requirement related to Testing, Qualification and Commissioning (TQC):

In case new safety features for nuclear installations will be applied that differ significantly from evolutionary LWRs or that use simplified, inherent, passive, or other innovative means to accomplish their safety functions (advanced reactors) a test programme must be implemented by the applicant / licensee or its designee to demonstrate the performance of the new safety features. It must be ensured by that program that the safety features will perform as predicted in the applicant's safety analysis report, to provide sufficient data to validate analytical codes, and that the effects of systems interactions are acceptable. The test program must include suitable qualification testing of a prototype simulating the most adverse design conditions. The test programme must be defined in writing and make provision for sign-offs as the test programme conditions are met.

A–14.9. UNITED KINGDOM–ONR

A–14.9.1. Question

Describe the challenges associated with regulatory assessment of novel/innovative design features (SSCs) in absent of previous application in nuclear industries. Provide information on how regulatory judgement was reached in view of the situation.

A–14.9.2. Response

Novel materials or new manufacturing methods for SSCs will need careful consideration to ensure that the necessary levels of integrity are established at build and through the life cycle of the design. Novel designs should be supported by appropriate research and development. Additional inspection and surveillance at build and through life may be necessary. Data used in analyses, and acceptance criteria, should be clearly conservative, taking account of uncertainties in the data and their contribution to the safety case.

ONR expectations on some key areas of engineering design associated with nuclear power plants are documented in the following Technical Assessment Guides:

- NS-TAST-GD-022 – Ventilation [A–247];
- NS-TAST-GD-056 – Nuclear Lifting Operations [A–248];
- NS-TAST-GD-057 – Design Safety Assurance [A–249];
- NS-TAST-GD-067 – Pressure Systems Safety [A–250].

ONR experience of SMR designs has been that there are differing levels of design maturity and in many cases, there is not sufficient information on how the novel or innovative technologies would be qualified for use in the UK as expected by ONR SAP [A–44] EQU.1. Areas of innovation in engineering design included the incorporation of compact heat exchangers and active magnetic bearings.

ONR inspectors have noted that when claiming the reliability of SSCs, the novelty of the SSC should be taken into account as per SAP [A–44] ERL.1, which expects that the Reliability claimed for an SSC should take into account its novelty, experience relevant to its proposed environment, and uncertainties in operating and fault conditions, physical data and design methods.

As previously stated, ONR guidance on ALARP and the goal setting regulatory regimes enables innovation and allows for alternative approaches. However, in this context it is important to recognise the level of design maturity, and the level of testing and qualification that would be expected to gain assurance that the risk associated with normal operations and fault conditions are adequately understood and accounted for.

A–14.10.   UNITED STATES OF AMERICA–NRC

A–14.10.1. Question

Describe the challenges associated with regulatory assessment of novel/innovative design features (SSCs) in absent of previous application in nuclear industries. Provide information on how regulatory judgement was reached in view of the situation.

A–14.10.2. Response

New designs, by their nature, challenge the staff's review because they employ some features and methodologies that have not previously been evaluated. Much of the technology is not 'off the shelf' and the technical staff assigned to the review may not be familiar with it. The NRC has established and refined a request for information (RAI) process for dealing with questions that arise in the course of the review. We have found that multiple rounds of RAIs are inefficient; therefore, we have instituted procedures (e.g. detailed teleconferences explaining the intention of RAIs when necessary) that have been effective in limiting the number of second round RAIs that we must issue.

Resolving issues with complicated new designs, however, does not lend itself to the RAI process, the questions are too intricate and tend to build one on the other. On-site audits and face to face meetings are often necessary to resolve complex problems. When new computer codes are used, the problems can become extremely complicated. Benchmarking, validation, and verification can pose challenges.

In its regulatory judgement, the staff is guided by the Code of Federal Regulations. Only when the reviewing staff has complete confidence that a new or innovative design can perform its safety function under the regulations will they give their approval.

A–15. CHALLENGES WHERE QUALIFICATION OF SSCS ARE ABSENT OR INCOMPLETE FOR SAFETY DEMONSTRATION OF SMRS

This Annex presents the responses provided by the following Member State regulatory bodies to the Question 15: "Describe the challenges where qualification of SSCs are absent or incomplete for safety demonstration of SMRs. Provide information on how regulatory judgement was reached in view of the situation."

A–15.1. ARGENTINA–ARN

A–15.1.1. Question

Describe the challenges where qualification of SSCs are absent or incomplete for safety demonstration of SMRs. Provide information on how regulatory judgement was reached in view of the situation.

A–15.1.2. Response

The qualification of SSCs consists in the demonstration that the items under study are capable of fulfilling their safety functions. This occurs during the entire range of operating conditions during the normal operating states of the plant, as well as, during the accidental conditions derived from Design Basis Accidents, seismic solicitations and the consideration of aging effects caused by environmental factors (such as conditions of vibration, irradiation, humidity or temperature).

As mentioned in the previous answers, the CAREM 25 Project developed a comprehensive and detailed methodology for establishing the engineering requirements for SSCs. Their objective was establishing an SSC Safety Classification system (based on the Safety Analysis) in order to set the design, manufacturing, assembly, testing, operation, and QA requirements among others that will apply to each SSC, demonstrating the functional safety of a design.

ARN established guidelines for developing an Equipment Qualification Program to achieve safety requirements. The Equipment Qualification Program must be considered from the early design stages with the objective of defining the equipment within its scope, establishing the qualification methods to be used, and establishing the measures for maintaining the qualification.

This guideline was developed following the state of the art in the matter and included references of IAEA, IEC, and IEEE.

The equipment qualification process consists of three marked phases:

(a) Design information necessary for qualification;
(b) Establishment of the qualification;
(c) Maintaining the qualification.

The Qualified equipment was designed and manufactured according to the requirements and regulations proposed by the RE and accepted by the Regulator. The certification issued by the manufacturer must establish that the component has demonstrated the ability to function within a postulated accident profile. This type of certification in many cases provides for the use of engineering and manufacturing procedures that follow the requirements of accepted standards. In the case of equipment and components whose qualification is not provided by a qualified manufacturer that can certify it, or are manufactured by the RE, different qualification methods can be used in order to demonstrate the suitability of the component under the established conditions of accidental and operation.

A–15.1.3. Follow-up Question

Were there any CAREM SSCs that had specific challenges with regard to qualification?

A–15.1.4. Response

In relation to the SSC safety classification of the CAREM 25 reactor, we can mention that the entire evaluation and review process was a great regulatory challenge.

Being an innovative classification methodology worldwide (it was recently consolidated with the appearance of the Approach on Defence in Depth as proposed by WENRA and INSAG and the IAEA SSG-30 [A–57] guide had not yet been issued). In addition, the regulations that were applied in the licensing of the other Argentine CN dates from the 1970s and 1980s). The ARN followed the Licensee process from a proactive perspective, counting, for the evaluation, with the support of external consultants and providing training to the personnel for the development and maintenance of the knowledge acquired.

As a result, the experience obtained in the described process facilitated the task in front of the tasks developed in the pre-licensing and then the signing of the Memorandum of Understanding (MOU) between the regulator and the licensee in view of the construction of the IV CN Argentina.

As an example of SSCs that presented a challenge, security variables monitoring systems can be mentioned.

A–15.2. CNSC–CANADA

A–15.2.1. Question

Describe the challenges where qualification of SSCs are absent or incomplete for safety demonstration of SMRs. Provide information on how regulatory judgement was reached in view of the situation.

A–15.2.2. Response

First, to define qualification, it is the demonstrating that the SSC is fit for purpose for normal operation and accident conditions. This includes but is not limited to environmental qualification, seismic, electro-magnetic interference and fire conditions, and radiation fields (impact on seals, metals and concrete). Considerations of design and installation includes qualifying the weld and the materials used (e.g. demonstrate pump welds, coatings, hard-facings and elastomers). The vulnerability of materials will impact the overall aging management of the facility and needs to be taken into account in the maintenance program. For more novel items, demonstration might require additional testing and modelling, and greater margins.

Qualification of SSCs is not a one-time discussion with a single answer of 'yes' or 'no'. Qualification of SSCs is a living discussion that is revisited as the body of knowledge grows. It starts with a discussion about whether there is sufficient information supporting the qualification claims for SSCs taking into account safety and control provisions that address uncertainties.

Considerations in deciding whether sufficient information exists to make regulatory decisions concerning 'sufficient qualification':

- Different SSCs will have differing levels of qualification data during the design process;
- The interfaces between SSCs means that these differing levels of qualification may present compounding/stacking of uncertainties;
- Does the vendor/applicant understand the long-term implications of the uncertainties? What are they doing about them? (long term R&D, data gathering from the first facility etc.);
- What interim safety and control measures will be proposed to address uncertainties?
- What would the regulator need to impose (licence conditions/hold points etc.) to further address any potential residual uncertainties?

In the CNSC pre-licensing vendor design review (VDR) process, it is too early to comment on specific qualification data to support SSCs (see below for some information on VDRs). Every design we have seen has varying degrees of available information from OPEX and/or R&D activities supporting qualification of SSCs. Even so-called 'proven' designs need to demonstrate that legacy qualification issues are being addressed.

It will be a licence applicant's job to support the safety case once the detailed design is proposed for a specific project. As a result, with vendors, we pay much more attention to the vendor's overall processes to develop the information to support qualification of SSCs in, for example, the R&D program. We also seek to understand the basis supporting qualification and fitness-for-service of the item.

An example of how adequate regulatory judgement can be reached is in Section 5.4 of REGDOC-2.5.2 [A–62] which provides requirements for use of proven engineering practices to demonstrate claims:

> "When a new Structure System and Component (SSC) design, feature or engineering practice is introduced, adequate safety shall be demonstrated by a combination of supporting research and development programs and by examination of relevant experience from similar applications. An adequate qualification program shall be established to verify that the new design meets all applicable safety requirements. New designs shall be tested before being brought into service and shall be monitored while in service so as to verify that the expected behaviour is achieved."

Section 11 of REGDOC-2.5.2 [A–62] provides the following guidance regarding alternative approaches:

> "The requirements in this regulatory document are intended to be technology neutral for water-cooled reactor designs. It is recognized that specific technologies may use alternative approaches.
>
> The CNSC will consider alternative approaches to the requirements in this document where:
>
> 1. the alternative approach would result in an equivalent or superior level of safety
> 2. the application of the requirements in this document conflicts with other rules or requirements

3. the application of the requirements in this document would not serve the underlying purpose, or is not necessary to achieve the underlying purpose

Any alternative approach shall demonstrate equivalence to the outcomes associated with the use of the requirements set out in this regulatory document."

Section 2.2 of REGDOC-1.1.2 [A–66] complements the above:

"This section should provide information pertaining to cases where the expectations contained in any of the various regulatory documents and other applicable codes and standards are not met. The safety significance of the deviations should be assessed and where necessary, a separate and complete justification should be provided for each deviation. This justification should include all the information necessary to assure the CNSC that any deviations from CNSC requirements and expectations will not negatively affect the facility's overall level of safety. This justification should be included in each of the applicable sections or documented in referenced documents provided with the application."

Ultimately, Canada's regulatory framework for design and procurement oversight is used in an integrated manner to ensure confidence in the site-specific configuration of a new reactor facility and addresses the requirements of IAEA SSR 2/1 [A–127]. It also provides flexibility for allowing an applicant to demonstrate how it meets Canadian regulations. From REGDOC-2.5.2 [A–62] which provides an applicant/reactor vendor with detailed expectations on how to establish design requirements and configuration information for SSCs and how to apply accepted practices for safety classification; to REGDOC-1.1.2 [A–66] and REGDOC-2.3.1 [A–144] which has specific information on procurement, manufacturing, construction and commissioning.

The development lifecycle of SMRs is like any other innovative technology. The development of the technology typically goes through a set of product development phases that follow the technology readiness scale[12].

Once the development of a new reactor is completed, it is expected that a First of a Kind (FOAK) reactor or a demonstration reactor will be constructed. The FOAK may need special construction and design adjustments to enable inspection, testing, or other means to substantiate safety claims. It is also expected that safety margins may need to be adjusted to compensate for the potential insufficient experimental data when licensing reactors that use new technologies. The pre-licensing processes can help in improving efficiencies in the licensing of new advanced technologies, particularly if the technologies are in the early phases of the development scale (Fig. A–11).

---

[12] The technological readiness scale is a scale of product development developed for NASA and the U.S. Department of Energy. This scale is now used in many technological development applications, including the Government of Canada's Innovation and Skills Plan
(https://www.ic.gc.ca/eic/site/080.nsf/eng/00002.html).

314

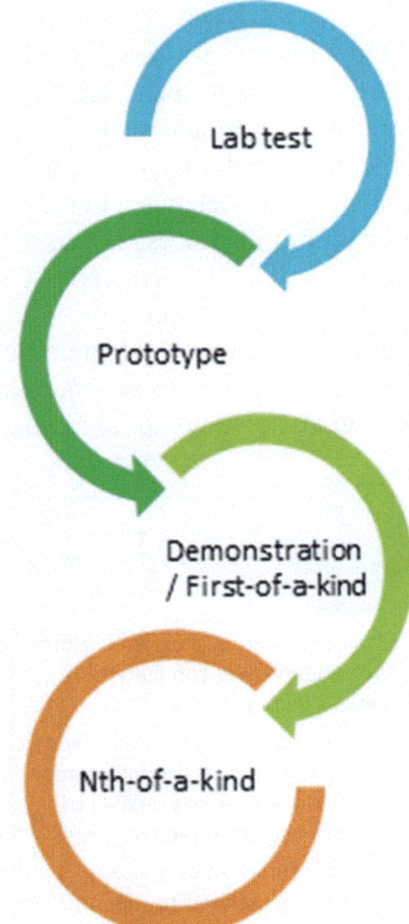

FIG. A–11. Development lifecycle of a new technology.

A FOAK can also mean new licensees or existing utilities employing new management systems. There are potential challenges with long lead items and supply chain through the introduction of new suppliers who may have varying experience with Canadian requirements. In some cases, engineering standards may not exist or they may be utilized differently. Licensees may have limited experience with use of new technology specifications. In construction, new approaches may be used which could result in varying quality of construction in new builds. In novel technologies, identifying key commissioning activities to provide greater confidence that they can be leveraged for future projects. Demonstration also includes future operation and maintenance performance of the licensee, not just the technology.

Some potential First of a Kind aspects that an applicant may consider in the demonstration needed to support a future fleet across Canada, includes:

- Manufacturing (pellets using TRISO particles, key nuclear components);
- Construction – use of modular construction techniques;
- Commissioning – includes potential First Plant only tests;
- Training and certification of staff;
- In-service OPEX collection (aging, performance of integrated systems);
- Operation and maintenance performance of the licensee;
- Security by design – minimization of security complement;
- Minimize need for offsite emergency response.

Vendor Design Reviews (VDR)

As an early feedback mechanism, the VDR process assesses that a reactor vendor is capable of demonstrating that design activities are systematic and quality assured and that the basis for design and safety analysis decisions are clearly documented, Fig. A–12. This review provides early identification and resolution of potential regulatory or technical issues in the design process, particularly those that could result in significant changes to the design or safety case. A challenge that the CNSC encounters in the conduct of this process is that much of the technologies reviewed are not yet proven; most designs are still at the conceptual stage, limited global operating experience. VDR results may also be used by an applicant in the licensing process. The CNSC expects a future applicant to be highly familiar with the technology it will eventually purchase for a proposed nuclear reactor facility. Potential applicants are encouraged to speak with vendors early on in the licensing process to discuss and resolve potential regulatory issues.

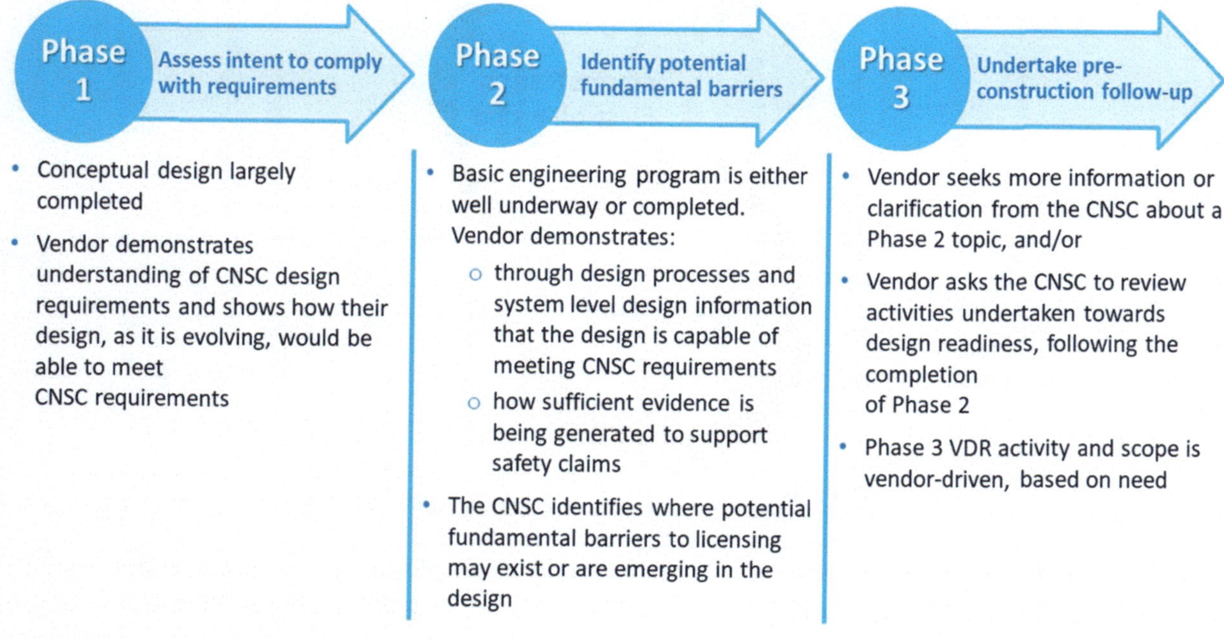

FIG. A–12. The VDR process progresses in phases that increases in information required by the vendor to demonstrate their claims.

A–15.2.3. Follow-up Question

Are they any SSCs for reactors under consideration that many have challenges with regard to qualification?

A–15.2.4. Response

One challenge could be whether a vendor submits creditable and sufficient information to support the qualification claims due to lack of OPEX on proposed SMR design. The information provided by vendors were often studies from decades ago, but the technology was not mature enough to be built and used at the time. SSC qualification continues to be work in progress.

A–15.3. CHINA–NNSA

A–15.3.1. Question

Describe the challenges where qualification of SSCs are absent or incomplete for safety demonstration of SMRs. Provide information on how regulatory judgement was reached in view of the situation.

A–15.3.2. Response

In Safety Regulations on Quality Assurance of Nuclear Power Plants (HAF003) [A–251], it is stipulated that "(…) the general program of quality assurance of nuclear power plants and the sub program of quality assurance of each kind of work must be formulated and effectively implemented in order to ensure the safety of nuclear power plants, including site selection, design, manufacturing, construction, commissioning, operation and decommissioning. All quality assurance programs follow the same principles". In HAF003 Regulations [A–251], relevant rules demand to control the process affecting the quality used in the manufacturing of nuclear power plant equipment. According to the requirements of relevant specifications, standards, technical specifications, codes or other special requirements, some measures must be formulated to ensure that these processes are completed by qualified personnel and in accordance with the approved standards and equipment. The corresponding inspection and test program is developed to inspect each step of work and verify the safety function of SSCs.

SSCs of SMRs should be manufactured in accordance with relevant quality assurance safety regulations. There may be some special requirements for SSCs of floating reactor (meeting the requirements of marine environment, etc.), so SSCs should meet the corresponding special requirements, and the corresponding provisions should be given in the corresponding quality assurance documents.

A–15.3.3. Follow-up Questions

Could you provide any further info on the SSCs for floating reactors that may have special requirements?

Please, could extend more about your experience? Could provide information about items important to safety qualification or another example.

A–15.3.4. Response

According to the marine environmental conditions, the corresponding safety analysis or environmental qualification requirements for floating reactor SSC are proposed. As the floating reactor has not completed the review of the project, there is no specific experience.

A–15.4. CZECH REPUBLIC–SÚJB

A–15.4.1. Question

Describe the challenges where qualification of SSCs are absent or incomplete for safety demonstration of SMRs. Provide information on how regulatory judgement was reached in view of the situation.

A–15.4.2. Response

Not applicable — because no SMR has been deployed in Czech Republic, there is currently no practical experience with absent or incomplete qualification for SSCs for this type of nuclear installation.

The system of selected equipment classification and conformity assessment is in line with IAEA standards. One of the more challenging aspects of SMR licencing would be the classification of the SSC as the current system of classification and conformity assessment is mainly adapted to the PWR so that the most detailed requirements are very specific and focused on this technology. The selected equipment is classified into safety classes 1 to 3 depending on the safety functions to the performance of which it contributes (graded approach applied in the legislation). The nuclear installation design shall, in line with the legislative requirements, specify the requirements for selected equipment in terms of the safety functions and classify selected equipment into safety classes accordingly.

The conformity assessment of the selected equipment with the technical requirements is performed prior to the use of the equipment. Both the Decree No. 329/2017 [A–22], on the requirements for nuclear installation design, and the Decree No. 358/2016 [A–23], on requirements for assurance of quality and technical safety and assessment and verification of conformity of selected equipment, contain very specific provisions for this classification that reflect PWR technology, as pressurized primary circuit, fuel cladding and assemblies (including specific pressure values and volumes).

The environmental qualification of systems, structures and components shall comply with the technical specifications set out by nuclear installation design, which shall also set the method of environmental qualification validity verification of selected equipment throughout the life cycle of nuclear installation. Prior to the use of selected equipment, the compliance of the equipment with all of these requirements is verified by the entitled person during conformity assessment (authorized or accredited person or manufacturer). SÚJB is authorized by Atomic Act [A–21] to perform inspections to verify if this assessment is in line with the legislative requirements, including prescribed scope and procedures for conformity assessment.

A–15.4.3. Follow-up Question

Answer describes current status, however, what is being considered to address classification and conformity assessment to address SMRs?

A–15.4.4. Response

Currently, no SMRs are planned to be deployed in the Czech Republic.

In the light of the above and given the lack of detailed information about various SMR designs and the uncertainty over whether and which type of SMR could be hypothetically deployed, detailed considerations in this regard have not been given.

A–15.5. FRANCE–ASN

A–15.5.1. Question

Describe the challenges where qualification of SSCs are absent or incomplete for safety demonstration of SMRs. Provide information on how regulatory judgement was reached in view of the situation.

A–15.5.2. Response

According to the article 2.5.1 of the Order [A–72], SSCs that are important for the protection of security, safety, public health and sanitation, nature and environment must be qualified to guarantee their capacity to realize their functions in situations to which they are necessary, in particular considering environmental conditions. Qualification is proportionate with what is at stake.

Also, the same article stipulates that qualification durability can be insured by provisions regarding studies, building, testing, controls and maintenance.

Finally, the licensee must present in the commissioning file of its installation its methodology to qualify SSCs and must list principal information about the results of the qualification.

For the Flamanville EPR, ASN enacted a resolution that contains additional requirements regarding SSCs' qualification. Indeed, ASN's resolution n° 2008-DC-0114 [A–252] requires from the licensee to provide information to ASN on a quarterly basis about the progress of SSCs' qualification, including eventual significant discrepancies. Also, it provides requirements about qualification of specific equipment.

ASN follows the qualification progress and considers that complete positive qualification results are necessary to authorize the commissioning of a basic nuclear installation. ASN controls, with the support of IRSN, the adequacy and the completeness of the licensee's qualification program for SSCs.

Also, the licensee has to provide elements to guarantee that it's SSCs have been designed manufactured, constructed, installed, commissioned, operated, tested, inspected and maintained in accordance with established processes that ensure design specifications, and the expected levels of safety performance are achieved. For example, for its EPR in Flamanville, EDF decided to use an I&C platform, the SPPA T2000, which is a 'conventional' industrial system, for functions linked to normal operations and for certain reactor protection operations in incident or accident situations.

ASN informed EDF that the safety of the 'SPPA T2000' platform could not be confirmed. ASN in particular asked EDF to provide additional justifications and examine a different design for the I&C of the Flamanville 3 EPR reactor. Since then, the licensee has been carrying out considerable work to comply with the ASN requests and finally, as requested by ASN, has implemented an I&C architecture modification designed to improve robustness and enable the SPPA-T2000 platform to be used for the Flamanville 3 EPR reactor.

A–15.6. JAPAN–NRA

A–15.6.1. Question

Describe the challenges where qualification of SSCs are absent or incomplete for safety demonstration of SMRs. Provide information on how regulatory judgement was reached in view of the situation.

A–15.6.2. Response

Not limited to the HTTR, it is generally required to conduct the design and construction based on the quality assurance, and there is no situation where qualification of SSCs are absent or incomplete.

A–15.6.3. Follow-up Questions

Please provide the reference to documents that govern SSC qualification.

Please, could extend more about your experience? Could provide examples, etc.

A–15.6.4. Response

Quality assurance activity for constructing facilities will be confirmed in the review process for the approval of design and construction plan.

A–15.7. RUSSIAN FEDERATION–ROSTECHNADZOR

A–15.7.1. Question

Describe the challenges where qualification of SSCs are absent or incomplete for safety demonstration of SMRs. Provide information on how regulatory judgement was reached in view of the situation.

A–15.7.2. Response

The products for which requirements have been established in order to ensure the nuclear facility safety (including the safety-significant systems), as well as the processes of product design, manufacture, installation and commissioning, must be assessed for conformity in accordance with the requirements of the federal nuclear safety regulations. The conformity assessment requirements are established in the format of:

- Review of technical documentation;
- Testing;
- Inspection;
- Acceptance;
- Decision to use foreign-made products at the nuclear facility;
- Mandatory certification of products;
- Registration.

However, NP-071-18 [A–149] does not apply to ships and other vessels with nuclear installations.

During the construction of the floating NPP Akademik Lomonosov, Rostechnadzor and Rosatom developed Decision No. 00-03-10/641 [A–150] that defines a temporary procedure for evaluation of the conformity of equipment, components, materials, and semi-finished products supplied to the nuclear-powered ships, floating structures, and support facilities mandatory requirements.

According to this Decision [A–150], works on the assessment of conformity of items with the mandatory requirements to be used on nuclear powered ships and floating structures, as well as at the support facilities, are carried out by the Russian Maritime Register of Shipping in accordance with the applicable Rules and Guidelines of the Register.

Rostechnadzor assesses the compliance of the items supplied to the nuclear-powered ships, floating structures, and support facilities with the mandatory requirements established in the regulatory legal acts in the field of the use of atomic energy in the form of federal state control (supervision) in accordance with its powers under [A–151] — Paragraph 2 of the decision.

According to Paragraph 52 of NP-022-17 [A–89], the safety-significant systems and components must undergo direct and full compliance testing at the commissioning stage, after modification and maintenance, and periodically throughout their lifetime. If direct and/or full verification is not possible, indirect and/or partial testing must be conducted.

A–15.7.3. Follow-up Question

Did not provide much info on how this topic was/is being approached.

What challenges were encountered, or do you expect to encounter with regard to qualification?

A–15.7.4. Response

In accordance with art. 40 of Federal Law No. 170-FZ [A–26], the requirements of the Russian Maritime Register of Shipping, along with federal regulations and rules in the field of the use of atomic energy use, must be applied at the stages of the life cycle of vessels and other floating crafts with nuclear reactors. In accordance with art. 22 of the Merchant Shipping Code of the Russian Federation No. 81-FZ of April 30, 1999 [A–28], and Resolution of the Government of the Russian Federation No. 121 of February 14, 2012 [A–253], the Russian Maritime Register of Shipping is a Russian organization that carries out the classification and examination of the vessels registered in the State Register of Ships, in the bareboat charter register, or in the Russian International Register of Ships.

According to art. 24 of the Merchant Shipping Code of the Russian Federation No. 81-FZ of 30.04.1999 [A–28], the Russian organizations authorized to classify and examine vessels issue rules of classification and construction of vessels, technical supervision over construction of vessels, rules of technical supervision over the manufacture of materials and products for vessels, and issue classification certificates confirming the compliance of vessels with these rules.

According to art. 24 of the Merchant Shipping Code of the Russian Federation No. 81-FZ of 30.04.1999 [A–28], vessels, except for small vessels used for non-commercial purposes, are examined by the Russian organizations authorized to classify and examine vessels for their compliance with the requirements of international treaties of the Russian Federation.

The classification certificate of a vessel is submitted by the Licensee to the regulatory body as part of the set of supporting documents and taken into account by the regulatory body when deciding whether to issue the license.

In addition, when justifying the safety of a vessel with a nuclear reactor, the requirements of the federal regulations and rules 'Standards for strength calculation of equipment and pipelines for ship nuclear steam generating installations with pressurized water reactors' (NP-054-04) [A–96] must be met. The compliance with the requirements of this document is checked during the examination of the safety case of a vessel with nuclear reactors.

A–15.8. SOUTH AFRICA–NNR

A–15.8.1. Question

Describe the challenges where qualification of SSCs are absent or incomplete for safety demonstration of SMRs. Provide information on how regulatory judgement was reached in view of the situation.

A–15.8.2. Response

There is no introduction of novel/innovative design features in the review of the HTTR.

The NNR has developed a position paper on Test, Qualification and Commissioning (TQC) for the PBMR (RG-0005 [A–34]) considering besides the overall approach on TQC, the uncertainties and the interrelationship between V&V and the test and qualification requirements associated with FOAK aspects of the design to clarify the requirements / expectations to the applicant.

Test and Commissioning are subsets of Qualification. The Test and Commissioning programme shall justify the data and assumptions of the design of SSCs and demonstrate their proper functioning. It must provide a structured approach to ensure that material, parts, components, subsystems and processes as utilised in the PBMR module, comply with the stated requirements.

The Test and Commissioning programme will include the testing of FOAK, and a commissioning programme for new technologies. It must be possible to demonstrate that such a programme of tests, and appropriate test methods will adequately qualify the FOAK, and ensure acceptance of the Plant as a whole. If the process for development of a testing programme is not acceptable to the Regulator, the Safety Case will be rejected.

As an example of NNR review comments on the high-level TQC process document, the salient points raised by the NNR in the review were:

- The qualification process for the various safety classes of SSC should be described, showing the interface between design and qualification;
- The relationship with V&V of the analytical models should be considered;
- The relationship of the TQC process with Safety Management must be clear;
- A description of the process for incorporating Environmental Qualification;
- Allocation of tests during all phases of the qualification process;
- Level of TQC detail for the Safety Case.

A–15.9. UNITED KINGDOM–ONR

A–15.9.1. Question

Describe the challenges where qualification of SSCs are absent or incomplete for safety demonstration of SMRs. Provide information on how regulatory judgement was reached in view of the situation.

A–15.9.2. Response

ONR has found that there is generally recognition amongst SMR proponents of the need to qualify SSCs, and prototypes or model equivalents are often proposed to achieve this goal. Generally speaking, these approaches are acceptable, however, the level of design maturity and development of some SMR designs means that there is limited information on the criterion that will be used to support the qualification of SSCs under the above methods and, as such, a meaningful assessment of the qualification process cannot be made.

ONR is aware of proposals to qualify certain SSCs for operational time scales which do not align with the intended operational life of the SMR units and may not fully recognise all foreseeable failure modes and consequences in the absence of relevant operational experience. ONR recognises that qualifying equipment for operational use upwards of 60 years is challenging. However, it expected that adequate arguments would be put forward to support how components would be qualified for the operational life, or arguments as to why a shorter qualified life is acceptable.

ONR expectations in this area are documented in the SAP [A–44] EQU.1, which states: "Qualification procedures should be applied to confirm that structures, systems and components will perform their allocated safety function(s) in all normal operational, fault and accident conditions identified in the safety case and for the duration of their operational lives."

A–15.10.   UNITED STATES OF AMERICA–NRC

A–15.10.1. Question

Describe the challenges where qualification of SSCs are absent or incomplete for safety demonstration of SMRs. Provide information on how regulatory judgement was reached in view of the situation.

A–15.10.2. Response

The same regulations that apply to existing LWRs and SMRs regarding SSCs, and the staff did not find any significant challenges in this area of review. Nuclear power plant systems and components important to safety should be designed, fabricated, erected, and tested to quality standards commensurate with the importance of the safety function to be performed. Important to safety SSCs are those SSCs that provide reasonable assurance that the facility can be operated with adequate protection to the health and safety of the public.

The NRC staff uses the term 'non-safety related' to refer to SSCs that are not classified as 'safety-related SSCs' as described in Title 10 of the Code of Federal Regulations [A–52] Part 50.2. However, among the 'non-safety-related' SSCs, there are those that are 'important to safety', as that term is used in the general design criteria (GDC) listed in Appendix A, 'General Design

Criteria for Nuclear Power Plants', to 10 CFR [A–52] Part 50, 'Domestic Licensing of Production and Utilization Facilities', and others that are not considered 'important to safety'. Generally, licensees apply augmented quality controls (a subset of the criteria in Appendix B to Part 50) to these 'important-to-safety' SSCs.

The NRC RG 1.26 [A–214] establishes an acceptable method for complying with these requirements by classifying fluid systems and components important to safety and applying corresponding quality codes and standards to such systems and components.

In GDC 1, the NRC requires, in part, that nuclear power plant SSCs important to safety be designed, fabricated, erected, and tested to quality standards commensurate with the importance of the safety functions they perform. Where generally recognized codes and standards are used, they shall be identified and evaluated to determine their applicability and adequacy and modified as necessary to assure a quality product in keeping with the required safety function.

In accordance with 10 CFR Part 50.55a(c)(1) [A–52], components that are part of the RCPB must meet the requirements for Class 1 components in ASME BPV Code [A–215] Section III, except as provided in 10 CFR Part 50.55a(c)(2) through (4). In accordance with 10 CFR Part 50.55a(d)(1), components classified as QG B must meet the requirements for Class 2 components in ASME BPV Code, Section III. In accordance with 10 CFR Part 50.55a(e)(1), QG C components must meet the requirements for Class 3 components in ASME Code [A–215] Section III.

An applicant's SSC classifications in conformance to RG 1.26 [A–214], and the applicable ASME BPV Codes and industry standards provides assurance that component quality will be commensurate with the importance of the safety functions of these systems. The staff found that the applicant satisfied the applicable design requirements pertinent to the qualification of SSCs.

A–16.   SHARED SAFETY SYSTEMS AND FEATURES AMONG UNITS

This Annex presents the responses provided by the following Member State regulatory bodies to the Question 16: "Describe the challenges associated with regulatory assessment of shared safety systems and features among units (Sharing of safety systems is precluded by the IAEA safety standards). Provide information on how regulatory judgement was reached in view of the situation."

A–16.1. ARGENTINA–ARN

A–16.1.1. Question

Describe the challenges associated with regulatory assessment of shared safety systems and features among units. Provide information on how regulatory judgement was reached in view of the situation.

A–16.1.2. Response

Not applicable.

A–16.2. CANADA–CNSC

A–16.2.1. Question

Describe the challenges associated with regulatory assessment of shared safety systems and features among units. Provide information on how regulatory judgement was reached in view of the situation.

A–16.2.2. Response

The regulatory experience for existing multiple-unit facilities in Canada should be applicable to multi-module SMR deployments.

A–16.2.1.1. Introduction / How do we regulate multiple-unit facilities in Canada

The Canadian Nuclear Safety Commission (CNSC) has decades of regulatory experience with multiple-unit facilities. This includes the facilities listed in Table A–12.

TABLE A–12. MULTIPLE-UNIT FACILITIES, CANADA

NPP	Description (# of Units, power)	In-service since
Pickering NGS A	4 units - 515 MW(e)	1971–1973
Pickering NGS B	4 units - 516 MW(e)	1983–1986
Bruce NGS A	4 units - 750 MW(e)	1977–1979
Bruce NGS B	4 units - 817 MW(e)	1985–1987
Darlington NGS	4 units - 881 MW(e)	1992–1993

There are architectural features that are common to all these stations. For example, these stations feature a vacuum building for common containment and a common services building which includes a central fuel handling area for on-line refuelling and used fuel pools. See Fig. A–13.

The Canadian licensing model for these facilities is one facility, one licence. A licence is issued for all activities concerning a reactor facility regardless of the number of units. If differences exist between units, they are reflected in the licensee's licensing basis documents, such as design manuals, operating manuals, etc. The CNSC can impose operating restrictions on individual units.

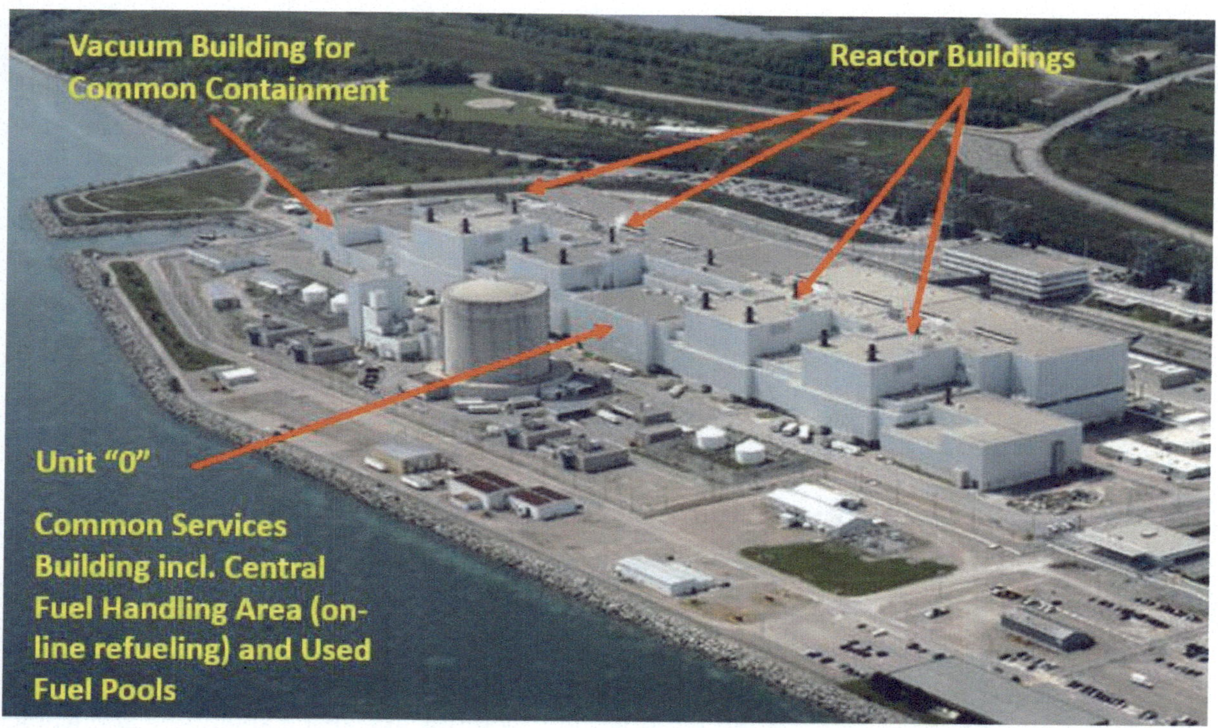

FIG. A–13. Image of Darlington Nuclear Generation Station with labelled with architecture features common to multiple-unit facilities in Canada.

A–16.2.1.2. Shared SSCs in multiple-unit facilities

Shared system, structure and component (SSC) features are designed to supplement unit specific DiD. Some of these features include:

- One main control room (MCR) with dedicated space allocated for each operating unit panels, including Unit 0 and fuel handling;
- Common containment, including one vacuum building;
- Common emergency coolant injection system (ECI) functions;
- Emergency power system to supplement unit-specific electrical supply architecture;
- Emergency service water to supplement unit specific water cooling systems.

A–16.2.1.3. Role of unit 0

In multiple unit facilities, Unit 0 SSCs play various roles:

- Supply common station needs (power, water, lighting, compressed air);
- Infrastructure for station-wide fuel handling systems;

- Ensure common station safety mitigation outside the dedicated unit systems, for example:
    - Used fuel pools (bays) environmental support systems;
    - Maintain station-wide confinement/containment envelope;
    - Fire protection and response.
- Provide supplemental support to each unit during all plant states including transients and accidents:
    - Executes and maintains containment 'button-up';
    - Central pressure relief and dousing on a unit pressure excursion event;
    - Emergency coolant injection supply function.
- For example, station containment system (see Fig. A–14) serves multiple functions, such as:
    - Allows fuelling machines to move between units and Unit 0 Common Services Area (new fuel rooms and irradiated fuel bays);
    - Very large containment volume reduces effects of pressure excursions;
    - Common emergency coolant recovery.

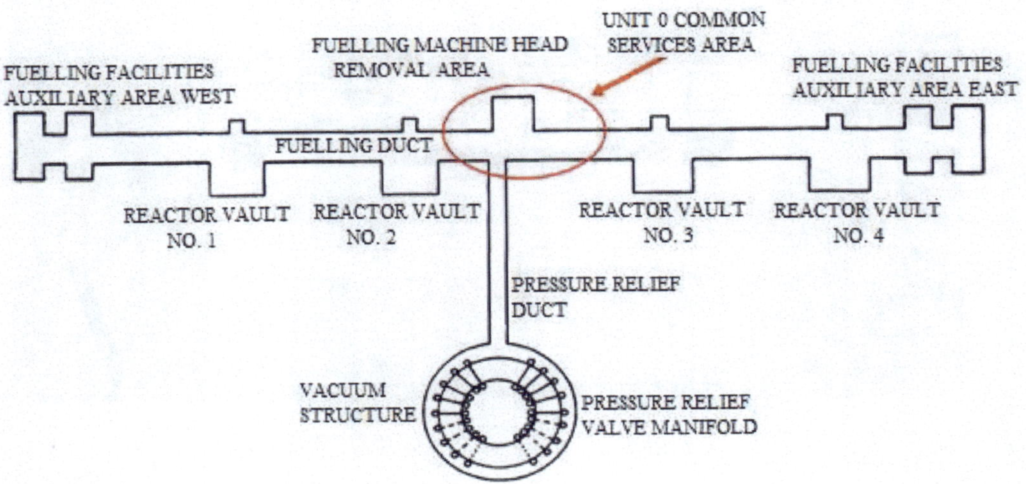

FIG. A–14. Diagram of Darlington Nuclear Generating Station containment system.

A–16.2.1.4. Unit 0 control room description

The Unit 0 control room fulfils many functions (see Fig. A–15), such as:

- Oversight of common station system operations, including:
    - Station-wide containment systems (including access control);
    - Common power supply systems and emergency power transfer.
- Station heavy water inventory quality assurance (upgrading, cleaning, etc.);
- Coordination of station wide emergency plan execution (e.g. fire response, providing emergency power);
- Support each unit during transients and accidents with Unit 0 systems.

There is significant use of (digital) automation in the MCR. Very little equipment is operated manually. The MCR is designed to give panel operators time to focus on big-picture situational awareness to always understand where they are inside the operating envelope. Operators are notified when automatic actions occur or do not occur within prescribed operating limits. Any manual operation or actions are driven by procedures based on diagnostic information such as alarm manuals and abnormal incident manuals (AIM). Operator responses to changing conditions are rehearsed and inter-unit communication is addressed in procedures and training.

For the conduct of control room operation, clear rules are in place to ensure safety in multiple configurations. Consideration must be taken that each unit can be in a different operating state, for example:

- Operating normally at full power;
- Maneuvering from one state to another;
- Shut down in guaranteed shutdown state (GSS) – outage;
- Unit transient or event.

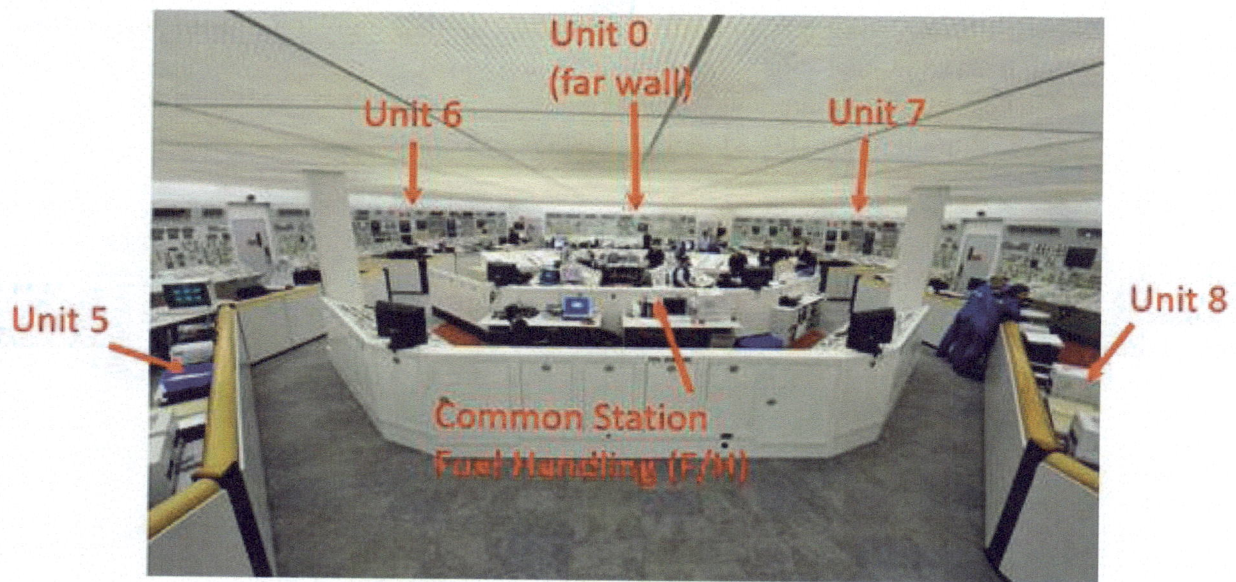

FIG. A–15. Labelled image of Bruce B MCR, Canada.

A–16.2.1.5. Minimum staff complement in multiple-unit facilities

Minimum staff complement applies to the entire facility. This topic is further discussed in REGDOC-2.2.5 [A–161]. Overall:

- CNSC expects a licensee to maintain a systematic analysis to determine the basis of the minimum staff complement while considering:
  - The most resource-intensive initiating events and credible failures considered in the Safety Analysis and the PSA;
  - Required actions;
  - Operating strategies;
  - Required interactions among personnel;
  - Staffing demands associated to the required tasks;
  - Staffing strategies under all operating conditions including normal operation, AOO, DBA and emergency conditions.
- Validation to show safe operation and response to the most resource-intensive conditions (including events that affect more than unit) under all operating states including normal operations, AOO, DBA and emergency conditions.

With a shared MCR, minimum shift complement is typically achieved by the following:

- Each unit has:
    - Dedicated ANO(s) and panel operators in MCR;
    - Field operators are also assigned (dedicated) to each unit;
    - Switching staff between units can be done but is not a regular practice (day-to-day configurations between units may be different);
- Under transient conditions, additional ANO and an ANO from a stable adjacent unit can provide support but dedicated ANO leads;
- Unit 0 has two certified control room operators (CRO) plus dedicated field operators;
- Fuel Handling has dedicated panel and field operators;
- Control room shift supervisor (CRSS) and shift manager (SM) are also licensed staff:
    - CRSS Coordinates MCR activities and is available to assist on incident unit upon request by the Unit ANO;
    - SM is responsible to maintain global oversight of station activities and leads execution of the emergency plan.
- Provisions, with shifting responsibilities and priorities, are also made for multiple unit transients:
    - Done through procedures and regular training exercises.

A–16.2.1.6. Probabilistic safety assessment for multiple-unit facilities

In multi-unit facilities, PSA has to reflect the station design, not just the unit design. This includes the following considerations:

- How do the units interact with each other in different station states?
- How are shared/common systems divided up in the PSA to reflect individual unit safety?
- Site based PSA versus unit based PSA?
    - External events (human-induced or natural)
    - Common-cause failures
    - What is the modelled release size and inventory?

A–16.2.1.7. Overall considerations

The following lessons learned were observed in multi-unit facilities in Canada over the years:

- Multiple unit facilities require a particularly strong configuration management program:
    - Every unit has some differences from others and the impacts of this must be understood for the whole facility.
- Be aware of the unit you are in when performing operations or maintenance! 'Wrong unit' work represents significant operational risk.
    - Units are colour-coded and equipment is tagged by unit;
    - Field/Control room communication always confirms correct unit during evolutions;
    - Procedure-use-and-adherence requires a check that the correct procedure is being used for the right unit (configuration can vary between units);
    - Where does human error fit into Safety Analysis?

A–16.2.3. Follow-up Question

Any conflicting issues when considering the single failure criteria, redundancy and shared systems?

A–16.2.4. Response

With regards to the single failure criteria, it is acceptable to demonstrate there are multiple means to achieve the safety function. It is not necessary to have redundant trains, as is found in currently operating LWRs.

With regards to sharing of SSCs, and in particular instrumentation, the CNSC does not permit sharing of process and safety system instrumentation.

With regards to sharing of SSCs in multi-unit facilities, it is possible to have shared SSCs, but the applicants/licensees must demonstrate that the safety functions can be achieved even with failure of a SSC in another unit. In other words, there is sharing of SSCs, but each unit must be able to operate independently of the others. See information and details in the answer from Canada to Question 7.

A–16.3. CHINA–NNSA

A–16.3.1. Question

Describe the challenges associated with regulatory assessment of shared safety systems and features among units. Provide information on how regulatory judgement was reached in view of the situation.

A–16.3.2. Response

Select the HTR-PM as an example. In HTR-PM design, helium purification system can be selected as an example of challenges associated with assessment of shared safety systems and features among units. Two normal purification lines are designed, each serves one unit of HTR-PM. Besides, a stand-by accident purification line, which is specially designed to remove the water/steam in the primary circuit after a water-ingress accident, serves for both units.

The authorized party (INET) had proved that, for design basis accidents (DBAs) and design extension conditions (DECs), relevant safety and regulatory requirements were met, even this shared line fails.

The use of TRISO particle fuel element and inherent safety design of HTR-PM can ensure there is no core meltdown or severe accident.

Furthermore, even in the DECs with extremely low probability, there is enough time, e.g. several days, to adopt appropriate measures to mitigate the consequence, which make it possible that some safety systems and features are shared among units in one power plants.

A–16.3.3. Follow-up Question

No regulatory activities in this context are mentioned. What are the regulatory requirements with regards to sharing of safety systems and features?

Any conflicting issues when considering the single failure criteria, redundancy and shared systems?

A–16.3.4. Response

At present, we do not have information about the design of safety systems and safety functions sharing, so we do not have the experience of review.

A–16.4. CZECH REPUBLIC–SÚJB

A–16.4.1. Question

Describe the challenges associated with regulatory assessment of shared safety systems and features among units. Provide information on how regulatory judgement was reached in view of the situation.

A–16.4.2. Response

Not Applicable — the standard system has not been modified as there are no SMRs to be deployed and therefore no challenges have been identified. Regulatory judgement would follow the standard practice.

In accordance with the IAEA safety standards each unit is expected to have its own systems and the legislation does not provide for sharing of the safety systems. Nuclear installation design shall, by means of physical separation, functional isolation, independence and redundancy of systems and by using of diverse means, to ensure reliable performance of the safety function of selected equipment in the event of malfunction of selected equipment due to a single failure and common-cause failures. At the same time, the design basis shall for external design events and scenarios which fall within the scope of design extension conditions ensure sufficient capacity and means for managing accident conditions and radiation accidents caused by external design events on-site with multiple nuclear installations expected to share support equipment and services.

A–16.5. FRANCE–ASN

A–16.5.1. Question

Describe the challenges associated with regulatory assessment of shared safety systems and features among units. Provide information on how regulatory judgement was reached in view of the situation.

A–16.5.2. Response

According to ASN's guide on pressurized water reactors design, sharing of important to safety SSCs and features must be limited and justified. In particular, the use of shared SSCs shall not:

- Negate the shutdown, the cooling and the residual heat removal of each unit;
- Lead to a lack of necessary cooling water or electric energy for each unit.

ASN controls that the licensee's safety demonstration and its general operating rules meet these objectives.

These objectives remain fully applicable for potential upcoming SMR projects.

A–16.6. JAPAN–NRA

A–16.6.1. Question

Describe the challenges associated with regulatory assessment of shared safety systems and features among units. Provide information on how regulatory judgement was reached in view of the situation.

A–16.6.2. Response

It is required for research reactor facilities that a safety facility should not lose the facilities safety when it is shared with more than 2 facilities or connected to each other.

Besides, safety facilities for the HTTR are not shared with other facilities.

A–16.7. RUSSIAN FEDERATION–ROSTECHNADZOR

A–16.7.1. Question

Describe the challenges associated with regulatory assessment of shared safety systems and features among units. Provide information on how regulatory judgement was reached in view of the situation.

A–16.7.2. Response

In general, the federal nuclear safety regulations pertaining to nuclear vessels allow integration of safety system and safety-significant system functions, provided that this will not lead to the violation of safety requirements for nuclear installations and to less reliable fulfilment of safety functions, in particular, in case of a multi-unit DBA or BDBA.

Thus, according to paragraph 51 of NP-022-17 [A–89], a multi-purpose use of a safety system and its components, as well as the integration of safety functions and normal operation functions, must not lead to the violation of safety requirements for nuclear installations and to less reliable fulfilment of safety functions.

Paragraph 98 of NP-022-17 allows sharing some components of the confining safety systems between several reactors, provided that the vessel design includes a rationale proving that an accident at one reactor will not affect the other reactor(s). According to Paragraph 99 of NP-022-17, the reactor on board a vessel must have double confinement — containment and safety enclosure. The safety enclosure may be combined with the hull structures.

A–16.7.3. Follow-up Question

Are there any specific considerations from regulators judgements on the floating NPP or BREST-300 regarding the sharing of safety systems? Can you provide further insights with regards to paragraph 1, in particular, examples when the sharing of systems is acceptable?

Any conflicting issues when considering the single failure criteria, redundancy and shared systems?

A–16.7.4. Response

Multipurpose use of safety systems and their components and the combination of safety functions with normal operation functions on vessels with nuclear reactors is allowed if the process of operation of the system and its functions it performs do not change; the performance of the safety functions also has priority over the performance of normal operation functions. For example, the primary circuit cleaning and cooling system can be used for both cooling down during normal operation of the reactor and cooling down in case of accidents. In the event of accidents, either complete prohibition is imposed on the operator's intervention in the system until it fully performs the safety functions, or the operator is left with the opportunity to manipulate the system controls in order to perform the safety functions.

When the system combines safety functions and normal operation functions, the system must be classified as a safety system and comply with the relevant requirements. In addition, the principle of diversity must be respected, that is, several systems based on different operating principles (for example, active and passive cooldown channels) must be provided to perform the safety functions.

In the limited space of a vessel, some systems or their components can be used for two reactors. Sharing means that the system has a channel design with redundant components within the channel. At the same time, there is a backup channel that can replace the failed safety channel of any of the reactors. Also, with appropriate demonstration, it is allowed to operate the channel of the system of one reactor for the other reactor or for both at once. If this is the case, it should be demonstrated that the operation of one channel is sufficient to perform the safety functions in the event of accidents on two reactors simultaneously.

In any case, the safety systems must comply with the principles of physical and functional separation, single failure, independence, and redundancy. Also, the principle of diversity must be implemented for the safety systems.

For land-based NPPs, in accordance with the requirements of clause 3.1.13 of the General Provisions for Ensuring the Safety of Nuclear Power Plants, NP-001-15 [A–254], the multipurpose use of safety systems and their elements must be justified. Combining safety functions with normal operation functions must not lead to a violation of NPP safety requirements and a decrease in the required reliability of safety functions performance.

The safety systems of one unit of a multi-unit NPP should be independent of the safety systems of another unit of the same NPP.

Research work to assess the new requirements for the use of combined systems in multi-reactor plants is planned.

A–16.8. SOUTH AFRICA–NNR

A–16.8.1. Question

Describe the challenges associated with regulatory assessment of shared safety systems and features among units. Provide information on how regulatory judgement was reached in view of the situation.

A–16.8.2. Response

The PBMR project did not reach a sufficiently mature stage for this topic to be developed.

A–16.9. UNITED KINGDOM–ONR

A–16.9.1. Question

Describe the challenges associated with regulatory assessment of shared safety systems and features among units. Provide information on how regulatory judgement was reached in view of the situation.

A–16.9.2. Response

Due to the potential for multiple reactor modules to be deployed in close proximity to each other and to supporting systems, SMRs designs may consider sharing safety systems and safety features for design extension conditions (particularly the latter), in features designed to enhance safety and grace periods.

ONR SAPs [A–44] state that facilities should have their own dedicated safety systems to protect against design basis faults and that such safety systems should not be shared between facilities (SAP para 135 and ST.6 Multi-Facility Sites). This arises from the strong design basis expectation that very high reliability safety systems are needed to protect against high consequence faults to demonstrate the adequacy of level 3 DiD (SAP EKP.3 Defence in Depth). This principle is also reinforced by safety system design SAPs ESS.18 to 20 which state that safety systems should exhibit failure independence between them, be dedicated to a single safety function and be physically separated / isolated. The single failure criterion (SAP EDR.4) and the need for design basis fault sequences to be protected with the relevant safety systems in their most onerous initial operating state (SAP FA.6).

Para 135 of ONR SAPs [A–44] is also clear that safety equipment designed to assist with controlling or mitigating accidents (i.e. at level 4 of Principle EKP.3) may be shared where this is justified to be in the interests of safety (e.g. if this provides a diverse, alternative means of restoring a lost safety function). Where equipment is shared, the safety case should demonstrate that the sharing does not increase either the likelihood or the consequences of an accident at any of the facilities.

In this context, ONR has participated in the IAEA activity to develop a TECDOC on the applicability of design safety requirements (SSR-2/1 Rev. 1 [A–127]) to small modular reactor technologies intended for near-term deployment, which covered both small modular light water reactors and high temperature gas-cooled reactors (HTGR).

As part of this activity, ONR helped review Requirement 33 of SSR-2/1 [A–127], and contributed to the development of a revised requirement for potential application to a multi-module reactor design, which was less prescriptive regarding the potential sharing of safety measures. This formulation included the caveat that where a safety measure is shared between reactor modules of a multi-module unit, the shared safety measure shall be functionally capable of fulfilling the safety requirements of each of these modules simultaneously, to protect against the consequences of events which have the potential to affect multiple modules.

A–16.10.   UNITED STATES OF AMERICA–NRC

A–16.10.1. Question

Describe the challenges associated with regulatory assessment of shared safety systems and features among units. Provide information on how regulatory judgement was reached in view of the situation.

A–16.10.2. Response

Staff did not find any significant challenges in this area of review. In 10 CFR [A–52] Part 52.1, it defines 'modular design' as a nuclear power station that consists of two or more essentially identical nuclear reactors (modules) and each module is a separate nuclear reactor capable of being operated independent of the state of completion or operating condition of any other module co-located on the same site, even though the nuclear power station may have some shared or common systems.

The provisions in Appendix A to 10 CFR [A–52] Part 50, 'General Design Criteria for Nuclear Power Plants' (GDC) in Criterion 5 require that SSCs important to safety shall not be shared among nuclear power units unless it can be shown that such sharing will not significantly impair the ability to perform their safety functions, including in the event of an accident in one unit, an orderly shutdown and cooldown of the remaining units.

The NRC staff determined that the applicant's design considers the risk and safety effects of the multimodule plant operation with shared systems to ensure the independence and protection of the safety systems of each unit during all operational modes. Staff also determined that the applicant's multimodule evaluation was adequate for the design certification (DC), since the applicant considered potential system interactions with other reactor modules, as specified in 10 CFR [A–52] Part 52.47(c)(3) and documented key assumptions in the Design Certification Application (DCA) to be confirmed in the combined license (COL) phase. The applicant's assessment is also technically adequate and consistent with the guidance in SRP [A–116] Section 19.0.

This Annex presents the responses provided by the following Member State regulatory bodies to the Question 17: "Describe the challenges associated with regulatory assessment of the adequacy of confinement function."

A–17.1. ARGENTINA–ARN

A–17.1.1. Question

Describe the challenges associated with regulatory assessment of the adequacy of confinement function.

A–17.1.2. Response

The CAREM 25 is an integrated type reactor, that is, with a compact primary system distribution. The design adopted for the containment is of the pressure suppressor type.

The AR 3.4.3 standard 'Confinement system in nuclear power reactors' [A–255] sets the requirements of the confinement barriers in order to meet the requirements of the AR 3.1.3 standard 'Radiological Criteria Relating to Accidents in Nuclear Power Reactors' [A–256] among other safety requirements like leak rate.

Regarding the assessment of the information related to the containment design of the CAREM 25 prototype reactor, some inconsistencies were detected in the inclusion of events for the design of the containment structures, including all openings, penetrations, and isolation systems. It was observed that they have been designed to preserve their structural and functional integrity in the face of postulated initiating events for a design basis. By mentioning only the PIE for DBA, it excludes DECs, thus resulting in an equivocal description of the role of Containment.

This observation was resolved, as well as the terminology used was homogenized. In this sense, it was verified that the information expansion details the characteristics such as the levels of DiD to which they belong, the sub-level of categorization, and the security class of the containment systems.

In relation to the final safe state that should be achieved through the action of active systems after 36 hours of the 'time or grace period', it was verified that the studies presented in the previous version of the CAREM reactor design report were expanded. These new studies demonstrating the ability of some active systems, already provided for by design, to adequately control the progressive increase in pressure that had been observed in various PIEs. The new calculation results show that after 36 hours and up to 76 hours, that is, for a period of 40 hours in addition to the grace period, it is possible to reach a safe end state. With the intervention of these systems and considering an initial pressure of 0.5 MPa, it is possible to reduce the pressure in the containment areas, avoiding exceeding the design pressure of the same.

In response to the observation of the ARN on the need to include a vent valve or containment relief, the CNEA, in order to avoid delays in the start of the construction of the nuclear module, included in the project a penetration to the containment for such a system in case its installation eventually becomes necessary or required.

A–17.1.3. Follow-up Question

Would there be any challenges in licensing a design without a leak tight containment/ pressure retaining containment in Argentina?

Please describe how regulatory judgement on the interface between the confinement function and requirements for external events e.g. aircraft impact, and control of radiological releases during normal operation and accident conditions was achieved.

A–17.1.4. Response

According to ARN Standard AR 3.4.3 [A–255], one or more confinement barriers for the reactor and primary pressure system must be provided, unless it can be proved by the licensee, that even without those barriers an accident condition that could imply a failure to the primary system, fulfil the ARN's requirements and criteria, regarding radiological releases in nuclear reactors.

The ARN required the licensee to describe in the Safety Report the compliance with ARN's Standards according to the design and criteria for Safety Structures, Systems, and Components, this includes external events. In the case of the confinement function, the design principles are described to achieve and comply with the different DiD levels, for normal and accidental conditions. Regarding aircraft impact, the site is in an air exclusion zone.

A–17.2. CANADA–CNSC

A–17.2.1. Question

Describe the challenges associated with regulatory assessment of the adequacy of confinement function.

A–17.2.2. Response

CNSC's REGDOC-2.5.2 [A–62] sets technology-neutral expectations for fulfilling the fundamental safety functions including:

- Confinement of radioactive material;
- Shielding against radiation;
- Control of operational discharges and hazardous substances, as well as limitation of accidental releases.

REGDOC-2.5.2 [A–62] also contains system-specific requirements for the containment system, containment isolation and access requirements. For reactors under 200 MW(th), CNSC's RD-367 [A–61] sets requirements for means of confinement. However, vendors engaged in the CNSC's pre-licensing vendor design review (VDR) process have typically chosen to apply REGDOC-2.5.2 [A–62] using a graded approach citing the benefits of considering applicable guidance in their design.

Experiences from new reactor technology assessments under the CNSC's Vendor Design Review Program

Overarching technological trends

CNSC staff have noted the following overarching technological trends in proposals from both large NPP vendors and vendors of SMRs and advanced reactor technologies:

- The concept of 'functional containment', where radionuclides are retained within multiple barriers with an emphasis on retention at their source (in the fuel);
- The concept of 'low-leakage containment', where there is continuous leakage from the containment during normal operation;
- Claims that a traditional concrete containment structure would not be necessary to fulfil fundamental safety functions and meet acceptance criteria for accident conditions;
- Potential novel safety classification and code classification that challenge traditional views that the containment is a Safety System. These types of proposals lead to greater interpretation of technical standards used in detailed design activities and a greater need for supporting R&D to establish reliability of SSCs.

There is flexibility in CNSC's framework to allow for alternative approaches to be proposed providing it is justified in the design and safety analysis documentation and supplemented with the appropriate research & development information and test results. An applicant must demonstrate that:

- The confinement function is fulfilled in all plant states with sufficient reliability taking into account the need for in service inspections, reliability testing and maintenance;
- Failure of the containment system would not impact the ability of the reactivity control and heat removal systems to perform their functions;
- Sufficient DiD is maintained at all times.

Some design developers claims that their design proposal does not need containment structures in the way that water-cooled reactors do. Although the reactor is located in the concrete structure, these types of designs rely primarily on the fuel particles/fuel matrix as the primary means of containment, with the reactor vessel pressure boundary serving to complement the confinement function. CNSC will verify that fuel qualification program activities are underway to provide results that will support the credibility of containment and confinement performance.

Low-leakage confinement structure is proposed as an alternative approach to the leak tight Canadian regulatory requirement. CNSC will verify that the future submissions will provide supporting evidence for this alternative approach.

CNSC noted that some designers' claims that the containment requirements need to be minimal given that over-pressurization accidents are not credible. Some designer claims that containment design follows the passive safety design principles; therefore, no engineered automatic actuation of containment isolation devices is required.

Some designs propose that the containment isolation by loop seals means. CNSC will verify that the containment isolation performance will be demonstrated by the designer via analytical assessments during the following design stages.

A–17.2.3. Follow-up Question

What are the metrics/criteria used to evaluate the adequacy/robustness of the fuel and the associated potential source term? What is the regulatory expectation regarding quality and reliability of the confinement barrier? Are potential common mode factors considered and how is DiD and diversity principles or multiple barrier concept applied.

A–17.2.4. Response

What are the metrics/criteria used to evaluate the adequacy/robustness of the fuel and the associated potential source term?

For many SMR designs, vendors are giving more credit to the fuel in the overall confinement scheme. The key metrics CNSC uses to assess the adequacy and robustness of the fuel are the dose acceptance criteria and safety goals. These are technology neutral performance-based criteria.

Furthermore, an applicant is expected to define the fuel conditions necessary to meet the dose acceptance criteria and safety goals. These are known as derived acceptance criteria and tend to be technology specific.

There are different considerations for solid and liquid fuel. For solid fuel, integrity is typically the acceptance criteria. For liquid fuel, such as molten salt fuel, it may be thermal physical condition and radio nucleus release limits.

While greater emphasis can be put on a single barrier, the concept of multiple barriers and principles of diversity are expected to be incorporated to address potential common modes of failure and maintain adequate DiD.

What is the regulatory expectation regarding quality and reliability of the confinement barrier?

CNSC will assess an individual barrier in terms of its overall role in the safety case presented by an applicant. This allows for flexibility in the design while maintaining sufficient quality and reliability to ensure the confinement function is met for all plant states. In addition, dose limits and safety goals must be met, as do expectations for robustness and emergency preparedness.

Are potential common mode factors considered and how is DiD and diversity principles or multiple barrier concept applied.

Yes, potential common mode factors considered in design and safety analysis. In the CNSC regulatory framework, there has to be systematic assessment of natural and man-made external events, as well as malevolent acts. As such, facilities need to be designed to withstand such common causes.

This is achieved through implementation of DiD. Such as having independent and overlapping provisions to ensure prevention, detection, correction, and mitigation of failure. This includes provision of a series of physical barriers to confine radioactive material.

A–17.3. CHINA–NNSA

A–17.3.1. Question

Describe the challenges associated with regulatory assessment of the adequacy of confinement function.

A–17.3.2. Response

Select the HTR-PM as an example. NNSA stated attitude about Containment of HTR-PM in the safety review principles for HTR-PM, Part 5:

> "For traditional PWR and BWR nuclear installations, they are particularly sensitive to reactor coolant loss accidents because of their fuel element form, high core power density and core residual heat. In order to maintain the cooling of fuel elements in the event of reactor coolant loss accident, a complex emergency core cooling system is set up. In this way, the containment not only plays the last barrier role of radioactive release to the environment, but also plays an important role in maintaining the total amount of coolant necessary after the accident and ensuring the long-term cooling of the core.
>
> While as for the HTR, the confinement of radioactive materials mainly depends on the coated granular fuel elements with high reliability. Because the coated particle fuel element can withstand high temperature and HTR has low core power density, after an accident, the core residual heat can be transferred to the heat removal system through natural mechanisms such as heat radiation and heat conduction, and then transferred to the final heat sink by the passive system, so HTR is not sensitive to the loss of reactor coolant. Even for the most serious accident conditions considered, the radioactive release of HTR is limited and has a large delay, which provides a long tolerance time for taking accident management measures. The above characteristics indicate that HTR can adopt containment which is very different from traditional PWR and BWR nuclear installations in principle (known as VLPC, ventilated low pressure containment, or containment).
>
> However, the rationality of adopting such containment concept must be proved by a complete safety evaluation, that is, it must meet the safety objectives determined for HTR-PM, and not reduce the overall defence level, including the defence of external events."

The containment of HTR-PM is not a normal safety barrier as LWR. As in case of loss of coolant accident, when the pressure in containment reaches 0.121 MPa or reaches 0.116 MPa in other compartments, it does not play a containment role, but allows the helium coolant to discharge directly to the environment. The reviewers believe that such containment is acceptable if there is sufficient test data to prove that the failure rate of HTR-PM fuel elements and the retention capacity of radioactive fission products meet the design requirements.

The quality of the spherical fuel element with coated particles is the key to ensure the safety of high temperature gas cooled reactor. Evaluation requires carrying out irradiation test under normal operation conditions, isothermal heating test under simulated accident conditions and oxidation corrosion test of irradiated fuel coated particles under high temperature air flow.

A–17.3.3. Follow-up Questions

The response articulates the approach to confinement including NNSA's acceptance of increased reliance on the spherical fuel element in the context of LOCA analyses. The response indicates that even for the most serious accident conditions considered, the radioactive release of HTR is limited and has a large delay. What is the most limiting/serious scenarios considered? It would appear that large air ingress e.g. was a break of the duct connecting the reactor vessel and SG postulated e.g. large air ingress with high fuel element temperatures?

Please describe how regulatory judgement on the interface between the confinement function and requirements for external events e.g. aircraft impact, and control of radiological releases during normal operation and accident conditions was achieved.

What are the metrics/criteria used to evaluate the adequacy/robustness of the fuel and the associated potential source term? What is the regulatory expectation regarding quality and reliability of the confinement barrier? Are potential common mode factors considered and how is DiD and diversity principles or multiple barrier concept applied.

A–17.3.4. Response

For the confinement function, the most important barrier is coating layer of TRISO coated particle, or the integrity of TRISO coated particles.

The most serious accident condition to challenge the TRISO coated particle is depressurized accident, which arouse the temperature up to the maximum fuel temperature limit.

For the air ingress accident (classified as beyond design basis accident), or in case of some of the accidents that can be classified as design extension conditions (required to be taken into consideration in design), there are enough grace time (for example 72 hours) to take action to stop air ingress accident, and the failure rate of coated particle and the release of fission product is limited.

In this context, the break of primary pressure boundary, the opening of reactor building, has no direct consequence of large release, although the integrity of primary boundary or reactor building can reduce the release.

Currently the acceptance criteria for confinement function can be described as probability safety goal as: the accumulated frequency for accident scenarios whose release is larger than 50 mSv in the site boundary must be less than 1E-6/reactor year.

The common mode factors for the DiD levels and multiple barriers need more investigation, although no obvious factor is found.

A–17.4. CZECH REPUBLIC–SÚJB

A–17.4.1. Question

Describe the challenges associated with regulatory assessment of the adequacy of confinement function.

A–17.4.2. Response

Not Applicable — a regulatory assessment of the adequacy of confinement function of an SMR has not been conducted as there has been no licence application for SMR yet.

In general, any nuclear installation design for nuclear installations with a nuclear reactor with thermal output exceeding 50 MW shall comprise the design of a containment safety system capable of preventing radioactive release and protection of the nuclear reactor against the effect of site characteristics (natural external events) and man-induced events. It shall consist of a hermetically sealed envelope enclosing a hermetically sealed space and ensure pressure and temperature control and handling and controlled removal of fission products, hydrogen and other substances produced by fission in order to prevent their release outside. Nuclear installation design shall set requirements for tightness, strength and functionality testing of the containment system and its individual parts during and after the construction of the nuclear installation (and for the heat removal system, integrity protection, means to allow entrance while maintaining its hermetical tightness and other systems and features).

The nuclear installation premises shall be designed (inter alia) so as to ensure optimization of the radiation protection, prevent a release of radioactive substance from systems, prevent release of radioactive substance outside the nuclear installation and create barriers preventing spread of radioactive material and contamination of persons and objects.

A–17.4.3. Follow-up Question

Would there be any challenges in licensing a design without a leak tight containment/ pressure retaining containment?

A–17.4.4. Response

The requirements pertaining the containment system contained in the Czech Republic legislation reflects, to certain extent, currently used technology (i.e. PWR). Therefore, licencing of a SMR design without a leak tight/pressure retaining containment can serve as a model example of such challenge.

A–17.5. FRANCE–ASN

A–17.5.1. Question

Describe the challenges associated with regulatory assessment of the adequacy of confinement function.

A–17.5.2. Response

According to the article 3.4 of the Order [A–72], safety demonstration must present how confinement function is insured. Confinement function must be insured by the positioning of one or several successive and sufficiently independent barriers between radioactive substances and people or environment and, if necessary, by a dynamic confinement system. The number and the efficiency and these provisions are proportionate to the importance of potential radioactive discharges. These regulatory requirements remain applicable for SMRs.

Chapter VI.3 of ASN's Guide n°22 [A–77] on pressurized water reactors design provides information to meet the regulatory requirements regarding confinement function.

For example:

- The confinement must be as efficient as possible, in order to meet the safety objectives presented in question 7. For this purpose:
  - Leak criteria shall be defined for containment buildings and its penetrations, and for other buildings of the nuclear island in which radioactive substances may be found;
  - Dynamic confinement system shall be equipped with appropriate filtration.

- Eventual radioactive leaks must be detected, and their consequences limited by appropriate means. In particular, the design shall include redundant and diversified SSCs to insulate circuits connected to the primary circuit. Eventual failures must be identified and equipment to detect them must be provided in the design;

- Design provisions shall enable to stabilise the corium in order to avoid foundation raft melt-through and to insure containment resistance against hydrogen explosion;

- In normal operation, SSCs must guarantee the control of pressure and temperature inside the containment building. Also, SSCs must enable the detection, the monitoring, and the treatment of radioactive substances that might be released inside the containment building;

- Appropriate static tightness shall be obtained in buildings in which SSCs contain or might contain radioactive substances. If necessary, SSCs must be installed to collect eventual discharges;

- The number of penetrations in the containment building must be as low as possible;

- Regarding ventilation systems, they have to be designed in a way that:
  - Reinforce as much as necessary static confinement provisions by creating a depression cascade phenomenon from low risk premises to higher risk premises to avoid radioactive discharges and to direct gaseous effluent to appropriate treatment systems before their release;
  - Maintain acceptable working conditions during normal operation, incidents and accidents;
  - Avoid explosive atmosphere;
  - Limit the risk of radioactive substance discharges in case of fire;
  - Maintain indoor conditions in premises that are compatible with SSCs' qualification.

- Containment building and its penetrations and insulating systems must be designed and built in a way that enable periodical testing.

ASN reviews licensee's design to control its compliance with regulatory requirements and to assess if it meets objectives defined in its guide.

During the instruction of the Flamanville's EPR design, ASN, with the support of IRSN, particularly focused on bypass of confinement issues.

A–17.6. JAPAN–NRA

A–17.6.1. Question

Describe the challenges associated with regulatory assessment of the adequacy of confinement function.

A–17.6.2. Response

For the HTTR, the release of radioactive materials outside the reactor facility is suppressed by maintaining a negative pressure inside the reactor containment vessel during normal operation and performing construction and maintenance so that the design leakage rate is not exceeded.

In the event of a design basis accident, the containment function of the reactor containment vessel and emergency circulation equipment with charcoal filter is assumed to be effective and evaluate the effective dose to the surrounding public.

In the event of a design extension condition (DEC), as an assumption that the design basis accident is exceeded which the containment function of the reactor containment vessel is lost, measures should be taken to reduce the effective dose to the surrounding public.

A–17.6.3. Follow -up Questions

What events are used to define the performance requirements of the reactor containment vessel in the HTTR? What measures are considered for DEC (to reduce the effective dose to the public in the case of the HTTR as credited)?

Please describe how regulatory judgement on the interface between the confinement function and requirements for external events e.g. aircraft impact, and control of radiological releases during normal operation and accident conditions was achieved.

A–17.6.4. Response

As for the DEC countermeasures for HTTR (when DBA countermeasures could not be expected) is achieved by:

- Monitoring dose in the reactor building and surrounding area;
- If higher dose detected, sealing the gaps or cracks outside of the reactor building to maintain airtightness and controlling the release of fission products from higher position of reactor building.

By above mentioned measures, the effective dose that public exposed could be reduced by reducing radioactivity concentrated in the area outside of facility compared with the case released from ground level.

A–17.7. RUSSIAN FEDERATION–ROSTECHNADZOR

A–17.7.1. Question

Describe the challenges associated with regulatory assessment of the adequacy of confinement function.

A–17.7.2. Response

According to NP-022-17 [A–89], confining safety systems must be provided to contain emergency radioactive substances and ionizing radiation within the boundaries specified in the vessel design. Confining safety systems must be provided for each vessel reactor and perform the assigned functions in design-basis accidents and beyond-design-basis accidents.

The containment must be designed to withstand an internal pressure caused by emergency coolant discharge from the reactor circuit following instantaneous pipeline rupture, taking into account the action of the system reducing internal containment pressure, and retain its functions in case of vessel flooding. During commissioning, the containment must be tested under the rated pressure. Further tests must be conducted under the pressure justified in the vessel design. The engineered features in the leak tight compartments must survive the tests without losing their serviceability. The vessel design must provide a methodology and technical means to verify the containment compliance with design parameters.

All components of the containment leak-tight circuit, via which in emergency the radioactive substances can leak beyond the leak-tight area boundary, must be equipped with the shut-off valves or with sealing means.

The vessel design must justify the acceptable levels of ionizing radiation beyond the biological shielding, and the acceptable size of containment leakage, at which the reference dose limits set for the vessel crew and special personnel, and the limits of radioactive substances release in the environment are not to be exceeded under normal operation and under abnormal operation, including design basis accidents.

Compliance of the actual containment tightness with the design one must be verified before the first criticality, and then tested in the course of operation at intervals established in the vessel design.

A–17.7.3. Follow-up Question

Please describe how regulatory judgement on the interface between the confinement function and requirements for external events e.g. aircraft impact, and control of radiological releases during normal operation and accident conditions was achieved.

A–17.7.4. Response

When reviewing safety documentation, the regulatory body examines how the normal operation systems and safety systems perform their functions, including confinement of radioactive substances in case of external initiating events, in particular, an aircraft crash.

For land-based NPPs, in accordance with the requirements of NP-001-15 [A–254], confining safety systems must be provided to confine radioactive substances and ionizing radiation in an accident within the boundaries specified in the NPP design.

The reactor and the systems containing radioactive substances, and the reactor components must be entirely located within the reactor containment to confine the radioactive substances released during design basis accidents. Controlled release of radioactive substances outside the reactor containment is allowed in severe accidents only to prevent the destruction of the containment, provided that measures are taken to ensure the radiation safety of the population (through the use of a release filtration system, shelter, evacuation of the population, or other measures).

The localizing safety systems must be provided for each NPP unit and perform their specified functions for design basis accidents, and beyond design basis accidents specified in NP-001-15 [A–254].

Detailed requirements are established by the federal regulations and rules, 'Rules for the Design and Operation of Confining Safety Systems for Nuclear Power Plants' (NP-010-16) [A–257].

A–17.8. SOUTH AFRICA–NNR

A–17.8.1. Question

Describe the challenges associated with regulatory assessment of the adequacy of confinement function.

A–17.8.2. Response

The question is what kind of confinement/containment will be required for the PBMR concerning internal events (leak tight or containment with early depressurisation function in case of larger helium coolant leaks). The concept is to base containment design on the technical evaluation of the potential source term/releases.

The NNR requires conservative design solutions because of the significant uncertainties of circuit contamination, dust and radioactivity mobilisation, and the resulting source term. Evidence must be provided that the selected design approach is ALARA in the event of leaks or breaks of the helium pressure boundary. The design approach preferred by PBMR is a confinement structure with an initial filtered depressurisation function to cope with the initial pressure pulse rather than a civil structure providing a high-pressure containment function.

Consideration of the dependencies between the confinement approach on the civil structures (as an additional barrier against radioactive releases and the related source term), Leak before Break application, Emergency Plan and associated Emergency Planning Zones is required. The licence applicant's strategy should also define the various aspects that must be considered to justify the confinement approach considering the possibility that the design and analysis to support the PBMR design approach could be inadequate to justify a departure from the LWR-industry norm of a high-pressure, leak-tight containment.

The NNR indicated that although not a full scope of analysis is required for the selected confinement design, it still needs to be demonstrated by qualitative safety justification, supported by some quantitative analysis for a full pressure design as part of the justification, that the selected design is more advantageous than the high-pressure containment approach.

The total releases associated with the preferred design approach for a confinement with initial depressurisation function will be governed by the efficiency and the reliability of the filtration function and the reliability of the isolation functions after depressurisation. The 'high pressure containment design' source term could potentially be dominated by contaminated helium leakages and containment bypasses during the high-pressure mode. This comparison between design options needs qualitative analysis and justification.

A qualified filter efficiency and reliability must be established. It is apparent that the feasibility of the filter function is crucial for the design decision and justification. The civil structures must cater for a limitation of the pressure pulse to the filter. The licence applicant's approach should recognise the importance of the development, verification and qualification steps of the filtration concept and identify the associated deliverables that will be produced to support the strategy to arrive at a plausible design approach as a basis for the safety case.

The deliverables must also provide information on the potential impact on the civil structures and how the design approach takes that into account. The strategy should address the steps needed as well as the aspects to be considered to arrive at a justified and feasible civil design approach. The analysis should also consider the consequences of smaller breaks, not only large breaks. The civil structures must limit the pressure pulse and temperature loads on, as well as the flow rates through, the filtration device. This will imply significant civil design and qualification efforts.

The consequences of different break sizes and locations are significant e.g. small breaks tend to impact the compartments (and connected pipe work) significantly concerning temperature loads, whereas the large breaks cause a major pressure pulse and huge loads on civil structures, doorways and HVAC isolation valves. The strategy should consider this.

The monitoring of releases during depressurisation needs to be addressed and the emergency preparedness implication clarified.

A–17.8.3. Follow-up Question

Please describe how regulatory judgement on the interface between the confinement function and requirements for external events e.g. aircraft impact, and control of radiological releases during Normal operation and accident conditions was achieved.

A–17.8.4. Response

Regulatory judgement on the interface between the confinement function and requirements for external events e.g. aircraft impact, and control of radiological releases during normal operation and accident conditions is achieved by attempting to reach agreement with the licence applicant about the strategies they will follow to resolve the key licensing issues (KLIs) of, in this case, KLI 5 'Containment', KLI 15 'Aircraft Crash / External Events', KLI 17 'Licensing Basis Events' and KLI 18 'Probabilistic Risk Assessment'. For that purpose, the PBMR licence applicant submitted strategy documents for each of these KLIs. Each of these strategy documents, in turn, identified deliverable documents supporting technical aspects claimed or undertaken in the respective KLI strategy documents. These KLI strategy documents and associated deliverable documents underwent review cycles by the NNR with a view of enhancing the chances of achieving consistency with regards to the interface between the confinement function and requirements for external events e.g. aircraft impact, and control of radiological releases during normal operation and accident conditions.

A–17.9. UNITED KINGDOM–ONR

A–17.9.1. Question

Describe the challenges associated with regulatory assessment of the adequacy of confinement function.

A–17.9.2. Response

ONR SAPs [A–44] note that confinement of radioactive material by the provision of multiple (and as far as practicable independent, physical barriers) is an important aspect of the implementation of DiD.

ONR SAPs ECV.1 to 10 provide expectations for containment and ventilation design, to confine radioactive material within the facility and to prevent its leakage or escape to the environment in normal operation and fault conditions, except in accordance with authorised discharge conditions. Potential challenges against these SAPs were reviewed as part of workshops held to develop guidance on DiD as applied to SMRs. ONR has also published a technical assessment guide NS-TAST-GD-020 on Civil Engineering Containments for Reactor Plants [A–258].

Some SMRs (i.e. HTGRs) are proposing to achieve the confinement function primarily through relying on the ceramic coated fuel (e.g. TRISO) to retain fission products in all accident conditions, rather than providing a pressure retaining, 'leak-tight' containment structure. Such designs may include a 'reactor building', and/or 'aircraft protection shell', to perform the other functions of a containment, such as protection against external events. In addition, liquid-metal cooled fast reactor technologies generally design for 'low-leakage' containments, with much lower design pressures than for a PWR type containment.

These approaches may challenge the interpretation of ONR guidance on the adequacy of the confinement function, as some aspects of the guidance are written assuming 'containment' (in a new nuclear reactor) to be a 'leak-tight' civil engineering structure. As such, some expectations may require specific interpretation for application to particular SMR technologies.

ONR has participated in the IAEA activity to develop a TECDOC on the applicability of design safety requirements (SSR-2/1 Rev. 1 [A–127]) to small modular reactor technologies intended for near-term deployment, which covered both small modular light water reactors and high temperature gas-cooled reactors (HTGR).

As part of this activity, ONR helped reviewed the relevant containment requirements in SSR-2/1 [A–127] (Requirements. 54-58), which have been written in the context of typical light-water reactor containment design, and identified the underlying purpose of the requirement, from a functional perspective. ONR considers that whilst the existing requirements as written in SSR-2/1 were not fully applicable to all SMRs, and should potentially be reformulated, it is important not to lose sight of the underlying functional intent of the requirements.

A–17.10.   UNITED STATES OF AMERICA–NRC

A–17.10.1. Question

Describe the challenges associated with regulatory assessment of the adequacy of confinement function.

A–17.10.2. Response

The NRC staff found that that applicants have typically satisfied the applicable design requirements pertinent to the adequacy of confinement function with minimal review challenges. This response is based on a small modular reactor (SMR) at a multi-module plant where all the modules are at the same site, in the same building and in a common pool of water.

For some small modular reactors, the staff has found that containments can be tested the same way the current LLWR containments are tested. The containment for this SMR is not a conventional large light water containment but is an American Society of Mechanical Engineers (ASME) Boiler and Pressure Vessel Code Class MC component (inclusive of all access and inspection openings, penetrations for emergency core cooling system trip/reset valves, and openings for electrical penetration assemblies). As permitted by ASME NCA-2134(c), the complete containment vessel is constructed and stamped as an ASME Class 1 vessel in accordance with ASME Boiler and Pressure Vessel Code [A–215] Section III, Subsection NB. All penetrations that are potential leakage pathways are either ASME Class 1 flanged joints capable of Type B testing or ASME Class 1 welded nozzles with isolation valves capable of Type C testing and are tested accordingly.

Because the potential vessel leakage pathways are testable containment penetrations, total containment leakage can be quantified via 10 CFR [A–52] Part 50, Appendix J, Type B and C tests, thus assuring that containment leakage does not exceed allowable leakage rate values.

The containment is built, tested and maintained as a leak tight code vessel. Comprehensive in-service inspections ensure that no new leakage pathways develop over the life of the containment system.

The containment is a small, high pressure, ASME Section III, Class 1 that is more comparable to typical reactor pressure vessels in design and dimensions than to typical containment structures. The post-accident containment atmospheric pressure is very high, and the containment volume is relatively small compared to LLWR containments. This causes the allowable leakage, which is measured in standard cubic feet per minute, to be extremely low, making it extremely difficult to be able to accurately measure the leakage rate from the containment.

Comprehensive preservice and in-service inspections and tests, applying ASME Class 1 criteria, ensure continued system leakage integrity. All surface areas and welds are accessible for inspection. All penetration pathways will be tested to Type B or C criteria at accident or design pressures to ensure that continued leakage integrity of the containment system is maintained. Therefore, the containment is designed to allow alternative testing and inspection that provide equivalent assurance that the allowable containment leakage is not exceeded during its service life, and therefore assures the performance of the overall containment system as a barrier to fission product releases.

This Annex presents the responses provided by the following Member State regulatory bodies to the Question 18: "Describe the challenges associated with regulatory assessment of the adequacy of reactivity control function."

A–18.1. ARGENTINA–ARN

A–18.1.1. Question

Describe the challenges associated with regulatory assessment of the adequacy of reactivity control function.

A–18.1.2. Response

In the CAREM 25 prototype reactor, the reactivity control to operational events and events in DiD level 2 and DiD level 3 is carried out using the First Shutdown System (FSS) and the Second Shutdown System (SSS). The ARN assessed and verified the information presented in the safety report to issue the Construction License.

The FSS operates the free fall of bars composed of neutron absorbent materials that will be inserted into the core of the reactor when its performance is required. It has two sets of absorbent bars, totalling 25 bars. The first of these, called the Rapid Shutdown System (RSS), must quickly cause the reactor to extinguish for all DBA in a period of no more than 2 seconds. The second set is in charge of controlling the reactivity of the reactor core in operation, called the Adjustment and Control System (ACS), if the action of the FSS is required, it must be inserted to guarantee the reactor's subcritical in all operating states. The ACS insertion time is around 20 seconds.

The FSS fulfils two functions, one of security and the other of adjustment and control:

- The RSS fulfils the safety function related to the extinction of the nuclear reaction and partially contributes to maintaining the shutdown condition. During operation, the SSS CRDs are retained outside the nucleus and upon a trigger signal, the system produces the complete and immediate insertion of the EEAA.

- The ACS fulfils control and security functions. The control is carried out during the operation by moving the CRDs between different extraction positions along their path. In the event of a trip signal, the system acts as part of the safety system, continuously lowering the EEAA, providing the necessary negative reactivity to keep the reactor off, considering the variations in reactivity that may occur during the different states of the plant in shutdown condition.

Each CRD of the RSS and the ACS are actuated by the rapid extinguishing and adjustment and control mechanisms, which are used to move the absorbent element vertically (or gradually) in order to adjust the reactivity of the nucleus.

Since the FSS is primarily made up of conventional mechanical parts, such as pipes, valves, pumps, filters, and heat exchangers, and other equipment commonly used in reactor systems, the evaluation of development focuses on three main issues:

(a)   Structural and functional verification of the kinematic chain in seismic conditions;

(b)   Validation and qualification of the position measurement system;
(c)   Validation of the conceptual design and qualification of the hydraulic mechanisms and drive system.

Point a) includes the verification of the fall times of the kinematic chain and the measurement of the main vibration modes in seismic conditions representative of the design base earthquake, which will be simulated in an experimental device built for this purpose.

Point b) includes the first stage of development in the laboratory to size the system and solve design aspects. System validation is performed under high-pressure conditions.

Point c) has been divided into three main phases by CNEA:

- A first phase allowed to obtain a preliminary design of the mechanism and define the general operating parameters;

- A second experimental phase under more controlled test conditions than the previous one (for which the Experimental Mechanism Testing Circuit, CEM) was built, the main objective of which was to determine the technical feasibility of the system to meet the functional requirements and evaluate the construction details. in which the technical feasibility of the Adjustment and Control System Mechanism (MSAC) has been verified. The circuit allowed different device variants to be tested until the current design was obtained, observing satisfactory behaviour for temperatures up to 80°C (circuit design limit) under turbulent flow conditions, which allowed inferring the behaviour under real working conditions, with the support of numerical simulations based on dynamic and stationary models developed specifically to represent the observed phenomena;

- For the next stage, a High-Pressure Circuit for Testing Mechanisms (CAPEM) has been assembled. CAPEM is a self-pressurized circuit and the working pressure and temperature are the operating conditions of the reactor. In the CAPEM, performance tests will be carried out separately from the MSAC and the Rapid Extinction System Mechanism (MSER).

In relation to the Second Shutdown System, it operates by discharging, through the action of gravity, a liquid solution with neutron poison. It was incorporated into the design to comply with criterion 9 of Standard AR 3.4.2, Rev. 1 [A–259]: "The set of extinguishing systems must have at least two independent extinguishing systems with adequate diversity, each with sufficient negative reactivity as to cause the extinction of the reactor without the action of the other being necessary."

Other systems related to event control in DiD N3 are also presented. Mention may be made of the Safety Injection System (SIS) and the design of the storage tank for irradiated fuels in order to ensure the subcriticality of the EECC stored in the pool in the presence of EP.

A–18.2. CANADA–CNSC

A–18.2.1. Question

Describe the challenges associated with regulatory assessment of the adequacy of reactivity control function.

A–18.2.2. Response

SMR vendors claim that their proposed designs are simpler, and that they incorporate enhanced engineered safety features and a high level of passive or inherent safety in the event of malfunctions and accidents. In relation to reactivity control these often related to significant negative coefficient of reactivity with power, significant thermal inertia, and sometime claiming that shutdown systems should not be classified as safety system required to manage level 3 DiD.

Novel approaches require quality assured and credible information that is supported by research and development in order to demonstrate claims of safety. Supporting information and data need to be demonstrated to be relevant and valid. In some cases, information from historical prototypic experimental reactors my not provide necessary pedigree to support modern first of a kind reactors. In some other cases, data extrapolation may lead in unquantified uncertainties that will need to be addressed by additional R&D work supported by modern QA practices. Some proponents claim that shutdown or reactivity control work by natural physical phenomenon such as convection, gravity, reactivity feedback and as such cannot fail and require no testing. However, long term behaviour of the core and reactivity feedback remain to be demonstrated, and natural phenomena are not a guarantee to operability (channels can get blocked, gravity driven control/shutdown rods can get stuck).

When estimating the effectiveness of the proposed reactivity control or shutdown functions in their submissions (for example, shutdown margins values), vendors should not over-rely on the calculation results, obtained from incompletely validated computer codes. Claims are to be fully justified in the context of a specific design and proven by a rigorous safety case. Uncertainties in core behaviour and computer code predictions need to be quantified to demonstrate safety margin.

The CNSC aims to be non-prescriptive in the means of achieving adequate reactivity control. This is achieved by deriving technology-neutral requirements from safety objectives.

When establishing the scope of CNSC staff's review of an application to construct a reactor facility, three levels of objectives are considered. These objectives are developed to assist in integrating individual reviews into an overall assessment of the adequacy of the licence application and are fully detailed in Appendix A of REGDOC-1.1.2 [A–66]. A summary of the relevant review objectives is provided for context.

The first-level objectives mirror subsection 24(4) of the Nuclear Safety and Control Act [A–13], which aims to ensure the applicant is qualified and will make adequate provisions in conducting the activity.

There are three second-level objectives regarding: design safety, construction program and qualifications. Of these, only the design safety objective applies to question eighteen: The design of an NPP to be constructed should make adequate provisions (not pose an unreasonable risk) for the protection of the environment, the health and safety of persons and the maintenance of national security and measures required to implement international obligations to which Canada has agreed.

Each of the three second-level objectives has third level objectives. Meeting these objectives means satisfying the relevant expectations outlined in REGDOC-2.5.2 [A–62] and/or RD-367 [A–61] (The expectations articulated in RD-367 that are not already captured in REGDOC-2.5.2 are planned to be integrated into the next revision of REGDOC 2.5.2) and other relevant

CNSC regulatory documents, such as REGDOC-2.4.1 [A–63]. The applicable third-level objectives to question 18 under the design second-level objective are:

- SO1: The NPP design captures all of the mitigation measures identified during the Environmental Assessment and ensures that operating performance meets all regulatory requirements concerning the radioactive and non-radioactive (hazardous substances) releases;
- SO2: The NPP design follows the ALARA principle; SO3: The NPP design complies with the dose acceptance criteria and safety goals;
- SO4: The NPP design complies with the DiD principle;
- SO5: The fundamental safety functions perform adequately in the NPP design;
- SO6: The NPP design provides adequate means to mitigate and manage accidents.

The CNSC has mapped each third-level objective to the applicable sections within REGDOC-1.1.2 [A–66], only once all applicable sections have been satisfactorily addressed will the objective then be met. Much of the material within REGDOC-1.1.2 takes the form of listing material that should be provided in the licence application such as specific systems, design descriptions, and the requirements the design were based on. Sections 6.6.1 'Reactivity control systems', 6.4.2 'Design of the reactor internals' and 6.4.3 'Nuclear design and core nuclear performance', all follow this style listing the required types of information to be submitted. REGDOC-1.1.2 also points to the relevant areas within the regulatory framework where the detailed topic-specific requirement are located, in this case to REGDOC-2.5.2 [A–62] which has superseded RD-337 'Design of New Nuclear Power Plants' [A–260]. The following content has been derived from REGDOC-2.5.2.

With regards to structural integrity

The requirements regarding the design of the reactor core is provided in section 8.1 REGDOC-2.5.2 [A–62] which requires that all foreseeable reactor core configurations for normal operation be considered. It also requires that:

> "The reactor core, including the fuel elements, reactivity control mechanisms, reflectors, fuel channel and structural parts, shall be designed so that the reactor can be shutdown, cooled and held subcritical with an adequate margin in operational states, DBAs and DECs.
>
> The anticipated upper limit of possible deformation or other changes due to irradiation conditions shall be evaluated. These evaluations shall be supported by data from experiments, and from experience with irradiation. The design shall provide protection against those deformations, or any other changes to reactor structures that have the potential to adversely affect the behaviour of the core or associated systems."

Section 8.1 also requires the core, associated structures and cooling systems to withstand all anticipated potential loadings, including static and dynamic loading, vibration, chemical, and radiation damage. The reactor design is also required to control the reactivity during operational states and to limit the maximum degree of positive reactivity and its maximum rate of increase to prevent failure of reactor systems.

With regards to reactor control

The control systems are addressed in section 8.1.2 REGDOC-2.5.2 [A–62]:

"The design shall provide the means for detecting levels and distributions of neutron flux. This shall apply to neutron flux in all regions of the core during normal operation (including after shutdown and during and after refuelling states), and during AOOs.

The reactor core control system shall detect and intercept deviations from normal operation with the goal of preventing AOOs from escalating to accident conditions.

Adequate means shall be provided to maintain both bulk and spatial power distributions within a predetermined range.

The control system shall limit the positive reactivity insertion rate to a level required to control reactivity changes and power manoeuvring.

The control system, combined with the inherent characteristics of the reactor and the selected operating limits and conditions, shall minimize the need for shutdown action.

The control system and the inherent reactor characteristics shall keep all critical reactor parameters within the specified limits for a wide range of AOOs.

In the design of the reactivity control devices, due account shall be taken of wear-out and of the effects of irradiation, such as burnup, changes in physical properties and production of gas."

With regards to means of shutdown

Means of shutdown is addressed in section 8.4 REGDOC-2.5.2 [A–62], which provides flexibility and has requirements that are technology neutral. It requires that:

"The design shall provide means of reactor shutdown capable of reducing reactor power to a low value, and maintaining that power for the required duration, when the reactor power control system and the inherent characteristics are insufficient or incapable of maintaining reactor power within the requirements of the operational limits and conditions.

The design shall include two separate, independent, and diverse means of shutting down the reactor.

At least one means of shutdown shall be independently capable of quickly rendering the nuclear reactor subcritical from normal operation in AOOs and DBAs, by an adequate margin, on the assumption of a single failure. For this means of shutdown, a transient re-criticality may be permitted in exceptional circumstances if the specified fuel and component limits are not exceeded.

At least one means of shutdown shall be independently capable of rendering the reactor subcritical from normal operation, in AOOs and DBAs, and maintaining the reactor subcritical by an adequate margin and with high reliability, for even the most reactive conditions of the core.

Means shall be provided to ensure that there is a capability to shut down the reactor in DECs, and to maintain the reactor subcritical even for the most limiting conditions of the reactor core, including severe degradation of the reactor core.

Redundancy shall be provided in the fast-acting means of shutdown if, in the event that the credited means of reactivity control fails during any AOO or DBA, inherent core characteristics are unable to maintain the reactor within specified limits.

While resetting the means of shutdown, the maximum amount of positive reactivity and the maximum rate of reactivity increase shall be within the capacity of the reactor control system.

To improve reliability, stored energy shall be used in shutdown actuation.

The effectiveness of the means of shutdown (i.e., speed of action and shutdown margin) shall be such that specified limits are not exceeded, and the possibility of recriticality or reactivity excursion following a PIE is minimized."

The term 'means' does not necessarily imply a dedicated system, but rather a combination of inherent behaviours coupled with sufficiently reliable SSCs (informed by safety classification) to achieve the shutdown state.

It should be noted that the word 'quickly' is interpreted by CNSC staff based on consideration of the characteristics of the specific technology such as speed and nature of event progression taking into account grace time.

The guidance section provides additional clarity on independence of means of shutdown, when a single fast-acting means of shutdown[13] would be considered[14] and some technology-specific considerations (such as bowing) that can impair a means of shutdown.

A–18.3. CHINA–NNSA

A–18.3.1. Question

Describe the challenges associated with regulatory assessment of the adequacy of reactivity control function.

A–18.3.2. Response

Reactivity control in the LWR SMR in China used control rods and boric acid which be added to reactor coolant.

For HTR, the authorized party (INET) claimed: "The flow rate of coolant is approximately proportional to nuclear power at normal operation, also approximately proportional to flow rate for water in steam generator. In this way, change of power can be achieved via flow rate adjustment, does not rely on control rod movement, which is used to compensate the xenon

---

[13] To have independence, the two means of shutdown do not share components. If complete separation is not possible since both means act within the core, separation of ex-core components is required.

[14] Safety analysis demonstrates that acceptance criteria are met if an AOO/DBA is coincident with failure of fast-acting means of shutdown.

dynamics. In the case of emergency trip of reactor during accident, trip of reactor is achieved by trip of helium circulator, as a consequence of core heating up by decay heat. Although control rod is supposed to drop at this time, but the real reactivity compensation is coming from temperature feedback. It is a special design features of HTR, from the viewpoint of reactivity control."

However, in the HTR-PM demonstration plant each reactor is equipped with two sets of independent reactivity control systems, i.e. control rod system and absorption ball shutdown system, which work according to different principles in the graphite side reflector near the active zone.

A–18.3.3. Follow-up Question

What are the regulatory requirements / expectations with regards to the assessment of the adequacy of reactivity control function? What challenges have been found in the context of SMRs, for example, maintaining core geometry in seismic events?

A–18.3.4. Response

The reactivity control function meets the requirements of the regulations HAF102 [A–68]. All the reactive control methods meet the seismic requirements.

A–18.4. SÚJB – CZECH REPUBLIC

A–18.4.1. Question

Describe the challenges associated with regulatory assessment of the adequacy of reactivity control function.

A–18.4.2. Response

Not applicable — as there was no licence application for SMR, no challenges associated with the regulatory assessment of the adequacy of reactivity control function could be identified.

According to the Czech legislation, any nuclear reactor shall be equipped with no less than two independent systems based on different technical principles for reactivity control and reactor shutdown that can shut reactor down in operational states and in the course of design basis accidents. These systems should keep reactor shut down in situations causing maximum core reactivity and ensure compliance with the set design criteria for nuclear fuel. The nuclear installation design shall also set measures that can ensure sub-criticality while managing design extension conditions and set reasonably practicable measures to ensure long-term sub-criticality of molten core in the event of a severe accident.

Nuclear installation design shall set requirements for core emergency cooling safety systems, which shall in the event of a design basis accident involving disruption of integrity of the nuclear reactor coolant pressure circuit and leakage of coolant from the primary circuit, ensure the heat removal from the core to the surrounding environment for a sufficiently long period of time so that there are no changes in geometry of fuel elements, fuel assemblies or inner part of the nuclear reactor that could affect core cooling efficiency.

A–18.5. FRANCE–ASN

A–18.5.1. Question

Describe the challenges associated with regulatory assessment of the adequacy of reactivity control function.

A–18.5.2. Response

According to the article 3.4 of the Order [A–72], the licensee must demonstrate that provisions enable to prevent criticality when this is not wanted. Nuclear chain reactions management must be presented in the safety report. This regulatory requirement remains applicable for SMRs.

ASN provided guidance in its guide on pressurized water reactors design that considers the return of experience from large PWRs, which may not be applicable to a significant proportion of SMRs regarding reactivity control function. Considering that no SMR project has been submitted to ASN up to the moment, no challenge has been identified.

A–18.6. JAPAN–NRA

A–18.6.1. Question

Describe the challenges associated with regulatory assessment of the adequacy of reactivity control function.

A–18.6.2. Response

Requirements for reactor reactivity control system and reactor shutdown system are not only imposed for HTTR, but also for research reactor facilities.

Concerning the reactivity control system, it is required to be able to control all reactivity changes that are expected during normal operation, and if control rods are used in the reactivity control system, they should not fall or pop out from the core, and the allowable design limit of fuel should not be exceeded even if an abnormal withdrawal occurs.

Concerning the reactor shutdown system, it is required to have two or more independent systems with control rods and other equipment that controls the reactivity, and during normal operation, abnormal transients and design basis accidents, it is required to shift the core to a non-critical state and to be able to maintain sub criticality. Also, control rods should be able to shift to non-criticality and maintain non-criticality at low temperature even if the one with the maximum reactivity value is not inserted.

A–18.6.3. Follow-up Questions

The reactivity control systems for the HTTR are not described and this should be provided for context / clarity? Were there any challenges associated with the regulator's assessment of their suitability?

A–18.6.4. Response

Reactor shutdown of HTTR is primarily achieved by control rod operation, and independent backup system for shutdown is installed by design requirement.

Concerning the reactivity control capability by control rods system, the design enables the system to shift to subcritical from high temperature operating state without exceeding fuel design limit and maintain subcritical, under the condition which a pair of control rods that have the maximum reactivity value stack at fully extracted position and unable to insert into core.

The backup system has following mechanism:

- Release the electric plug installed in the lower part of hopper which stores boron carbide pellets;
- Boron carbide pellets are dropped into core by gravity;
- The (negative reactivity of) pellets shut down the reactor.

Thus, the design of backup shutdown system satisfies the requirement which backup system can shift to and maintain subcritical from high temperature operating state.

A–18.7. RUSSIAN FEDERATION–ROSTECHNADZOR

A–18.7.1. Question

Describe the challenges associated with regulatory assessment of the adequacy of reactivity control function.

A–18.7.2. Response

According to the requirements of Paragraph 57 of NP-022-17 [A–89], the vessel reactor core must be designed so that neither in normal operation nor under abnormal operating conditions, including design-basis accidents, the core components are not distorted to the extent that can disturb normal performance of reactivity control elements and upset the reactor scram, or degrade the fuel rod cooling so that the fuel rod damage limits are exceeded.

In compliance with the requirements of Paragraph 19 of NP-029-17 [A–93], the values of the reactivity coefficients must be negative within the operating nuclear core temperature range under normal operation conditions and in case of any operational occurrences including design-basis accidents.

As required by Paragraph 60 of NP-022-17 [A–89], the design of the reactor and reactivity control elements must render impossible an inadvertent reactivity variation under the inclinations (rotations), capsizing, vibration, impacts, and other design-basis dynamic loads.

According to Paragraph 61 of NP-022-17 [A–89], the reactivity control elements must be capable of bringing the vessel reactor to a subcritical state from any reactor power level within the timeframe specified in the vessel reactor design.

In compliance with the requirements of Paragraph 21 of NP-029-17 [A–93], activation (switch-off) of the circulation pumps of the reactor coolant circuit and/or the emergency cooldown system with the reactor shut down must not upset its sub-critical state in case of any initiating event considered in the reactor design.

In compliance with the requirements of Paragraph 25 of NP-029-17 [A–93], the characteristics of the nuclear core and the reactivity controls must prevent reactivity increase at any section of their travel in the course of insertion into the nuclear core in any combination of their positions.

In compliance with the requirements of Paragraph 78 of NP-029-17 [A–93], in case of the ship capsizing the core controls must be inserted into the core, particularly in case of total blackout.

In compliance with the requirements of Paragraph 79 of NP-029-17 [A–93], sub-criticality of the nuclear core after lifting of the core control devices with all other CPS controls inserted into the core must be at least 0.01 ($K_{eff} \leq 0.99$) for the campaign moment and the core state with the maximum effective multiplication factor.

A–18.7.3. Follow-up Question

Similar questions to previous responses from Rostechnadzor: where there any specific challenges in the regulatory assessment of SMRs in this regard (floating NPP and land-based e.g. BREST-300 LFR). How is uncertainty accounted for in the context of K_{eff} values given?

A–18.7.4. Response

According to Russian approaches to nuclear safety regulation, BREST-300 does not fall into the category of SMR facilities, and hence is subject to the licensing process adopted for the average NPPs.

Uncertainties in the K_{eff} estimation are accounted for in a conservative way for the design-basis accidents, and in a realistic manner for the beyond-design-basis events. The requirements concerning the adequacy of reactivity control functions at a floating nuclear power plant are set out in the federal nuclear safety regulation NP-029-17 [A–93].

Requirements for reactivity control are set out in more detail in paragraph 2.3.3 of the 'Nuclear Safety Rules for Reactor Installations of Nuclear Power Plants' NP-082-07 [A–261]. The section contains the requirement that the method and errors in determining the reactivity must be justified in the reactor design.

The approaches for BREST-OD-300 are similar. The specifics are determined by the concrete implementation of the control channels and the characteristics of the core (for example, a small value of the effective fraction of delayed neutrons).

A–18.8. SOUTH AFRICA–NNR

A–18.8.1. Question

Describe the challenges associated with regulatory assessment of the adequacy of reactivity control function.

A–18.8.2. Response

Requirements for reactivity control

Positive control over the nuclear reaction is a fundamental requirement of nuclear safety. Operational safety limits are to be established from safety analyses. Specific variables are measured by the reactor protection system and the shutdown systems are activated if predetermined set points are reached.

It needs to be demonstrated that the most important postulated reactivity events can be detected, monitored and controlled either by automated control and protection systems, operator intervention or by inherent design characteristics.

The design principles being used for the instrumentation must be clear. For example, how will redundancy and diversity of individual instruments be implemented and how will this mitigate against faults in individual components?

Reactivity Control and Shutdown Capability:

(a)   At least two diverse and independent systems for reactivity control and shutdown must be provided.

    (i)   At least one of these systems must be, on its own, capable of quickly rendering the reactor subcritical by an adequate margin from operational states and in Category A and Category B events, on the assumption of a single failure;

    (ii)   At least one of these systems must be, on its own, capable of rendering the reactor subcritical from operational states and in Category A and Category B events in terms of RD-0018 [A–107], and of maintaining the reactor subcritical by an adequate margin and with high reliability, even for the most reactive conditions of the core. This system shall be designed such that a single failure in any component will not impair the functioning of the entire system.

(b)   The physically possible reactivity effects must be compared with the capacity of the control equipment and sufficient margins must be provided by the design to avoid any unintended re-criticality or power increase.

Analyses must be performed to demonstrate that both initial and long-term sub-criticality are ensured for the bounding PIEs identified for the PBMR. For this purpose uncertainty analyses on the shutdown capability are necessary.

Note: Safety-important characteristics of a shutdown system to be considered in the respective analyses are for example:

- The effectiveness and net effectiveness;
- The shutdown rate;
- The highest possible positive reactivity rate which may be caused by a wrong operation of reactivity actuators;
- The reactivity requirements for these systems depend on the following properties of the nuclear core design:
  - Excess reactivity of the core;
  - Reactivity coefficients of power, moderator temperature, fuel temperature;
  - Influence of xenon poisoning;
  - Effectiveness of burnable absorbers (if existent), possible water ingress due to failures in the power conversion system.

(c)   If a reactivity control and shutdown system provides both operational control and safety functions, the shutdown function must be provided passively with fail-safe design.

(d)   If the shutdown system to be actuated first is not sufficient for cold shutdown, this system and the core must be designed in such a way that after a shutdown and until the second shutdown system becomes effective the reactor is in the sub-critical state when cooled down.

(e)   For a system designed for short term shutdown, a minimum effectiveness must be specified and monitored to ensure sufficient reactivity and shutdown effectiveness during all operational states. Reactor operation must be automatically prohibited for control rod positions below the equivalent minimum rod positions for scram.

(f)   Scrams must be initiated automatically by the RPS with high reliability if one or more safety-important parameter reaches the specified limits. A manual activation must also be possible.

(g)   The long term sub-criticality must be assured at temperature levels appropriate for long term stable state conditions and verified by conservative analysis.

(h)   Instrumentation shall be provided and tests shall be specified to ensure that the shutdown means are always in the state required for the given plant condition.

Examples of further considerations

Challenges that may arise include an insufficient consideration of uncertainties and of adverse conditions in the safety analyses as well as a questionable categorisation of initiating events related to reactivity control.

Both long term and short term reactivity changes must be discussed.

Uncertainties on temperature coefficients must be taken into account. These will affect the reactivity requirement to shut down and will need to be taken into account at the design stage. The tests on the reactor will only serve to verify that the design calculations are adequate.

Group rod withdrawal from zero power without scram was intended to be analysed at a relatively later stage. As this may well be a bounding reactivity accident, this was considered not acceptable.

It is not just uncertainties on control rod worth that will affect the excess reactivity of the first reactor. This will also be affected by uncertainties on temperature coefficients, etc.

The large core, distance from core to flux detectors and the random packing of fuel pebbles raise the question of whether the operator will remain informed of possible changes in the core make-up. The implication could be that core parameters such as fuel temperatures, localised power densities, etc. can deviate from allowable values without the operator's knowledge and could, in accident conditions, lead to higher than forecast releases. The longer-term reactivity changes that the core may experience must be identified and it must be shown how these will be detected before any serious deviation from licensing conditions.

Since the average flux from all detectors is being used to indicate reactor power, there would appear to be the possibility that a failure of this single system would leave the Reactor Protection System (RPS) and the power controller without any indication of power level. A description of the provisions for redundancy and diversity for the RPS (and the Nuclear Instrumentation System providing input to it) must be provided.

Will the compacted core also be more reactive, in addition to the control rods being less effective?

For the HTR-Modul (earlier German High Temperature Gas-cooled Reactor project) design earthquake, the reduced neutron leakage of the compacted pebble bed was the dominating effect compared to the movement of the core relative to the control rod positions.

Reliability of reactor protection and other I&C important to safety, including:

- Design for inherent safety or for failure to a defined safe state;
- Use of redundancy, diversity and segregation to meet claimed reliability;

- Compliance with the single failure criterion;
- Where high reliability required, consideration of common cause failure;
- Operability in presence of faults or internal / external hazards;
- Design for testability and maintainability such that ongoing reliability assured.

Reactor protection system requirements, including:

- Methods used for requirements specification;
- Avoidance of complexity of technology;
- Impact of demand profile on system operation;
- Automatic initiation of safety actions;
- Demonstration of adequacy of plant safety monitoring / initiating variables;
- Identification of potential failure modes using formal analysis;
- Facilities for self-supervision, and extent of diagnostic coverage;
- Actions taken upon failure detection;
- Periodic testing of complete system and of all functions performed;
- Allowance made for unavailability of equipment (e.g. due to maintenance, testing, non-repairable failures or unrevealed failures);
- Avoidance of unnecessary interfaces with other systems;
- Use of isolation facilities to prevent propagation of failures;
- Protection functions not reliant on external power sources;
- Design margins known and shown to be adequate;
- Development processes include requirements traceability;
- Operating personnel alerted to all demands for protection and actions initiated.

Use of software (or other complex equipment) for reactor protection functions:

- Demonstration of production excellence (QA, best technical design practices, comprehensive testing (prior to and following installation) and dynamic testing);
- Independent confidence building (complete, preferably diverse, verification of validated production software and independent assessment of test programme).

Further reactivity control related requirements are covered under I&C requirements.

A–18.9. UNITED KINGDOM–ONR

A–18.9.1. Question

Describe the challenges associated with regulatory assessment of the adequacy of reactivity control function.

A–18.9.2. Response

ONR has recently considered the definition of shutdown and adequacy of shutdown systems in the context of DiD for SMRs. The work was prompted as ONR encountered novel reactor shutdown systems (such as a 'freeze plug' in some Molten Salt Reactor designs) during ANT project activities.

In addition to its own guidance (e.g. SAPs [A–44] and TAGs [A–46]) ONR considered specifically the IAEA SSR-2/1 [A–127] Requirements 44, 45, and 46 for the assessment of the adequacy of shutdown systems and considered primarily land-based reactors. ONR focused the core of the assessment of the adequacy of a plant shutdown capability on the hazard and the reliability within the five levels of DiD identified in the IAEA formulation.

The expectation in the UK is that the successful operation of design basis safety measures should reduce the sequence frequency of a failure to shutdown to below ~1 × 10^{-7}/year. UK operational experience and RGP was also considered part of the work and as the basis for guidance, recognising the conflicting requirements of design simplification and the need to incorporate lessons learned from past nuclear events.

Classification of SSCs was also considered when assessing the adequacy of a shutdown system referring to a specific ONR TAG TAST-GD-094 [A–209] which provided ONR's expectations regarding the licensee's / requesting party's (RP's) arrangements for identifying and categorising safety functions and identifying and classifying SSCs. The TAG also provides guidance that covers the factors and RGP that should be taken into account when categorising safety functions and classifying SSCs.

ONR noted that many of the SMR designs place great reliance on passive decay heat removal systems during shutdown as well as during operation. ONR concluded that a robust demonstration of the effectiveness of the passive systems is required, together with an extensive substantiation of reliability claims, and a robust demonstration of sub-criticality where the core is relocated (e.g. in case of MSRs).

It is ONR's view that it is necessary to demonstrate that means of safe shutdown can be provided before core damage occurs in the event of failure of the primary shutdown method and that transient re-criticality is prevented due to cooldowns, xenon decay, and plant recovery. Some SFR vendors may argue that in the event the control rods could not be inserted, the core thermal expansion could be claimed as a mean for reactor shutdown as this tends to increase neutron leakages therefore reducing the power of the reactor. ONR judges that this is only a temporary situation as the reactor would cool down at a later stage and return to power. ONR has considered that there is also the potential for the distorted core geometry to have characteristics that may not have been anticipated.

For reactor designs which require safety functions consistent with existing reactor technologies and have similar potential radiological consequences and system reliabilities, two independent and diverse shutdown systems are expected. For new reactor designs which are able to demonstrate safety by other means, inherent reactivity feedback mechanisms may be justifiable as an alternative to one of the shutdown systems. However, given the novelty and uncertainty associated with such a claim, the levels of substantiation required from the reactor vendor and regulatory scrutiny required from ONR will need to be high. ONR's initial expectation remains that two independent and diverse shutdown systems should be provided.

For lead-cooled reactors, corrosion has been identified as a key issue for shutdown systems relying on control rods, as it may hinder their insertion. ONR's view is that reliability of control rods in lead-cooled reactors might be order of magnitude inferior compared to current LWR solutions. Therefore, ONR expects a robust demonstration of reliability, and an independent and diverse shutdown system.

Finally, ONR has considered the rupture of the primary piping in the helium-cooled and graphite-moderated HTGR, which represents a design basis fault that should not result in significant safety consequences. In general, in such a loss-of-coolant event, the reactor would be shut down by means of control rods, and the decay heat would be removed passively. However, ONR concluded that the potential for an anticipated transient without action of the primary shutdown system needs to be considered. Reactor vendors tend to rely on the negative reactivity feedback of the fuel and argue that if air can be excluded, fuel damage can be avoided. A major concern is graphite oxidation damage to the fuel and core should a major air ingress take place through the breached primary pressure boundary. In the case of fuel operating in an oxidizing environment, maintenance of fuel integrity would require operation at temperatures significantly below normal (there are indications in the region of 600°C although data on design limits is sparse).

The maintenance of the temperature below design limits is a crucial factor in this event and is likely to require a diverse shutdown of the reactor, together with sufficient diversity and redundancy of the cooling systems. The successful demonstration of passive systems may reduce requirements for emergency power to operate electric power-driven pumps. More generally, the possibility of damage or melting of the control rods during a fault should be considered. This is true especially in high temperature versions of HTGR.

A–18.10.   UNITED STATES OF AMERICA–NRC

A–18.10.1. Question

Describe the challenges associated with regulatory assessment of the adequacy of reactivity control function.

A–18.10.2. Response

The NRC staff did experience a significant challenge regarding the effects of boron volatility and redistribution for the SMR design that was reviewed. This challenge for the staff was, however, ultimately resolved using analytical methods and its current guidance. The following considers this challenge along with the review of the fuel rod and fuel assembly design and the nuclear and thermal-hydraulic design.

The SMR applicant's fuel assembly design contains multiple fuel rods and burnable absorber rods, guide tubes, and instruments tubes within their fuel assembly. The design used soluble boron through the chemical and volume control system (CVCS) that was not safety related and control rods as the two-independent means for reactivity control. The CVCS and control rod assemblies were each capable of holding the reactor subcritical under cold conditions. The design relied on the control rod drive system to prevent and mitigate DBEs. The control rods were capable of reliably controlling reactivity changes under conditions of normal operation, including AOOs, and with appropriate margin for stuck rods, such that specified acceptable fuel design limits (SAFDLs) were not exceeded. The CVCS was capable of reliably controlling the rate of reactivity changes resulting from planned, normal power changes to assure acceptable fuel design limits were not exceeded. The staff then concluded that the design was designed so that core coolability will always be maintained, even after severe postulated accidents, thereby meeting the related requirements of applicable regulations.

Regarding the challenge that the NRC staff evaluated on the effects of boron volatility and redistribution during long-term shutdown passive cooling operation — during this mode of operation, boron free steam will enter the downcomer and containment which can potentially challenge reactor core shutdown margin and could lead to a return to power. The loss of boron from the reactor coolant via steaming from the core region can challenge reactor core shutdown margin and potentially lead to a return to power following a postulated accident or anticipated operational occurrence. The NRC reviewed calculations analysis provided by the applicant showing that the reactor remained subcritical and that specified acceptable fuel design limits are not exceeded. The staff evaluated the technical basis for the applicant's approach and conducted confirmatory calculations and independent assessments to determine its acceptability and found it to be like that of the existing light water reactor fleet and therefore minimized the challenges associated with regulatory assessment.

**A–19.   ADEQUACY OF NUCLEAR HEAT REMOVAL FUNCTION UNDER
NORMAL OPERATION AND ACCIDENT CONDITIONS**

This Annex presents the responses provided by the following Member State regulatory bodies to the Question 19: "Describe the challenges associated with regulatory assessment of the adequacy of nuclear heat removal function under normal operation and accident conditions."

A–19.1. ARGENTINA–ARN

A–19.1.1. Question

Describe the challenges associated with regulatory assessment of the adequacy of nuclear heat removal function under normal operation and accident conditions.

A–19.1.2. Response

The CAREM 25 reactor heat removal safety function includes the heat generated in the reactor core and in the pool of irradiated fuel elements, for all operating states and under normal operating and accident conditions.

CAREM 25 reactor prototype reactor design is a light water integrated reactor. The core, steam generators, primary coolant, and steam dome (whole high-energy primary system) are contained inside a single pressure vessel and the cooling flow in the reactor is achieved by natural circulation through the core.

Some of the auxiliary systems are the chemical/volume control system, suppression pool cooling and purification system, shutdown cooling system (with two functions: to cool RPV water, during standard shutdown and refuelling and to heat RPV water during plant start-up). Components cooling system — closed external circuit, fuel pool cooling, and purification system, control rod drive — hydraulic system, among others.

The residual heat removal system (RHRS) is designed to reduce, in case of loss of heat sink accidents, the pressure on the primary system and remove the decay heat. The system operates condensing steam from the primary system in heat exchangers. While the top header of the reactor is connected to the vessel steam dome, the lower header is connected to the reactor vessel at level that is below the reactor water level. The condensers are located in a cold-water pool inside the containment building. The steam line inlet valves remain always open, while the outlet valves are normally closed. Hence, the tube bundles are filled with condensate. When the system is triggered, the outlet valves open automatically. The water drains from the tubes and steam from the primary system enter the tube bundles and condense on the cold surface of the tubes. The condensate is returned to the reactor vessel forming a closed circuit driven by natural circulation, removing heat from the reactor coolant. During the condensation process, the heat is transferred to the water of the pool by a boiling process. This evaporated water is then condensed in the suppression pool of the containment.

In the case of LOCA, the Emergency Injection System prevents the exposure of the core. The system features two redundant borate water accumulators connected to the RPV. The tanks are pressurized, thus when the pressure in the reactor vessel reaches a triggering low pressure during a LOCA, the RPV is flooded after the break of the rupture disks, preventing the uncovering of the core for a long period. The residual heat removal system is also triggered to help to depressurize the primary system, in case of small LOCAs.

In case of large unbalance between the core power and the power removed from the RPV, overpressure protection of the RPV is achieved through two 100% capacity safety relief valves. The blow-down pipes from the safety relief valves discharge in the suppression pool. The design also features a manually operated relief valve system to depressurize the primary system in case of failure of RHRS.

A–19.2. CANADA–CNSC

A–19.2.1. Question

Describe the challenges associated with regulatory assessment of the adequacy of nuclear heat removal function under normal operation and accident conditions.

A–19.2.2. Response

CNSC's REGDOC-2.5.2 [A–62] sets technology-neutral expectations for fulfilling the fundamental safety functions of cooling including:

- Removal of heat from the fuel in the reactor and spent fuel storage. (e.g. maintaining fuel within design limits to prevent degradation of fuel in the short and long term);
- Removal of heat from other structures systems and components that form or support maintenance of a multiple layer barrier to releases.

REGDOC-2.5.2 [A–62] also contains system-specific requirements for Reactor Coolant Systems, Emergency Core Cooling Systems and Emergency Heat Removal Systems.

The following clauses of REGDOC-2.5.2 are applicable to the reactor coolant, emergency core cooling and emergency heat removal systems:

5.4 Proven engineering practices;
7.1 Safety classification of structures, systems, and components;
7.6 Design for reliability;
7.6.1 Common-cause failures;
7.6.2 Single failure criterion;
7.6.3 Fail-safe design;
7.6.4 Allowance for equipment outages;
7.6.5 Shared systems;
7.7 Pressure-retaining structures, systems and components;
7.8 Equipment environmental qualification;
7.12 Fire safety;
7.13 Seismic qualification and design;
7.14 In-service testing, maintenance, repair inspection, and monitoring;
8.2 Reactor coolant system;
8.2.4 Removal of residual heat from reactor core;
8.5 Emergency core cooling system;
8.7 Heat transfer to an ultimate heat sink;
8.8 Emergency heat removal system;
8.13 Radiation protection;
9.0 Safety analysis.

In addition to REGDOC-2.5.2 [A–62], the general CNSC review criteria are also based on REGDOC 2.4.1 [A–63]. These criteria were successfully applied to natural and forced circulation configured water-cooled and non-water-cooled reactors during pre-licensing design reviews using a graded approach.

Experience gained from new reactor technology assessments under the CNSC's Vendor Design Review Program.

Overarching technological trends

CNSC staff have noted the following overarching technological trends in proposals from both large NPP vendors and vendors of SMRs and advanced reactor technologies:

- In many cases, less reliance on specific dedicated systems called[15] emergency core cooling system (ECCS) and emergency heat removal system (EHRS);

- More reliance on 'always-on/available' structures, systems and components (SCCs) that will be designed to have the capacity to support the removal of heat from the core and then from the facility under both anticipated transient conditions and emergency conditions. (i.e. not dedicated to emergencies);

- These SSCs may also play a greater role in meeting fundamental safety functions of:
  o control of reactivity through passive heat transfer at predictable rates;
  o confinement of radioactive material (integrated as a barrier to release);
  o control of operational discharges and hazardous substances, as well as limitation of accidental releases (e.g. trapping of tritium that migrates through reactor components);
  o monitoring of safety-critical parameters to guide operator actions – plant parameters may be inferred from measurements from cooling systems where direct measurement and control may not be pragmatic.

- Potential novel safety classification and code classification proposals for these SSCs that challenge traditional views that these SSCs would be classified as Safety Systems. There is a trend by industry to seek to reduce the costs of these systems through use of commercially available components while maintaining an appropriate level of reliability commensurate with the safety function. These types of proposals lead to greater interpretation of technical standards used in detailed design activities and a greater need for supporting R&D to establish reliability of SSCs.

- In a number of cases, systems being used for normal operation are also being credited for accident conditions, leading to a greater reliance on fewer systems to perform multiple functions. This raises questions about whether there is sufficient diversity and redundancy in SSCs that support cooling functions important to safety.

CNSC staff observations concerning technological proposals for different plant states:

- Normal operation
  o Direct cycle concepts — Heat removal approach appear to be remaining generally consistent with traditional direct cycle technologies used in NPPs;

---

[15] Vendors may have alternative nomenclatures for ECCS and EHRS equivalent systems.

368

- o Indirect cycle concepts: the primary means of fulfilling the heat removal function is the reactor coolant system transferring heat to a secondary coolant loop (traditional approach) and, for some designs, to a tertiary loop (some advanced designs). Although, coolants and flow types may vary, this is a common approach for existing nuclear power plants and small modular reactors (SMR).
- Anticipated plant transients (e.g. AOOs)
  - o Developers are proposing 'always-on' passive[16] means of heat removal that couple specific systems with inherent physics and reactor coolant thermal hydraulic behaviours that reduce reactor power and aid with reducing heat generation and maintaining sufficient heat transfer;
  - o In many cases, the traditional secondary coolant loop may continue some heat removal functions (e.g. turbine bypass) but are not being credited as contributing to safety functions;
  - o Questions tend to arise about the capacity of proposed heat removal systems such as whether they are sized only for decay heat removal or can begin performing their cooling functions at higher reactor power (reactor in a low power critical state).
- Accident conditions (DBA, DEC, BDBA)
  - o ECC and EHR functions have typically been fulfilled by separate systems. Developers are proposing that that acceptance criteria could be met using either a single system or systems with shared components and crediting natural circulation;
  - o 'Always on' core cooling and heat removal functions are being credited to take overheat removal for a period of time ranging from 72 hours (e.g. where auxiliary coolant makeup storage tanks are used) to indefinite time (e.g. passive systems such as Direct Reactor Auxiliary System (DRACS));
  - o Questions tend to arise about, for instance:
    - Robustness of SSC performance against combinations of internal and external events, particularly in BDBA plant states where some barriers to releases may have been compromised;
    - Whether there is sufficient redundancy against failures of these systems or for performance of periodic maintenance and in-service inspections;
    - The capacity and long term performance of such systems such as whether they are sized only for decay heat removal or can begin performing their cooling functions at higher reactor power (e.g. reactor failure to shutdown scenarios).

The following are some of the more specific challenges CNSC staff encountered in their review of the adequacy of nuclear heat removal function under normal operation and accident conditions:

(a) Adequacy of a FOAK design for passive cooling, and the need for experimental verification;
(b) Reliability of cooling systems which operation is based on natural circulation;
(c) Determination of driving force for natural circulation to effectively cool the fuel and prevent dry-out following accidents resulting in the reactor pressure vessel's depressurization;
(d) Single failure requirement for passive and semi-passive cooling systems;
(e) Computer code applicability and validation for SMRs;

---

[16] The degree of passive behaviour will vary from 'always on' connected systems to passive systems available on standby but require initiation logic.

(f)   Determination of thermal-hydraulic instabilities in the cooling systems under different conditions;

(g)   Operational challenges with novel steam generators such as helical-coil steam generators;

(h)   Effects of flow oscillations in the secondary side on the primary side coolant;

(i)   Diversity and redundancy of important cooling systems' valves;

(j)   Modelling of passive valves;

(k)   Use of one system (passive or active) to mitigate consequences of failure of more than one level of DiD;

(l)   For some SMR designs, use of unacceptable design standards (i.e. use of obsolete or cancelled design standards).

There is flexibility in CNSC's framework to allow for novel approaches to be proposed providing it is justified in the design and safety analysis documentation and supplemented with the appropriate research and development information and test results. An applicant must demonstrate that:

- The heat removal function is fulfilled in all plant states with sufficient reliability taking into account the need for in service inspections, reliability testing and maintenance;
- Sufficient DiD is maintained at all times.

CNSC Staff note that a potential challenge for technology developers and future licensees will be the demonstration of reliability of heat removal systems, including the establishment of periodic testing requirements and procedures for cooling systems that are always removing heat during normal operation. This issue becomes more complex when the systems are relying on passive and inherent behaviours under changing operational conditions that may involve chemistry related effects. CNSC staff also note an increased use of high fidelity (and coupled) modelling software for prediction of heat generation, thermal hydraulic behaviours and heat transfers. In Vendor Design Reviews, the validation and verification of these software tools are expected to be demonstrated along including where verification using experiments may be necessary. This is particularly important to give credibility to the safety and operational margins that will be available for the FOAK facility.

A–19.3. CHINA–NNSA

A–19.3.1. Question

Describe the challenges associated with regulatory assessment of the adequacy of nuclear heat removal function under normal operation and accident conditions.

A–19.3.2. Response

Using the HTR-PM as an example:

(a)   In the HTR-PM, under normal operation and AOOs, which would not result in the scram of the reactor, the heat removal function is guaranteed by the forced circulation of the primary helium.

(b)   Safety analysis of the HTR-PM guaranteed that, after reactor scram due to AOOs or accident conditions (including the DEC), even in the case of failure of all active cooling systems and complete loss of coolant, the decay heat can be removed from the core by means of physical processes, such as heat conductivity and radiation. In other words, the HTR-PM is designed with inherent safety. Emergency cooling system is not necessary for the HTR-PM.

(c)   Due to the excellent fission-product-retention capability of the TRISO particles, the radioactivity in the primary circuit is maintained at very low levels. Inherent safety design ensures that specified design limits will not be exceeded in any operational state.

(d)   Nevertheless, helium purification system is designed for the clean-up of reactor coolant. Two pressure relief valves are installed to protect the pressure boundary of the reactor coolant systems. Besides, negative pressure ventilation system is designed to filter the gas before it is released to the environment in a depressurized loss of forced cooling (DLOFC) accident or in condition of pressure relief valve open.

(e)   Reactor cavity cooling system has the capability to transfer heat to an ultimate heat sink – atmosphere. It is designed as a passive system. But even this system totally fails, the design limit for the fuel temperature will not be exceeded due to the inherent safety design.

For the removal of accident residual heat, NNSA has focused on the analysis and test verification of heat conduction of fuel ball and pebble bed reactor core, including the equivalent heat conduction test of pebble bed conducted by the authorized party. Capability and regular test requirements of passive residual heat removal system were also key reviewed. In addition, the most important is the calculation of the maximum temperature of the fuel particles and the reliability and sensitivity analysis.

The challenge of safety review mainly comes from the insufficient operating experience and test data. In addition, there are some disadvantages in the use of passive, inherent characteristics of residual heat removal, that is, a certain temperature gradient is required to ensure the export of heat. In other words, the core will keep very high temperature for a long time. This is not in compliance with the current LWR safety requirements, so a reasonable demonstration is needed.

A–19.4. CZECH REPUBLIC–SÚJB

A–19.4.1. Question

Describe the challenges associated with regulatory assessment of the adequacy of nuclear heat removal function under normal operation and accident conditions.

A–19.4.2. Response

Not applicable — as there was no licence application for SMR, no challenges associated with the regulatory assessment of the adequacy of heat removal function have been identified.

The nuclear installation design shall set requirements for cooling systems —primary and secondary — and emergency cooling systems in order to ensure heat removal from the core for a sufficiently long time so that the design limits are not exceeded, and consequences are not grave (no changes in geometry of the core) as well as safety system providing the removal of residual heat from the core (heat from decay of fission products and accumulated heat of components).

The design of systems, structures and components of the primary circuit system in nuclear installation design shall provide the operators with diverse and alternative means and enable to carry out organisational measures for emergency cooling of the core and removal of a residual heat and depressurization of the primary circuit and prevention of core melting under high pressure in accident conditions.

The provisions of the Decree No. 329/2017 [A–22], on the requirements for nuclear installation design, are, as mentioned elsewhere, relatively specific and reflect PWR technology (design with fuel cladding, elements and assemblies). These requirements would therefore have to be probably adapted to different type of SMR technology should it be deployed in Czech Republic.

A–19.5. FRANCE–ASN

A–19.5.1. Question

Describe the challenges associated with regulatory assessment of the adequacy of nuclear heat removal function under normal operation and accident conditions.

A–19.5.2. Response

According to the article 3.4 of the Order [A–72], the licensee must present in the safety report how heat removal is insured.

ASN provided guidance in its guide on pressurized water reactors design that considers the return of experience from large PWRs, which may not be applicable to a significant proportion of SMRs regarding heat removal function. Considering that no SMR project has been submitted to ASN up to the moment, no challenge has been identified.

A–19.6. JAPAN–NRA

A–19.6.1. Question

Describe the challenges associated with regulatory assessment of the adequacy of nuclear heat removal function under normal operation and accident conditions.

A–19.6.2. Response

The cooling system of the HTTR can reliably remove the heat generated in the nuclear reactor during normal operation, ensure its soundness during normal operation and abnormal conditions, not exceed the allowable fuel design limit, and the design is such that decay heat and other residual heat can be removed without impairing the soundness of the reactor coolant pressure boundary.

In the event of a DBA, in order to prevent a large amount of fuel damage and prevent and mitigate the diffusion of radioactive materials, an engineering safety facility for decay heat removal, etc. is installed, which has a sufficient capacity in order to cope with a single failure and have multiplicity and independence. The engineering safety facility is designed so that it can be operated by the emergency power supply facility even when the commercial power supply is lost.

A–19.6.3. Follow-up Question

Please provide a brief description of regulatory judgement made on the engineering safety facility including the suitability multiplicity and independence claimed.

A–19.6.4. Response

For redundancy and independence of engineered safety features, the following aspects are confirmed in the review.

The design has:

- High reliability for its operation and performance, with engineered safety features that operate as designed whenever needed;
- The system to be (always) controlled under the condition that assumes single failure exists;
- Backup electric power or other drive source anytime available and that enables system operation, under the condition that external power is not available.

A–19.7. RUSSIAN FEDERATION–ROSTECHNADZOR

A–19.7.1. Question

Describe the challenges associated with regulatory assessment of the adequacy of nuclear heat removal function under normal operation and accident conditions.

A–19.7.2. Response

The current Russian regulations pertaining to designing ships and other vessels with nuclear reactors, including small-power nuclear plants with SMRs, include the requirements for a system removing the fuel heat both in normal and in abnormal operating conditions. Thus, NP-022-17 [A–89] sets the following requirements:

- The nuclear power installation of a vessel must have safety systems designed to perform fundamental safety functions, including emergency heat removal from the reactor and from the spent fuel storage facilities (Paragraph 46 of NP-022-17);
- The nuclear power installation must have systems to remove residual heat from the core during normal operation, in case of a scram, in the process of refuelling, and during outages. The systems must remain operable during and after any design-basis accident (Paragraph 62 of NP-022-17);
- The protective safety systems must include systems providing emergency heat removal from the reactor to the ultimate heat sink; each system must consist of several independent trains (Paragraph 94 of NP-022-17);
- Measures must be taken to prevent criticality and unacceptable pressure in the primary circuit when the emergency heat removal system is started up and in the course of its operation (Paragraph 95 of NP-022-17);
- The vessel design must demonstrate safe management of nuclear fuel. The SAR section dealing with nuclear fuel management must contain a list of potential operational occurrences, initiating events for DBAs and BDBAs, and safety justification. The spent fuel storage facility must be provided with systems of heat removal to the ultimate heat sink to prevent damage of nuclear fuel and avoid release of radioactive substances into the vessel premises or in the environment (Paragraph 115 of NP-022-17);
- Decay heat removal must be provided for the entire refuelling cycle (Paragraph 164 of NP-022-17).

A–19.7.3. Follow-up Question

General question still applies (other SMRs e.g. BREST-300).

A–19.7.4. Response

For land-based power units, the requirements for heat removal systems are set in NP-001-15 [A–254] and NP-082-07 [A–261] and do not depend on the type of reactor.

According to Russian approaches to nuclear safety regulation, BREST-300 does not belong to the category of SMR facilities, and hence is subject to the licensing process adopted for the conventional NPPs.

The RITM-200 design incorporates standard heat removal systems so that there are no challenges associated with their assessment.

A–19.8. SOUTH AFRICA–NNR

A–19.8.1. Question

Describe the challenges associated with regulatory assessment of the adequacy of nuclear heat removal function under normal operation and accident conditions.

A–19.8.2. Response

Requirements for heat removal

A coherent heat removal pathway for all PIE considered for the design that will ensure that the maximum acceptable fuel temperatures are not exceeded is required. In addition, for events that are beyond the design basis, the intent is that adequate heat removal will be available such that maximum acceptable fuel temperatures are not exceeded.

Sufficient cooling must be provided for all parts of the core to remove heat such that the temperature limits for the fuel and structural components are not exceeded.

Examples of thermal-hydraulic scenarios are:

- Heat removal by forced cooling systems during operational transients and PIE with nuclear power transients;
- Instant or delayed core cool-down using the provided heat removal systems;
- Pressurized Loss of Forced Cooling (PLOFC);
- Depressurized Loss of Forced Cooling (DLOFC);
- Escalation from PLOFC to DLOFC;
- Recovery from DLOFC to PLOFC;
- Recovery from PLOFC events to either Core Conditioning System (CCS) operation or normal operation;
- Long term heat removal;
- Temporary unavailability of the RCCS during PLOFC or DLOFC events.

Monitoring systems and instrumentation of the reactor core and the cooling circuits must be provided such that it can reliably identify the cooling conditions inside the reactor core and the core structure ceramics (CSC). The monitoring systems must provide sufficient confidence that the actual in-core conditions are monitored during the operational lifetime of the reactor.

The thermal-hydraulic instrumentation of the core and the cooling circuits must be able to detect:

(a)   The temperature of the coolant at the core inlet and outlet;
(b)   The mass flow rate of the coolant.

These safety-important parameters must be measured continuously during normal operation and AOO with core coolant mass flow. Sufficient redundancy must be provided for each measuring device. This requirement applies to operation of the main heat sink and of auxiliary heat sinks such as CCS and core barrel conditioning system (CBCS).

For purposes of design verification also the radial and azimuthal distribution of the core outlet temperature must be measured.[17]

The instrumentation must be able to provide sufficient information to demonstrate that potential coolant bypass flows around the pebble bed do not lead to excessive fuel and structural temperatures.[18]

In case of PIE sequences that lead to possible reverse flow of coolant through the core, (e.g. after PLOFC) due to either natural convection or depressurisation at the core inlet, the thermal-hydraulic core instrumentation must be able to detect the temperature distribution and possible impact to metallic core structures including control rods with adequate accuracy.[19]

In the case of passive heat removal sufficient instrumentation must be available for measurement or demonstrably reliable monitoring of the maximum core barrel and reactor vessel temperatures at representative positions.

In the course of PIE sequences without forced heat removal from the reactor core the produced heat is transported by passive mechanisms only within the core and from the core boundaries via the core structures, the core barrel and the RPV to the reactor cavity cooling system located in the reactor cavity. The addressed passive heat transfer mechanisms comprise natural convection, conduction and radiation. For the analysis of such PIE different requirements apply depending on the coolant pressure and composition in the reactor. At atmospheric pressure in a He atmosphere the convective heat transfer in a pebble bed reactor core is insignificant and the transport of heat within the core may be reduced to a heat conduction problem based on an effective thermal conductivity. In this case the highest temperatures are calculated in the core, however, the temperatures of other SSC (e.g. control rods, top reflector suspensions) may be underestimated.

---

[17] This instrumentation will have to be located in the CSC assembly and needs not necessarily be replaceable. As part of the overall instrumentation and monitoring concept, it is expected that this instrumentation will be used to calibrate alternative methods to assess the temperature distribution at the core outlet as long as it will be available.

[18] This instrumentation will have to be located in the CSC assembly and needs not necessarily be replaceable. As part of the overall instrumentation and monitoring concept, it is expected that this instrumentation will be used to calibrate alternative methods to assess the distribution of core bypasses as long as it will be available. In this context specific attention will be given to the fact that core bypasses will increase with reactor lifetime due to neutron dose induced shrinkage of reflector blocks.

[19] The concept to monitor the temperatures in the upper part of the core assembly is expected to be part of the overall instrumentation and monitoring concept required.

A–19.8.3. Follow-up Question

But were there any specific challenges linked to the regulator's assessment?

A–19.8.4. Response

Amongst the deliverables for resolution of KLI 21 'Heat Removal', were documents with the titles:

- 'Reactor Unit Temperature Limits Report', which summarises the design temperature limits of the fuel and the structural components of the PBMR reactor unit and RPV. Respecting these limits will ensure that the components do not fail their structural function and that fission product releases from fuel at elevated temperatures comply with regulatory licensing criteria.

- 'Reactor Unit Heat Generation Categories and Heat Removal Systems Table', which defines heat generation categories for the PBMR core, lists and describes available heat removal systems, and defines enveloping combinations of heat generation categories and heat removal systems for analysis.

- 'PBMR Demonstration Plant System Heat Removal Capability', which demonstrate for the enveloping combinations defined in the second document above by means of supporting analyses that sufficient heat can be removed from the core and the reactor unit to comply with the temperature limits established in the first document above.

Regulatory assessments identified the following as issues to be resolved:

- The important case of heat removal under chemical attack (In certain conditions associated with air ingress, carbon can oxidise. The reaction is complex, with a number of different reactions present that are both exothermic and endothermic. The net effect is an exothermic reaction that, if unconstrained, could lead to very high temperatures in the core that may lead to additional fuel failures and a release of radioactivity);
- The safety case must demonstrate heat removal for DiD level 4 events [REF];
- Demonstration that cases considered are enveloping and bounding;
- Heat removal from used/spent fuel tanks should be considered;
- Temperature limits for the core as well as core structure ceramics must be adequately justified.

A–19.9. UNITED KINGDOM–ONR

A–19.9.1. Question

Describe the challenges associated with regulatory assessment of the adequacy of nuclear heat removal function under normal operation and accident conditions.

A–19.9.2. Response

Some SMR designs claim not to require forced circulation for heat removal, during normal operations or accident conditions (and thus include no pumps in the primary circuit). Some designs claim to have effectively eliminated the possibility of certain loss of coolant accidents (LOCAs), for example, by using an integrated approach, guard vessel, or shared pool within which the reactor sits. Other designs claim that sufficient decay heat removal can be provided by natural phenomenon, even following a LOCA.

ONR SAPs [A–44] EHT.1 to 5 cover heat transport systems. The expectations associated with these SAPs are sufficiently high-level and technology neutral to avoid any obvious challenges in their application to SMRs.

ONR has participated in the IAEA activity to develop a TECDOC on the applicability of design safety requirements (SSR-2/1, Rev. 1 [A–127]) to SMR technologies intended for near-term deployment, which covered both small modular light water reactors and high temperature gas-cooled reactors (HTGR).

As part of this activity, ONR helped review Requirements 47–53 of SSR-2/1 [A–127]. It was noted that some requirements may not be fully applicable to SMRs as written, some may require a technology specific interpretation and some additional requirements may also be needed. For example, for HTGRs, the isolation of leaks is more important for preventing long-term oxidation of the core (air ingress scenario), or reactivity control (water ingress scenario), than in maintaining the coolant for heat removal purposes.

A–19.10.   UNITED STATES OF AMERICA–NRC

A–19.10.1. Question

Describe the challenges associated with regulatory assessment of the adequacy of nuclear heat removal function under normal operation and accident conditions.

A–19.10.2. Response

Staff did not find any significant challenges in this area of review. The reactor coolant system (RCS) provides for the circulation of the primary coolant. The SMR applicant's design relied on natural circulation flow for the reactor coolant and did not include reactor coolant pumps or an external piping system. The applicant's design had two safety-related passive heat removal systems, the decay heat removal system (DHRS) and the emergency core cooling system (ECCS).

In traditional LWR designs, residual heat removal (RHR) systems are used to cool the reactor coolant system (RCS) following a shutdown. In the SMR applicant's design, safety-related RHR following accidents was accomplished using the passive DHRS, while the cooldown following a routine shutdown is performed by using normal feedwater and secondary-side systems followed by the containment flood and drain system.

The staff's review of the DHRS to determine whether safety-related class 1E power was required and found that the applicant's proposed design did not rely on safety-related power. Because the actuation valves open on de-energization (the only powered component of the system required to change state) and the system is then driven by natural, passive forces, the system did not rely on power to operate. For this reason, the staff determined the DHRS was able to fulfil its design bases without any need for safety-related power.

The staff found that the applicant satisfied the applicable design requirements pertinent to the DHRS. DHRS acts as a robust system to transfer residual heat from the reactor core at a rate such that specified acceptable fuel design limits and the design conditions of the reactor coolant pressure boundary are not exceeded for events in which the RCS is not breached.

This Annex presents the responses provided by the following Member State regulatory bodies to the Question 20: "Describe the challenges associated with regulatory assessment of worker dose. Also consider radioactive discharges to the environment and the dose to the public."

A–20.1. ARGENTINA–ARN

A–20.1.1. Question

Describe the challenges associated with regulatory assessment of worker dose. Also consider radioactive discharges to the environment and the dose to the public.

A–20.1.2. Response

The design of the different systems that will integrate the CAREM 25 prototype reactor includes various aspects that will allow the operation of the reactor at a reduced cost in doses, both individual and collective, for the plant's operation and maintenance personnel, as well as for the general population. In accordance with the standards, guidelines, and regulations in force in the Argentine Republic (Nuclear Regulatory Authority) and international recommendations. Operations at the CAREM 25 reactor that means or may mean the exposure of people, will be subject to the following basic criteria that make up the Dose Limitation System for normal operation:

- Justification: Any practice that means exposing people to ionizing radiation must produce a net positive benefit.
- Limitation: The resulting doses of all practices for the design of the CAREM-25 plant, with the exception of those due to natural radiation and medical treatments or practices, will not exceed the applicable dose limits.
- Optimization: People's exposures should be kept as low as reasonably achievable, taking into account economic and social factors (ALARA principle).

The radiological protection systems of the plant, such as shields, ventilation systems, and others, are defined based on the optimization criteria, considering a cost of the collective dose unit of not less than USD $ 10 000/man Sv, provided that the values thus obtained do not exceed the individual limits imposed by the Regulatory Authority.

The Annual Dose Limit adopted for the design of the CAREM 25 plant according to IAEA GSR Part 3 [A–160], is more restrictive than the values established in the Radiation Protection Requirements, AR 10.1.1, Rev. 3 [A–262].

Regarding the assessment of the information related to the radiological protection of the CAREM 25 prototype reactor, some of most relevant that could have affected the start of construction findings were related to analysing if there are operational aspects that require the provision of spaces, facilities or aspects specific to be considered in the design. For example, the space necessary for manoeuvring and eventual temporary deposit of large components. Regarding the calculation of shields, it was observed that the contribution of the delayed photons of the fission products had to be considered in the calculation.

The information requested by ARN was delivered and contributed to the approval and issuance of the reactor construction license.

A–20.2. CANADA–CNSC

A–20.2.1. Question

Describe the challenges associated with regulatory assessment of worker dose. Also consider radioactive discharges to the environment and the dose to the public.

A–20.2.2. Response

Canada has established technology neutral criteria for the regulatory assessment of worker doses, radioactive discharges to the environment and dose to public. Albeit existing criteria, the CNSC acknowledges some challenges with respect to review of this topic in advanced reactor facilities and SMRs.

The CNSC has noted that some proposed SMR designs have claims of lower doses to workers, in comparison to existing/operational reactor facilities, through novel features intended to enhance human performance. Some claims include enhancements due to increased use of automation resulting in reduced human error of interfaces with the plant and improved supervisory functions through better-quality information. It is also claimed that use of very small reactor facilities in remote locations could present unique advantages at minimizing doses to the public as a result of an accident but may also present unique challenges in accident response. Additionally, there are some proponents of very small designs investigating the long-term feasibility of fully autonomous plant operations with remote monitoring and intervention. Since these designs are novel in nature, claims remain to be supported by adequate evidence. The designs are typically at a conceptual stage and evolving as research and development advances.

More broadly, because advanced reactor technologies make use of new fuel and coolant technologies, consideration must be given to unique radiation sources and pathways. Where little operational experience exists, there may be a challenge in determining if the applicant has identified all radiation sources expected to occur over the lifetime of the reactor facility (including construction, operation and decommissioning) and under accident conditions, with account taken of both contained and immobile sources, out-of-core criticality, and potential sources of airborne radioactive material. Where significant uncertainties exist in a proposal, the CNSC expects the applicant to propose appropriate safety and control measures to address those uncertainties. This may mean research and development to validate claims or additional engineered features in the design.

In review of an application for a licence to construct, CNSC staff consider the following to ensure the applicant makes adequate provisions for incorporating radiation protection into the design of the reactor facility:

- Application of the ALARA principle, and the expected occupational radiation exposures during normal operation and AOOs, including measures to avoid and restrict exposures;
- Minimizing the number and locations of radiation sources;
- The capability for monitoring all significant radiation sources in all activities throughout the lifetime of the SMR;
- Preventing accidents with radiological consequences and minimizing radiological consequences of any accident.

The applicant must demonstrate that it meets the following provisions regarding the ALARA principle:

- Occupational radiation exposures during normal operations will be kept below regulatory dose limits and ALARA, social and economic factors taken into account (e.g. avoiding the need for workers to be in areas where they are exposed to radiation for long periods of time have been duly taken into account in the design of the SMR).
- Occupational radiation exposures during AOOs and accident conditions will be kept below regulatory dose limits and ALARA, social and economic factors taken into account.
- Radiation doses to workers resulting from the operation of the SMR will be reduced wherever practicable by means of engineered controls and RP measures.
- The ALARA principle will be applied in a systematic manner by the licensee for all phases of the SMR's life cycle, including construction, operation, maintenance, and decommissioning of the SMR, to further reduce occupational exposures wherever practicable.

The CNSC expects the applicant to include reference dose data to demonstrate that adequate RP measures have been incorporated into the design of the reactor facility in order to achieve RP objectives and the ALARA principle, described above. The dose data should be reliable and verifiable, and representative of the proposed reactor facility. The CNSC has experienced that this can be a challenge for applicants of advanced reactor technologies, to provide detailed dose assessments where source terms have not been fully identified and/or characterized.

Furthermore, REGDOC-2.5.2 [A–62] sections 4.1.1, 6.4, 7.3.1, 8.11, 8.11.2, 8.13.1, 10.2 includes further expectations for the application of the ALARA principle.

IAEA SSR-2/1 [A–127] Req. 81: Design for radiation protection, states:

> "Provision shall be made for ensuring that doses to operating personnel at the nuclear power plant will be maintained below the dose limits and will be kept as low as reasonably achievable, and that the relevant dose constraints will be taken into consideration."

To address this requirement, CNSC expectations in REGDOC-2.5.2 [A–62] elaborate that the reactor facility layout should provide for efficient operation, inspection, maintenance, and replacement to minimize radiation exposures. The reactor facility design should limit the amount of activated material and its build-up. The design needs to account for frequently occupied locations and support the need for human access to locations and equipment. Access routes should be shielded where needed. Issues such as avoiding the need for workers to be in areas where they are exposed to radiation for long periods of time or facilitating means to reduce the time in which workers are required to be in such an area (e.g. improved access), should have been duly taken into account in the design.

Additionally, to expectations regarding RP in design of the reactor facility, the applicant is required to describe a radiation protection program, required by section 4 of the Radiation Protection Regulations. Section 11.6 of REGDOC-1.1.2 [A–66], requires the following:

> "It should also demonstrate that the radiation protection program is based on a risk assessment that takes into account the location and magnitude of all radiation hazards in the plant and that addresses matters such as the following:

1. classification of work areas and access control
2. local rules and supervision of work
3. monitoring of individuals and the workplace
4. work planning and work permits
5. protective clothing and protective equipment
6. facilities, shielding and equipment
7. optimization of protection
8. source reduction
9. training
10. arrangements for response to emergencies"

GSR Part 3 [A–160] Req. 12: Dose limits states:

"The government or the regulatory body shall establish dose limits for occupational exposure and public exposure, and registrants and licensees shall apply these limits."

The CNSC's effective dose limits for a nuclear energy worker is set at 50 mSv in any one year and 100 mSv in five consecutive years. The dose limit for pregnant nuclear energy workers is 4 mSv from the time the pregnancy is declared, for the balance of the pregnancy. Licensees must ensure that all doses are kept below the regulatory dose limits and ALARA, social and economic factors being taken into account.

The Radiation Protection Regulations are based on the recommendations of the International Commission on Radiological Protection and the IAEA's GSR Part 3 [A–160]. Regarding doses to members of the public, the Radiation Protection Regulations, the prescribed limit for the general public is 1 mSv per calendar year. Dose acceptance criteria for a proposed exclusion zone is also available in REGDOC-1.1.1 [A–64] as the following:

"The applicant shall consider the following criteria (for an operating unit) in determining the size of the proposed exclusion zone:

- committed whole-body dose for average members of the critical groups who are most at risk at or beyond the exclusion zone boundary, is calculated in the deterministic safety analysis for a period of 30 days after the analysed event

- under normal operating conditions, the effective dose at the exclusion zone boundary to a person who is not a nuclear energy worker shall not exceed 1 mSv over the period of one calendar year

- under anticipated operational occurrence (AOO) conditions, the effective dose at the exclusion zone boundary to a person who is not a nuclear energy worker shall not exceed 0.5 mSv over the release time due to the AOO

- under design-basis accident (DBA) conditions, the effective dose at the exclusion zone boundary to a person who is not a nuclear energy worker shall not exceed 20 mSv over the release time due to the DBA

- demonstration that the dispersion model used for the dose calculations is representative of the actual site".

A–20.3. CHINA–NNSA

A–20.3.1. Question

Describe the challenges associated with regulatory assessment of worker dose. Also consider radioactive discharges to the environment and the dose to the public.

A–20.3.2. Response

The occupational exposure management of nuclear power plants in China generally follows two national standards, namely, the Basic Standard for Protection Against Ionizing Radiation and Safety of Radiation Sources (GB18871-2002) [A–263] and the Regulations on Radiation Protection of Nuclear Power Plants (GB6249-2011) [A–264]. In the safety review principles for small PWR nuclear power plants radiation protection objective was also specified, that is "(…) ensure that the radiation exposure in the small PWR nuclear power plant under all operating conditions or due to the planned discharge of any radioactive substance in the small PWR nuclear power plant is kept below the specified limit value and as low as reasonably acceptable so as to reduce the radioactive consequences of any accident."

According to nuclear safety regulations of China, the average effective dose of radiation workers in nuclear power plants in five years should not exceed 20 mSv, the annual equivalent dose of eye crystal should not exceed 150 mSv, and the annual equivalent dose of limbs or skin shall not exceed 500 mSv.

A–20.3.3. Follow-up Question

Were challenges encountered with regard to assessing worker dose for the HTR and other SMRs being considered?

A–20.3.4. Response

SMR personnel dose did not encounter problems, SMR personnel dose are in line with the regulatory requirements.

A–20.4. CZECH REPUBLIC–SÚJB

A–20.4.1. Question

Describe the challenges associated with regulatory assessment of worker dose. Also consider radioactive discharges to the environment and the dose to the public.

A–20.4.2. Response

Not applicable — no challenges associated with regulatory assessment of worker dose or radioactive discharges to the environment and the dose to the public have been identified as there was no SMR licence application.

Current system is in line with EU legislation and various international documents (WENRA, IAEA and ICRP) and describes exposure limits and regulates discharges from workplaces in order to limit the exposure of member of the public (model group is being represented by representative person) so that the adequate radiation protection of public is ensured. In this regard, the SMR would be regulated in the same way as any other nuclear installation (category IV workplace). An authorised limit is laid down by the Office in the licence for activities performed in exposure situations. It is a quantitative indicator which is a result of radiation protection optimisation for individual activities involving radiation or an individual source of ionising radiation and is usually lower than the dose constraint.

In particular licence holders (performing activities in planned exposure situations) are obliged to restrict exposure of an individual so that the total exposure resulting from a combination of exposures from these activities is justified, optimised and does not exceed in total the exposure limits. They are also obliged to ensure monitoring of the addition of the doses from all work activities of their exposed workers to compare with the limits for workers' exposition. They are obliged to optimise a radiation protection to keep the doses levels as low as reasonably achievable, taking all economic and societal aspects into account together with the extent of exposure, its likelihood and the number of natural persons exposed to radiation. In order to optimise public exposure, they shall apply the dose constraints in the Atomic Act [A–21] and in the monitoring programme they shall determine dose constrains for workers to optimize their exposure (over a certain period). The licence holder shall provide for the monitoring of discharges and the surrounding area in accordance with the monitoring programme, including accidental monitoring. Monitoring programme is obligatory part of the documentation for the activity to be licenced (including siting licence).

According to the Decree No. 329/2017 [A–22], on the requirements for nuclear installation design, the nuclear installation design shall set requirements for equipment of the nuclear installation also from radiation protection perspective. In general, a nuclear installation design shall set requirements for technical means and conditions for ensuring radiation protection so as to satisfy the requirements of implementing decrees on radiation protection and on radiation situation monitoring. It shall set requirements with means enabling control of discharges and handling of radioactive waste. The nuclear installation premises shall be designed, inter alia, so as to ensure optimization of the radiation protection, ensure that equipment which is frequently attended or maintained is preferably located in locations with a favourable radiation situation and to provide sufficient means and number of points with enough capacity to measure the contamination of persons and objects.

Among other things, the nuclear installation shall also be equipped with a system for monitoring of external exposure and contamination of persons, systems for accounting discharges of radioactive material into atmosphere and water flows and a laboratory for measuring the activity of gas and liquid samples. The nuclear installation shall also be equipped with mobile and stationary monitoring means (stationary monitoring systems shall ensure the monitoring of dose rates, activity concentration in atmospheres and systems, surface contamination, etc. in all states of nuclear installation). The abovementioned requirements are not narrow in application and should provide a sound basis also for SMR deployment.

A–20.3.3. Follow-up Question

What do you expect to encounter with regard to worker dose?

A–20.3.4. Response

Currently, no SMRs are planned to be deployed in the Czech Republic. Given the lack of detailed information about various SMR designs and the uncertainty over whether and which type of SMR could be hypothetically deployed, there are no specific expectations with regard to the worker dose.

A–20.5. FRANCE–ASN

A–20.5.1. Question

Describe the challenges associated with regulatory assessment of worker dose. Also consider radioactive discharges to the environment and the dose to the public.

A–20.5.2. Response

Limit exposure for workers and public are defined in the Public Health Code [A–165] and Labour Code [A–166] (see the answer from France to Question 7 for more information). These limits remain applicable to SMRs. The licensee is responsible for its worker's radiation protection and must monitor their dose.

Also, according to the article 4.2.1 of the Order [A–72], the licensee must monitor the effects of its installation on the environment. At this effect, the licensee monitors radiation level, but also chemical quality inside and outside its installation. For example, samplings are realised in neighbouring water and plants.

ASN conducts inspections to control licensee's compliance with these regulatory requirements.

A–20.6. JAPAN–NRA

A–20.6.1. Question

Describe the challenges associated with regulatory assessment of worker dose. Also consider radioactive discharges to the environment and the dose to the public.

A–20.6.2. Response

The radiation dose limits for radiation workers are stipulated for normal and emergency work for both HTTR and for research reactor facilities.

Regarding the exposure dose to the general public, the target value is 50 μSv annually outside the environmental monitoring area during normal times, and it is designed to be less than 5 mSv per accident at the reactor facility outside the environmental monitoring area during an accident.

A–20.6.3. Follow-up Question

What challenges were encountered, or do you expect to encounter with regard to worker dose?

A–20.6.4. Response

As for radiation exposure of emergency workers under DEC event, it is confirmed that the estimated effective dose for the most affected worker does not exceed 5 mSv.

Therefore, there is no particular problem with radiation exposure.

A–20.7. RUSSIAN FEDERATION–ROSTECHNADZOR

A–20.7.1. Question

Describe the challenges associated with regulatory assessment of worker dose. Also consider radioactive discharges to the environment and the dose to the public.

A–20.7.2. Response

The basic Russian regulations governing radiation safety of the personnel and the public are NRB-99/2009 [A–169] and OSPORB-99/2010 [A–170] developed on the basis of the IAEA recommendations in GSR Part 3 [A–160].

A–20.7.3. Follow-up Question

What challenges were encountered, or do you expect to encounter with regard to worker dose?

A–20.7.4. Response

In the Russian Federation, the main regulatory documents governing the radiation safety of personnel and the public are NRB-99/2009 [A–169] and OSPORB-99/2010 [A–170].

The NRB-99/2009 [A–169] radiation safety regulations are applied to ensure human safety in all conditions of exposure to ionizing radiation of artificial or natural origin.

The requirements and standards established by the Regulations are mandatory for all legal entities and individuals, regardless of their subordination and form of ownership, whose activities may cause exposure of people to radiation, as well as for the administrations of the constituent entities of the Russian Federation, local authorities, citizens of the Russian Federation, foreign citizens, and persons without citizenship living in the territory of the Russian Federation.

In general, the approaches to radiological safety assurance for personnel and the public laid down in NRB-99/2009 [A–169] and OSPORB-99/2010 [A–170] with due regard for the recommendations of the IAEA Standard GSR Part 3 [A–160] are suitable for protecting the personnel and the public against the potential radiological impact of the floating nuclear power plant.

There is no intention to make amendments in the existing regulatory framework regarding radiological safety of personnel and the public.

A–20.8. SOUTH AFRICA–NNR

A–20.8.1. Question

Describe the challenges associated with regulatory assessment of worker dose. Also consider radioactive discharges to the environment and the dose to the public.

A–20.8.2. Response

A description of the facility and equipment design, planning and procedures, management policy and organisational structure, Radiation Protection programme, and the techniques and practices employed in ensuring occupational exposure is within regulatory limits and ALARA is required.

Also required is a description of the dose expected for site workers during normal operation (including all planned normal operational modes, maintenance and inspection, and AOO). It must include a description of the best estimate radiological source, DiD measures that provide worker protection in case of the failure of another radiation feature, and measures to minimise worker dose. Operational programs (Dose Assessment, ALARA, Radiation Protection), and design features intended to minimise dose to the worker must be described.

The Act 47(1999) [A–30] /RG 1/ and the Regulation No. R388 [A–31] form the general basis for radiation protection requirements.

The following 'principal radiation protection and nuclear safety requirements' apply to actions authorised by, or seeking authorisation in terms of a nuclear installation licence, a nuclear vessel licence or a certificate of registration, cf. section 3 of /RG 1/:

The dose to an individual arising from normal operating conditions must not exceed the limits specified in Annex 2 of /RG 1/.

(a)   For Occupational exposure the following applies:

The occupational exposure of any worker must respect the following limits:

- Section 1.1.1: An (average) effective dose of 20 mSv per year averaged over five consecutive years;
  The start of the averaging period shall be coincident with the first day of the relevant annual period starting from the date of entry into force of the Regulations, with no retroactive averaging.
- Section 1.1.2: A (maximum) effective dose of 50 mSv in any single year;
- Section 1.1.3: An equivalent dose to the lens of the eye of 150 mSv in a year;
- Section 1.1.4: An equivalent dose to the extremities (hands and feet) or the skin of 500 mSv in a year.

There is provision in section 1.1.5 for the dose limit to be varied temporarily in special circumstances by the Regulator.

Similar but different limits as above are specified for each of 'Apprentices and students' and 'women'.

(b)   For Emergencies the following applies:

In the case of an emergency or as a responder to an accident, a worker who undertakes emergency measures may be exposed to a dose in excess of the annual dose limit:

- Section 1.4.1: For the purpose of saving life or preventing serious injury;
- Section 1.4.2: If undertaking actions intended to avert a large collective dose; or
- Section 1.4.3: If undertaking actions to prevent the development of catastrophic conditions.

Under any of the situations referred to in 1.4.2 or 1.4.3 above, all reasonable efforts must be made to keep the worker's dose below twice the maximum annual dose limit. Regarding situations referred to in 1.4.1 above, every effort shall be made to keep doses below ten times the maximum annual dose limit. In addition, workers undertaking interventions which may result in their doses approaching or exceeding ten times the annual dose limit may only do so when the benefits to others clearly outweigh their own risk.

For exposure of visitors and non-occupationally exposed workers at sites the following applies:

The annual effective dose limit for visitors to the sites and those not deemed to be occupationally exposed is 1 mSv. The annual dose equivalent limit for individual organs and tissues of such persons is 10 mSv.

Design features for radiation protection

A description of the design features of the equipment and the facility has to be provided that ensures radiation protection. The description should give information on the shielding provided for each of the radiation sources identified, describe the features for occupational radiation protection and the instrumentation for fixed radiation area monitoring and airborne radioactive isotopes continuous monitoring, and the criteria for their selection and placement, and address, if necessary, provisions for any equipment decontamination.

The radiation protection principles applied in the design need to be stated. For example:

"(a) No person shall receive doses of radiation in excess of the authorized dose limits as a result of normal plant operation;
(b) The occupational exposures in the course of normal operation shall be ALARA;
(c) Dose constraints shall be used to avoid inequities in the dose distributions;
(d) Measures shall be taken to prevent any workers from receiving doses near the dose limits year by year;
(e) All practicable steps shall be taken to prevent accidents with radiological consequences;
(f)   All practicable steps shall be taken to minimize the radiological consequences of any accident."

Radiation dose targets have to be stated.

It should be demonstrated, for the overall design, that suitable provision is made in the design, layout and use of the plant to reduce doses and radioactive releases from all sources. Such provisions should include the adequate design of systems, structures and components so that exposures in all activities throughout the lifetime of the plant are reduced or, where no significant benefit accrues from the activities concerned, eliminated. Reference to the chapter of the SAR on description and conformance to the design of plant systems on this subject may be appropriate.

At the detailed level, at least the following points must be addressed to an acceptable level:

- Radiation protection aspects related to other SAR chapters;
- Detailed description of the relevant source terms, also by reference to other SAR chapters;
- Detailed description of the scenarios leading to exposure of workers;
- Definition and description of tasks and worker profiles with respect to the dose assessment;
- Detailed description of the scenarios leading to exposure of public;

- Detailed definition, description and discussion of all relevant parameters used for dose assessment;
- Consideration of liquid and gaseous pathways;
- Consideration of direct radiation;
- Transparent and comprehensive description of assumptions and conditions with respect to SSC and systems relevant for the respective scenarios;
- Clear definition and description of the radiological zoning;
- Consideration of committed doses where appropriate.

A–20.8.3. Follow-up Question

Were challenges encountered with regard to assessing worker dose for the PBR?

A–20.8.4. Response

The consideration of potential worker dose led to limitations on design options, for example:

- The complex design of pipework of the PBMR DPP caused by the Brayton Cycle (helium turbine and compressors) imposes a multitude of potential leak and break locations with the potential for subsequent air ingress event that could potentially lead to graphite and fuel element corrosion. Since early operator actions or action imposing high worker doses are not acceptable, reliable safety functions such as inertisation or fast core cooling would be required. The associated safety concept and associated design solutions were still under development.

- The only mitigation measure to cope with air ingress events appeared to be methods to close pipe breaks, although this was assumed by the licence applicant to be no longer possible in the light of the clarification of allowable worker doses. The use of other measures or systems such as isolation valves, nitrogen injection, post event clean-up system were considered but not adopted. Whilst it is not necessary to provide detailed discussion of other potential mitigation measures on a strategy level, it was an important result of regulatory communications with the licence applicant that a strategy relying solely on closing of pipe breaks is not acceptable. Therefore, a description in sufficient detail was required of other mitigation methods to constitute an acceptable strategy to cope with air ingress events.

Other concerns related to worker dose included for example the following consideration:

The lack of progress with fuel qualification was of concern as fuel qualification and performance is the basis of the PBMR safety case and directly related to source term and confinement issues. State-of-the-art fuel at the time of licensing tends to leak certain metallic and other fission products during normal operation as a result of diffusion phenomena under elevated temperature conditions as well as due to the failure of a small percentage of coated particles that have inevitably been damaged during the fuel manufacturing process. The relevant uncertainties must be conservatively accounted for in the contamination and source term analyses. This in turn must be considered for worker doses during maintenance activities as well as the potential dose to the public during normal operation and events.

A–20.9. UNITED KINGDOM–ONR

A–20.9.1. Question

Describe the challenges associated with regulatory assessment of worker dose. Also consider radioactive discharges to the environment and the dose to the public.

A–20.9.2. Response

ONR assessment of worker doses from nuclear power plants is carried out against UK Law. For Radiation Protection, the Ionising Radiations Regulations 2017 (IRR 17) [A–265] are the main legal requirements. Document L121 [A–266] includes Approved Code of Practice (ACOP) and statutory Guidance on practical implementation of the regulations. It should be noted that in the UK, assessment of public dose for normal operation is the remit of the Environment Agency (EA). Assessment of worker dose in normal operation and accident conditions is undertaken by ONR.

ONR assessment of worker dose is supported by the SAPs [A–44], Technical Assessment Guides (TAGs) [A–46], and the Health and Safety Executive document 'Reducing Risks, Protecting People' (R2P2) [A–267]. In this context publications from authoritative international sources such as the IAEA, WENRA and NEA are viewed as RGP by ONR and are used to provide a benchmark against which designs can be compared. The following TAGs provide the principal expectations of ONR regarding radiation protection. These are:

- NS-TAST-GD-002 - Radiological Shielding [A–268];
- NS-TAST-GD-004 - Fundamental Principles [A–269];
- NS-TAST-GD-005 - Guidance on the demonstration of ALARP (As Low as Reasonably Practicable) [A–43];
- NS-TAST-GD-038 - Radiological Protection [A–270];
- NS-TAST-GD-041 - Criticality Safety [A–271];
- NS-TAST-GD-043 - Radiological Analysis Normal Operation [A–272];
- NS-TAST-GD-045 - Radiological Analysis Fault Conditions [A–273].

Additional Guidance can be obtained from Technical Inspection Guides e.g. ONR-INSP-GD-054 'The Ionising Radiations Regulations 2017' [A–274].

ONR expects that worker doses are reduced to As Low As Reasonably Practicable (ALARP), and a hierarchy of controls are put in place to limit dose to workers (UK Ionising Radiations Regulations 2017 Regulation 9(2) – IRR 17) [A–265].

The Numerical Targets and Legal Limits are set in the SAPs [A–44]. The Basic Safety Level (BSL) and Basic Safety Objective (BSO) translate the Tolerability of Risk [A–275] framework developed to guide inspectors in their decision making. ONR policy state that the BSLs indicate which doses/risks new facilities need to meet and also provide benchmarks for existing facilities. It must be recognised that the BSO doses/risks have been set at a level where ONR considers it not to be a good use of its resources or taxpayers' money, nor consistent with a targeted and graded regulatory approach, to pursue further improvements in safety. In contrast, facility operators and owners have the overriding duty, irrespective of whether the BSOs are met, to consider on a case by case basis whether they have reduced risks ALARP. On these bases, it will be inappropriate for operators (etc.) to use the BSOs as design targets, or as surrogates to denote when ALARP levels of dose or risk have been achieved.

Targets 1, 2 and 3 of the SAPs [A–44] are the most relevant for Radiological Protection. It should be noted that these are linked to the legal dose limits in IRR 17 [A–265]. Target 1 deals with occupational exposure during normal operations.

Target 1 - Normal operation – any person on the site

Employees working with ionising radiation. The target for effective dose in a calendar year:

- Basic safety level (BSL): 20 mSv — This is also the legal dose limit for employees under IRR regulation 12(1);
- Basic safety objective (BSO): 1 mSv.

Other employees on the site. The target for effective dose in a calendar year:

- Basic safety level (BSL): 2 mSv;
- Basic safety objective (BSO): 0.1 mSv.

Target 2 - Normal operation – any group on the site (e.g. welders, scaffolders)

Average effective dose in a calendar year to defined groups of employees working with ionising radiation are:

- Basic safety level (BSL): 10 mSv;
- Basic safety objective (BSO): 0.5 mSv.

Target 3 - Normal operation – any person off the site

The target for effective dose in a calendar year for any person off the site from sources of ionising radiation originating on the site are:

- Basic safety level (BSL): 1 mSv — This is also the legal dose limit for members of the public under IRR regulation 12(1);
- Basic safety objective (BSO): 0.02 mSv.

ONR has identified a number of challenges in the context of worker dose and accident source terms for SMRs, which are in some cases common to many disciplines and technical areas. These have been as follows:

- Lack of OPEX (applies also to answer from United Kingdom to question 22, on accident source term);
- Lack of analysis of source term for normal operation and faults (also question 22);
- Fuel handling. Shielding along fuel route and maintenance of remote and difficult to access plant;
- Coolant activation. Unavailability of data on achievable coolant impurity levels, corrosion of surfaces, mobility and resultant coolant activity levels;
- Mobility of source term (dust). Particular to HTGRs;
- Structural activation of components near the reactor core in compact designs;
- Fission product release rates from novel fuel (also question 22);
- Plate out and clean up rates of activation and fission products from coolant in novel designs (also question 22);

- Performance of novel heating, ventilation and air conditioning (HVAC) and containment/confinement systems in removing airborne activity (also question 22).

A–20.10. UNITED STATES OF AMERICA–NRC

A–20.10.1. Question

Describe the challenges associated with regulatory assessment of worker dose. Also consider radioactive discharges to the environment and the dose to the public.

A–20.10.2. Response

Staff did not find any significant challenges in this area of review. ALARA means to make every reasonable effort to maintain exposures to radiation as far as practicable and below the dose limits of Title 10 of the Code of Federal Regulations (10 CFR) [A–52] Part 20, 'Standards for Protection Against Radiation'. This includes accounting for the state of technology and the economics of improvements, including the use of procedures and engineering controls based on sound radiation protection principles, in relation to benefits to public health and safety.

For the SMR design reviewed by the NRC, the ALARA principles were applied during the design process to identify and describe design features and specifications intended to limit and minimize the amount of radiation exposure from operating modules of the SMR to workers constructing or installing additional modules; and radiation exposure to occupational workers during plant operation, AOOs, maintenance and inspection activities, and accidents. Operational program elements were used to complement design features and specifications to limit and minimize radiation exposure.

Most nuclear plant worker occupational radiation exposure (ORE) results from the operation and maintenance of systems that contain radioactive material, radioactive waste handling, in-service inspection, refuelling, abnormal operations, and decommissioning work activities. The design of the SMR addressed and included these activities through the plant physical layout, selection of materials, shielding, and chemistry control.

During the design process, ALARA design reviews were periodically conducted. To the extent that the experience is relevant to the SMR design, the design was based on experience and lessons learned from operating reactors. Examples of facility design features in the SMR design that ensured that the design was ALARA include: (1) the separation of radioactive components into individual shielded compartments; (2) the use of remote operating equipment, where possible, to reduce radiation exposure; and (3) the minimization of field run piping to the extent practicable.

Operators of the SMR will be required to describe the operational program to maintain exposures to ionizing radiation as far below the dose limits as practical, ALARA.

A–21. INTERPRETATION OR DEFINITION OF CORE DAMAGE AND SEVERE ACCIDENTS

This Annex presents the responses provided by the following Member State regulatory bodies to the Question 21: "Describe the challenges associated with interpretation or definition of core damage and severe accidents."

A–21.1. ARGENTINA–ARN

A–21.1.1. Question

Describe the challenges associated with interpretation or definition of core damage and severe accidents.

A–21.1.2. Response

The so-called and postulated Severe Accidents (SA) can be caused by initiating events in conjunction with system failures and human errors. In order to simulate the plant response in the case of SA, reactor and containment models were developed using MELCOR code.

At CAREM 25 prototype reactor, due to the intrinsic characteristics of its design, certain typical events of the current generation of reactors, such as large LOCA, LOFA, ejection of control rods and boron dilution in the coolant, do not have a possibility of occurrence by design. It has also been shown that in CAREM 25 a larger diameter pipe break can be mitigated with a safety-related system without the need for SIS intervention.

The postulated event of a total loss of electricity supply (blackout) does not have the impact or the severity that it has in said reactors, as it is a reactor with passive cooling systems, which allows the plant to be kept in a safe state during the period of grace and complying with the simple failure principle applied to safety systems.

After the ARN assessment, it is important to highlight the strategy proposed for the multiple failure events control within DiD Sublevel 3B, that through autonomous powered systems it is possible to maintain a secure state beyond the grace period. These so-called 'Safe State Extension' systems are classified as related to safety, they present a higher level of protection than current reactors, in order to avoid severe accidents.

The large refrigerant inventory, the high ratio of the volume of steam to that of liquid, negative coefficients of reactivity, and natural circulation, with a self-adaptive response to power changes, are the relevant characteristics of the CAREM 25 that make the observed robust behaviour before the analysed initiating events.

See the answer from Argentina to Question 22 to more information.

A–21.2. CANADA–CNSC

A–21.2.1. Question

Describe the challenges associated with interpretation or definition of core damage and severe accidents.

A–21.2.2. Response

In Canada, CNSC staff acknowledge that the definition of core damage as defined in REGDOC-2.5.2 [A–62], and its associated safety goals may not be pragmatically applicable, as written, to all reactor designs. For example, some concepts such as liquid fuel reactors or using fuels with very high temperature failure tolerance are seeking to demonstrate that the fuel is no longer a significant defining aspect of core failure. As a result, technology developers may be seeking to define 'severe accident' differently, such as failure of a core vessel or even making claims that the concept of a severe accident does not exist for their design.

Some designers claim that the Canadian regulatory framework for SMRs needs to take into account designs that include extensive use of passive features. As per these claims, these features will prevent many of the traditional accident events and potential scenarios from causing any core damage or from releasing radioactive materials from the reactor containment. Their concerns was that the traditional safety analysis methodologies may be difficult to employ, and alternate approaches should be recognized as applicable to or acceptable for safety cases. CNSC note that uncertainties presented by alternative and innovative features can affect the confidence on the outcomes of safety analyses.

Some design developers claims that the risk metrics of quantitative safety goals in REGDOC-2.5.2 [A–62], core damage frequency (CDF), small release frequency (SRF) and large release frequency (LRF) are not appropriate for their design. They propose the use of a Non-LWR PSA standard, which contains technology inclusive metrics, such as source term parameters, and radiological doses.

CNSC REGDOC-2.5.2 [A–62] defines core damage and severe accident as:

- "Core damage: core degradation resulting from event sequences more severe than design-basis accidents
- Severe accident: an accident more severe than a design-basis accident and involving severe fuel degradation in the reactor core or spent fuel pool."

Canada's definition of core damage frequency (CDF) and large release frequency (LRF) are surrogates of the two qualitative safety goals established for societal risk (latent or cancer deaths) and individual risk (prompt deaths), respectively.

The definitions of CDF, small release frequency (SRF), and LRF are based on water-cooled technologies. Per REGDOC-2.5.2 [A–62], SRF and LRF are defined by the release magnitude of a specific radionuclide (10^{15} becquerels of iodine-131, and 10^{14} becquerels of caesium-137, respectively). However, in RD-367 [A–61], the release criteria were more broadly written to be technology neutral, as they do not refer to particular isotopes (caesium or iodine) but to doses necessary for triggering evacuation or relocation of the population. The design developers should define the core degradation states for their specific design. CNSC consider that the impacts of core degradation need to be understood and documented as part of demonstrating technical safety objectives have been met. A beyond-design-basis accident may or may not involve core degradation.

Canadian design requirements permit the proposal of other surrogates under the condition that the underlying objectives in the above requirements continue to be met. For example, for sodium reactors, a high temperature operational constraint (e.g. 800°C) with a sufficient degree of conservatism has been proposed as a surrogate to a formal definition core damage.

At this stage, CNSC staff are willing to engage in a case by case discussion with technology developers and potential applicants for licences with the understanding that they are expected to support safety claims with suitable evidence commensurate with safety importance including analysis that may lead to a better understanding of severe accidents.

The design developers should identify the set of DECs relying on both probabilistic and deterministic methods, engineering judgment, operational experience and the results of research and analysis. These DECs need to be used to improve the safety of the nuclear facility further by enhancing the plant's capabilities to withstand, without substantial radiological releases, as well as accidents that are either more severe than DBAs or that involve other additional failures.

CNSC expectations[20] for DEC are articulated in REGDOC 2.5.2 [A–62] and expand on considerations from IAEA SSR 2/1 [A–127] Req 20. Provisions for DECs can vary greatly between designs. CNSC is noting that technology developers are claiming to have increasing challenges identifying DECs because of the specific robust design provisions being proposed. CNSC is asking that the proposed alternative approach should be supported by adequate evidence and would result in an equivalent level of safety. Section 8.6.12 of REGDOC-2.5.2 indicates that following onset of core damage, the containment boundary shall be capable of contributing to the reduction of radioactivity releases to allow sufficient time for the implementation of offsite emergency procedures. The following requirements for the containment system are further described:

"The ability of the containment system to withstand loads associated with design extension conditions (DECs) shall be demonstrated in design documentation, and shall include the following considerations:

- various heat sources, including residual heat, metal-water reactions, combustion of gases, and standing flames
- pressure control
- control of combustible gases
- sources of non-condensable gases
- control of radioactive material leakage
- effectiveness of isolation devices
- functionality and leak tightness of airlocks and containment penetrations
- effects of the accident on the integrity and functionality of internal structures"

Section 8.6.12 of REGDOC-2.5.2 [A–62] also expands on expectations in IAEA SSR-2/1 [A–127] 5.27, for extension of the capability of features to mitigate a severe accident and maintain the integrity of containment:

"The design authority shall demonstrate that complementary design features have been incorporated that will:

- prevent a containment melt-through or failure due to the thermal impact of the core debris
- facilitate cooling of the core debris
- minimize generation of non-condensable gases and radioactive products
- preclude unfiltered and uncontrolled release from containment"

---

[20] The term expectations refers to the broad combination of regulatory requirements and guidance.

CNSC's expectations regarding the treatment of severe accidents is described in REGDOC-1.1.2. [A–66]. Overall, an applicant should demonstrate that severe accident management has been considered during the design of the reactor facility, including SMRs. Furthermore, the applicant should demonstrate provisions for mitigating the consequences of severe accidents will be effective in:

- Terminating core degradation as early as possible in the accident sequence;
- Maintaining the integrity of the containment;
- Achieving a stable and controlled state of the reactor core or core debris;
- Minimizing the release of radioactive material into the environment.

The following CNSC expectations address issues related to severe accidents in REGDOC-1.1.2 [A–66]:

- Section 5.9.6 provides expectations around the determination of the adequacy of design provisions for severe accidents and the acceptability of proposed severe accident management guidelines;
- Section 7.6 is focused on the safety analysis related to BDBA and severe accidents;
- Section 9.6 provides expectations around the determination of an adequate accident management plan.

To address claims of low probability or physical impossibility, compared to the SSR-2/1 (Rev. 1) [A–127] 5.31 which indicates "The design shall be such that the possibility of conditions arising that could lead to an early radioactive release or a large radioactive release is 'practically eliminated'". The following guidance regarding the practical elimination of events is described in section 7.3.4 of REGDOC-2.5.2 [A–62]:

"To demonstrate practical elimination as extremely unlikely with a high degree of confidence, the following should be considered:

- The degree of substantiation provided for the demonstration of practical elimination should take account of the assessed frequency of the situation to be eliminated and of the degree of confidence in the assessed frequency.

- Practical elimination of an accident should not be claimed solely based on compliance with a probabilistic cut-off value. Even if the probability of an accident sequence is very low, any additional design features, operational measures or accident management procedures to lower the risk further should be implemented to the extent practicable.

- The most stringent requirements regarding the demonstration of practical elimination should apply in the case of an event with the potential to lead directly to a severe accident, i.e. from level 1 to level 4 for DiD. For example, demonstration of practical elimination of a heterogeneous boron dilution event in a pressurized water reactor (PWR) would require a detailed substantiation.

- The necessary high confidence in low likelihood should, wherever possible, be supported by means such as:
    o multiple layers of protection
    o application of the safety principles of independence, diversity, separation, redundancy

        o   use of passive safety features
        o   use of multiple independent controls

- It should be ensured that the practical elimination provisions remain in place and valid throughout the plant lifetime, for example, through in-service and periodic inspections.

In each case, the demonstration should show sufficient knowledge of the accident sequence analysed and of the phenomena involved, substantiated by relevant evidence.

To minimize uncertainties and to increase the robustness of a plant's safety case, demonstration of practical elimination should preferably rely on the criterion of physical impossibility, rather than the second probabilistic criterion (extreme unlikelihood with high confidence)."

A–21.3. CHINA–NNSA

A–21.3.1. Question

Describe the challenges associated with interpretation or definition of core damage and severe accidents.

A–21.3.2. Response

For HTR-PM, there is no core melting, and for all considered DBA and BDBA, the fuel temperature will not reach the limit of 1620°C. In other words, there will be no large-scale fuel failure.

In addition, we are also considering the requirements of the fourth and fifth level of DiD for some SMRs with low temperature and low pressure, such as the pool reactor, which can be designed so that the core will not melt in any credible accident.

A–21.3.3. Follow-up Question

Please could you identify the most limiting scenario considered and the considerations in the judgement that no core melting demonstration was acceptable in the DEC used to define that the fuel temperature will not reach the limit of 1620°C?

A–21.3.4. Response

The current value for possible failure rate of TRISO coated particles in modular HTGR is in the order of 1E-4, even for most serious accident of depressurization accident in the viewpoint of maximum fuel temperature, or of air ingress accident in the viewpoint for additional coated particle failure mechanism. Therefore, it can be claimed that there is no core damage and no severe accident for modular HTGR.

A–21.4. CZECH REPUBLIC–SÚJB

A–21.4.1. Question

Describe the challenges associated with interpretation or definition of core damage and severe accidents.

Not applicable — as there was no licence application for SMR, no challenges associated with the interpretation of core damage and severe accidents have been identified.

The general legislative requirement for core damage and severe accidents are in the Decree No. 329/2017 [A–22], on the requirements for nuclear installation design. These requirements are general enough to allow deployment of various SMR technologies — however in practice the interpretation or definition of core damage and severe accidents would be based on a particular type of the SMR design.

The severe accident is defined as an accident conditions involving serious damage of nuclear fuel either due to serious damage to and irreversible loss of the structure of the core of the nuclear reactor or the system for nuclear fuel storing due to damage to fuel assemblies as a result of nuclear fuel melt.

In the context of ensuring compliance with requirements for the application of the DiD principles, the designer of the nuclear installation needs to set out technical and organisational measures for managing abnormal operation, all design basis accidents and design extension conditions, including severe accident.

In order to manage design extension conditions, the design of the nuclear reactor installation shall include reasonably practicable technical and organisational measures such that:

- Severe accidents leading to early radiation releases or large radiation releases, are practically eliminated;
- Those severe accidents that cannot be practically eliminated but could lead to radiation releases are managed in such a manner that no protective measures stricter than urgent protective actions (sheltering prophylaxis and evacuation) and restriction of the use of food and water and feeding stuff are necessary.

Nuclear reactor installation design shall also set out and evaluate reasonably practicable measures for managing a postulated severe accident corresponding to the type of the nuclear reactor so that the damaged/melting core and the nuclear fuel in storage are cooled, and the melt from the damaged/melting core is contained, the subsequent fission chain reaction is prevented and the safety objectives of the design (practical exclusion of early and large radiation releases) are complied with.

When ensuring compliance with the principles for the safe use of nuclear energy (basic safety functions) for design basis external events and their very likely combinations and external design events and scenarios which (due to their frequency of incidence and severity) fall within the scope of design extension conditions, the external design events and the corresponding scenarios falling within the scope of design extension conditions shall be assessed and the nuclear installation design shall propose reasonably practicable measures focused to extreme events.

Practically eliminated are (according to this decree on the design) those condition, state or event, the occurrence of which is considered physically impossible, or which are, with a high degree of confidence, very unlikely.

Certain legislative requirements might be considered technologically specific (reflecting the currently used technology in Czech Republic i.e. PWR). One example might be that the nuclear installation design shall specify the characteristics of the fuel system and the core and the operating conditions for the nuclear reactor so that the fuel elements and assemblies remain in place under design basis accident conditions and do not suffer damage that would prevent insertion of reactivity control system components into the core, the functioning of other systems for reactivity control and the reactor shutdown, effective cooling down of the core or subsequent handling of the fuel assemblies.

A–21.5. FRANCE–ASN

A–21.5.1. Question

Describe the challenges associated with interpretation or definition of core damage and severe accidents.

A–21.5.2. Response

According to ASN's guide on PWR design, the list of events for severe accidents must rely on deterministic and probabilistic considerations, eventually confirmed by expert opinions. Taking into account these events aim at defining provisions to limit, in terms of scope and durations, the consequences of severe accidents. The examination of these events must take into account:

- The installation's environment and siting;
- The capacity of important to safety SSCs to accomplish their functions considering the conditions induced by the situation;
- The practicability of the actions required to manage the accident, and the needed time to realise them;
- The input of probabilistic safety assessments.

Also, ASN's guide stipulates that in case of severe accident, the reactor must be durably brought into a state in which the subcriticality is insured, the heat is removed, and the core is cooled. Also, radioactive substances' containment must be insured. The licensee must define technical acceptance criteria in order to translate in operational terms the objectives previously described.

Finally, ASN's guide stipulates that the radiological consequences must be assessed in compliance with the article 3.7 of the Order [A–72] (see the answer from France to Question 22). The assessment of radiological consequences contributes to demonstrate the provisions' adequacy with the safety objectives.

ASN, with the support of IRSN, reviews this item of the safety demonstration through instructions.

A–21.6. JAPAN–NRA

A–21.6.1. Question

Describe the challenges associated with interpretation or definition of core damage and severe accidents.

A–21.6.2. Response

It is not required to implement severe accident measures or conduct its effectiveness evaluation, because the scale of the research reactor facility, the amount of radioactive material contained in it, and the risk of radiation are smaller than those of the commercial reactor facility.

However, for reactor facilities with a certain output (heat output of 500 kW) or more, and for gas-cooled or sodium-cooled research reactor facilities, etc., it is required to evaluate and take measures against accidents (so-called DEC) that may cause excessive radiation exposure (effective dose evaluation value exceeds 5 mSv per occurrence accident).

For the HTTR, as the DEC event, the strictest DBA is selected from the viewpoint of graphite oxidation and radioactive material release. Thus, events which superimposed the reactor shutdown, cooling, and radioactive material confinement functions (depending on the assumption, there may be multiple superpositions) were selected, evaluated it, and took necessary measures.

A–21.6.3. Follow-up Questions

What are the criteria used to assess radiological releases and their acceptability? Are they technology neutral?

A–21.6.4. Response

As for research and test reactor facility, there is no specific limitation value for the amount of released radioactive material, however, in the light of preventing excessive exposure of public, it is required that the dose evaluation value in DEC event should not exceed 5 mSv per an occurrence of accident.

A–21.7. RUSSIAN FEDERATION–ROSTECHNADZOR

A–21.7.1. Question

Describe the challenges associated with interpretation or definition of core damage and severe accidents.

A–21.7.2. Response

The existing design of the floating NPP (Akademik Lomonosov) employs reactors with pressurised water coolant and moderator, and classical fuel assemblies, and fuel rods. So, we face no challenges when defining severe accidents.

In accordance with definition No. 22 in the federal rules and regulations NP-029-17 [A–93], "A severe accident implies accident conditions, which are more severe than a design basis accident and involve significant core damage". Thus, the definition of a severe accident in NP-029-17 is in harmony with the definition in the IAEA Safety Glossary [A–206].

Following the requirements of NP-022-17 [A–89], the system of technical and organisational measures encompasses measures intended to ensure the vessel safety, as well as the measures to provide the safety of vessel crew, special personnel and passengers, and also the measures to ensure the safety of the public when the vessel is in a mooring location or at a shipyard. The indicative lists of beyond design basis accidents specific to each reactor type are defined in the federal rules and regulations in the field of the use of atomic energy that establish requirements for the safety analysis report. The final list of BDBA (including severe accidents) is given in the SAR. The list shall encompass the representative scenarios in order to identify the measures and actions essential for managing such accidents. Representativeness of the scenarios is ensured by considering the severity of vessel condition and potential availability or unavailability of the safety systems and special technical features designed to cope with beyond design basis accidents.

The safety analysis report must contain a realistic (non-conservative) analysis of these beyond design basis accidents, including the assessment of the probability of the BDBA courses and consequences. The analysis of beyond design basis accidents presented in the safety analysis report provides an input to develop an action plan to protect the workers (personnel) and the public in case of an accident, and to develop the BDBA management procedures.

The vessel design must provide technical features and organisational measures intended to prevent and minimise consequences of accidents at nuclear power installation, so as:

- To ensure that limits set in the design for design basis accidents at the nuclear power installation are not be exceeded, owing to the inherent safety of the reactor and the use of the safety systems;
- To limit the consequences of beyond design basis accidents at the nuclear power installation, owing to the use of dedicated technical features for managing beyond design basis accidents, and the use of other technical features capable of fulfilling the required functions under the existing conditions, and due to the implementation of organisational measures, including the measures for coping with beyond design basis accidents.

For the beyond design basis accidents that are not excluded by the inherent safety of the reactor and the design philosophy, whatever the probability of these events, organisational accident management measures must be developed, including measures to minimise radiation exposure of the vessel crew and special personnel, the public and the environment.

A–21.8. SOUTH AFRICA–NNR

A–21.8.1. Question

Describe the challenges associated with interpretation or definition of core damage and severe accidents.

A–21.8.2. Response

The capability of the design and proposed Severe Accident Management actions to mitigate unlikely event sequences that are beyond Design Basis, and which have the potential to lead to significant releases shall be evaluated and described. Despite the claimed low probability of occurrence of these sequences, safety submissions shall demonstrate the ability of the design to meet the following objectives:

i. To decrease the probability or mitigate the consequences of complex sequences involving multiple failures beyond those considered in the deterministic design basis;

ii. To mitigate Severe Accidents with core degradation. The confinement system shall be able to mitigate the consequences of a degraded core.

Amongst others, the following acceptance criteria for radiological safety of new reactor concepts have to be fulfilled:

- Exclusion of off-site emergency measures (such as evacuation);
- Reduction of emergency planning zones;
- Limitation of emergency measures.

The mitigation measures shall be designed based on the representative Severe Accidents to be addressed as BDBAs. Generally, Severe Accidents with highest contribution to Core Damage frequency shall be selected as BDBAs and mitigated as appropriate until the probabilistic safety targets are met. However, care must be taken not to discard Severe Accidents of lower contribution to Core Damage frequency but involving potential significant releases.

A–21.8.3. Follow-up Question

Are you able to share any insights from the regulatory assessment of the PBMR?

A–21.8.4. Response

The following provides some indication of the challenges associated with interpretation or definition of core damage and severe accidents in the context of the PBMR:

The probabilistic risk assessment (PRA) for the PBMR is fundamentally different to that for the LWR. The concept of core damage frequency cannot be used. Extensive use of passive features must be modelled. Events of very low frequency have to be addressed.

PRA studies performed for LWRs are classically used to:

(a) Identify the sequence of events that can lead to core damage and estimate the core damage frequency (level 1);
(b) Identify the ways in which radioactive releases from the plant can occur and estimate their magnitude and frequency (level 2);
(c) Estimate the risk to public health (and other societal risks) such as the contamination of land or food (level 3).

The PRA for the PBMR seeks to achieve the same overall goal but is structurally different to PRAs carried out for LWRs. This follows directly from the differences in the design and safety philosophy.

As a result:

- The concept of core damage and large early release end states is not considered;
- The PRA is fundamentally a challenge-response analysis of the fission product barrier — the fuel particle coating;
- Extensive use is made of passive systems for which failure probabilities are correspondingly small and therefore difficult to justify from operational or test data.

The following therefore needs to be addressed:

- The reliability of passive systems in particular for their long-term response;
- The modelling of fuel degradation as a function of time and temperature;

- The urgency for developing the PRA depends upon the intention to employ risk-informed methodologies and the way in which these interact with the design process.

Since early PBMR PRA proposals discarded the most severe accidents, the PRA's representativity of the actual risk associated with the PBMR was questioned. As a result, the NNR required the following action related to PRA work:

Provide assessments of the probability of occurrence of severe core damage states and the risk associated with large off-site releases.

A–21.9. UNITED KINGDOM–ONR

A–21.9.1. Question

Describe the challenges associated with interpretation or definition of core damage and severe accidents.

A–21.9.2. Response

ONR provides a definition of the term 'severe accidents' in the SAPs as follows:

"(…) severe accidents are defined as those fault sequences that could lead either to consequences exceeding the highest off-site radiological doses given in the BSLs of Numerical Target 4 (i.e. 100 mSv, conservatively assessed) or to an unintended relocation of a substantial quantity of radioactive material within the facility which places a demand on the integrity of the remaining physical barriers. A substantial quantity of radioactive material is one which if released could result in the consequences specified in the societal risk Target 9."

The consequences specified in Target 9 are 100 or more fatalities, either immediate or eventual. Further guidance is provided for ONR inspectors in the severe accident analysis (SAA) technical assessment guide (TAG [A–46]).

This definition has been designed to be technology neutral, to permit application to the full range of nuclear facilities in the UK, including pressurised water reactors, advanced gas-cooled reactors, and non-reactor facilities (e.g. fuel cycle). By adopting this approach, the challenge from advanced reactor technologies to ONR guidance is reduced. For example, a leak of molten fuel salt from a molten salt reactor would still be considered a severe accident under this definition, as would a substantial release of radioactivity from ceramic coated fuel (as the SAPs [A–44] definition makes no reference to fuel or core melt).

ONR does not provide a target 'core damage frequency' (CDF) in the SAPs; the numerical targets are based on dose consequence, accident frequency and risk of death from exposure to ionising radiation. The facility risk must be reduced to as low as reasonably practicable (ALARP).

The ONR NS-TAST-GD-030 (on PSA) [A–174] does however refer to CDF, alongside large release frequency. However, it is important to note that whilst discussion of 'fuel damage' may not be fully applicable to, for example, molten salt reactors, 'core damage' remains possible, as the core is typically taken to include the structural components, control rods, and heat removal means, failure of which could potentially result in significant radiological releases.

Some potential challenges associated with advanced technologies relate to the definition of an appropriate degraded plant state as the starting point for severe accident analysis, and selection of an appropriate metric for PSA studies.

Development of internal guidance on appropriate selection of a degraded plant state for severe accident analysis of advanced reactor technologies (particularly high-temperature gas-cooled reactors using ceramic coated fuel), has been identified as an area of focus in the ANT project's guidance review activities.

A–21.10. UNITED STATES OF AMERICA–NRC

A–21.10.1. Question

Describe the challenges associated with interpretation or definition of core damage and severe accidents.

A–21.10.2. Response

The NRC staff did not find any significant challenges in this area of review. The SMR design incorporated several innovative design features that provide enhanced capabilities for mitigating an extended loss of electrical power compared to currently operating nuclear reactor plants. These features include the use of passive safety systems capable of maintaining core cooling, containment, and spent fuel cooling functions. Each unit sits in a large reactor pool, which serves as the ultimate heat sink (UHS) for the facility. These features are intended to enable the design to mitigate beyond-design-basis external events (BDBEEs) using only installed plant equipment for an extended duration (greater than or equal to the first 72 hours following the event) without the need for alternating current (ac) power. Although the regulation governing mitigation of beyond-design-basis events (10 CFR [A–52] Part 50.155) does not apply to applicants for design certification, the applicant did voluntarily seek the NRC's approval of its proposal in the DCA to use installed design features for mitigation of BDBEEs.

In Technical Report (TR)-0816-50797, 'Mitigation Strategies for Loss of All AC Power Event' [A–276] Revision 3, Section 5.0, the applicant described that, to develop a mitigation strategy, the baseline coping capability of the plant design must be determined. The determination is made by evaluating the status of the three key safety functions (core cooling, containment, and spent fuel pool (SFP) cooling) stated in 10 CFR [A–52] Part 50.155(b)(1)(i) during the integrated plant response to a loss of all ac power. Specifically, the staff determined that the design capacities and capabilities of the permanently installed SSCs in the design, as described in the final safety analysis report, are capable of providing adequate core cooling, containment, and SFP cooling consistent with the requirements of 10 CFR Part 50.155(b)(1)(i) and (c)(1) for 72 hours following a BDBEE.

A–22.   ADEQUACY OF THE CLAIMS ABOUT THE ACCIDENT SOURCE TERMS

This Annex presents the responses provided by the following Member State regulatory bodies to the Question 22: "Describe the challenges associated with regulatory assessment of the adequacy of the claims about the accident source terms."

A–22.1. ARGENTINA–ARN

A–22.1.1. Question

Describe the challenges associated with regulatory assessment of the adequacy of the claims about the accident source terms.

A–22.1.2. Response

The information about sources was submitted in chapter 12 'radiological protection' and assessed by ARN.

The information for the calculation of the source term for the Probabilistic Safety Analysis level II, in general, is not available in basic design stages. Thus, for example, for the calculation of the source term, a detail of volumes, containment structures, and location of ventilation ducts, passages, and doors that are plausible to generate a bypass or containment failure is required, not compatible with conceptual engineering or basic. Given that, the aforementioned information will not be available until detailed engineering is completed.

For the calculation, according to the requirements of construction license, CNEA followed the international recommendation that is to use a 'best estimate' approach in modelling, both deterministic (progression of a severe accident) and probabilistic (event trees), of a severe accident.

No modelling of the response has been made in the case of severe accidents (yes for single failure PIE, and multiple failure PIE) or the progression of the severe accident inside the reactor. Therefore, there are no specific level II simulations for CAREM, and that is why the term source was conservatively estimated from the existing bibliography [Severe Accident Risks: An Assessment for Five U.S. Nuclear Power Plants, NRC (1990) NUREG/CR 1150 [A–277]], considering similar contentions. The results were presented grouped into a single accidental sequence whose annual probability is the sum of all those contributing to core damage and whose source term is the most severe, which is assumed to contain early failure.

To determine the accidental sequences to be evaluated, an analysis of the progression of the severe accident was carried out: in this analysis, an evaluation of the accidental sequences with core damage was carried out using deterministic physical models and their uncertainties for determining the probability of containment failure and size of the source term. Three PIEs were analysed, which were verified with data from other reference studies, interpolated to CAREM 25.

For the estimation of the annual probability of occurrence of the source term, the value of the total annual probability of central damage obtained in the APS level 1 is used and it is assumed that the probability of failure of the containment is equal to 1. Then, the Annual frequency of occurrence is the highest possible for internal events. It is important to mention that this frequency (the highest) is assigned to the worst source term of the analysed sequences.

As a result of the evaluation by ARN, it should be noted that together, both the reactor and the containment of the CAREM 25 reactor have characteristics that allow assuming a similar or better response than current reactors in the event of a severe accident. In particular, it can be mentioned as an intrinsic characteristic of thermal inertia, which provides additional time for the implementation of additional measures and procedures. In addition, specific design and operational measures are implemented to further increase the level of safety. For this, devices are used to contain the molten material inside the pressure vessel, such as external spraying of it, or to limit the concentration of hydrogen, such as catalytic recombinators.

The level 3 Probabilistic Safety Analysis (APS N3) corresponds to the evaluation of the consequences on people and property, outside the plant, as a result of a PSA. In this stage, the analysis of the transport and dispersion of radioactive material in the atmosphere, its transfer through it, and the calculation of the dose in the individuals of the public are carried out. In Argentina, this is required for the licensing of a nuclear installation and is clearly expressed in Standard AR 3.1.3 [A–256], sections 20, 21, and 22.

The input data for APS N3 are the radioactive inventory of the core at the time of reactor extinction, the source term, its probability of occurrence, site meteorological data, and population distribution.

The Argentine acceptability criterion defines the individual radiological risk as to the probability that, during a certain time interval and as a consequence of a postulated event in a nuclear power reactor, an individual located in his environment, who is accidentally exposed to ionizing radiation, receive a given effective dose and die.

For the verification of the AR 3.1.3 [A–256] standard, the individual risk assessment associated with an nth source term was used as the calculation methodology. The environment of the installation was divided into sectors, placing it in the centre and the risk was calculated in each of them, and finally, the maximum value was taken in the entire domain. This calculation methodology presupposes that a possible critical group can be located in each sector of the environment and that is why the risk in each of them is evaluated. Finally, the critical group will be the one located.

A–22.1.3. Follow-up Question

There is a lot of information on the severe accident source term, but is there information available on the DBA source term? (as a metric for level 3 DiD)

A–22.1.4. Response

Yes, the information about sources was submit to ARN in chapter 12 'radiological protection' of Safety Report.

In CAREM 25 by design, level 3 of DiD aims to control both single postulated initiating events (DBA) and multiple failure postulated events (DEC), in order to avoid by design damage to the fuel elements and the pressure boundary, maintaining the effectiveness of the barriers, including containment. In both cases, the objective is to limit the release of radioactive material and prevent the escalation of such events to severe accident conditions.

For the CAREM 25 licensing the requirements called for the development of a preliminary PSA levels 1 to 3, with the objective of demonstrating that the probabilistic standard (standard AR 3.1.3 [A–256]) regulation is met.

The basic acceptance criterion of safety analysis is a requirement of the ARN and is set in the AR 3.1.3 [A–256]. This requirement establishes that, for the licensing of a NPP, compliance with risk acceptability criteria for the public must be ensured, showing that "No accidental sequence — with radiological consequences for the public — should have an annual probability of occurrence that, plotted as a function of the effective dose ... results in a point located in the unacceptable zone of the Criterion Curve". This criterion is based on the evaluation of the radiological risk —probabilistic concept — for the licensing of a NPP based on severe accidents and those postulated events that can produce a dose in the critical group. Basically, this norm stipulates that the product of frequencies and consequences of severe accident sequences need to be below certain threshold. The 'Criterion Curve' of AR 3.1.3 [A–256] standard, establishes a limit to the risk from any accident situation through a probabilistic quantification.

A–22.2. CANADA–CNSC

A–22.2.1. Question

Describe the challenges associated with regulatory assessment of the adequacy of the claims about the accident source terms.

A–22.2.2. Response

The CNSC acknowledges that there are some challenges associated with regulatory assessment of the adequacy of the claims about accident source terms. For example, claims that with a smaller reactor core, there is correspondingly smaller potential radioactivity that could be released. These claims do not always acknowledge considerations of the composition of the fuel and fission product inventories, fuel enrichment level, re-fuelling approach (or even no refuelling), uncertainties in fuel behaviour under accident conditions, release mechanisms, and/or the energies involved.

There have been claims made for reactor technologies such as molten salt reactors, that noble gases would not be released in an accident scenario due to their continuous extraction and management. This might not be realistic to assume when for a number of designs, the fission gases are piped into pressurized tanks for long term storage. This collection system presents vulnerabilities in a design if not qualified properly for external and internal events. It should also be noted that in new designs with limited operational experience, potential for corrosion control issues could influence the nature and extent of accident consequences. To address potential uncertainties regarding accident source terms, CNSC expects applicants to propose appropriate safety and control measures to address the uncertainties.

In Canada, CNSC staff review the source term as early as in the review of the application for a licence to prepare site. Except for acceptance criteria regarding core damage frequency as described in the answer from Canada to Question 21 of this survey, the current expectations regarding accident source term are technology neutral and applicable to SMRs and advanced reactor technologies of all types.

As described in Appendix F.2.2.2 of REGDOC-1.1.1 [A–64]:

"The applicant shall describe:

- the source term (for example, list of radionuclides, magnitude and timing of the release)
- a description of the process followed to arrive at the final list of radionuclides

- where applicable, a justification of the basis for screening out radionuclides that are not included."

More specifically, in review of an application for a Licence to Construct, the review of the deterministic safety analysis and the level 2 PSA includes the review of the postulated source terms. As per REGDOC-1.1.2 [A–66]:

> "The summary results of the probabilistic analyses carried out for the plant should be described in this section, and it should be demonstrated that these results meet the expectations for safety goals contained in section 4.2.2 of RD-337 [A–260] (now REGDOC-2.5.2 [A–62]). The results should be presented in a manner that clearly conveys the quantitative risk measures taken, and the aspects of the plant design and operation that are the most important contributors to these risk measures. This section should identify and refer to the completed plant PSA as a separate document, which should accompany the application."

The CNSC review of source term analysis should address the following elements:

- The source terms estimation;
- The fission product grouping according to the release characteristics;
- Fission product release from fuel;
- Fission product transport inside containment;
- Releases outside containment.

CNSC expectations for accident source term are articulated in REGDOC-2.4.1 [A–63] and REGDOC-2.4.2, Probabilistic Safety Assessment (PSA) for Nuclear Power Plants [A–182] and expand on IAEA SSR 2/1 [A–127] Req.34.

REGDOC-2.4.1. describes how deterministic safety analysis is used to demonstrate safety goals are met by predicted source term and doses during severe accidents.

For normal operation:

> "During the design phase, the normal plant operation is analysed as a separate class of event. This allows sources of radiation or releases of radioactive materials to be assessed in various modes of operation or transition between modes."

For AOOs and DBAs:

As per Canadian expectations, the design developer should predict source term and doses during accidents as a support of demonstration that the specific design meet the dose acceptance criteria.

For severe accidents:

The assessments for BDBAs are aimed to meet risk criteria such as safety goals related to frequency of severe core damage and significant releases of radioactivity.

> "These calculations should demonstrate, for example, that:
>
> - containment failure will not occur in the short term following a severe accident (see REGDOC-2.5.2 [A–62]);

- the public is provided a level of protection from the consequences of NPP operation, such that there is no significant additional risk to the life and health of individuals".

As support of defining specification of the complementary design features for DECs, the reference source term should be calculated for a set of representative accident scenarios based on the best-estimate models. This should take into account the uncertainties of key parameters and the possible changes in governing physical processes.

REGDOC-2.4.2 [A–182] indicates:

"Perform a level 1 and level 2 PSA for each NPP.

Considerations shall include the reactor core and other radioactive sources such as the spent fuel pool. Multi-unit impacts, if applicable, shall be included.

For radioactive sources outside the reactor core, the licensee may, with the agreement of persons authorized by the Commission, choose an alternate analysis method to conduct the assessment."

A–22.3. CHINA–NNSA

A–22.3.1. Question

Describe the challenges associated with regulatory assessment of the adequacy of the claims about the accident source terms.

A–22.3.2. Response

The most challenging aspect for HTR-PM source term analysis was the insufficient understanding of the mechanisms by which radioactive fission products could be released from fuel elements, the primary system and containment, especially under some accident conditions. Thus, for accident source term assessment, very conservative assumptions were chosen.

In order to assess the source term more accurately under accident conditions for HTR-PM, we focus on the following aspects:

(a) For depressurized loss of forced cooling (DLOFC) accident, i.e. depressurization of primary circuit following rapid loss of primary coolant through the break, the transport behaviour and mechanism of graphite dust must be deeply studied. Necessary tests may be arranged to know the dust suspension during various gas flow conditions. Moreover, we need to pay more attention to the distribution of graphite dust in primary system and the absorption of radionuclides by graphite dust during normal operation because these features of dust influence the accident source term to a great extent.

(b) For extreme high temperature up to 1600°C in reactor core under DLOFC accident, diffusion rate of fission product from fuel element must be considered more cautiously. Maybe more experiments are needed to determine the diffusion coefficients for some key nuclides such as Cs-137, I-131, Sr-90 and Ag-110m.

(c) For water ingress accident, i.e. water/steam enter into reactor core from secondary circuit, the interaction mechanisms between steam and gaseous fission products in fuel elements are not clear and need to be investigated more comprehensively. Both experimental study and theoretical simulation must be strengthened to get reasonable release fraction from fuel elements by steam 'wash', especially iodine isotopes release fraction.

A–22.4. CZECH REPUBLIC–SÚJB

A–22.4.1. Question

Describe the challenges associated with regulatory assessment of the adequacy of the claims about the accident source terms.

A–22.4.2. Response

Not applicable — challenges associated with regulatory assessment of the adequacy of the claims about the accident source terms have not been identified (no SMR designs have been assessed in this regard).

In general, the related legislative requirements are formulated as a general ones and are not specific for a certain type of the SMR technology. The establishment of the emergency planning zone (that is established on basis of the analysis and evaluation of radiation extraordinary event — see the answer from Czech Republic to Question 24) shall, inter alia, contain:

- Description of the radiation accident and its scenario considered in the analysis and evaluation of radiation extraordinary event;
- Description of a nuclear installation;
- Description of the time course of the leakage of radioactive substances or the spread of ionising radiation;
- List of released radionuclides and the estimation of their activity in individual periods of time of the leakage.

The nuclear installation design shall comprise the design of containment system capable of preventing radioactive release outside the nuclear installation. It shall also provide for means to protect the primary circuit against over-pressurisation to ensure that there is no radioactive release outside the nuclear installation and into the operational and working space, with the exception of justified and time-limited discharges into systems or spaces inside the containment of the nuclear reactor designed for this purpose, if it is necessary for coping with accident conditions (abnormal operation shall be managed without intervention by these means).

With regard to the technical means for ensuring radiation protection, the nuclear installation premises shall be designed so as to, among other things, prevent a release of radioactive substance from systems, prevent contamination of the workplace and dispersion of radioactive substance into the atmosphere of the workplace, prevent release of radioactive substance outside the nuclear installation and create barriers preventing spread of radioactive material and contamination of persons and objects.

A–22.4.3. Follow-up Question

What challenges do you expect to encounter with regard to accident source term?

A–22.4.4. Response

Currently, no SMRs are planned to be deployed in the Czech Republic. Given the lack of detailed information about various SMR designs and the uncertainty over whether and which type of SMR could be hypothetically deployed, there are no specific expectations with regard to the accident source term.

A–22.5. FRANCE–ASN

A–22.5.1. Question

Describe the challenges associated with regulatory assessment of the adequacy of the claims about the accident source terms.

A–22.5.2. Response

According to the article 3.7 of the Order [A–72], the safety demonstration must include an assessment of potential consequences of incidents and accidents. This assessment must be reasonably pessimistic and must present the retained hypotheses.

Hence, the licensee must determine the inventories of radionuclides that are produced, and the amount of these radionuclides that could be discharged in the environment.

ASN, with the support of IRSN, reviews licensee's assessment of source term and controls the compliance of the potential consequences' evaluation with the safety objectives.

A–22.6. JAPAN–NRA

A–22.6.1. Question

Describe the challenges associated with regulatory assessment of the adequacy of the claims about the accident source terms.

A–22.6.2. Response

In order to make the exposure assessment during severe DBAs, the operating power, fuel burnup, fuel damage rate, emission rate of radioactive noble gas and iodine, deposition rate, and leakage rate from the reactor containment vessel immediately before the accident were set, and the amount and type of radioactive materials released in an accident are assessed.

A–22.6.3 Follow-up Questions

What challenges do you expect to encounter with regard to accident source term?

Please extend more about your experience. Could provide references of the regulatory standards and/or guidelines, examples, etc.?

A–22.6.4. Response

In HTTR, it is confirmed that the fuel in reactor core maintains integrity without causing meltdown and the confinement feature is maintained under the assumption of DEC event.

Therefore, radioactive substances that are assumed to be released in DEC event are focused on iodine and rare gases, which are the same as in a DBA; there was no specific challenge for the approach on assessing the release of radioactive substances.

The concept for the radioactive substances that are assumed to be released in DBA is described in [A–168], and this concept is also used as reference in gas cooled HTTR.

A–22.7. RUSSIAN FEDERATION–ROSTECHNADZOR

A–22.7.1. Question

Describe the challenges associated with regulatory assessment of the adequacy of the claims about the accident source terms.

A–22.7.2. Response

The vessel regulations do not use the 'source term' concept. Nevertheless, the federal nuclear safety regulations establish requirements for source term estimation.

According to Paragraph 14 of NP-022-17 [A–89], the SAR must present the results of deterministic and probabilistic safety analyses. The safety analyses must be conducted for all design-basis operational states of the vessel and take into consideration all vessel areas accommodating nuclear material, radioactive substances, and radioactive waste, where normal operating conditions can be disrupted. A deterministic safety analysis for design-basis accidents must be conducted on the basis of a conservative approach. The software used in the analysis must be certified.

According to Paragraph 15.3.2 of [A–106], accident analysis must provide information on the mathematical models and software used to perform the analysis, in particular, to assess the source term for emergency release.

The SAR must describe the model of analysed processes and list the key physical phenomena driving each process.

The mathematical models describing the fission product transport in the core, circuits and systems of the nuclear power installation must take into account the physical and chemical processes affecting the concentration of radioactive substances in the nuclear installation circuits and premises where the radioactive substances can be released under the analysed accident scenario. The minimum set of these processes must include:

- Natural deposition on the internal surfaces;
- Desorption from the internal surfaces to the steam and gas medium;
- Radioactive decay;
- Leaking with the steam and gas medium through loose areas into the adjacent rooms and the environment, due to the pressure difference;
- Leaking into the environment after pressure balancing due to free convection caused by the temperature difference and different composition of the environment in the room and in the atmosphere;
- Purification of the steam and air medium as it goes through the passive condensation devices (bubblers);
- Purification of the steam and air medium by the sprinkler system;
- Purification of the steam and air medium by the special ventilation system;
- Chemical reactions in water, leading to a change in the physical and chemical properties of the fission products;
- Chemical reactions in the steam and gas phase and on the surfaces leading to a change in the physical and chemical properties of the fission products;

- Water purification to remove the radioactive products. The mathematical models must take into account the behaviour of the aerosols and fission products grouped on the basis of their physical and chemical properties. The considered groups must include:
  - Noble radioactive gases;
  - Volatile (organic and inorganic) forms of iodine.

The mathematical models must use only verified values of coefficients characterising the modelled physical processes (diffusion, desorption, elimination, etc.). Applicability of any new coefficient and credibility of its value must be demonstrated.

The applied mathematical models must use verified values for the adopted weight ratio of radioactive iodine in a molecular form, in the form of organic compounds, and in an aerosol form.

If the model does not take account of individual processes, the conservatism of the analysis must be demonstrated.

A–22.7.3. Follow-up Question

What challenges were encountered, or do you expect to encounter with regard to accident source term?

A–22.7.4. Response

In essence, the approaches to estimation and justification of the accident source term established in the regulations NP-022-17 [A–89] and NP-023-2000 [A–106] cover the specifics of the floating NPP.

There is no intention to make amendments in the existing regulatory framework regarding assessment of accident source term.

A–22.8. SOUTH AFRICA–NNR

A–22.8.1. Question

Describe the challenges associated with regulatory assessment of the adequacy of the claims about the accident source terms.

A–22.8.2. Response

The most important credible form of the PBMR radioactive releases is the accidental gaseous release following loss of coolant events. While early releases are related to the aerosol and dust load of the He coolant and the remobilisation of the dust deposited in the helium pressure boundary (HPB) during operation, the additional source terms of the fuel and of corrosion must be considered for delayed releases.

For these, PBMR-specific source term related phenomena investigations are undertaken, and the appropriate evaluation tools used or developed covering the transport of fission products (FP) within the building and atmospheric dispersion, analysis of the mechanisms relevant for graphite dust production, radioactivity and deposition/remobilisation. Enveloping assumptions are difficult to formulate since high uncertainties have been identified considering the assumptions made for the PBMR. There is some concern internationally concerning the potential for higher accident fuel-temperature ranges for the PBMR than anticipated and, the results are directly related to the licensing criteria and the demonstration of the ALARA principle.

The source term analysis may be divided into analyses under the following five subsection headings[21]:

A–22.8.2.1. Releases from coated fuel particles (CP) and Fuel Spheres

In the PBMR design, the CP form the primary barrier for the fission products generated in the fuel. The following release paths have to be considered for the CP:

- Releases from CP manufacturing defects;
- Releases due to the temperature dependent diffusion rates of individual FP;
- Releases from CP failures caused by irradiation;
- Releases from additional CP failures due to high temperature accidents.

Due to the very large number of CP in the reactor, the assessment of CP failure rates is susceptible to statistical evaluation of manufacturing data and results of irradiation tests and subsequent heating tests. In addition, there must be consideration of temperature dependent diffusion of certain isotopes through intact CP and the diffusion of all releases from the CP and of FP generated outside the CP due to uranium contamination of the matrix material. Contrary to LWR fuel, where design limits of fuel can be determined as discrete limit values, CP failures rates increase with temperature and design limits must be derived failure rate/temperature curves. In addition, temperature related diffusion factors and fraction curves must be considered for source term analysis.

A computer code has been used to model the significant increase of FP release with the operational fuel temperature. The variation of the FP releases with the temperature supports the need for a thorough analysis considering the PBMR may feature regions of increased temperatures caused by increased local power, reduced local coolant flow or pebble bed density.

A–22.8.2.2. Generation and depletion of Fission Products and Dust in the Helium Pressure Boundary (HPB)

Fission and activation products diffuse from the fuel spheres into the coolant and are transported through the HPB. Partly they plate out at colder surfaces or are removed by the purification system.

Apart from the fission products, graphite dust is generated by the friction between the fuel spheres through the pebble bed, the fuel discharge and supply system and by friction with graphite blocks. The dust generated is as contaminated as the fuel spheres surfaces themselves or as the graphite block surfaces. The fission products and the graphite dust are transported through the HPB by the coolant and partly depleted at the HPB surfaces especially in eddy zones and low flow areas. Depleted dust is further contaminated by FP isotopes adsorbed from the coolant while losing activity by decay.

---

[21] The source term analysis applies analogously to used fuel storage systems and other concentrations of radionuclides.

A preliminary evaluation of available past data revealed significant uncertainties which may amount to orders of magnitude, thus indicating that contaminated dust remobilised might be the main source of releases from the reactor.

A–22.8.2.3. Releases from the Helium Pressure Boundary (HPB)

In case of a HPB break, the coolant inventory (operational gaseous contamination, suspended dust and additional contaminants) is released into the building. The break, depending on size and location, may prompt the remobilization of parts of the dust depleted in the HPB and also be released into the building (initial source term). If the HPB break cannot be closed, there may be a delayed release of fission products caused by additional CP failures, increased diffusion rates due to elevated accident temperatures and corrosion products.

It is clear that the analysis will not resolve the issue of high uncertainties on dust contamination and releases. Hence, it will be very important to monitor the production of dust, contamination and deposition during operation of the PBMR and put in place mitigation measures in case that acceptable limits are exceeded. As for the delayed release from the CP, the assumptions on the fraction released need to be reviewed from the break.

A–22.8.2.4. Releases from the Building

In case of leaks, the contaminated dust and gaseous releases from the HPB may be released to the environment through the pressure relief shafts. (Depending on the pressure build up in the building, the designer plans to implement a dust filtration system in the initial phase. After depressurisation the designed is planning the use of a post-event clean-up system. Both systems are intended to minimize releases).

The phenomena to be analysed are similar to the phenomena encountered in other reactor types baring the difference in inventory composition and concentration in the building atmosphere.

The regulatory review comprises a review of the building retention assumptions, filtration measures and independent calculations of the enveloping cases to determine the dose and risk to the workers and the input for atmospheric dispersion calculations. The results are measured against the basic licensing requirements (BLR) [A–107] on dose to the personnel as the major basis for the licence.

A–22.8.2.5. Environmental Dispersion

The atmospheric dispersion is expected to be similar to that of other reactor types, apart from differing isotope inventory. The assessment comprises a review of the environmental dispersion analysis to be submitted by the licensee and independent calculations of the enveloping cases to determine radiation doses and risk to the public. The results are then measured against the BLR [A–107] on dose and risk to the public as a major basis for the nuclear licence.

Scoping calculations are performed to demonstrate the wide range of possible dose results dependent on the assumptions for weather conditions. The results demonstrated the need for a careful definition of conservative weather conditions for the analysis of design basis events.

A–22.9. UNITED KINGDOM–ONR

A–22.9.1. Question

Describe the challenges associated with regulatory assessment of the adequacy of the claims about the accident source terms.

A–22.9.2. Response

ONR does not prescribe the accident source term applicable to nuclear reactor facilities. ONR expects vendor/licensees to develop the relevant source term by appropriate methods, which should be supported and substantiated with evidence.

A–22.9.3. Follow-up Question

What challenges does the ONR anticipate with regard to accident source term?

A–22.9.4. Response

The challenges in this area had been presented in question 20 as they are generally assessed by the same specialists within ONR (Radiological protection inspectors):

- Lack of OPEX;
- Lack of analysis of source term for normal operation and faults;
- Mobility of source term (dust). Particular to HTGRs;
- Fission product release rates from novel fuel;
- Plate out and clean up rates of activation and fission products from coolant in novel designs;
- Performance of novel HVAC and containment/confinement systems in removing airborne activity.

A–22.10. UNITED STATES OF AMERICA–NRC

A–22.10.1. Question

Describe the challenges associated with regulatory assessment of the adequacy of the claims about the accident source terms.

A–22.10.2. Response

The NRC staff did not find any significant challenges in this area of review. As stated in the NRC regulations under 10 CFR [A–52] Part 50.2, 'Definitions' an accident source term refers to "(…) the magnitude and mix of the radionuclides released from the fuel, expressed as fractions of the fission product inventory in the fuel, as well as their physical and chemical form, and the timing of their release." For the SMR design reviewed by the NRC, applicant are required to develop source terms for deterministic accidents for the SMR design that are similar to those used in safety and siting assessment for LLWRs, as described in Chapter 15, 'Transient and Accident Analysis' of NUREG-0800 [A–116]. The DBAs described in the SMR methodology are the main steam line break (MSLB) outside containment, rod ejection accident (REA), fuel handling accident (FHA), steam generator tube failure (SGTF), and the failure of small lines carrying primary coolant outside containment. The SMR methodology also described an iodine spike design-basis source term (DBST), which is a surrogate accident to bound potential accidents with release of the reactor coolant into the containment vessel. In addition, the methodology provided source term and accident assessment methodology for a core damage event (CDE) in which significant core damage is assumed to occur in accordance with the description of the postulated accident fission product release in Footnote 3 to the NRC regulation 10 CFR [A–52] 52.47(a)(2)(iv).

For large LWRs, the accident associated with the siting and safety analysis regulatory requirements with respect to radiological consequences has historically been a postulated LOCA, in which a break in the reactor coolant system (RCS) piping results in the inability of the emergency systems to maintain core cooling with subsequent damage to the reactor core, without damage to the reactor vessel itself and with the containment remaining intact. In general, currently operating power reactors were originally licensed by using the LOCA dose analysis source term described in Atomic Energy Commission Technical Information Document TID-14844, 'Calculation of Distance Factors for Power and Test Reactor Sites' dated March 23, 1962 (ADAMS Accession No. ML021720780) [A–278], which is also listed as a reference in 10 CFR [A–52] 100.11, 'Determination of Exclusion Area, Low Population Zone, and Population Center Distance' for the siting requirements for power reactors licensed before January 10, 1997. In 1995, the NRC published NUREG-1465, 'Accident Source Terms for Light-Water Nuclear Power Plants' [A–279] which described revised accident source terms for LWRs. NRC Regulatory Guide (RG) 1.183, 'Alternative Radiological Source Terms for Evaluating Design Basis Accidents at Nuclear Power Reactors' [A–280] provides guidance on acceptable use of alternative source terms based on NUREG-1465 in DBA radiological consequence analyses in licensing actions for power reactors. The DBA LOCA source terms in TID-14844 [A–278] and RG 1.183 [A–280] are not intended to reflect a specific LOCA scenario, but each is intended to represent a conservative surrogate accident based on a spectrum of break sizes up through the double-ended guillotine break of the largest RCS piping. The radiological consequence analysis of this accident is intended to evaluate the performance of the containment and release mitigation systems and to evaluate the proposed siting of the facility.

The SMR design did not include large RCS piping; therefore, the accident scenario that would result in a fission product release to containment consistent with the regulatory requirements would not be the same as for the LLWR LOCA. Instead, to address the regulatory requirements, the SMR applicant proposed a methodology to develop a core damage source term (CDST) based on several severe accident scenarios that result in core damage, taken from the design-specific probabilistic risk assessment (PRA). This CDST is the surrogate radiological source term for a CDE.

A–23.   PROVISION FOR INSPECTION OF REACTOR INTERNALS, CIVIL STRUCTURES AND ALL THE SSCS, INCLUDING INNOVATIVE INSPECTION APPROACHES

This Annex presents the responses provided by the following Member State regulatory bodies to the Question 23: "Describe the challenges associated with regulatory assessment of the provision for inspection of reactor internals, civil structures and all the SSCs, including innovative inspection approaches."

A–23.1. ARGENTINA–ARN

A–23.1.1. Question

Describe the challenges associated with regulatory assessment of the provision for inspection of reactor internals, civil structures and all the SSCs, including innovative inspection approaches.

A–23.1.2. Response

The ageing management program, the in-service inspection program (ISI), the RPR surveillance program, etc. These are some of the requirements that the RE of the CAREM 25 project must present for the issuance of the Commissioning License and the Operating License.

During the licensing process and from the design stage, an important exchange is taking place between both the regulator and the Responsible Entity work areas, promoting fluid communication between the parties, in order to facilitate the discussion and management of possible objections that arise.

For example, the RPR surveillance program presented in the technical specification that the RE sent to the manufacturer of the RPR was evaluated in a previous stage. This evaluation allowed detecting some inconsistencies, which allowed establishing new requirements that were added to the ET. It also works in a similar way with aging management, in order to establish guidelines that avoid future inconsistencies.

ARN's experience in licensing other diverse national projects, which present problems with old or outdated regulations, becomes very important in new projects.

The licensing scheme proposed by ARN for the CAREM 25 reactor construction project, aims to accompany the development of the project so as to detect, early, aspects that could affect the licensing process in the future.

A–23.2. CANADA–CNSC

A–23.2.1. Question

Describe the challenges associated with regulatory assessment of the provision for inspection of reactor internals, civil structures and all the SSCs, including innovative inspection approaches.

A–23.2.2. Response

CNSC expects SSCs important to safety to be designed so they can be inspected over the lifetime of the plant to standards commensurate with the safety significance of their functions.

For discussion on cases where standards for periodic inspection are either not fully applicable or absent, please refer to the answer from Canada to Question 12.

For SSCs important to safety that cannot be designed to support the desirable testing, inspection, or monitoring schedules, one of the following approaches needs to be taken:

- Proven alternative methods, such as surveillance of reference items, or use of verified and validated calculation methods shall be specified;
- Conservative safety margins are applied, or other appropriate precautions to compensate for possible unanticipated failures.

SMRs may present a number of challenges associated with regulatory assessment of the provision for inspection of reactor internals, civil structures and other SSCs. In many cases, due to the compact design of integral SMRs and difficulty in accessing requisite components, vendors are opting for alternative methods and/or conservative safety margins in lieu of a typical periodic testing regime. Canadian experience with the CANDU, NRU and SLOWPOKE reactors have demonstrated that it is possible to design a reactor core and it's support systems with limited access for in-service inspection and operate it safely; however, it is important to have leak detection or other monitoring systems in place that could identify potential issues before safety was compromised.

Innovative approaches can be developed to ensure in-service inspection capability in challenging conditions, even after new degradation phenomena were identified during plant operation. The objective with advanced reactors is to identify these inspections approaches before first operation.

Some vendors have proposed sealed core configurations where reactor internals would not be inspected over the life of the plant. In such cases, the design is expected to include adequate safety margins and account for projected degradation and aging mechanisms for the entire life of the plant.

Innovative inspection approaches have also been proposed by existing licensees in Canada, so this challenge is not necessarily unique to SMRs. CSA-N287, the industry standard governing In-service inspections for containment structures in Canada includes provisions for use of remote operating vehicles, video equipment, and telephoto lenses, provided that the clarity and resolution of the photographic images is acceptable.

A–23.3. CHINA–NNSA

A–23.3.1. Question

Describe the challenges associated with regulatory assessment of the provision for inspection of reactor internals, civil structures and all the SSCs, including innovative inspection approaches.

A–23.3.2. Response

According to the requirements of NNSA, a series of inspections and tests are carried out to ensure the integrity of the structure and pressure boundary of nuclear safety related equipment. The comprehensive inspection and test is called pre-service inspection before the equipment is put into operation. Equipment components may be affected by temperature, stress, irradiation, hydrogen adsorption, corrosion, vibration, wear and other factors, resulting in changes in material properties of components, so in-service inspection is required during the operation of nuclear power plant.

The licensee must work out an in-service inspection program in accordance with the requirements of NNSA and the relevant documents provided by the manufacturer. It must specify the pre-service inspection completed before operation and all inspections and tests which will be carried out during the operation life of the nuclear power plant unit. Its main contents include the selection of inspected parts and areas, the determination of inspection type and inspection cycle, as well as inspection methods and techniques. The in-service inspection program including implementation documents and inspection results report should be submitted to NNSA.

The reactor internals and SSCs of small reactors will be fully inspected before putting into operation. However, the accessibility required by the in-service inspection of some SSCs cannot be met due to the integrated compact design adopted in the design of the small reactor. The operation organization applies for exemption for the in-service inspection of these SSCs according to the reliability of the equipment and the safety impact after the defect occurs.

A–23.3.3. Follow-up Question

It would be interesting to learn about compensatory measures in situations where it is difficult or impossible to inspect SSCs.

A–23.3.4. Response

Based on the reliability of the equipment and the safety impact of the defects, the operating unit applies for a waiver of in-service inspection of these SSCs.

A–23.4. CZECH REPUBLIC–SÚJB

A–23.4.1. Question

Describe the challenges associated with regulatory assessment of the provision for inspection of reactor internals, civil structures and all the SSCs, including innovative inspection approaches.

A–23.4.2. Response

Not applicable — no challenges associated with regulatory assessment of the SSCs have been identified as no SMR is currently in operation or under construction in Czech Republic.

Current legislative requirements contained in the Atomic Act [A–21] and the implementing legislation related to inspection activities are relatively general and technologically neutral and provide sufficient flexibility for the deployment of various SMR technologies. However, the main challenges associated with the selected equipment and other SSCs are mainly to be expected if novel or very innovative technologies are used in a situation when there is no or very limited experience with the application in nuclear sector. These new features would certainly create a challenge and would require a regulatory judgement and potentially an innovative inspection approach from the regulatory perspective. As no SMR is planned to be deployed in Czech Republic, only a general description of the legislative requirements is provided.

The licence holders are obliged to continuously monitor the state of the nuclear installation and its systems, structures and components in terms of the implementation of the controlled ageing process in accordance with the controlled ageing programme (from the commencement of construction to decommissioning of the nuclear installation). Starting with the application for licence for construction of a nuclear installation the licencing documentation contains the controlled ageing programme (preliminary, pre-operational, operational).

During the commissioning and operation of selected equipment plans and programmes for commission and operation of the selected equipment shall be drawn up, implemented, and maintained; activities shall be performed in accordance with commissioning programmes for the selected equipment; activities shall be planned and performed in accordance with internal rules and other operating documentation, including plans and programmes for operating checks of selected equipment, the pre-operational and operational managed ageing programme, and plans and programmes for checks during maintenance, repairs, or modifications to selected equipment, and with technical requirements specified in the legislation.

During the operation of selected equipment, within the scope of an implemented controlled ageing process for selected equipment, its condition shall be systematically monitored and the impact of ageing and the effect of degradation mechanisms that could lead to defects and a reduction of the technical safety level of the selected equipment shall be determined. Verification of conformity of the selected equipment with technical requirements shall be documented by a comprehensive set of documentation of the verification of conformity that contains, among other things: an operating check programme, plans for operating checks, operational controlled ageing programme, accompanying technical documentation for selected equipment, documentation applicable to the preparation and performance of repairs and maintenance of selected equipment and records of operating checks and other tests. An operating check programme shall in particular contain a list of individual selected equipment broken down by selected equipment type and operating check programmes of individual pieces of selected equipment.

The ageing management process shall, inter alia, involve definition of the rules and criteria for the selection of systems, structures and components subject to ageing management process, identification of the degradation mechanisms, definition of monitored parameters, definition of acceptance criteria, monitoring and determination of the development of the degradation mechanisms and the impacts of ageing and ensuring of the early detection and monitoring of degradation mechanisms and impacts of ageing.

A–23.5. FRANCE–ASN

A–23.5.1. Question

Describe the challenges associated with regulatory assessment of the provision for inspection of reactor internals, civil structures and all the SSCs, including innovative inspection approaches.

A–23.5.2. Response

According to the article 2.5.1 of the Order [A–72], and as mentioned in question 15, important to safety SSCs qualification's durability relies, among other things, on maintenance.

Based on a failure mode and effects analyses and to guarantee the capacity of SSCs to perform their functions if needed, the licensee must:

- Realise an adequate maintenance of the SSCs;
- Monitor and inspect the ageing of the installation, especially for critical components that are difficult to replace;
- Test the important to safety SSCs. The periodicity of the tests is defined in the general operating rules.

Regarding pressure equipment, specific requirements and in service follow-up are realised by ASN, licensee and approved bodies. To control the compliance of a nuclear pressure equipment, ASN assesses using and maintenance recommendations provided by the manufacturer.

ASN doesn't formally approve the licensee's maintenance program but are assessed as part of the general operating rules attached in the commissioning file. However, the licensee must provide sufficient information to explain its approach to define the maintenance program.

Regarding maintenance, periodic testing and qualification, ASN conducts inspections to control the compliance with regulatory requirements and licensee's general operating rules. Also, ASN controls licensee's oversight on activities realised by subcontractors as required by the article L.593-6-1 of the Environmental Code [A–24].

ASN assesses installations' ageing through inspections, but also specific instructions, especially during the periodic safety reviews required by the article L.593-18 of the Environmental Code [A–24]. At this occasion, the licensee must reassess the state of its installation vis-a-vis applicable requirements.

A–23.6. JAPAN–NRA

A–23.6.1. Question

Describe the challenges associated with regulatory assessment of the provision for inspection of reactor internals, civil structures and all the SSCs, including innovative inspection approaches.

A–23.6.2. Response

It is required by the Ordinance concerning the installation and operation of research reactors, which was enforced in April 2020, for operators to establish a facility management implementation plan that includes inspections detail and frequency. Therefore, it will be checked after the establishment of the plan.

A–23.6.3. Follow-up Questions

What challenges do you expect to encounter with regard to inspection of SSCs?

Please, could extend more about your experience? Could provide references of the regulatory standards and / or guidelines, examples, etc.

A–23.6.4. Response

We have not reviewed any plan for inspection because concrete plan for certain facility has not been submitted.

A–23.7. RUSSIAN FEDERATION–ROSTECHNADZOR

A–23.7.1. Question

Describe the challenges associated with regulatory assessment of the provision for inspection of reactor internals, civil structures and all the SSCs, including innovative inspection approaches.

A–23.7.2. Response

In developing the designs for ships and other vessels with nuclear installations, including the designs of small nuclear power plants with SMRs, the requirements of the current regulatory legal acts of the Russian Federation provide for the fulfilment of the requirements related to the assessment of performance of the safety-significant systems as well as the requirements for the management of their resources. However, these activities are mostly carried out by the specialists of the operating organization or other external organizations that hold the required Rostechnadzor licenses. In particular, the provisions of NP-022-17 [A–89] stipulate compliance with the following requirements:

- The systems and components important for safety must undergo direct and full compliance testing during commissioning, after modifications and maintenance, and periodically throughout their lifetime. If direct and/or full verification is not possible, indirect and/or partial testing must be conducted. (Paragraph 52, NP-022-17).

- Prevention of abnormal operation is achieved by keeping the safety-significant systems and components in good operating condition, due to timely flaw detection, implementation of preventive measures, service life monitoring, organization of effective technical maintenance, and documenting of the results of work (Paragraph 9, NP-022-17).

- Actuation of the protective safety systems must not result in any failures of the normal operation systems and/or components. The permissible number of actuations for the protective safety systems (including spurious actuations) within the service life of the ship nuclear installation must be substantiated in the design on the basis of their impact on the remaining life of the systems and mechanisms of the nuclear installation (Paragraph 96, NP-022-17).

- Location of the reactor and associated systems and/or components on the ship must be substantiated in the ship design with due regard for the reactor specifics; in this case access to the equipment in the course of maintenance and safety of the ship crew and special personnel must be ensured (Paragraph 110, NP-022-17).

A–23.7.3. Follow-up Question

What challenges were encountered, or do you expect to encounter with regard to inspection of SSCs?

A–23.7.4. Response

In general, the existing inspection process established to verify the technical condition and availability of safety-significant systems of the marine nuclear power installations is applicable to the floating NPP and covers its specifics.

There is no intention to make amendments in the existing regulatory framework regarding inspection of the technical condition and availability of the safety-significant systems, including the reactor internals.

A–23.8. SOUTH AFRICA–NNR

A–23.8.1. Question

Describe the challenges associated with regulatory assessment of the provision for inspection of reactor internals, civil structures and all the SSCs, including innovative inspection approaches.

A–23.8.2. Response

The inspections programme must define the requirements that must be met in order to maintain the reliability of the important to safety SSC at the levels assumed in the SAR.

The traditional LWR approach to inspection and testing requirements has consisted of a somewhat loose collection of periodic operational surveillance tests (OTS), in-service inspection/testing requirements (e.g. ASME XI), and monitoring inputs to 'Predictive' maintenance programmes (RCM).

These approaches have not always produced coherent surveillance programmes commensurate with the SSC importance to safety and investment protection. Such nuclear power plant surveillance requirements are generally focused on LWR practices and may not be applicable to the PBMR. The LWR surveillance Programme requirements originated in an era without the analytical tools that are currently available, i.e. for probability analysis, and failure modes/frequencies.

International licensing practice has been evolving to replace those requirements with criteria based on more logical methodologies, which provides the opportunity for SMR proposals to develop an integrated Programme that modifies the traditional approach by also applying modern risk based analytical techniques.

It is necessary that adequate surveillance requirements be developed to describe and schedule inspections and tests that predict, prevent or mitigate the effects of failure mechanisms, and which provide continued assurance of the reliability and integrity of all applicable SSC. It follows that the Programme must include inspections and tests that might previously have been grouped under the subheadings ISI/T, OTS, performance monitoring, condition monitoring programmes, and maintenance programme inspections.

Requirements:

The technical surveillance programme shall demonstrate that the failures, system disturbances, and transients that are described and analysed in the SAR have been taken into account in the development of the requirements. These reference sources must be identified in the bases for the inspections and tests and must include the assumptions made on the operating duty of SSC arising from different system operating configurations.

The programme for the construction stage shall provide details applicable to technical surveillance performed during installation and commissioning up to fuel loading, and proposals for future stages.

The programme shall describe the following:

- The bases for the inspections and tests and maintenance, including any special requirements related to FOAK SSC (e.g. helium pressure boundary);
- The means of compliance with requirements;
- The frequency of inspection and tests;
- The initial conditions, required operations and return to service requirements;
- Special tools and equipment requirements;
- Equipment and personnel safety;
- Control of cleanliness and consumables;
- Special process qualification requirements;
- The skills, training and personnel qualification requirements;
- Treatment of test & inspection results, and follow-up actions;
- Measures to ensure operability after repair and replacement of SSC.

A–23.8.3. Follow-up Questions

Please discuss any anticipated challenges with regard to inspection of SSCs.

Please, could provide information on how the requirements were assessed in the case of the PBMR?

A–23.8.4. Response

The online refuelling would possibly introduce dynamic loads and stresses on the core structure ceramics requiring the need for routine inspections of the structures to determine integrity of these internal structures.

Furthermore, maintenance outages were planned for every five years. This in itself would require that the internal structures must maintain its integrity over long operating periods.

The dust source term as well as activation products would present worker dose challenges during inspection activities.

The contamination of the turbine due to the direct cycle also introduced concerns with worker dose during outages.

It was required that the DPP should have adequate provision for monitoring and inspections as well as instrumentation. In light of these challenges, the NNR developed RG-0005 [A–34] on Testing, Qualification and Commissioning of the PBMR DPP.

A–23.9. UNITED KINGDOM–ONR

A–23.9.1. Question

Describe the challenges associated with regulatory assessment of the provision for inspection of reactor internals, civil structures and all the SSCs, including innovative inspection approaches.

A–23.9.2. Response

The ONR SAPs [A–44] provide a number of general and high level expectations for examination, inspection, maintenance and testing (EIMT) of SSCs in SAPs EMT 1 to 8 and on ageing and degradation in SAPs EAD 1 to 5.

However, some of the characteristics of SMR technologies have highlighted the potential challenges to these SAPs, for example, reactor designs with sealed cores; lack operating experience that could inform the plant performance and ageing and degradation predictions and monitoring/inspection issues due to opaque coolant.

Structural Integrity expectations related to metal SSCs are included in SAPs EMC.1 to EMC.34 and include specific principles on inspection, testing and maintenance of reactor internals that supplement the more general principles outlined in SAPs EMT 1 to 8. They would be applied to all metal SSCs, including rector internals if metallic. They are supported by the technical assessment guides (TAGs [A–46]) on Integrity of Metal Structures, Systems and Components and Examination, Inspection, Maintenance and Testing of Items Important to Safety. A general expectation is that inspection should be carried out with a capability and regularity that is commensurate with their safety classification in accordance with SAP ECS.3.

SMR technologies may require additional inspection requirements due to constraints to access for inspection due to novel component designs. Furthermore, different coolant properties may require novel inspection technologies. It should be noted that, where classification dictates that inspection is required, ONR would expect components to be designed to facilitate sufficiently reliable inspection over the plant lifetime. Inspection programmes should take account of anticipated degradation mechanisms, and for components important to safety ONR would expect speculative inspections where novelty of design gives rise to significant uncertainty regarding the potential for degradation.

Metal SSC 'Highest reliability components' (see the answer from United Kingdom to Question 10) may require additional though life inspection commensurate with their reliability claim, including qualified volumetric examinations.

For cores containing graphite SAPs EGR 1 to 15 and the ONR TAG on Graphite Reactor Cores are relevant. This guidance provides reference to the degradation mechanism of graphite oxidation in CO_2, resulting in weight loss and challenges to core integrity. However, for SMRs, further OPEX and information on HTGR or MSR specific degradation mechanisms such as helium migration, molten salt flow erosion and high temperature degradation will be required in order to adequately justify inspection, testing and maintenance requirements. Furthermore, novel inspection technologies and approaches may be required to inspect in these environments.

High level civil engineering expectations on inspection, testing and monitoring of civil engineering structures are described in SAPs ECE 1 to 26 and in the ONR technical assessment guide NS-TAST-GD-020 [A–258]. It should be noted that the expectations on containment inspections, monitoring and testing are based on Advanced Gas Reactor pressure vessels and PWR type containment. However, independently of the type of the reactor the inspection, monitoring and testing regime needs to demonstrate that the structure continues to meet its safety functional requirements.

Particular challenges for civil engineering structures are the lack of access for inspections, reliability of monitoring techniques and use of novel materials.

ONR has participated in the IAEA activity to develop a TECDOC on the applicability of design safety requirements (SSR-2/1, Rev. 1 [A–127]) to small modular reactor technologies intended for near-term deployment, which covered both small modular light water reactors and high temperature gas-cooled reactors (HTGR). SSR-2/1 Req. 31 on Ageing Management remains applicable to these technologies and it is in line with the intent of the ONR SAPs.

A–23.10.   UNITED STATES OF AMERICA–NRC

A–23.10.1. Question

Describe the challenges associated with regulatory assessment of the provision for inspection of reactor internals, civil structures and all the SSCs, including innovative inspection approaches.

A–23.10.2. Response

The differences in the sizes and designs (SMRs to LLWRs) of the components and lack of operating experience made applying the existing regulations, codes, and standards challenging. In some areas during the staff's recent SMR design review, ASME Code [A–215] Section XI was a good fit, and with others it was more challenging with the NRC staff evaluating each component to determine if existing regulations, codes, and standards were appropriate.

Nuclear power plants periodically use in-service inspection (ISI) and in-service testing (IST) to assess the structural and leak-tight integrity of the reactor coolant pressure boundary (RCPB) throughout the operating lifetime of the facility. As required by 10 CFR [A–52] Part 50.55a(g)(3), reactor designs certified on or after July 1, 1974, are required to provide access to enable the performance of ISI of ASME Code [A–215] Class 1 RCPB components. Typically, a design should be developed that implements an ISI program consistent with the provisions of ASME Code [A–215] Section XI, Division 1, as supplemented by augmented ISI requirements in 10 CFR 50.55a. However, based on the specific attributes of a reactor design, additional augmented ISI may need to be proposed, and designed, to support compliance with applicable general design criteria, which requires the RCPB to be "(…) designed, fabricated, erected, and tested so as to have an extremely low probability of abnormal leakage, of rapidly propagating failure, and of gross rupture."

The SMR applicant's design of the RCPB incorporated provisions for access to enable the performance of ISI examinations in accordance with 10 CFR 50.55a(g)(3) and ASME Code, Section XI. The final ISI program is required to meet the latest ASME Code, Section XI, edition and addenda incorporated by reference 18 months before the date scheduled for initial loading of fuel.

Suitable equipment will be developed and installed to facilitate the remote inspection of these areas of the RCPB that were not readily accessible to inspection personnel. The staff concluded that the description of the preservice inspection (PSI) and ISI programs was acceptable and met the inspection and testing requirements of applicable general design criteria and 10 CFR 50.55a. This conclusion was based on the applicant meeting the requirements of ASME Code, Section XI, Division 1.

The reactor coolant system (RCS) also provided for the circulation of the primary coolant. The applicant's design relied on natural circulation flow for the reactor coolant and did not include reactor coolant pumps or an external piping system. The RCS included the reactor vessel and integral pressurizer, the reactor vessel internals, the reactor safety valves, RCS piping inside the containment vessel. The staff concluded that the applicant had met the requirements of 10 CFR [A–52] Part 50.55a, specific to the RCPB, for the construction of systems, structures, and components important to safety to quality standards by ensuring that RCPB components, as defined by 10 CFR 50.55a, were classified properly as ASME Code [A–215] Section III, Class 1 (Quality Group A) components. This SMR design did not describe whip restraints for any Class 1 piping. This was deemed acceptable as the design did not include any significant length of Class 1 piping that would require a whip restraint.

This Annex presents the responses provided by the following Member State regulatory bodies to the Question 24: "Describe any changes in the approach to establishing emergency planning zones for SMRs. Describe the challenges associated with regulatory assessment of the adequacy of the claims about the emergency planning zones associated with the project."

A–24.1. ARGENTINA–ARN

A–24.1.1. Question

Describe any changes in the approach to establishing emergency planning zones for SMRs. Describe the challenges associated with regulatory assessment of the adequacy of the claims about the emergency planning zones associated with the project.

A–24.1.2. Response

Onsite and off-site emergency plan to respond in case of nuclear or radiological emergency at NPPs, is required to the operator by the Regulatory Body. This plan must contemplate the response actions to be taken within the installation as well as the necessary to be implemented offsite.

The Emergency Plan includes all aspects related to the necessary strategy to control, mitigate and limit the consequences in the event of an emergency and establishes the automatic measures for the protection of the population and the actions to be implemented by the response organizations.

The main important issues are:

- ARN advises the Executive Branch and national, provincial and municipal organizations on the issues of their incidence, including radiological and nuclear emergencies;
- ARN must provide protection against harmful effects of ionizing radiation, even under emergencies;
- ARN approves the procedures and emergency plans, including emergency plans developed by local, provincial and national authorities, as well as training plans and training plans for members of the public near the NPPs;
- ARN coordinates the representatives from response organizations in relation to the protection actions necessary in case of a nuclear accident.

Urgent protective actions and other actions that must be performed once declared the emergency in the NPP are stablished in the approved emergency plans. For that purpose, zones are predefined as following:

- Precautionary action zone (PAZ): Is the area enclosed by the 3 km radius from the NPP;
- Urgent protective action planning zone (UPZ): Defined as the area enclosed between the 3 km to 10 km from NPP. It is being considerate, under IAEA post-Fukushima Daiichi NPP recommendations, to extent this zone and include the 360° around the NPP.

In addition to these zones, JOEN considers the following zone for the application of other measures:

- Extended complementary planning zone : This zone covers areas beyond the UPZ zone and is limited by the results of radiological monitoring. In the extended complementary planning zone other measures are defined, other than those applied in PAZ and UPZ. Among them, the instruction to reduce accidental ingestion, restriction of consumption of certain foods, decontamination, etc.

For the issuance of the construction license for the CAREM 25 reactor, the ARN established as a requirement, the assessment of the viability of the Emergency Plan considering that the site is next to the Atucha nuclear power plant.

In the Regulatory Framework of Argentina, the Emergency Plan is a requirement to obtain the operating license and applies to CAREM 25 at that stage. The same requirements established to the operational NPPs will be considered to CAREM 25 reactor.

From the beginning of the construction and commissioning project of CAREM 25, the authorities of the CNEA contacted those of Nucleoeléctrica Argentina S.A. (NA-SA) to agree on measures of mutual coordination and assistance between the two organizations at the Atucha Nuclear Site. As a result, CNEA and NA-SA signed a Framework Agreement on November 11, 2009, in order to establish the aforementioned formal relations of mutual cooperation and assistance, including Site logistics and community relations.

During the construction stage, the aforementioned agreement covers the preparation of the site and the construction of the CAREM 25 facilities, without the introduction of nuclear material.

Emergencies that may involve site personnel are those that could occur at the Atucha NPPs, emergencies due to physical protection issues, and conventional accidents on the premises itself.

During the construction period, all the personnel that carries out their tasks in the site of the construction of CAREM 25, will submit to the authority of the NA-SA during the management of nuclear emergencies, for this purpose it was agreed that the CAREM 25´s site is incorporated into the NA-SA emergency plan.

Regarding effective means of communication, it is worth mentioning that the participation of CNEA personnel performing tasks on the premises during the Atucha 1 emergency drills, allowed the identification of mutual coordination measures in the implementation of warning, transport, and training systems.

In order to adequately fulfil their own physical security needs and not constitute an additional risk to the NA-SA nuclear power plants, coordination measures were agreed between the CNEA and the NA-SA.

A–24.1.3. Follow-up Question

Did you have any challenges with regard to demonstrating an acceptable EPZ for CAREM?

A–24.1.4. Response

No, we don´t had significant challenges. For the CAREM 25 licensing, the ARN regulation for nuclear power plants was applied, which is in line with what is established by the IAEA. The particular case in which the EPZ is limited according to the source term and risk of a specific design of reactor was not studied.

A–24.2. CANADA–CNSC

A–24.2.1. Question

Describe any changes in the approach to establishing emergency planning zones for SMRs. Describe the challenges associated with regulatory assessment of the adequacy of the claims about the emergency planning zones associated with the project.

A–24.2.2. Response

Technology developers have claimed capability to reduce EPZ size, taking into account technology improvements. In some cases, reactor vendors are making the claim for a 0 metre emergency planning zone.

Before describing the challenges associated with the assessment of claims about emergency planning zones, first the terms 'exclusion zone' and 'emergency planning zone' will be defined, and a schematic is shown in the Fig. A–16.

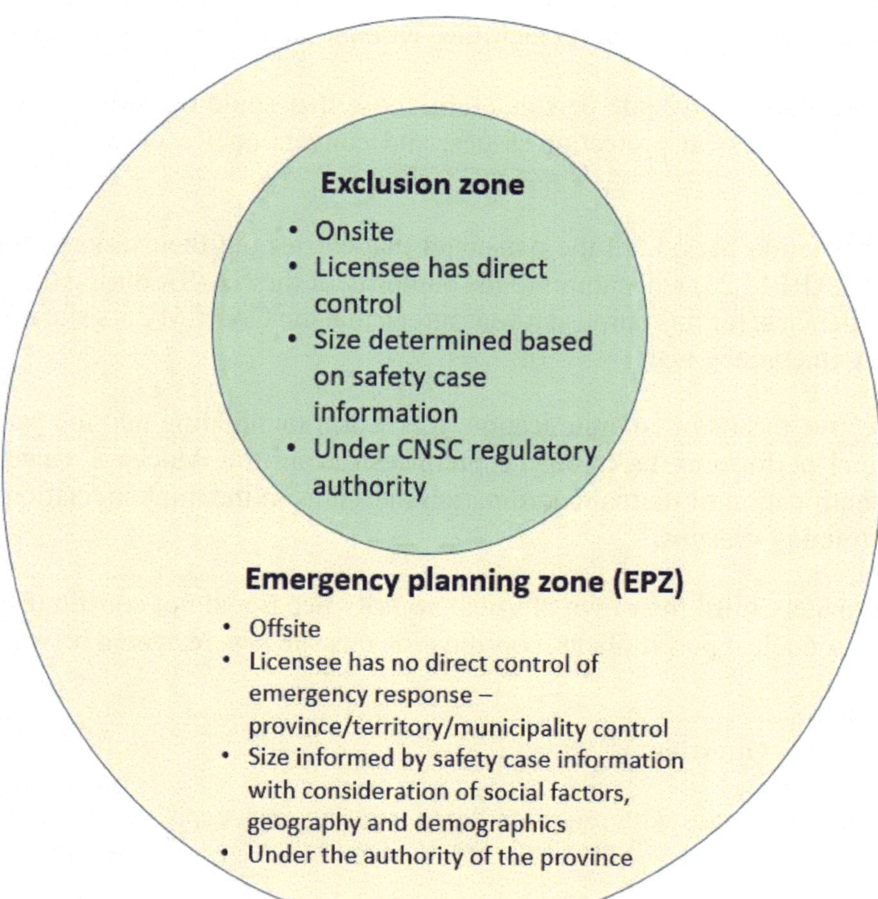

FIG. A–16. Planning zones – Canada.

Exclusion zone: Per Class I Nuclear Facilities Regulations [A–60] – Section 1, an 'exclusion zone' is an area of land within or surrounding a nuclear facility whereby permanent dwellings are not allowed and over which the licensee retains the legal authority to exercise control. REGDOC-1.1.1 [A–64] gives further details on the exclusion zone.

EPZ: The EPZ is defined as the area in which, to protect public health, safety, and the environment, it may be required the implementation of operational and protective actions during a nuclear emergency. An EPZ is normally controlled by an external emergency planning authority and encompasses emergency measures to be implemented and executed by that authority.

In Canada, there are no regulatory nor legislative requirements for the size of the EPZ and, therefore there are no restrictions on minimum EPZ size. Hence, EPZ and other emergency planning actions should be commensurate to the risks associated with the activity or technology. The results from safety analyses (i.e. the PSA) in combination with the protection strategy used by offsite planners ought to determine the size of the EPZ. This approach is consistent with the documented IAEA methodologies.

CNSC regulatory framework in areas such as physical design of reactor facilities (REGDOC-2.5.2 [A–62] and RD-367 [A–61]) and safety analysis (REGDOC-2.4.1 [A–63] and REGDOC-2.4.2, Safety Analysis: Probabilistic Safety Assessment (PSA) for Nuclear Power Plants [A–182]) provides requirements and guidance for applicants on key areas to support the methodology for determining the EPZ size). This guidance is also applicable to new reactor designs they are intending for deployment in Canada. The Class I Nuclear Facilities Regulations [A–60] require a licence application for a nuclear reactor to demonstrate that the design is suitable for specific site and regional characteristics. Composite bounding designs submitted as a bounding approach are possible; however, the applicant is limited to the projected releases as set in the environmental assessment and confirmed at the time of the construction licence review.

The mechanistic source term approach described here provides an opportunity for these design developers to realistically assess the radiological consequences of an accident and may allow reduced EPZs and smaller plant sites. Some designers proposed mechanistic source term approach evaluation as a support of EPZ definition.

The site evaluation plays an important role in the postulated initiating events identification for the specific site. The CNSC sets out requirements and guidance on-site evaluation for new NPPs. In addition to the documents mentioned above, further information on-site evaluation is contained in REGDOC-1.1.1 [A–64]. In the construction licence application, the estimates of releases and disturbances used in risk modelling are confirmed when the design and safety features of the NPP have been confirmed. The licensee is then expected to re-evaluate risk modelling as operating experience is gained over the facility lifetime. CNSC staff then review re-evaluated risk models as necessary.

As per REGDOC-2.10.1 [A–217], applicants and licensees need to provide offsite emergency authority with key information for the offsite planning. This information includes various results of safety analyses, which planners utilize for the establishment of the EPZs and response plans. Also, applicants and licensees need to work with and support relevant offsite organizations, such as municipalities and provincial governments, to develop an effective onsite and offsite emergency response plan. All these information are inputs to the decision on the size of the EPZ.

In addition to regulatory framework of the CNSC, the Canadian Standards Association (CSA) Group also maintains standards supporting emergency planning. For instance, CSA N1600-16 [A–281] General Requirements for Nuclear Emergency Management Programs is in-line with the IAEA standards, as listed in Table A–13.

TABLE A–13. GENERAL REQUIREMENTS FOR NUCLEAR EMERGENCY MANAGEMENT PROGRAMS

IAEA	CSA N1600 [A–281]	CSA N1600 definition
Precautionary Action Zone	Automatic Action Zone	A pre-designated area immediately surrounding a reactor facility where pre-planned protective actions would be implemented by default on the basis of reactor facility conditions [Source: Modified IAEA Safety Guide GS-G-2.1 [A–282]]
Urgent Protective Action Planning Zone	Detailed Planning Zone	A pre-designated area surrounding a reactor facility, incorporating the automatic action zone, where pre-planned protective actions are implemented as needed on the basis of reactor facility conditions, dose modelling, and environmental monitoring, with the aim of preventing or reducing the occurrence of stochastic effects. [Source: GS-G-2.1 [A–282]]
Extended Planning Distance	Contingency Planning Zone	A pre-designated area surrounding a reactor facility, beyond the detailed planning zone, where contingency planning and arrangements are made in advance, so that during a nuclear emergency protective actions can be extended beyond the detailed planning zone as required to reduce potential for exposure. Note: Contingency planning and arrangements in the contingency planning zone would be less detailed and have less specificity than the planes in the detailed planning zone. [Source: Modified IAEA EPR-NPP Public Protective Actions [A–283]]
Ingestion and Commodities Planning	Ingestion Control Zone	A pre-designated area surrounding a reactor facility where plans or arrangements are made to: a) Protect the food chain b) Protect drinking water supplies c) Restrict consumption and distribution of potentially contaminated produce, wild grown products, milk from grazing animals, rainwater, animal feed d) Restrict distribution of non-food commodities until further assessments are performed Note: Wild-grown products can include mushrooms and game. [Source: Modified IAEA EPR-NPP Public Protective Actions [A–283]]

A–24.2.3. Follow-up Question

Do you anticipate any challenges with regard to demonstrating an acceptable EPZ for SMRs under consideration?

A–24.2.4. Response

Yes, to date there is insufficient information on accidents and malfunctions and accident releases to support EPZ sizing. See information regarding in the answer from Canada to Question 22 on accident source term.

A–24.3. CHINA–NNSA

A–24.3.1. Question

Describe any changes in the approach to establishing emergency planning zones for SMRs. Describe the challenges associated with regulatory assessment of the adequacy of the claims about the emergency planning zones associated with the project.

A–24.3.2. Response

Small reactors are small in size, low in core power and good in safety, which creates conditions for reducing the emergency planning area technically. And NNSA stated attitude about EPZ of SMR in Ref. [A–70], Part7.

> "The design of small PWR should ensure that the effective dose and thyroid equivalent dose which individuals (adults) on the boundary of the site may receive are lower than the general optimization intervention level of concealment and iodine protection in case of accident. In other words, it should be designed to make sure the off-site emergency can be not necessary."

A–24.3.3. Follow-up Questions

Please state the criteria used for EPZ determination and were there any challenges with regard to demonstrating the off-site response was/is not necessary.

Could you explain in more detail how the change developed and how it is based?

How is it applied to floating plants?

A–24.3.4. Response

As the HTR-PM and ACP100 are located at the same site as other large commercial PWRs, the size and supervision of their EPZs has been covered by that of large commercial PWRs.

A–24.4. CZECH REPUBLIC–SÚJB

A–24.4.1. Question

Describe any changes in the approach to establishing EPZs for SMRs. Describe the challenges associated with regulatory assessment of the adequacy of the claims about the emergency planning zones associated with the project.

A–24.4.2. Response

Not applicable — there have been no changes in the approach to establishing the EPZ for SMRs. As there was no assessment of any SMR design, no challenges associated with regulatory assessment of the adequacy of the claims about the emergency planning zone associated with the nuclear installation project have been identified.

In general, the requirements related to radiation extraordinary event management (including EPZ) contained in Czech legislation are general in nature and thus not limited to certain kind of technology used. The radiation extraordinary event analysis that is to be conducted by the licence applicant is the cornerstone in this area (it is necessary for the categorization of potential events, determining threat category, establishment of the emergency planning zone, etc.).

The area of the emergency planning zone is defined as an area in which (based on radiation extraordinary event analysis and assessment) the requirements for preparation for taking urgent protective action, other measures to protect the general public as a result of the expected exceedance of reference levels and other measures to protect the general public apply. It is established as a circle — the centre corresponds to the centre of the smallest circle, which includes the projection of the floor plan of all the building with nuclear reactor (or all buildings with nuclear reactor in case of multi-unit NPP) and the radius of the circle equals to the distance at which the need for planning the introduction of urgent protective measures is not eliminated for a radiation accident with a frequency of occurrence higher than or equal to 1×10^{-7}/year.

According to the legislative requirements, when radiation extraordinary event analysis and assessment are prepared, the possibility of simultaneous occurrence of a radiation extraordinary event on two and more nuclear reactors located on the nuclear installation grounds shall be taken into account.

A–24.4.3. Follow-up Question

Please state the dose criteria used for EPZ determination.

Do you anticipate any challenges with regard to demonstrating an acceptable EPZ for SMRs under consideration?

A–24.4.4. Response

The EPZ is defined in the Atomic Act [A–21] as the area surrounding the nuclear installation grounds in which (based on radiation extraordinary event analysis and assessment) the requirements for preparation for taking urgent protective action, other measures to protect the general public as a result of the expected exceedance of reference levels and other measures to protect the general public apply. The urgent protective action comprises in particular sheltering, iodine prophylaxis and evacuation (according to the Article 104 of the Atomic Act).

The requirements for establishing EPZs are in the Article 4 of the Decree No. 359/2016 Coll. [A–284], on details of ensuring radiation extraordinary event management. The area of the emergency planning zone shall be established as a circle. The centre of the circle corresponds to the centre of the smallest circle, which includes the projection of the floor plan of the building with nuclear reactor or, where appropriate, all buildings with nuclear reactors located on the nuclear installation grounds. The radius of the circle equals to the distance at which the need for planning the introduction of urgent protective measures is not eliminated for a radiation accident with a frequency of occurrence higher than or equal to 1×10^{-7}/year. Radiation accident in this context means a radiation extraordinary event that cannot be handled by forces and means of the operators or shift personnel of the person whose activities gave rise to the radiation extraordinary event.

According to the Article 107 of the Decree No. 422/2016 Coll., on Radiation Protection and Security of a Radioactive Source [A–285], urgent protective measures shall be always implemented if doses absorbed in organs could exceed, over less than 2 days in any individual, the levels specified in Annex 29 to this decree (absorbed dose that is assumed or expected to be received over the course of less than 2 days [Gy]: Whole body 1; Lungs 6; Skin 3; Thyroid gland 5; Lens of the eye 1.5; Gonads 1). A justified urgent protective measure refers to: 1. sheltering (if the averted effective dose is greater than 10 mSv over the period of sheltering lasting no longer than 2 days); 2. stable iodine administration (if the internal contamination by radioactive iodine is imminent and the averted committed equivalent dose in the thyroid gland caused by iodine radioisotopes is greater than 100 mSv); or 3. evacuation (if the sum of the effective dose so far received in an emergency exposure situation when taking into account the effect of the already implemented protective measures and the effective dose, which could be averted, is greater than 100 mSv over the first 7 days). For reducing accidental exposure of the public in a radiation accident, the values of the selected directly measurable quantities (operational intervention levels) are determined in Annex 9 of the Decree 359/2016 Coll. [A–284], above which the introduction of urgent protective measures shall be taken into account (value of photon or ambient dose equivalent rate measured at a distance of 1 m above the contaminated ground equal to: for evacuation 1 mSv/h; for sheltering 0.1 mSv/h; for use of iodine prophylaxis in releases containing radioactive iodine 0.1 mSv/h).

Currently, no SMRs are planned to be deployed in the Czech Republic. Given the lack of detailed information about various SMR designs and the uncertainty over whether and which type of SMR could be hypothetically deployed, there are no specific expectations with regard to demonstrating acceptable EPZs for SMRs.

A–24.5. FRANCE–ASN

A–24.5.1. Question

Describe any changes in the approach to establishing emergency planning zones for SMRs. Describe the challenges associated with regulatory assessment of the adequacy of the claims about the emergency planning zones associated with the project.

A–24.5.2. Response

Since the application for a construction license, the applicant shall provide all relevant information for the offsite authorities assess or make an informed decision on the EPZ, such as the source term and accident sequences. The calculations need to be included in the safety case. This is applicable for any kind of basic nuclear installation, including SMRs.

The internal emergency plan must be attached to the commissioning file.

A–24.6. JAPAN–NRA

A–24.6.1. Question

Describe any changes in the approach to establishing emergency planning zones for SMRs. Describe the challenges associated with regulatory assessment of the adequacy of the claims about the emergency planning zones associated with the project.

A–24.6.2. Response

At nuclear facilities, only the EPZ was previously set up in order to prepare for the protective measures to be implemented at the initial stage of emergencies, and a radius of about 1500 m was used as a guideline EPZ for research reactors where the thermal output is in the range 10–50 MW. (However, about 200 m in radius was set for the HTTR).

After that, based on the lessons learned from the Fukushima Daiichi NPP accident, a nuclear disaster countermeasure priority area was set, and instead of EPZ, Precautionary Action Zone (PAZ) and Urgent Protective action planning Zone (UPZ) has been applied respectively.

At present, for research reactors with a heat output of more than 10 MW and less than or equal to 100 MW, in accordance with the threat category of IAEA Safety Standard GS-G-2.1 [A–282], only the UPZ is set because on-site events that could give rise to severe deterministic health effects off the site are not postulated, and as a guideline, a radius of approximately 5 km is set as the UPZ.

The nuclear disaster countermeasure priority area for each facility category was established after discussions by the study team set up within the NRA. However, due to the diversity of design and nuclear fuel materials etc. used in research reactors, there was a proper discussion about the suitability of establishing a nuclear disaster countermeasure priority area by conforming to the threat category based only on thermal output.

A–24.6.3. Follow-up Questions

Please state the dose criteria used for determination of the various planning zones.

What was the EPZ change for HTTR based on?

Please, provide information about the discussions of the study team of NRA and the challenges they faced.

A–24.6.4. Response

Emergency Planning Zone (EPZ)

The radius was set by following:

The preliminary EPZ was set by estimating the release amount of radioactive substances outside the said EPZ resulting in 10 mSv per day of Gamma exposure for external whole-body dose and 100 mSv per day of equivalent dose (by iodine) for paediatric thyroid — they were the minimum value in radiation protection index of Japan.

The radius with margin was set by evaluating the amount above exceeded the total amount of radioactive substances to be released on the maximum credible accident — that was evaluated in the safety review for siting — to a great degree.

Precautionary Action Zone (PAZ)/Urgent Protective action planning Zone (UPZ)

In light of the lessons learned from the Fukushima Daiichi NPP accident, the concept of PAZ and UPZ was introduced as the priority area that should take countermeasures for emergency preparedness against nuclear disaster (called 'nuclear disaster countermeasure priority area'), replacing the preceding EPZ.

Guideline of the PAZ for commercial power reactor was set to 'approximately 5 km radius from commercial reactor facility' considering following issues:

The area with dose level that cause deterministic effect (bone marrow dose of 1 Gy for external acute exposure) — which IAEA set as a standard for determining the PAZ — was estimated approximately within 3 km radius from nuclear facility, by probabilistic analysis.

Also, IAEA standards established the maximum radius for PAZ to be set within 3–5 km radius from nuclear facility (5 km is recommended).

Guideline of UPZ for commercial power reactor was set to 'approximately 30 km radius from commercial reactor facility' considering the following 4 issues comprehensively:

(a)   In the Fukushima Daiichi NPP accident, the point that reached 1000 μSv/h of dose — OIL (operational intervention level) that requires immediate evacuation or sheltering preferably in substantial buildings which IAEA set — were almost remained within the boundary of the NPP;

(b)   In the said accident, the point that reached 100 μSv/h of dose — OIL that requires temporary relocation which IAEA set — were almost within 30 km from the NPP;

(c)   Using source term analysis on severe accident and dose estimation by the result of the analysis, compared with the IAEA criteria, the area that required evacuation or sheltering remained within approximately 10 km and area that required iodine thyroid blocking remained approx. 30 km from nuclear facility;

(d)   Maximum radius for UPZ was to be set between 5—30 km in the IAEA standards.

As for setting the nuclear disaster countermeasure priority area for research reactors, dose criteria have not been used directly. Threat category was decided based on thermal output, as provided in IAEA standard, and maximum radius corresponding to each thermal output range in the same standard was set as the nuclear disaster countermeasure priority area.

Guideline for the EPZ for research reactors has been determined on the basis of its thermal output, however, EPZ was individually reviewed and set for each facility that has special condition, such as:

- JRR-4 (EU / LW swimming pool type research reactor);
- HTTR (LE-UO$_2$ coated fuel particle / graphite moderated / helium gas cooled high temperature engineering test reactor);
- FCA (EU and Pu fuel / horizontal, split-table type critical assembly);
- Toshiba NCA (LEU /LW heterogeneous critical assembly).

As a result of individual review, approx. 200 meter radius for HTTR had been set considering its design and operational characteristics.

However, HTTR does not currently set the PAZ and the UPZ is set to radius of 5 km from the facility as nuclear disaster countermeasure priority area. Please refer to the following response for the details.

Former guideline for nuclear disaster countermeasure priority area of research reactors had been set based on the thermal output of each reactor as in IAEA standard, however, categorization method and thermal output range for each category had been different from IAEA standard.

Therefore, the 'Study team on Nuclear Emergency Preparedness Measures' had reviewed the concept for setting nuclear disaster countermeasure priority area and concluded following basic concept:

Threat category of research reactors was decided based on thermal output, which is consistent with IAEA standards. The maximum radius corresponding to each thermal output range in the same IAEA standards, was introduced as the guideline of nuclear disaster countermeasure priority area for the relevant facility.

In the IAEA standards, it is to be applied to the reactor which is in continuous operation in certain thermal output.

On the other hand, some research reactors may operate under different conditions depending on its research and test purpose, and the operation may not be long and continuous.

Considering such characteristics, on the basis of application of IAEA standards, thermal output limitation — stipulated in the operational safety program for each research reactor under the assumption of continuous operation in constant thermal output — is used in determining threat categories.

As a result, the UPZ is set to 5 km radius and the PAZ has not been set for HTTR.

The Study team also pointed out the needs for consideration on the diversity in research reactors.

And the team discussed on the importance of considering the characteristics of reactor type and location of each facility as well - for swift and effective protective action - when establishing nuclear disaster countermeasure priority area based on the concept for the guideline said above.

This context is also provided in Guideline for Emergency Preparedness and Response (NRA EPR guide) [A–286].

A–24.7. RUSSIAN FEDERATION–ROSTECHNADZOR

A–24.7.1. Question

Describe any changes in the approach to establishing EPZs for SMRs. Describe the challenges associated with regulatory assessment of the adequacy of the claims about the EPZs associated with the project.

A–24.7.2. Response

According to Paragraph 165 of NP-022-17 [A–89], an action plan must be developed and available for implementation to protect the workers (personnel) in case of emergency occurring in the course of the designed vessel operation, when the vessel is on its permanent base or in a temporary mooring location, or in an authorized port, including shipyards (before fuelling).

The requirements for the planning of the measures for actions and protection of the personnel in case of a nuclear or radiological emergency on the vessel are set in NP-079-18 [A–104]. According to these requirements, the action plan section addressing the personnel protection measures in case of a nuclear or radiological emergency on a floating nuclear plant vessel must be developed with due regard of the provisions of the federal rules and regulations in the field of the use of atomic energy 'Standard content of the action plan for protection of personnel in the event of accident at nuclear power plant' (NP-015-12) [A–287] approved by Rostechnadzor Order No. 518 of September 18, 2012 (registered by the Ministry of Justice of the Russian Federation on February 12, 2013 under reference No. 27011, Bulletin of Federal Agency Regulations No. 16, 2013).

NP-015-12 [A–287] sets requirements for establishing EPZs. Thus, the size of the zones, prediction of external and internal exposure doses to the personnel in the event of a severe BDBA (the BDBA is defined by the nuclear plant designer) must be established assuming the worst weather conditions. The contamination zones, doses of external and internal exposure in case of a beyond design basis accident must be calculated, considering the following:

(a) The rationale for the choice of the BDBA.

(b) The scenario of the selected BDBA. The radionuclide composition and activity of radionuclide release at the early phase of the accident, starting with its onset (until confinement (elimination) of the source term).

(c) Analysis of the radiological consequences of the selected BDBA and the calculation results for the predicted doses of external and internal exposure. To estimate the radiological consequences, the external and internal doses are calculated using the conservative approach.

(d) The potential number of the affected personnel who may need medical assistance, considering the severity and type of exposure.

(e) Assessment of the radiological situation in the premises of the nuclear power installations, Main Control Room, Emergency Control Room, at the site, and in the buffer zone.

(f) The key calculation data for the zones, their boundaries, and characteristics (contamination zones, zones for protective action planning, zones for planning the actions on mandatory evacuation of the personnel, contamination levels on the zone boundaries) under assumed meteorological conditions.

(g) The key measures to protect the personnel based on the calculation of consequences predicted for the selected beyond design basis accident.

(h) Conclusions: The calculated sizes of the zones in case of a radiological emergency are charted on the zoning plan of potential hazardous contamination, indicating the prevailing winds ('wind rose'), average speed of the surface wind, average air temperature, and symbols.

A–24.7.3. Follow-up Question

Have you had any challenges with regard to demonstrating an acceptable EPZ for the floating SMR or other SMRs under consideration?

Please, provide information on how the emergency is managed in the different ports in relation to the emergency management tasks with other organizations, drills, etc.

How is it applied to floating plants?

A–24.7.4. Response

Owing to the experience Rostechnadzor gained in reviewing the emergency response and personnel protection plans for nuclear and radiological accidents (including those on vessels and other watercrafts with nuclear reactors), there have been no difficulties in reviewing the emergency response plans of the floating NPP.

There is no intention to make amendments in the existing regulations regarding prevention and management of emergencies.

A–24.8. SOUTH AFRICA–NNR

A–24.8.1. Question

Describe any changes in the approach to establishing emergency planning zones for SMRs. Describe the challenges associated with regulatory assessment of the adequacy of the claims about the emergency planning zones associated with the project.

A–24.8.2. Response

At the time of the PBMR project (up to 2010), the NNR requirements for emergency planning were contained in RD-0014 [A–113].

In addition to the information contained in RD-0014, the following is stated here:

According to RD-0018 [A–107], emergency or remedial measures must be considered where there is a potential for the off-site annual individual effective doses to the public to be more than 1 mSv. According to RD-0014 [A–113], urgent protective measures must be taken if the intervention levels (10/50 mSv) can be achieved. It is noted that the filtered depressurisation function is designed for breaks >10 mm diameter. It was the NNR position that for high frequent leaks and breaks (e.g. $P \approx 10^{-2}$/a) the off-site annual individual effective doses to the public will not be more than 1 mSv (emergency preparedness level not achieved), that dose constraints for higher frequency category B events be defined so that the intervention level for sheltering (10 mSv) according RD-0014 will not be challenged, and that the intervention level for evacuation (50 mSv) will not be achieved even for low frequency category B events. The strategy must recognise these limitations.

PBMR (Pty) Ltd was required to motivate that the source term on which the EPZ for the Demonstration Power Plant is based, taking into account the chosen confinement design performance, does not require an extension of the established Emergency Planning Zones, and does not contradict the EP requirements described in the Safety Concept Requirements Specification.

PBMR (Pty) Ltd was therefore required to provide a deliverable document for this purpose that addresses the following:

- Derivation of the Source term assumed for EPZ;
- Rules used for the Analysis i.e.:
    - Use of deterministic analysis methods;
    - Use of best estimate methods for calculating source terms and meteorology;
    - Assumptions for availability of SC-H and SC-M SSC safety functions;
    - Assumptions for mitigation actions such as food ban, iodine prophylaxis, sheltering or evacuation;
    - Characteristics assumed for the reference population for deterministic effects;
    - Characteristics assumed for the reference population for stochastic effects.

Amongst others, the following acceptance criteria for radiological safety of new reactor concepts had to be fulfilled:

- Exclusion of off-site emergency measures (such as evacuation);
- Reduction of emergency planning zones;
- Limitation of emergency measures.

The mitigation measures shall be designed based on the representative Severe Accidents to be addressed as BDBAs.

A–24.8.3. Follow-up Question

Were there challenges with regard to demonstrating an acceptable EPZ for the PBMR?

A–24.8.4. Response

Any potential challenges with regard to demonstrating an acceptable EPZ for the PBMR were masked by the main challenge of not a yet sufficiently mature safety analysis to identify such challenges.

A–24.9. UNITED KINGDOM–ONR

A–24.9.1. Question

"Describe any changes in the approach to establishing emergency planning zones for SMRs. Describe the challenges associated with regulatory assessment of the adequacy of the claims about the emergency planning zones associated with the project."

A–24.9.2. Response

No changes are proposed to the approach for establishing emergency planning zones to specifically address SMRs.

Regulation of emergency planning in the UK is governed by Radiation (Emergency Preparedness and Public Information) Regulations 2019 (REPPIR 2019) [A–288] supported by an Approved Code of Practice and Guidance [A–266]. A review of REPPIR 2019 and the implications for SMRs in the UK (supported by information provided to ONR by vendors as part of the SMR engagement project and the Advanced Modular Reactor (AMR) Feasibility & Development (F&D) Project) was conducted in 2020 as part of the ONR's ANTs project.

REPPIR 2019 [A–288] is based upon a risk framework, with the level of preparedness and the planning zones required being proportionate to the likelihood and consequences of an accident with off-site consequences.

Review of REPPIR 2019 revealed the following key challenges relevant to SMRs. If the regulations are deemed to apply then the operator must conduct a hazard evaluation and consequence assessment, including sensitivity studies. Information provided by some vendors includes claims that no postulated credible events would result in radiological consequences large enough to require offsite emergency planning. Some vendors explicitly identify the goal to limit any detailed planning zone to within the site boundary.

Two categories of planning zones are defined within REPPIR 2019 [A–288], a Detailed Emergecy Planning Zone (DEPZ) and an Outline Planning Zone (OPZ). The extent of the OPZ is defined by REPPIR 2019 based upon the nature of the site. For operating nuclear power plants, the OPZ is defined to be 30 km with no consideration of reactor power. The extent of the DEPZ is informed by the consequence assessment and the REPPIR 2019 risk framework.

However, it is identified within the REPPIR 2019 Guidance that operating reactors are expected to have at a minimum a DEPZ in-line with international guidance and standards produced by IAEA. Current international guidance identifies a DEPZ of at least 0.5 km for reactors of over 100 MW thermal power (GS-G-2.1 [A–282]).

A–24.9.3. Follow-up Question

Do you anticipate any challenges with regard to demonstrating an acceptable EPZ for SMRs under consideration?

A–24.9.4. Response

As per previous response, Regulation of emergency planning in the UK is governed by Radiation (Emergency Preparedness and Public Information) Regulations 2019 (REPPIR 2019) [A–288] supported by an Approved Code of Practice and Guidance [A–266]. A review of REPPIR 2019 and the implications for SMRs in the UK (supported by information provided to ONR by vendors as part of the SMR engagement project and the Advanced Modular Reactor (AMR) Feasibility & Development (F&D) Project) was conducted in 2020 as part of the ONR's ANTs project. REPPIR 2019 is based upon a risk framework, with the level of preparedness and the planning zones required being proportionate to the likelihood and consequences of an accident with off-site consequences.

The review of REPPIR 2019 revealed some key challenges relevant to SMRs and these have been provided previously:

- If the regulations are deemed to apply then the operator must conduct a hazard evaluation and consequence assessment, including sensitivity studies. Information provided by some vendors includes claims that no postulated credible events would result in radiological consequences large enough to require offsite emergency planning. Some vendors explicitly identify the goal to limit any detailed planning zone to within the site boundary.
- Two categories of planning zones are defined within REPPIR 2019, a Detailed Emergency Planning Zone (DEPZ) and an Outline Planning Zone (OPZ). The extent of the OPZ is defined by REPPIR 2019 based upon the nature of the site. For operating nuclear power plants, the OPZ is defined to be 30 km with no consideration of reactor power. The extent of the DEPZ is informed by the consequence assessment and the REPPIR 2019 risk framework. However, it is identified within the REPPIR 2019 Guidance that operating reactors are expected to have at a minimum a DEPZ in-line with international guidance and standards produced by IAEA. Current international guidance identifies a DEPZ of at least 0.5 km for reactors of over 100 MW thermal power (GS-G-2.1 [A–282]).

A–24.10.   UNITED STATES OF AMERICA–NRC

A–24.10.1. Question

Describe any changes in the approach to establishing emergency planning zones for SMRs. Describe the challenges associated with regulatory assessment of the adequacy of the claims about the emergency planning zones associated with the project.

The NRC's existing emergency preparedness program for nuclear power plants is focused on LLWRs. Based on the challenges of the applicants SMR designs, the NRC is proposing to amend its regulations to create an alternate emergency framework for SMRs and other new technologies. The alternative requirements and implementing guidance would adopt a risk-informed, performance-based, technology-inclusive, and consequence-oriented approach. The alternative requirements would include a scalable approach for determining the size of the plume exposure pathway emergency planning zone around each facility. The NRC is interested in addressing specific emergency preparedness policy issues such as:

- How planning activities should apply to the performance-based approach;
- How should hazard analysis be applied to the performance-based approach;
- What specific factors or technical considerations are needed when applying the scalable emergency planning zone approach.

A–25. STAFFING LEVELS OF MULTI-UNITS WITHIN A FACILITY

This Annex presents the responses provided by the following Member State regulatory bodies to the Question 25: "Describe any changes in the approach to establishing staffing levels of multi-units within a facility. Describe the challenges associated with regulatory assessment of the adequacy staffing levels of multi-units within a facility."

A–25.1. ARGENTINA–ARN

A–25.1.1. Question

Describe any changes in the approach to establishing staffing levels of multi-units within a facility. Describe the challenges associated with regulatory assessment of the adequacy staffing levels of multi-units within a facility.

A–25.1.2. Response

Not applicable.

A–25.1.3. Follow-up Question

Do you anticipate having any challenges with regard to demonstrating acceptable staffing levels for prototype and commercial deployment of CAREM?

A–25.1.4. Response

Yes, it's an area that in the future will surely present challenges in staff training and qualification as CAREM 25 reactor is a FOAK NPP.

Regulatory Standards AR 0.11.1. [A–289] and AR 0.11.2. [A–290] set the criteria and procedures to provide Individual Licenses and Specific Authorizations (two kinds of conceptually different documents, which imply certifications) to the personnel who apply for licensable functions in nuclear installations. These regulatory standards also establish terms and conditions according to which the ARN may issue these Individual Licenses and Specific Authorizations. In addition, Regulatory Standard AR 0.11.3 [A–291] establishes criteria on retraining of personnel for this type of installations.

The applicant for an Individual License, Specific Authorization or for the renewal of the latter must fulfil a number of requisites concerning qualification, working experience, training, retraining and psychophysical aptitude, depending on the installation and the function.

In case of licensing of staff of CAREM 25 the working experience in the area of reactor operation that the personnel in charge of carrying out this task should have, could be presented as a challenge, as it's a FOAK reactor.

A–25.2. CANADA–CNSC

A–25.2.1. Question

Describe any changes in the approach to establishing staffing levels of multi-units within a facility. Describe the challenges associated with regulatory assessment of the adequacy staffing levels of multi-units within a facility.

A–25.2.2. Response

In Canada, minimum staff complement applies to the entire facility. The CNSC has many years of experience regulating multi-unit facilities in Canada. The regulatory experience for existing multiple-unit facilities should be applicable to multi-module SMR deployment. Regulatory requirements and expectations are documented in REGDOC-2.2.5 [A–161].

A number of SMR developers are seeking to develop technologies that reduce the need for onsite human support in a facility, such as:

- Instrumentation and control architectures to replace the need for field surveillance by onsite personnel;
- Reactor safety characteristics that reduce the need for human intervention or provide for long response times by plant operators [22] under anticipated plant operating conditions.

The approach to the regulatory assessment of the adequacy of the staffing levels of multi-unit facilities can be described as follows:

- CNSC expects a licensee to conduct and maintain a systematic analysis to determine the basis of the minimum staff complement while considering:
  - o The most resource-intensive events and credible failures considered in the Safety Analysis and the PSA;
  - o Required actions;
  - o Operating strategies;
  - o Required interactions among personnel;
  - o Staff numbers, competencies, qualifications and workload demands associated with the required tasks;
  - o Staffing strategies under all operating conditions including normal operation, AOO, DBA and emergency conditions.
- Validation to demonstrate safe operation and response to the most resource-intensive conditions (including events that affect more than unit) under all operating states including normal operations, AOO, DBA and emergency conditions.

Several regulatory challenges exist in the assessment of an application for new technologies and for potentially different concepts of operation. While Canada has experience with the current technologies, the lack of this experience for the newer technologies may pose a challenge for determining staffing levels. For staffing considerations, this will require detailed task analysis and novel approaches for validation (such as task simulations in mock-ups or simulators and task network modelling) in the absence of operating experience or analogous information for these new technologies. In addition, the safety characteristics of the new reactor designs will be a key factor to determine the required human actions, which are likely to differ from those needed in the current NPPs.

Currently NPPs exist mainly to produce electrical power. Future SMRs may produce electrical power but a facility may also carry out additional missions such as district heating or desalination. This would add additional tasks to what the control room operators do in current NPPs. Whether a single facility with multiple missions would have the same control room

---

[22] Response time by plant operators.

operator controlling all those process applications, or whether a separate operator would be assigned to each type of process operation, would need careful examination and analysis.

Additionally, for multi-mission facilities, if load following is proposed, consideration would be needed for how such an operation would be staffed, given the impacts on the multiple missions. Careful analysis and demonstration that this would not adversely affect safe operations would be required.

For any reactors operated remotely, especially those in regions that are not immediately accessible, staffing would need to address security and emergency response personnel and where they would be located. In the current situation, emergency response personnel (who are part of the minimum staff complement) are located on-site or close by. For remote communities the response capability might be limited, so the number of emergency response personnel, where they are located, how they are counted in the minimum staff complement, and their training/qualification would need careful consideration.

Currently Canadian NPPs have an operator(s) dedicated to refuelling, either on-line or off-line. Novel technologies may involve novel refuelling methods which might be done only occasionally, or even rarely. This has implications for the staffing approach to handling such refuelling, and the maintenance of appropriate qualification of the staff for doing such work.

Several of the SMR vendors have indicated that additional reactor units could be added as the facility need arises. Questions exist concerning the staffing and management of several units in different stages of operation. Whether one operator would monitor multiple units or whether each unit would have its own dedicated operator remains to be decided. Additionally, methods would need to be developed to demonstrate that acceptable situation awareness will exist for staff handling multiple units that are potentially in different operational states.

A–25.3. CHINA–NNSA

A–25.3.1. Question

Describe any changes in the approach to establishing staffing levels of multi-units within a facility. Describe the challenges associated with regulatory assessment of the adequacy staffing levels of multi-units within a facility.

A–25.3.2. Response

The organization of nuclear installation shall be determined according to the scale of the power plant, number of existing units and planned units, engineering characteristics, etc., and according to the relevant requirements of nuclear safety regulations and nuclear industry standards. As the SMRs built in China are all located at site with other big reactors, any staffs and organizations are sharing.

The organizational requirements for individual SMR and Floating Reactors are under study.

The control room staffing rules are based on the experience of existing large-scale light water reactors at present, which rely on operator monitoring and active safety system control for normal and abnormal operation and accident mitigation. Advanced reactor design, advanced automation technology, and the elimination or significant reduction of the use of active safety systems are introduced into small reactors design. These optimizations may change the role, responsibility, composition and scale of operators. In the design of small reactor, measures of accident prevention and human factors engineering are used to reduce the task burden of control room operators, and the active safety system is not relied on excessively for normal, abnormal or accident operation. In addition, the number of small reactor's total system is far less than that of large light water reactors. It can significantly extend the response time of personnel after the accident, thus reducing the necessary control actions of operators.

In this case, each small reactor applicant is required to carry out task analysis for all design accidents, identify appropriate staffing, and determine the functions assigned to the control room operator. The introduction of advanced reactor design and increased use of automatic control systems can have a significant impact on accident analysis, and ultimately affect the role, responsibility, composition and scale of the staff needed to control plant operations. Because of the design differences between the small reactor and the previously approved LWR, the small reactor may need fewer operators to perform the same task. For multiple reactor units, task analysis needs to consider operating multiple units in different operation modes. Not only the actions needed to operate the unit should be defined, but also the interaction with other field maintenance and support organizations of multiple units should be carried out in the accident analysis.

This situation will bring more complex work to nuclear safety supervisors, who need to understand the whole design, operation concept of each reactor type and the role and responsibility of operators in the accident. The applicants should establish staffing guidelines to better define the scope of tasks that operators need to perform in a comprehensive accident analysis in order to help auditors work.

A–25.4. CZECH REPUBLIC–SÚJB

A–25.4.1. Question

Describe any changes in the approach to establishing staffing levels of multi-units within a facility. Describe the challenges associated with regulatory assessment of the adequacy staffing levels of multi-units within a facility.

A–25.4.2. Response

Not applicable — no challenges associated with regulatory assessment of the adequacy staffing levels of multi units within a facility have been identified. There have been no changes implemented in the approach to establishing staffing levels of multi-units within a single facility. SÚJB has experience with existing multiple unit facilities on the same site (Dukovany power plant has 4 × VVER-440 and Temelín power plant 2 × VVER-1000). Each unit is licenced as a separate nuclear installation in line with the definition of nuclear installation (facility or plant comprising a nuclear reactor). If an SMR multi-unit facility would be deployed, the adequacy of staffing levels of multi-unit plants would be done in the same way.

According to the Atomic Act [A–21] a licence holder for the construction of a nuclear installation is obliged to ensure that already reached level of safety of another nuclear installation located in the site for a nuclear installation under constructions is sited does not degrade (human resources included).

The licence holders are obliged to ensure and make use (inter alia) of the human resources, including suitable working environment, which are essential for ensuring and increasing the level of safety. Adequacy of staffing levels is continuously assessed during commissioning of a nuclear installation to ensure the readiness for further operation. Adequate staffing and system of education, training and exercise and qualification shall be described in the licensing documentation.

At present — in case of currently operated NPPs — the minimum number of shift personnel is specified in the licence document 'Limits and Conditions for Safe Operation'. These minimum numbers are verified as part of emergency exercises, their scenarios can combine accident conditions on several units, accidents covering the whole site, etc. Any changes related to changes in the number of employees, or the volume of the activities affecting nuclear safety are subject to safety assessment. The assessment of the category 'Human Factor' includes assessment of the ability to ensure the performance of activities important to safety from the perspective of the sufficiency of persons, their substitutability, working mode (accumulation of activities, overload, stress...), the potential for human error, qualification, ability to maintain and disseminate knowledge and expertise, etc. In the case of evaluation that there could be deterioration, the change is either not implemented or such measures are taken to eliminate the negative effects.

The issue of the sufficiency of human resources is covered by SÚJB inspections in the area of the Integrated Management System. This area also forms a natural part of other types of inspections, which may give rise to other activities in this area.

A–25.4.3. Follow-up Question

What challenges have you had with regard to staffing levels for the SMRs under consideration?

A–25.4.4. Response

As there are currently no SMRs planned to be deployed in the Czech Republic and given the lack of detailed information about various SMR designs and the uncertainty over whether and which type of SMR could be hypothetically deployed, there are no specific expectations with regard to staffing levels for SMRs.

A–25.5. FRANCE–ASN

A–25.5.1. Question

Describe any changes in the approach to establishing staffing levels of multi-units within a facility. Describe the challenges associated with regulatory assessment of the adequacy staffing levels of multi-units within a facility.

A–25.5.2. Response

In French NPPs, some staff can be shared between two different units in the same plant.

The article 2.5.5 of the Order [A–72] stipulates that important to safety activities must be realised by trained staff with necessary skills and qualifications.

Furthermore, the article 2.5.2 of the same Order [A–72] stipulates that important to safety activities must be realised in conditions that enable to satisfy beforehand the requirements associated to these activities. An appropriate staffing level is one of these conditions.

Although the regulation doesn't require a numerical objective for staff, ASN assesses and reviews the licensee's staffing level through inspections and instructions and gives attention to the risks related to sharing of staff, in particular regarding organisational and human factors.

A–25.6. JAPAN–NRA

A–25.6.1. Question

Describe any changes in the approach to establishing staffing levels of multi-units within a facility. Describe the challenges associated with regulatory assessment of the adequacy staffing levels of multi-units within a facility.

A–25.6.2. Response

The HTTR does not share personnel with other reactor facilities within Oarai Research Centre.

It has been confirmed that the engineers required for the HTTR have been secured and that they have the necessary technical capabilities for operation.

A–25.6.3. Follow-up Questions

What factors are taken into account in determining staffing levels?

What challenges have you had with regard to staffing levels for the HTTR?

A–25.6.4. Response

As for HTTR DEC events, we assumed following 3 multiple events led by common cause such as earthquake, and estimated the required reaction time is about 1 hour for countermeasures:

- Sealing the gaps or cracks from the outside of the reactor building;
- Reaction time for setting mobile power supplies and establishing observation team under SBO condition;
- Stopping leakage from Spent Fuel pool caused by siphoning – by opening the vent valves for feed-water pipes.

Considering assumptions above, it is confirmed that the required personnel are stationed.

A–25.7. RUSSIAN FEDERATION–ROSTECHNADZOR

A–25.7.1. Question

Describe any changes in the approach to establishing staffing levels of multi-units within a facility. Describe the challenges associated with regulatory assessment of the adequacy staffing levels of multi-units within a facility.

A–25.7.2. Response

According to Paragraph 133 of the federal rules and regulations in the field of the use of atomic energy 'General safety provisions for nuclear power installations of ships and other vessels' NP-022-17 [A–89], the vessel design must provide a rationale for the size of the vessel crew and/or the number of special personnel required for the safe operation of the nuclear power installation. Any changes in the staffing level of the vessel crew and/or special personnel in the course of the plant operation must be agreed with the principal designer and approved by the operating organisation for each vessel (or vessel design), considering its purpose and specifics.

A–25.7.3. Follow-up Questions

What factors are taken into account in determining staffing levels?

What challenges have you had with regard to staffing levels for the floating SMR and other SMRs under consideration?

A–25.7.4. Response

The staffing level is estimated with due account of all credible relevant factors, including process management specifics, expected exposure of personnel, operating conditions of the nuclear facility, etc.

A–25.8. SOUTH AFRICA–NNR

A–25.8.1. Question

Describe any changes in the approach to establishing staffing levels of multi-units within a facility. Describe the challenges associated with regulatory assessment of the adequacy staffing levels of multi-units within a facility.

A–25.8.2. Response

The PBMR project did not reach the stage where this topic was developed as all attention was focused on the envisaged first PBMR demonstration power plant.

A–25.8.3. Follow-up Questions

What challenges would have anticipated with regard to staffing levels for the PBMR?

Could provide references of the regulatory standards and / or guidelines on staffing?

A–25.8.4. Response

In early versions of the PBMR SAR, it was stated that the PBMR Demonstration Power Plant (DPP) design is characterized by the use of automated control, monitoring and protection systems that bring the plant back to normal conditions or to a safe shutdown state without the immediate need for operator action. Consequently, higher levels of automation and first of a kind systems may influence the number of staff and the knowledge, skills and abilities required for control and mitigation actions.

The NNR review comments reminded the licence applicant that use of first of a kind systems does not, in itself, provide justification for less staff, and any claims will need to be supported by a comprehensive staffing analysis.

Generic staffing scenarios are described in the PBMR SAR based on existing practices within the nuclear industry and it was stated that the required staffing for the main control room and other operational tasks is, however, subject to verification upon completion of task analyses, workload analyses and Human Reliability Analysis.

The NNR draft 'Specific Nuclear Safety Regulations: Nuclear Facilities' under development contains the following requirements with regard to staffing:

"8. Operations

…

(3) Staffing

(a) The adequacy of staffing of the facility for safe operation (including response to accidents), addressing competence, experience levels and suitability for safety work shall be justified, verified annually, and documented.

(b) Changes to the number of staff, which might be significant for safety, shall be justified in advance, carefully planned and evaluated after implementation.

(c) A long-term staffing plan shall exist for activities that are important to nuclear safety.

(d) The authorisation holder shall ensure that for all functions and processes necessary for the lifecycle of a nuclear facility, including technical and engineering support, licensing and other supporting processes, that sufficient resources are available to respond to all foreseeable circumstances, including normal operations, abnormal and emergency conditions."

A–25.9. UNITED KINGDOM–ONR

A–25.9.1. Question

Describe any changes in the approach to establishing staffing levels of multi-units within a facility. Describe the challenges associated with regulatory assessment of the adequacy staffing levels of multi-units within a facility.

A–25.9.2. Response

Generally speaking, and in the context of GDA and in the area of human factors (HF), ONR HF specialists assess whether the RP has taken a systematic approach to understanding the human contribution to safety. This will include assessment of the feasibility and acceptability of the human based safety claims (HBSCs) and that the human contribution to risk has been reduced as far as is reasonably practical. The scope of work necessary to achieve this includes, but is not limited to, demonstrating suitability and sufficiency in the following areas:

- The RP's organisational HF capability;
- The applied codes, standards, methods and guidance;
- HF integration (HFI) into all risk important areas, systems, structures and components;
- HF input into: design (including analysis and testing) build, operation, EIMT and decommissioning;
- The HFI programme;
- Consideration of operational experience and research;
- Human reliability analysis (HRA) (including all normal and fault states and demonstration of task feasibility).

ONR expects that assumptions relating to the future operating organisation and to nuclear safety, are captured for adoption / validation by a future licensee (including ongoing activities). These assumptions do not need to be fully developed, however, there needs to be sufficient information for ONR to judge their credibility. ONR organisational capability and staffing level expectations for licensees are documented in LC36: Organisational capability [A–292].

LC36 states that "The licensee shall provide and maintain adequate financial and human resources to ensure the safe operation of the licensed site" and that "(…) the licensee shall make and implement adequate arrangements to control any change to its organisational structure or resources which may affect safety". In this context, ONR requires licensees to provide for the classification of changes to the organisational structure or resources according to their safety significance and that the arrangements shall include a requirement for the provision of adequate documentation to justify the safety of any proposed change.

ONR has not formally assessed staffing levels in the context of SMR design assessment or permissioning activities. However, ONR is currently considering areas of HF uncertainty relating to novel designs, operational, and organisational features of SMR concepts. The aim is to develop of a set of regulatory questions that can be used to guide ONR's assessment of these uncertainties (in-line with Para 27 of the SAPs [A–44]). These questions would also be used to aid early engagement with vendors to provide advice and guidance.

It is anticipated that areas of specific interest will include: impact of inherently safe / passively safe engineering; single control room operation of multi-units; off-site control rooms; unmanned operation; significantly reduced staffing levels; changes to refuelling concepts (return to factory); automated manufacture; management of phased deployment models; where units can be on-grid whilst other are being built and commissioned; and off-site severe accident response. It is not expected that new assessment methods / tools will be required.

A–25.9.3. Follow-up Question

Do you anticipate any challenges with regard to demonstrating acceptable staffing levels for SMRs under consideration?

A–25.9.4. Response

ONR is not currently assessing SMRs or staffing levels of multi-unit sites but has planned an activity within the Advanced Nuclear Technologies project for Human Factors specialists to identify areas of specific regulatory uncertainty in the context of novel design, operational, and organisational features of AMR/SMR concepts. These would then be used to develop of a set of regulatory questions to potentially guide ONR's assessment of uncertainties (in-line with Para 27 of the SAPs [A–44]).

A–25.10.   UNITED STATES OF AMERICA–NRC

A–25.10.1. Question

Describe any changes in the approach to establishing staffing levels of multi-units within a facility. Describe the challenges associated with regulatory assessment of the adequacy staffing levels of multi-units within a facility.

A–25.10.2. Response

While the SMR design that the NRC staff reviewed was similar to existing LWR fleet in many ways, having multiple units coupled with advances in control technologies brought about challenges in the request to minimize staffing requirements. In this instance, an applicant for a combined operating license (COL) would submit its plans for a corporate-level, technical, and onsite organizational structure to support, design, construct, test, operate, and maintain the nuclear plant. A description of the corporate-level management and technical support organization and the onsite operating organization are deferred to a COL applicant.

Plant procedures established at the COL stage would include: (1) administrative procedures that provide for administrative control over safety-related activities for the operation of the facility, (2) operating procedures and emergency operating procedures (EOPs) used to ensure that routine operating, off-normal (i.e. abnormal), and emergency activities are conducted in a safe manner, and (3) procedures for other safety-related plant operating activities, including related maintenance activities, that the operating program or EOP program does not cover.

This Annex presents the responses provided by the following Member State regulatory bodies to the Question 26: "Describe any changes in the safeguards approach. Describe the challenges associated with regulatory assessment of the adequacy safeguards approach."

A–26.1. ARGENTINA–ARN

A–26.1.1. Question

Describe any changes in the safeguards approach. Describe the challenges associated with regulatory assessment of the adequacy safeguards approach.

A–26.1.2. Response

In order for the ARN to grant authorization to enter the site of the necessary nuclear material, the RE must comply in a timely manner with the requirements and procedures for Safeguards, established in the AR 10.14.1 [A–293].

However, before the beginning of construction, the RE submitted to ARN the Design Information Questionnaire (DIQ) in order to be presented to IAEA. In 2019 the DIQ was updated according to some design changes of the facility.

A–26.1.3. Follow-up Question

What challenges do you expect to encounter in the next stages of licensing?

A–26.1.4. Response

The ARN doesn't expect to find relevant challenges for the next stages of licensing. In general lines, the tasks that are expected to be carried out for the licensing of safeguards are framed within the existing regulatory regulations.

A–26.2. CANADA–CNSC

A–26.2.1. Question

Describe any changes in the safeguards approach. Describe the challenges associated with regulatory assessment of the adequacy safeguards approach.

A–26.2.2. Response

CNSC REGDOC-2.13.1 Safeguards and Nuclear Material Accountancy [A–294] is a modern document, implemented in 2018, that consolidates all safeguards requirements into a single document that are essential for compliance with Canadian regulations and to meet international commitments.

Pursuant to its obligations under the NPT (Treaty on the Non-Proliferation of Nuclear Weapons), Canada entered two safeguards agreements with the IAEA:

- "Agreement Between the Government of Canada and the International Atomic Energy Agency for the Application of Safeguards in Connection with the treaty on the Non-Proliferation of Nuclear Weapons";

- "Protocol Additional to the Agreement between Canada and the International Atomic Energy Agency for the Application of Safeguards in Connection with the Treaty on the Non-Proliferation of Nuclear Weapons."

The main objective of these agreements is to allow the IAEA to provide assurances to Canada and the international community that all nuclear materials in the country remain in peaceful-use activities on an annual basis.

The objective is achieved by:

- The timely detection of any diversion of significant quantities of nuclear material from peaceful-use activities to other uses, such as the manufacturing of nuclear weapons or other nuclear explosive devices or for purposes unknown;
- Deterrence, by risk of early detection, of such diversion.

In the CNSC regulatory framework, the safeguards and non-proliferation area covers all the activities required to meet the obligations arising from the Canada/IAEA safeguards agreements as well as all other measures arising from the Treaty on the Non-Proliferation of Nuclear Weapons. The General Nuclear Safety and Control Regulations [A–129] Sections 12(1)(i) and 30(1) require licensees to take all the necessary measures to facilitate Canada's compliance with the applicable safeguards agreement.

REGDOC-2.13.1 [A–294] sets out requirements and guidance for safeguards programs for applicants and licensees who, inter alia, possess nuclear material, carry out certain specified work related to nuclear fuel-cycle research and development, and/or carry out specified activities related to nuclear manufacturing. REGDOC-2.13.1 provides a flexible regulatory framework and is technology and facility neutral. REGDOC-2.13.1 can be applied equally to every nuclear reactor facility type, whether a research reactor, SMR, or full-scale NPP. REGDOC-2.13.1 indicates that licensees must have a safeguards program in place to cover the following specific areas:

- Nuclear material accountancy and control;
- Access and assistance to the IAEA;
- Support for safeguards equipment;
- Provision of Operational and design information.

The same way the SMR technologies vary significantly, so do the fuel designs and the types and characteristics of fissile materials used in SMRs. Thus, materials and fuels may include:

- Fuels and fissile material compositions already in use in operating NPP fleets;
- For molten salt reactors: liquid fuels;
- For liquid metal and high-temperature gas reactors: metallic and graphite-based fuels.

Different types and characteristics of the fuel require new safeguards approaches for implementing nuclear material accountancy and control measures.

From the safeguards perspective, some SMR deployment models will present technical and logistical challenges to safeguards inspectors. For instance, fleets of smaller SMR facilities spread across a large remote region make the traveling for physical inspections possibly very complex. Hence, these types of deployments will necessitate alternative but equally rigorous safeguards approaches. An equipment based approach can be used to address remote access challenges and to reduce the demand for on-site and in person safeguards inspections.

In the regime of safeguards by design (SBD), vendors are encouraged to communicate their designs to the CNSC and the IAEA at an early stage so that safeguards measures, requirements and equipment installations can be integrated into the facility design and construction phases without undue burden. One challenge with this process is that early design concepts can change a lot during the design process and therefore an iterative approach is required.

The CNSC's pre-licensing vendor design review process (VDR) provides reactor vendors with early feedback on whether their proposal meets Canadian requirements for the implementation of safeguards in the design. Focus area 15 of the VDR — robustness, security and safeguards — will confirm whether the documentation submitted by the vendor is consistent with Canada's overall safeguards approach and to facilitate Canada's implementation of its safeguards agreements with the IAEA. The vendor may later use this information to potential customers who may apply for a CNSC licence to build and operate their proposed design in Canada.

The safeguards measures applied are based on the design and operation of the facilities. The CNSC has engaged in discussion with SMR vendors who have provided early stage design information to the CNSC. This has allowed the CNSC to consult with the IAEA at an early stage so that the vendors can work to incorporate safeguards requirements into their design and construction plans without any unexpected requirements or retrofits being necessary.

Taking into account that there are differing levels of maturity and design development, and that most of the new designs have novel features and operational concepts, it is recognized that issues related to new fuel configurations, fuel handling methods, and methods of operation will have to be addressed for material accountancy and control on a case by case basis.

Based on activities and interactions with licensees and vendors, the CNSC has the following recommendations:

- The continued publication of IAEA Nuclear Energy Series guidance documents, like No. NP-T-2.9 International Safeguards in the Design of Nuclear Reactors (2014) [A–295] and No. NP-T-2.8 International Safeguards in Nuclear Facility Design and Construction (2013) provides a valuable resource [A–296];
- A DIQ completion guideline would be useful to applicants;
- A communication protocol detailing how information should flow between the IAEA, CNSC, applicant, and/or designer would be beneficial to ensure requirements are understood and questions are answered in the most efficient manner;
- A step by step framework and associated timelines for engagement with the IAEA on how to progress through the creation of a material balance area, key measurement points, and facility safeguards approach would be useful.

In conclusion, Canada's safeguards regulatory framework is adequate and robust for ensuring safeguards are applied to any reactor technology. The nuclear material accountancy and control requirements and system are adequate for SMRs, however additional guidance documents could make the process of implementing safeguards more user friendly.

A–26.3. CHINA–NNSA

A–26.3.1. Question

Describe any changes in the safeguards approach. Describe the challenges associated with regulatory assessment of the adequacy safeguards approach.

A–26.3.2. Response

Safeguards are not included in the scope of the NNSA's duties.

A–26.3.3. Follow-up Question

What challenges do you expect to encounter?

A–26.3.4. Response

Safeguards are not included in the scope of the NNSA's duties.

A–26.4. CZECH REPUBLIC–SÚJB

A–26.4.1. Question

Describe any changes in the safeguards approach. Describe the challenges associated with regulatory assessment of the adequacy safeguards approach.

A–26.4.2. Response

Not applicable — there have been no changes implemented in the safeguards approach and no challenges associated with regulatory assessment of the adequacy of the safeguard approach have been identified since no SMR is being constructed or planned to be constructed in Czech Republic.

In general, for the non-proliferation purposes, safeguarded installation is defined. The definition covers any kind of nuclear installation with a nuclear reactor. Thus, any SMR regardless of its design is considered a safeguarded installation and is subject to all relevant safeguards legal requirements that are based upon the Euratom safeguards system — the tripartite agreement between the EU member countries, the European Atomic Energy Community and the International Atomic Energy Agency in implementation of the Treaty on the Non-Proliferation of Nuclear Weapons and Additional Protocol to the Agreement. The Czech Republic is a state with so called 'monistic approach' to international treaties (agreements) and therefore, according to the article 10 of the Constitution of the Czech Republic, international treaties become (with some exceptions) an inherent part of the Czech legislative system.

For abovementioned reasons the design requirements are not specific — the design of safeguarded installations shall comply with the technical requirements concerning safeguards of the IAEA arising from international treaties binding on the Czech Republic. These design requirements include technical requirements to ensure independent power supply and lightning or for building and its modifications that allows for the effective IAEA control of the safeguarded installation.

A–26.4.3. Follow-up Question

What challenges do you expect to encounter?

A–26.4.4. Response

Currently, no SMRs are envisaged to be deployed in the Czech Republic. Given the lack of detailed information about various SMR designs and the uncertainty over whether and which type of SMR could be hypothetically deployed, there are no concrete expectations with regard to regulatory assessment of the safeguards approach.

A–26.5. FRANCE–ASN

A–26.5.1. Question

Describe any changes in the safeguards approach. Describe the challenges associated with regulatory assessment of the adequacy safeguards approach.

A–26.5.2. Response

Nuclear safety and nuclear safeguards are controlled by distinctive authorities.

A–26.6. JAPAN–NRA

A–26.6.1. Question

Describe any changes in the safeguards approach. Describe the challenges associated with regulatory assessment of the adequacy safeguards approach.

A–26.6.2. Response

For the HTTR, based on the idea of Safeguards by Design, the safeguards approach was discussed with the IAEA from the design stage, and the approach based on its characteristics has been agreed and applied. Specifically, the design of equipment relating safeguards — for example, unattended spent fuel block flow monitoring system is installed in the door valve — and safeguards approach was examined in consideration of the feature that the HTTR does not perform fuel exchange for a long period and does not open the pressure vessel during the period compared to other research reactors and light water reactors.

Regarding the interface of nuclear safety, security and safeguards related to research reactor facilities, the prime responsibility rests with operators to harmonize them, and the NRA makes each operator aware of this and urging them to respond. Also, from July 2018, when a license application for nuclear safety and nuclear security is made by an operator, or safeguards equipment is installed/updated at facilities by IAEA, the department in charge of each part (Nuclear Safety, Nuclear Security and Safeguards) shares information with the others to eliminate mutual adverse effects as much as possible.

A–26.7. RUSSIAN FEDERATION–ROSTECHNADZOR

A–26.7.1. Question

Describe any changes in the safeguards approach. Describe the challenges associated with regulatory assessment of the adequacy safeguards approach.

A–26.7.2. Response

The approaches to accounting and control have been standardized in the Russian Federation. They are applicable to any nuclear facility and captured in the following federal rules and regulations in the field of the use of atomic energy:

- Basic Rules for Nuclear Material Accounting and Control, NP-030-12 [A–90];
- Requirements for Organization of Material Balance Areas, NP-081-07 [A–91].

A–26.7.3. Follow-up Question

What challenges were encountered?

Provide some examples.

A–26.7.4. Response

In general, the approaches to accounting and control of nuclear materials and the approaches to arrangement of material balance areas established by the federal nuclear regulation NP-030-19 [A–297] cover the specifics of nuclear material accounting and control of the floating NPP and associated land infrastructure.

No changes are being planned in the current regulatory framework regarding accounting and control of nuclear materials and arrangement of material balance zones.

A–26.8. SOUTH AFRICA–NNR

A–26.8.1. Question

Describe any changes in the safeguards approach. Describe the challenges associated with regulatory assessment of the adequacy safeguards approach.

A–26.8.2. Response

Not available.

A–26.8.3. Follow-up Question

What challenges do you expect to encounter?

A–26.8.4. Response

The main challenges in a pebble bed reactor with online refuelling relates to fuel accountancy.

A–26.9. UNITED KINGDOM–ONR

A–26.9.1. Question

Describe any changes in the safeguards approach. Describe the challenges associated with regulatory assessment of the adequacy safeguards approach.

A–26.9.2. Response

The safeguards approach eventually agreed will be discussed with all relevant stakeholders as required during the evolution of that particular reactor type (including vendor/operator, ONR Safeguards (and security and safety colleagues), IAEA, etc.) i.e. during planning, design, construction and commissioning. Direct discussion with the vendor/operator will go through a number of stages as the project matures:

(a)   Ensuring vendor/operator understands safeguards requirements and how they will impact on the facility (and any associated services) covering three main areas:
(i)   Operator's nuclear materials accountancy and control system;
(ii)   Arrangements for reporting these nuclear materials to ONR and the IAEA;
(iii)   Agreeing arrangements for any on-site verification equipment.
(b)   Continuing interactions as the design of the particular plant becomes more fixed. The relative emphasis of the various contributors to verification in the safeguards approach will depend on the ease of physical verification and whether indirect verification may be needed e.g. the several 100 000s of pebbles in PBR.

It is likely that a different safeguards approach will be agreed for each of the reactor types as more detailed interactions proceed and as the plant design becomes more fixed. Early interactions, certainly well before any formal safeguards deadlines, will also be undertaken with other stakeholders including our ONR Safety, Security and Transport colleagues and the Environment Agency, the Scottish Environmental Protection Agency and Natural Resources Wales as required.

With regards to formal safeguards deadlines, the operator is obliged to provide ONR with preliminary design information on any new nuclear facility as early as possible after the decision has been taken to construct, or authorise construction of a facility, in the form of the relevant basic technical characteristics (BTC) as detailed in Regulation 3(2) of the EU Exit Regulations 2019 [A–298]. The Regulations also require the operator to produce a BTC document for each qualifying nuclear facility using the relevant questionnaire shown in Part 1 of Schedule 1 of the Regulations. Such design information must be provided at least 200 days before qualifying nuclear material is introduced into the new facility.

A–26.10. UNITED STATES OF AMERICA–NRC

A–26.10.1. Question

Describe any changes in the safeguards approach. Describe the challenges associated with regulatory assessment of the adequacy safeguards approach.

A–26.10.2. Response

Regarding security of controlled materials, many of the security aspects are proprietary and plant specific. The NRC staff did not, however, find any significant challenges in this area of review.

SMR applicants are required to comply with the reactor security requirements in 10 CFR [A–52], Part 73, Physical Protection of Plants and Materials. A portion of the security requirements are performance-based, so the amount of security infrastructure (i.e. personnel and physical infrastructure) needed for an SMR will likely not be identical to that at the current fleet of LLWR in the USA. The staff encourages SMR design vendors to consider safety and security requirements together in the design process such that security issues can be effectively resolved through facility design and engineered security features, and formulation of mitigation measures, with reduced reliance on human actions.

The NRC also has other requirements that address transport and storage of spent fuel away of the reactor, information security, cyber security, materials control and accountability, and by-product material security requirements for sealed sources. These requirements should be independent of the reactor type or size.

A–27. SECURITY APPROACH

This Annex presents the responses provided by the following Member State regulatory bodies to the Question 27: "Describe any changes in the security approach. Describe the challenges associated with regulatory assessment of the adequacy security approach."

A–27.1. ARGENTINA–ARN

A–27.1.1. Question

Describe any changes in the security approach. Describe the challenges associated with regulatory assessment of the adequacy security approach.

A–27.1.2. Response

In order for the ARN to grant authorization to enter the site of the necessary nuclear material, the RE must comply in a timely manner with the requirements and procedures for Physical Protection, established in the AR 10.13.1 [A–299].

Before the beginning of construction, the RE and NA-SA signed the specific agreement No. 6 which established cooperation in the areas of physical protection, fight against fire, conventional security and relations with the community and communications. This agreement was part of the mandatory documentation submitted to the ARN to obtain the construction license and covered the following areas:

Physical Protection

- Coordination in communications and use of the Response Force;
- Standardization of directives for surveillance personnel;
- Communications to staff about the rules of entry and exit to Atucha Nuclear Site;
- Training of surveillance personnel with similar patterns and joint exercises of both entities (GN, Surveillance, Supervisors);
- Coordination measures to require external assistance;
- Establishment of communication codes to identify violations or possible violations in Physical Protection Systems;
- CNEA participation in the establishment of the perimeter fence;
- Any other need or activity that tends to improve the Physical Protection and Site Security systems.

Fight Against Fire

- Training and joint training of firefighters from both entities;
- Communication and maintenance of mutual records on existing capacities;
- Cooperation in the establishment of the Site Fire Station;
- Support for Lima Town volunteer firefighters in a coordinated manner between both entities;
- Mutual cooperation in the event of fire, providing support of its own elements to the affected entity; Possibility of using the fire systems of the entities or facilities not affected by those affected. Systems standardization to facilitate it;
- Standardization in the use of materials and equipment to prevent the spread of fires to other entities or neighbouring fields.

Community relations and communications

- Establish an Annual Plan of common community relations and communications activities;
- Keep each other informed about problems or incidents that occurred in the respective facilities that may lead to affect relations with the community or communications;
- Transfer to the corresponding party any request for reports or private notes or public that is requested by external entities;
- Invite the parties to any activity that can be used to improve the level of communications or relationship with the community.

Periodically, the ARN requests that RE submit updated information on the physical protection system and the possible changes that are generated.

A–27.1.3. Follow-up Question

What challenges do you expect to encounter in the next stages of licensing?

Please provide some examples.

A–27.1.4. Response

The ARN doesn't expect to find relevant challenges for the next stages of licensing. In general lines, the tasks that are expected to be carried out for the licensing of security are framed within the existing regulatory regulations.

In particular, an exhaustive review will be carried out related to the activities of transportation of EECC, transitory deposit and the reactor itself to find and solve possible issues.

A–27.2. CANADA–CNSC

A–27.2.1. Question

Describe any changes in the security approach. Describe the challenges associated with regulatory assessment of the adequacy security approach.

A–27.2.2. Response

The Nuclear Security Regulations (NSR) [A–16] and associated regulatory documents define nuclear security requirements and guidance for the licensing, construction and operation of nuclear facilities (including high-security nuclear facilities) and for the production, use, transport and/or storage of nuclear substances, including sealed sources and category I, II and III nuclear material. In addition, the NSR ensure that Canada continues to achieve conformity with measures of control and international obligations related to nuclear security to which Canada has agreed.

Developers of SMR technologies are seeking alternative approaches to security, such as security by design, in order to reduce the need for security personnel. One of the concerns is that current security requirements are not sufficiently flexible to address design approaches that could allow for the reduction in security personnel staffing.

Security approaches typically involve a multi-layered DiD system that includes a combination of measures including engineered features, administrative measures and the use of highly qualified security personnel. The NSR permit a measure of flexibility in the use of alternative approaches while ensuring security will remain commensurate with the proposed activities while taking into account the design basis threat (DBT) and the facility-specific threat and risk assessment (TRA). The NSR permit the application of a graded approach, particularly as they apply to the security requirements for nuclear substances (sealed sources and category I, II and III nuclear material). For example, sabotage scenarios would need to be considered taking into account all features and consider where nuclear substances are located, or other vital areas in the facility.

Section 3 of the NSR requires the licence applicant to perform and submit a site-specific TRA and substantiate it in a security program document and site security plan, including information on how it would meet requirements. The licensee would use the TRA in concert with the DBT to assess the threat environment that is applicable to the licensed location and then implement the required security measures accordingly. This would also include how security by design would be credited.

CNSC has received feedback from industry stakeholders that new approaches to site security may need to be considered for SMRs because the threats and risks to these units may be completely different from those faced by existing facilities. For example, should nuclear material not be stored onsite, other than in the reactor, the need to have security measures specific to preventing the theft of nuclear material could be significantly reduced, subject to the performance objectives in-place. The use of passive systems may also potentially eliminate the need for various systems that are traditionally vulnerable to sabotage.

This information is being considered as part of the review and update of the NSR. It is important to note that the use of 'security by design' is possible under the existing regulations and that a graded approach to security can be applied to meet requirements based on security risk-informed considerations.

Since the last major revision to the NSR [A–16] in 2006, security threats, operational experience and technological advancements have evolved, and there is a need to ensure that nuclear facilities and nuclear substances are well-protected with clear and robust requirements, while providing licensee with the flexibility to meet said requirement with alternate means.

Technology, which is embedded in many SMR designs, continues to have a major impact on nuclear security. New technology can present new challenges for the security of nuclear facilities and nuclear substances, including the introduction of new vulnerabilities such as those posed cyber security threats, while at the same time, provide opportunities to better protect nuclear security infrastructure against threats.

While the revision of the NSRs are still in progress, the CNSC recognizes the need to have a flexible regulatory approach that could consider the potential radiological consequences and health impacts. As other regulatory criteria could potentially change such as acceptable radiological consequences on-site and off-site and minimum permissible dose to the public and environment, these changes could have impacts on the security design at the facility. It is also important to note that public acceptance for consequence-based NSRs, particularly for sabotage events, will also play a role in how NSRs are revised.

On November 2019, during the World Institute of Nuclear Security (WINS) Workshop in Ottawa, the President of the CNSC emphasised the importance of integrating security considerations in the design phases of SMRs. It was also confirmed the CNSC's readiness to regulate these new technologies and highlighted the recent memorandum of cooperation with the US Nuclear Regulatory Commission (US NRC) to modernize the regulation of SMRs. The audience was issued with the following four challenges:

(a) Set the path forward toward the effective integration of safety, security and safeguards requirements for SMRs;
(b) Drive the evolution of prescriptive security requirements to a goal-oriented, graded approach commensurate with the risks of SMRs;
(c) Imagine the best next steps in international harmonisation;
(d) Develop concrete recommendations toward modern security requirements.

A–27.3. CHINA–NNSA

A–27.3.1. Question

Describe any changes in the security approach. Describe the challenges associated with regulatory assessment of the adequacy security approach.

A–27.3.2. Response

As the SMRs built in China are all located at site with other big reactors, we haven't met these challenges. We are also considering the organizational requirements for individual SMR and floating reactors.

A–27.3.3. Follow-up Question

What challenges were encountered?

Please provide some examples.

A–27.3.4. Response

As the SMR currently being built in China are all located on-site along with other large reactors, there is no security problem at present. Security regulation of individual SMR and floating reactors is still under study.

A–27.4. FRANCE–ASN

A–27.4.1. Question

Describe any changes in the security approach. Describe the challenges associated with regulatory assessment of the adequacy security approach.

A–27.4.2. Response

In France, nuclear safety and security are ruled by different laws and regulations and are controlled by distinct authorities.

Legal requirements on nuclear safety are found in the Environmental Code [A–24] and the compliance with these requirements is controlled by ASN. The security of nuclear materials and installations is ruled by the Code of Defence [A–300] and subsequent regulations. It is controlled by the Department of Nuclear Security of the High Official for Defence and Security (HFDS, *Haut Fonctionnaire de Défense et de Sécurité*) within the ministry in charge of energy.

Both ASN and HFDS receive technical support from IRSN.

The security of new reactor projects is assessed in two aspects:

- The protection and control of nuclear materials;
- The consideration of malicious acts in the demonstration of nuclear safety.

Regarding protection and control of nuclear materials, the article L. 1333-2 of the Code of Defence [A–300] provides that the importation, exportation, elaboration, possession, transfer, use and transportation of nuclear materials is subject to authorisation or declaration, and control.

To obtain this authorisation, the operator has to submit an application, which includes a security study, to the ministry in charge of energy. This application is reviewed by HFDS.

Regarding the integration of security in the safety demonstration, the Order [A–72] provides that malicious acts shall be included in both the internal (article 3.5) and external (article 3.6) hazards to be considered in the demonstration of nuclear safety.

Moreover, ASN resolution n° 2015-DC-0532 [A–226] relative to the safety analysis report of basic nuclear installations specifies that the safety report shall include a separate classified part presenting:

- The triggering events that could result from given malicious acts in spite of the protective measures implemented;
- The accident situations that could result from these triggering events;
- The study of these accident situations, their consequences and the justification that the emergency provisions to limit their consequences are sufficient.

The classified part of the safety report is examined by ASN. ASN may ask HFDS' opinion on the content of the report in terms of consistency with the security study.

The nuclear power plants that are currently operating in France have been designed at a time when the major threat identified was nuclear proliferation. In 2005, the IAEA Convention on the Physical Protection of Nuclear Material was amended to take into account the possibility that terrorist acts may be intended to result in radiological consequences. Additional protection measures had therefore to be implemented to secure the original design of existing NPPs.

For new reactors, it is essential that the interfaces between nuclear safety and security must be identified early in the design phase. For instance, the design of the EPR reactor that is under construction in Flamanville includes increased redundancy, diversification and physical separation of main safeguard systems and power supplies, and a concrete shell able to withstand the impact of airplane crashes.

These interfaces between safety and security must be taken into account at the earliest stage of new reactor projects, including SMRs.

A–27.5. CZECH REPUBLIC–SÚJB

A–27.5.1. Question

Describe any changes in the security approach. Describe the challenges associated with regulatory assessment of the adequacy security approach.

A–27.5.2. Response

Not applicable — no changes in the security approach have been made as no SMR has been deployed in Czech Republic. For the same reason challenges associated with regulatory assessment of the adequacy security approach have not been identified.

At the same time, there are certain aspects that could be highlighted in the current legislative system governing nuclear security. The provisions of the Atomic Act [A–21] and the Decree No. 361/2016 [A–301], on security of nuclear installation and nuclear material, that are relevant to in this context (i.e. nuclear security) are a combination of prescriptive and performance-based approach. Some of these provisions are relatively very detailed, including very specific description of the requirements. Most probably the biggest challenges could be envisaged in the area of delineation and physical demarcation of guarded area, protected area, inner area or vital area, as certain provisions, especially in the implementing decree, are quite specific. It includes, for example, very detailed description of the technical measures for delineation (height of fences, CCTV etc.).

At the same time, the graded approach should be applied — the nuclear material classification also reflects the risk of unauthorized diversion. In this regard, the legislative system is flexible enough to allow to combine the individual security areas in exceptional and justified cases (in such cases, impeding devices and intrusion detection systems shall be strengthened to ensure same level of physical protection — that shall be demonstrated by efficiency assessment).

Currently the design basis threat is established for NPPs in operation and research reactors (hypothetically a new one would be established for SMR of a particular design to reflect its specificities).

Other provisions that are performance-based are linked to the application of the design basis threat concept — these provisions should be general enough (thus flexible) to reflect various technological solutions used by different SMRs. More general and thus more flexible in this context (SMR deployment) are also provisions governing cybersecurity, categorization of sensitive activities, physical security and response.

A–27.6. JAPAN–NRA

A–27.6.1. Question

Describe any changes in the security approach. Describe the challenges associated with regulatory assessment of the adequacy security approach.

A–27.6.2. Response

The NRA revised the related guides in March 2019 to strengthen measures against internal threats to research reactor facilities by introducing trustworthiness checks and so on.

Regarding the interface of nuclear safety, security and safeguards related to research reactor facilities, from July 2018, in case the license application for nuclear safety and nuclear security is made by the operator or safeguards equipment is installed/updated at facilities by IAEA, the department in charge of the NRA shares the result of confirmation about the effect on the other regulation with the others to eliminate mutual adverse effects as much as possible.

A–27.6.3. Follow-up Question

Please provide further information on the security provisions for the HTTR (i.e. what kind of response force).

Did you have any challenges with regard to security for the HTTR?

A–27.6.4. Response

In regard to the concrete information of HTTR security issue such as response force, we would like to refrain from describing the detail for security reason.

As for the challenges with regard to security, no special issue is recognized for HTTR in Japan.

A–27.7. RUSSIAN FEDERATION–ROSTECHNADZOR

A–27.7.1. Question

Describe any changes in the security approach. Describe the challenges associated with regulatory assessment of the adequacy security approach.

A–27.7.2. Response

There are several physical protection aspects about transportation of the floating plant with a fully loaded core, including the need for providing guarded patrol along the entire route of its movement to the site. However, the resolutions of the Government of the Russian Federation (RF GR No. 456 of July 19, 2007 [A–302]; RF GR No. 646 of May 27, 2017 [A–303]) basically establish standard approaches to physical protection applicable to any nuclear facility. The specific features of physical protection for the floating nuclear plants are established in the federal regulations 'Requirements for the Physical Protection of Vessels with Nuclear Reactors, Nuclear Service Vessels, Vessels Transporting Nuclear Material, and Floating Nuclear Plants' NP-085-19 [A–304].

A–27.7.3. Follow-up Question

What challenges have you had with regard to security for the floating SMR and other SMRs under consideration?

A–27.7.4. Response

In general, the physical protection approaches established by the government orders [A–302] and [A–303], and elaborated in the regulation NP-085-19 [A–304] cover the specifics of the security approach for the floating NPP.

There is no intention to make amendments to the existing regulatory framework regarding the security approach.

A–27.8. SOUTH AFRICA–NNR

A–27.8.1. Question

Describe any changes in the security approach. Describe the challenges associated with regulatory assessment of the adequacy security approach.

A–27.8.2. Response

Not available.

A–27.8.3. Follow-up Question

What are your future plans and security and any anticipated challenges?

A–27.8.4. Response

Not available.

A–27.9. UNITED KINGDOM–ONR

A–27.9.1. Question

Describe any changes in the security approach. Describe the challenges associated with regulatory assessment of the adequacy security approach.

A–27.9.2. Response

In March 2017, ONR introduced the SyAPs [A–45]. This was a significant move away from ONR's prescriptive security methodology and towards a more outcome-focused approach. This new regulatory 'philosophy' applies to all dutyholders and those who wish to build new facilities whatever the design.

When SyAPs was originally conceived and drafted, it was intended for them to be technology-neutral, so the regulatory expectations detailed in SyAPs are as relevant and applicable to SMRs as they are to other regulated nuclear facilities. As a top-level document, SyAPs provide the foundation for introducing outcome-focused regulation for all security disciplines. It also places greater emphasis on security leadership and management.

In assessing SMRs, and especially advanced and novel designs, ONR has started to take an approach based on experience within generic design assessment (GDA). In terms of a security 'approach', these technologies and builds do offer opportunities to reduce security risk and equally may present new risks. The adoption of SyAPs enables a flexible and risk-based approach that is applicable to SMRs. Using the GDA security framework for thinking and assessing, we might identify any changes to our approach in the context of SMRs:

- First, within our security approach, we examine the reactor (nuclear island) design focused on critical safety systems identified through 'Safety Case' work. This is conducted through the perspective of what and where a malicious actor might target to achieve sufficient damage that could lead to a release of radioactivity. We would also examine the fuel in terms of waste — as a target — and its value for theft to be used for an improvised weapon. This approach has been described as 'secure by design' that seeks to remove a security risk or reduce it by safety design and modifications (not security). This security approach, which draws from safety-based analysis, requires joint working across a number of specialisms (focused first on fault analysis) with Security experts. SMRs, that may offer benefits to security design, require to be examined through joint working between Security and Safety teams to exploit the opportunities within 'secure by design'. This approach, while not new although highlighted with SyAPs, is underdeveloped and requires more sophisticated thinking that demands closer cross-specialism working. In terms of SyAPs, it tends to suit a 'goal setting' or 'outcome-focused' approach that allows such innovation and risked-based approach rather than proscribing a security solution. This change presents an opportunity to innovate although a complex one to deliver.

- Next, is the methodology used to evaluate security risks and that is based on IAEA good practice, developed to meet UK regulatory needs. This includes vital area identification and categorisation for theft that both provide RGP applicable to SMRs and other reactor designs. However, there is a change in our approach to 'Cyber Security Risk Assessments' that aims to identify where a 'cyber' threat might singularly, or in combination, enable theft and sabotage. This methodology is relevant to SMRs although elements are still developing.

- Thirdly, there is no real change in applying a security regime should security risks not be 'designed out'. With an outcome focused approach (the 'what'), ONR does not prescribe a security template (the 'how') with fences, barriers and means to detect and delay an intruder. In terms of the commercial and safety benefits of a small and modular build approach employing novel technologies and specific structural design (e.g. mostly underground), ONR's outcome-based approach is highly relevant allowing innovation and flexibility within the final security solution.

Challenges associated with regulatory assessment of the adequacy security approach

SyAPs [A–45] is an approach that offers principles and guidance as to how dutyholders in Great Britain might meet regulatory expectations. As stated, it is aligned with SMR regulation. It has required a culture change in regulatory approach. First, as it affects industry, as they present their own innovative solutions to meet the regulatory expectations or 'outcomes'. Then, secondly, the way in which ONR Security inspectors assesses the adequacy of the proposed arrangements. The adequacy is more subjective now there are no 'model standards' to provide a benchmark, however, it is based on RGP. That RGP includes both the methodology used within security analysis, adopting 'secure by design' and then in applying and justifying a security regime that meets regulatory 'outcomes' rather than ALARP.

The challenges might be summarised as:

- Assessing a novel approach for example when a vendor, or requesting party, chooses to integrate safety and security (and other 'hazards') into what is an 'engineering-centric' approach. Such nuclear innovation work, while welcomed, requires a degree of translation between ONR and our guidance and the vendor's thinking. The outcome-based and non-prescriptive approach offer flexibility but poses analytical demands as the specific means to achieve the goal are not prescribed.

- The site design, based on a smaller modular approach, presents security benefits. For example, a partially underground plant clearly has some benefits in terms of protection. A smaller site may limit the scope for 'DiD' in traditional thinking in terms of fences. However, under an outcome-focused approach, how they (the RP or vendor) create the necessary effect, such as layering and adding complexity through depth, is their choice. They have the freedom of action as long as solutions meet the necessary outcome and security posture expected.

- Over time there may be several SMR locations and obviously this creates additional security requirements. These are to be expected. Maybe there will be more transport security with SMRs and increased demand on armed Police to respond effectively.

- Implementation of an outcome-based approach and mind-set offers flexibility but with it brings challenges in developing the necessary management and security capacity and capability as the 'required solution' is not prescribed. ONR would be looking for an effective 'intelligent customer' capability in any vendor. That requires the whole team from CEO to security manager to be actively and intelligently involved demanding a major cultural shift for some organisations. Equally, the Regulator needs to develop and maintain capability to assess and challenge.

- There must be better Safety and Security joint working so security personnel understand the design and can influence it during its early stages. Security and Safety are not the same, but the relationship is complementary and can be misunderstood. Security teams should include safety experts given the technical challenges of the vital area identification process and especially understanding safety systems interdependencies.

The Key Security Plan Principles (KSyPPs) in SyAPs [A–45] also identify key factors to be considered during all stages of the design of a nuclear facility, including the requirement to incorporate security into the very initial stages of the design.

Reference materials

- New nuclear reactors: Generic Design Assessment Guidance to Requesting Parties for the UK HPR1000. ONR-GDA-GD-001. Revision 4. October 2019. ONR [A–305];
- Security Assessment Principles for the Civil Nuclear Industry. 2017 Edition, Version 0. March 2017, ONR [A–45].

Technical Assessment Guides [A–46]:

- Categorisation for Theft. CNS-TAST-GD-6.1 [A–306];
- Target Identification for Sabotage. CNS-TAST-GD-6.2 [A–307];
- Physical Protection System Design (OS). CNS-TAST-GD-6.3. Revision 0. March 2017. ONR [A–308];
- Effective Cyber and Information Risk Management. CNS-TAST-GD-7.1. Revision 0. March 2017 [A–309];
- Protection of Nuclear Technology and Operations. CNS-TAST-GD-7.3 [A–310];
- Physical Protection of Information. CNS-TAST-GD-7.4 [A–311];
- Guidance on the Security Assessment of Generic New Nuclear Reactor Designs. CNS-TAST-GD-11.1. Revision 0. June 2017. ONR [A–312].

A–27.10. UNITED STATES OF AMERICA–NRC

A–27.10.1 Question

Describe any changes in the security approach. Describe the challenges associated with regulatory assessment of the adequacy security approach.

A–27.10.2. Response

The NRC staff did not find any significant challenges in this area of review. Some of the features included in the SMRs' designs are conceived as safety/security improvements as they will serve also to reduce their vulnerability to physical threats. One common feature to many SMR designs is the integrated cooling system, resulting in a compact reactor coolant boundary. In these designs, the large break LOCAs may not need to be postulated, which result in a safety enhancement. This is different from the fleet of power reactors currently in operation.

Other features common to many SMRs is the larger number of passive physical barriers and greater simplicity in shutdown systems. These features such as RPVs and containment vessels located entirely underwater or below grade, the reactor building located partially or completely below grade, and fewer safe shutdown systems, may simplify the physical protection specific features.

SMR designs incorporating security considerations from the conceptual phase through design implementation, may be equipped with features which deter and delay adversarial actions. The below grade installation of near term SMRs provide additional security benefits, such as minimizing aircraft impact, limiting access to vital areas and limiting the communication ability of adversaries. These same features may provide an excellent means of enhancing security system effectiveness against radiological sabotage. Application of the traditional multi-layered defensive approach of deterrence, detection, assessment, delay, and interdiction can be used effectively for physical protection of SMRs.

REFERENCES TO THE ANNEX

[A–1]   ARGENTINA, The National Law on Nuclear Activity, Law 24,804, Buenos Aires (1997)

[A–2]   ARGENTINA, Law of Confidentiality of Information and Products, Law No. 24,776, Buenos Aires, (1996)

[A–3]   ARGENTINA, Creation of the National Atomic Energy Commission, Decree No. 10,936, Buenos Aires (1950)

[A–4]   ARGENTINA, Validation of decrees, Law No. 14,467, Buenos Aires, (1957)

[A–5]   ARGENTINA, Regulations for the use of radioisotopes and ionizing radiations Decree No. 842/58, Buenos Aires, (1958)

[A–6]   ARGENTINA, Creation of the Nuclear Regulatory Authority, Decree No. 1,540/94, Buenos Aires, (1994)

[A–7]   ARGENTINA, Approval of the Regulations of Law 24,804, Decree No. 1,390/1998, Buenos Aires (1998)

[A–8]   ARGENTINA, Declaration of national interest the construction of the CAREM 25 nuclear reactor, Decree No. 1107/06, Buenos Aires (2006)

[A–9]   ARGENTINA, Activities to allow the life extension of the EMBALSE nuclear power plant Law No. 26,566, Buenos Aires (2009)

[A–10]   ARGENTINA, Regime of control of sensitive export and war equipment, Decree No. 603/92, Buenos Aires (1992)

[A–11]   ARGENTINE NUCLEAR REGULATORY AUTHORITY, Licensing of Relevant Nuclear Installations, AR 0.0.1, Rev.1 ARN, Buenos Aires (2002)

[A–12]   CANADIAN NUCLEAR SAFETY COMMISSION, Canadian National Report for the Convention on Nuclear Safety - Seventh Report, Ottawa (2016)

[A–13]   CANADA MINISTRY OF JUSTICE, Nuclear Safety and Control Act (S.C. 1997, c.9), Nuclear Safety and Control Act, Ministry of Justice, Ottawa (1997).

[A–14]   CANADIAN NUCLEAR SAFETY COMMISSION, Small Modular Reactors: Regulatory Strategy, Approaches and Challenges, Discussion Paper DIS-16-04, Ottawa (2016)

[A–15]   CANADIAN NUCLEAR SAFETY COMMISSION, Strategy for Readiness to Regulate Advanced Reactor Technologies, CNSC, Ottawa (2019)

[A–16]   CANADIAN NUCLEAR SAFETY COMMISSION, Nuclear Security Regulations, SOR/2000-209, CNSC, Ottawa (2000)

[A–17]   CANADA MINISTRY OF JUSTICE, Nuclear Liability and Compensation Act (NLCA), Ottawa (2017)

[A–18]   CANADA MINISTRY OF JUSTICE, Impact Assessment Act, Ottawa (2019)

[A–19]   CANADA MINISTRY OF JUSTICE, Canadian Environmental Assessment Act, Ottawa (2012)

[A–20]   CANADIAN NUCLEAR SAFETY COMMISSION, Regulatory Fundamentals, REGDOC-3.5.3, Version 2, Ottawa (2021)

[A–21]   CZECH REPUBLIC, Atomic Act, No. 263/2016 Coll., Prague (2016)

[A–22]   STATE OFFICE FOR NUCLEAR SAFETY, Requirements for the Design of a Nuclear Installation, Decree No. 329/2017, SÚJB, Prague (2017)

[A–23]   STATE OFFICE FOR NUCLEAR SAFETY, Requirements for Assurance of Quality and Technical Safety and Assessment and Verification of Conformity of Selected Equipment, Decree No. 358/2016, SUBJ, Prague (2016)

[A–24]   FRANCE, Environmental Code (Code de l'environnement) https://www.legifrance.gouv.fr/codes/texte_lc/LEGITEXT000006074220

[A–25]   JAPAN, Reactor Regulation Act, Act No. 166 of 1957, Tokyo (1957)

[A–26]   RUSSIAN FEDERATION, Federal Law on Atomic Energy Use, No. 170-FZ (1995)

[A–27]   INTERNATIONAL ATOMIC ENERGY AGENCY, Governmental, Legal and Regulatory Framework for Safety, IAEA Safety Standards Series No. GSR Part 1 Rev 1, IAEA, Vienna (2016).

[A–28]   RUSSIAN FEDERATION, Merchant Marine Code of the Russian Federation, Federal Law No. 81-FZ (1999)

[A–29]   NATIONAL NUCLEAR REGULATOR, Basic licensing requirements for the PBMR, LG-1037, NNR, Centurion (2000)

[A–30]   SOUTH AFRICA, National Nuclear Regulator Act, Act No 47 of 1999, Pretoria (1999)

[A–31]   NATIONAL NUCLEAR REGULATOR, Regulations on Safety Standards and Regulatory Practices (SSRP), R388, NNR, Centurion (2006)

[A–32]   NATIONAL NUCLEAR REGULATOR, Design Authorisation Framework, PP-0008, NNR, Centurion, (2009)

[A–33]   NATIONAL NUCLEAR REGULATOR , Quality and Safety Management Requirements for Nuclear Installations, RD-0034 Rev 0, NNR, Centurion (2008)

[A–34]   NATIONAL NUCLEAR REGULATOR, Guidance on Testing, Qualification and Commissioning of the PBMR DPP, RG-0005 Rev 0, NNR, Centurion (2010)

[A–35] NATIONAL NUCLEAR REGULATOR -, Guidance on the Verification and Validation of Evaluation and Calculation Models used in Safety and Design Analyses, RG-0016 , NNR, Centurion (2015)

[A–36] NATIONAL NUCLEAR REGULATOR, Authorisations for Nuclear Installations, PP-0009, NNR, Centurion (2016)

[A–37] NATIONAL NUCLEAR REGULATOR, Manufacturing of Components for Nuclear Installations, PP-0012, NNR, Centurion (2020)

[A–38] NATIONAL NUCLEAR REGULATOR, Emergency Technical Basis for New Nuclear Installations, PP-0015, NNR, Centurion (2018)

[A–39] NATIONAL NUCLEAR REGULATOR, Conformity Assessment of Pressure Equipment in Nuclear Service, PP-0016, NNR, Centurion (2020)

[A–40] UNITED KINGDOM, The Energy Act (2013)

[A–41] UNITED KINGDOM, Health and Safety at Work Act (HSWA) (1974)

[A–42] OFFICE FOR NUCLEAR REGULATION, Risk-informed regulatory decision making, ONR, Bootle (2017), http://www.onr.org.uk/documents/2017/risk-informed-regulatory-decision-making.pdf

[A–43] OFFICE FOR NUCLEAR REGULATION, Guidance on the demonstration of ALARP (As Low as Reasonably Practicable), NS-TAST-GD-005, ONR, Bootle (2020)

[A–44] OFFICE FOR NUCLEAR REGULATION, Safety Assessment Principles (SAPs) for Nuclear Facilities. 2014 Edition Revision 1, ONR, Bootle (2020)

[A–45] OFFICE FOR NUCLEAR REGULATION, Security Assessment Principles (SyAPs) for the Civil Nuclear Industry, Ver 0., ONR, Bootle (2017)

[A–46] OFFICE FOR NUCLEAR REGULATION, Technical Assessment Guides (TAGs), http://www.onr.org.uk/tagsrevision.htm

[A–47] INTERNATIONAL ATOMIC ENERGY AGENCY, Integrated Regulatory Review Service (IRRS) to the United Kingdom of Great Britain and Norther Ireland, IAEA-NS-IRRS-2019/06 - Rev. 1, Bootle (2019),

https://www.gov.uk/government/publications/nuclear-and-radiological-safety-review-of-the-uk-framework-2019

[A–48] UNITED KINGDOM, Nuclear Installations Act (NIA), London (1965)

[A–49] UNITED KINGDOM, Nuclear Installations Regulations (NIR), London (1971)

[A–50] OFFICE FOR NUCLEAR REGULATION, Licence Condition Handbook, ONR, Bootle (2017)

[A–51]   OFFICE FOR NUCLEAR REGULATION, Licensing Nuclear Installations, ONR, Bootle (2019)

[A–52]   UNITED STATES OF AMERICA, Title 10 of the Code of Federal Regulations (10 CFR)

[A–53]   UNITED STATES, The Nuclear Energy Innovation and Modernization Act (NEIMA), Washington DC (2017-2018)

[A–54]   U.S. NUCLEAR REGULATORY COMMISSION, Rulemaking Plan On Risk-Informed, Technology-Inclusive Regulatory Framework for Advanced Reactors (RIN-3150-AK31; NRC-2019-0062), SECY-20-0032, US NRC, Washington DC (2020)

[A–55]   UNITED NATIONS, Vienna Declaration on Nuclear Safety, Vienna (2015)

[A–56]   U.S. NUCLEAR REGULATORY COMMISSION, Guidance on the format and content of the Safety Report, RG 1.206, US NRC, Washington DC (2007)

[A–57]   INTERNATIONAL ATOMIC ENERGY AGENCY, Classification of Structures, Systems and Components in Nuclear Power Plants, SSG-30, IAEA, Vienna (2014)

[A–58]   INTERNATIONAL ATOMIC ENERGY AGENCY, Ageing Management and Development of a Programme for Long Term Operation of Nuclear Power Plants, SSG-48, IAEA, Vienna (2018)

[A–59]   CANADIAN NUCLEAR SAFETY COMMISSION, Computer programs used in the design and safety analysis of nuclear power plants and research reactors, G-149, CNSC, Ottawa (2000)

[A–60]   CANADIAN NUCLEAR SAFETY COMMISSION, Class I Nuclear Facilities Regulations, SOR/2000-204, CNSC, Ottawa (2000).

[A–61]   CANADIAN NUCLEAR SAFETY COMMISSION, Design of Small Reactor Facilities, RD-367, CNSC, Ottawa (2014)

[A–62]   CANADIAN NUCLEAR SAFETY COMMISSION, Design of Reactor Facilities: Nuclear Power Plants, REGDOC-2.5.2, CNSC, Ottawa (2014)

[A–63]   CANADIAN NUCLEAR SAFETY COMMISSION, Deterministic Safety Analysis, REGDOC 2.4.1, CNSC, Ottawa (2014)

[A–64]   CANADIAN NUCLEAR SAFETY COMMISSION, Site Evaluation and Site Preparation for New Reactor Facilities, REGDOC-1.1.1, CNSC, Ottawa (2018)

[A–65]   CANADIAN NUCLEAR SAFETY COMMISSION, Supplemental Information for Small Modular Reactor Proponents, REGDOC-1.1.5, CNSC, Ottawa (2018)

[A–66]   CANADIAN NUCLEAR SAFETY COMMISSION, Licence Application Guide: Licence to Construct a Nuclear Power Plant, REGDOC-1.1.2, CNSC, Ottawa (2019)

[A–67]   CANADIAN NUCLEAR SAFETY COMMISSION, Licence Application Guide: Licence to Operate a Nuclear Power Plant, REGDOC-1.1.3, CNSC, Ottawa (2017)

[A–68]   NATIONAL NUCLEAR SAFETY ADMINISTRATION, Nuclear Power Plant Design Safety Regulations, Standard Document, HAF102, NNSA, Beijing (2016)

[A–69]   NATIONAL NUCLEAR SAFETY ADMINISTRATION, Accompanying Guides to HAF102, NNSA, Beijing (2016)

[A–70]   NATIONAL NUCLEAR SAFETY ADMINISTRATION, The safety review principles for small PWR nuclear power plants (Trial), NNSA, Beijing (2016)

[A–71]   NATIONAL NUCLEAR SAFETY ADMINISTRATION, External events selected in the design of floating nuclear power plant (Trial), NNSA, Beijing (2018)

[A–72]   NUCLEAR SAFETY AUTHORITY, Order of 7 February 2012 Setting the General Rules Relative to Basic Nuclear Installations, ASN, Montrouge (2012)

[A–73]   NUCLEAR SAFETY AUTHORITY, Codifying the provisions applicable to basic nuclear installations, the transport of radioactive substances and transparency in nuclear matters,´ Decree nr 2019-190,ASN, Montrouge (2019)

[A–74]   NUCLEAR SAFETY AUTHORITY, Criticality Risk Control in Basic Nuclear Installations, Resolution n° 2014-DC-0462, ASN, Montrouge (2014)

[A–75]   NUCLEAR SAFETY AUTHORITY, Prevention of Risks Resulting from the Dispersion of Pathogenic Microorganisms (Legionella and Amoeba) by the Cooling Installations of the Secondary Circuit of Nuclear Power Reactors Pressurized Water, Resolution n° 2016-DC-0578, ASN, Montrouge (2016)

[A–76]   NUCLEAR SAFETY AUTHORITY, On Setting out Requirements to be Met by Électricité De France – Société Anonyme (Edf Sa) at Flamanville Nuclear Site (Manche) for the Flamanville 3 Reactor (INB No. 167) Commissioning Tests, Resolution n° 2013-DC-0347, ASN, Montrouge (2013)

[A–77]   NUCLEAR SAFETY AUTHORITY, Design of Pressurized Water Reactors, Guide n°22, ASN, Montrouge (2017)

[A–78]   NUCLEAR REGULATORY AUTHORITY OF JAPAN, NRA Order Prescribing Standards for the Location, Structure, and Equipment of Research and Test Reactors and their Auxiliary Facilities, NRA Order No 5 of 2013, NRA, Tokyo (2013)

[A–79]   NUCLEAR REGULATORY AUTHORITY OF JAPAN, Review standard for the fire protection of commercial reactors and their auxiliary facilities, NRA Decision No. 2033110 of 2020, NRA, Tokyo (2020)

[A–80]   NUCLEAR REGULATORY AUTHORITY OF JAPAN, Guide for Evaluation on Volcanic Hazards, NRA Decision No 1912182 of 2019, NRA, Tokyo (2019)

[A–81]   NUCLEAR REGULATORY AUTHORITY OF JAPAN, Guide for Evaluation on Tornado Hazards, NRA Decision No 1812177 of 2018, NRA, Tokyo (2018)

[A–82] NUCLEAR REGULATORY AUTHORITY OF JAPAN, Guide for Evaluation on External Fires, NRA Decision No 13061912 of 2013, NRA, Tokyo (2013)

[A–83] NUCLEAR REGULATORY AUTHORITY OF JAPAN, Guide for Evaluation on Internal Flooding, NRA Decision No 20033110, NRA, Tokyo (2020)

[A–84] NUCLEAR REGULATORY AUTHORITY OF JAPAN, Guide for Evaluation on Internal Fires, NRA Decision No 1707195 of 2017, NRA, Tokyo (2017)

[A–85] NUCLEAR REGULATORY AUTHORITY OF JAPAN, Guide for Evaluation on Surveys for geological features and structure of Site and its vicinity, NRA Decision No 20033110 of 2020, NRA, Tokyo (2020)

[A–86] NUCLEAR REGULATORY AUTHORITY OF JAPAN, Guide for Review on Design-Basis Earthquake and Seismic-resistance Design, NRA Decision No 20033110 of 2020, NRA, Tokyo (2020)

[A–87] NUCLEAR REGULATORY AUTHORITY OF JAPAN, Guide for Review on Design-Basis Tsunami and Tsunami-resistance Design, NRA Decision No 20033110 of 2020, NRA, Tokyo (2020)

[A–88] NUCLEAR REGULATORY AUTHORITY OF JAPAN, Guide for Review on Foundation Grounds and Slope Stability Assessment, NRA Decision No 1306194 of 2013, NRA, Tokyo (2013)

[A–89] FEDERAL ENVIRONMENTAL, INDUSTRIAL AND NUCLEAR SUPERVISION SERVICE, General Safety Assurance Provisions for Ships and Other Floating Craft with Nuclear Reactors, NP-022-17, Moscow (2017)

[A–90] FEDERAL ENVIRONMENTAL, INDUSTRIAL AND NUCLEAR SUPERVISION SERVICE, The Basic Rules of Nuclear Material Record and Control NP-030-12, Moscow (2012)

[A–91] FEDERAL ENVIRONMENTAL, INDUSTRIAL AND NUCLEAR SUPERVISION SERVICE, Requirements for Organization of Material Balance Areas, NP-081-07, Moscow (2007)

[A–92] FEDERAL ENVIRONMENTAL, INDUSTRIAL AND NUCLEAR SUPERVISION SERVICE, Ensuring of Safety during Decommissioning of Nuclear Facilities. General Provisions, NP-091-14, Moscow (2014)

[A–93] FEDERAL ENVIRONMENTAL, INDUSTRIAL AND NUCLEAR SUPERVISION SERVICE, Nuclear Safety Rules for Ships and other Floating Craft with Nuclear Reactors, NP-029-17, Moscow (2017)

[A–94] FEDERAL ENVIRONMENTAL, INDUSTRIAL AND NUCLEAR SUPERVISION SERVICE, Federal Standards and Rules in the Field of Use of Atomic Energy, NP-061-05, Moscow (2005)

[A–95] FEDERAL ENVIRONMENTAL, INDUSTRIAL AND NUCLEAR SUPERVISION SERVICE, Safety Rules in Transportation of Radioactive Materials, NP-053-16, Moscow (2016)

[A–96] FEDERAL ENVIRONMENTAL, INDUSTRIAL AND NUCLEAR SUPERVISION SERVICE, Strength Calculations Standards for Equipment Components and Pipelines of Ships Nuclear Steam Generating Installations with WWER-type Reactors, NP-054-04, Moscow (2004)

[A–97] FEDERAL ENVIRONMENTAL, INDUSTRIAL AND NUCLEAR SUPERVISION SERVICE, Requirements to Quality Assurance Programs for Nuclear Facilities, NP-090-11, Moscow (2011)

[A–98] FEDERAL ENVIRONMENTAL, INDUSTRIAL AND NUCLEAR SUPERVISION SERVICE, Safety in Radioactive Waste Management. General Provisions, NP-058-14, Moscow (2014)

[A–99] FEDERAL ENVIRONMENTAL, INDUSTRIAL AND NUCLEAR SUPERVISION SERVICE, Basic Rules on Accounting and Control of Radioactive Materials and Radioactive Waste in an Organization, NP-067-16, Moscow (2016)

[A–100] FEDERAL ENVIRONMENTAL, INDUSTRIAL AND NUCLEAR SUPERVISION SERVICE, Rules of Nuclear Materials Transfer to the Category of Radioactive Materials or Nuclear Waste, NP-072-13, Moscow (2013)

[A–101] FEDERAL ENVIRONMENTAL, INDUSTRIAL AND NUCLEAR SUPERVISION SERVICE, Rules of Physical Protection of Radioactive Materials and Radiation Sources during Transportation, NP-073-11, Moscow (2011)

[A–102] GOSATOMNADZOR, Requirements to Justification of the Possibility to Extend Design Life-time of Nuclear Facilities, NP-024-2000, Moscow (2000)

[A–103] FEDERAL ENVIRONMENTAL, INDUSTRIAL AND NUCLEAR SUPERVISION SERVICE, Rules on Safety Ensuring During Decommissioning of Ships and other Vessels with Nuclear Installations and Radiation Sources, NP-037-11, Moscow (2011)

[A–104] FEDERAL ENVIRONMENTAL, INDUSTRIAL AND NUCLEAR SUPERVISION SERVICE, Requirements to Planning of Measures for Actions and Protection of Personnel at Nuclear and Radiological Accidents on Vessels and other Watercraft With Nuclear Reactors, NP-079-18, Moscow (2018)

[A–105] FEDERAL ENVIRONMENTAL, INDUSTRIAL AND NUCLEAR SUPERVISION SERVICE, Regulations on Order of Investigation and Accounting of Operational Violations of Ships with Nuclear Installations and Radiation Sources, NP-088-11, Moscow (2011)

[A–106] GOSATOMNADZOR, Requirements for Safety Analysis Report of Nuclear Power Installations of Ships, NP-023-2000, Gosatomnadzor, Moscow (2000)

[A–107] NATIONAL NUCLEAR REGULATOR, Basic Licensing Requirements for the Pebble Bed Modular Reactor, RD-0018Rev 1, NNR, Centurion (2012), http://www.nnr.co.za/acts-regulations/regulatory-documents

[A–108]   NATIONAL NUCLEAR REGULATOR, Requirements on the risk assessment and compliance with principal safety criteria for nuclear installations, RD-0024 Rev 0, NNR, Centurion (2008)

[A–109]   NATIONAL NUCLEAR REGULATOR, Guidance on the Verification and Validation of Evaluation and Calculation Models used in Safety and Design Analyses, RD-0016 Rev 0, NNR, Centurion (2015)

[A–110]   NATIONAL NUCLEAR REGULATOR, Licensing Guide On Safety Assessments of Nuclear Power Reactors, LG-1038, NNR, Centurion (2001)

[A–111]   NATIONAL NUCLEAR REGULATOR, Fuel qualification requirements for Pebble Bed Modular Reactor, LD-1096 Rev 0, NNR, Centurion (2010)

[A–112]   NATIONAL NUCLEAR REGULATOR, Qualification Requirements for the Core Structure Ceramics of the Pebble Bed Modular Reactor, LD-1097 Rev 0, NNR, Centurion (2006)

[A–113]   NATIONAL NUCLEAR REGULATOR, Emergency Preparedness and response requirements for nuclear installations, RD-0014 Rev 0, NNR, Centurion (2005)

[A–114]   NATIONAL NUCLEAR REGULATOR - LG-1041: Licensing guide on safety assessments for nuclear power plants Rev 0, NNR, Centurion (2002)

[A–115]   NATIONAL NUCLEAR REGULATOR, Guidance for licensing submissions involving computer software and evaluation models for safety calculations, LG-1045 Rev 0, NNR, Centurion (2006)

[A–116]   U.S. NUCLEAR REGULATORY COMMISSION, Standard Review Plan for the Review of Safety Analysis Reports for Nuclear Power Plants: LWR Edition (SRP), NUREG-0800, NRC, Washington DC (2009)

[A–117]   U.S. NUCLEAR REGULATORY COMMISSION, Population Related Siting Considerations for Advanced Reactors, SECY-20-0045, USNRC, Washington DC (2020)

[A–118]   U.S. NUCLEAR REGULATORY COMMISSION, Insurance and Liability Regulatory Requirements for Small Modular Reactor Facilities, SECY-18-0076, USNRC, Washington DC (2011)

[A–119]   U.S. NUCLEAR REGULATORY COMMISSION, -: Options for Physical Security for Light-Water Small Modular Reactors and Non-Light-Water Reactors, USNRC, Washington DC (2018)

[A–120]   U.S. NUCLEAR REGULATORY COMMISSION, License Structure for Multi-Module Facilities Related to Small Modular Nuclear Power Reactors, SECY-11-0079, USNRC, Washington DC (2011)

[A–121]   U.S. NUCLEAR REGULATORY COMMISSION, Proposed Variable Annual Fee Structure for Small Modular Reactors, SECY-15 0044, USNRC, Washington DC (2015)

[A–122] U.S. NUCLEAR REGULATORY COMMISSION, Decommissioning Funding Assurance for Small Modular Nuclear Reactors, SECY-11-0181, USNRC, Washington DC (2011)

[A–123] U.S. NUCLEAR REGULATORY COMMISSION, Recommendations on Issues Related to Implementation of Risk Management Regulatory Framework or RMRF, SRM-SECY-15-0168, USNRC, Washington DC (2015)

[A–124] U.S. NUCLEAR REGULATORY COMMISSION, An Approach For Using Probabilistic Risk Assessment in Risk-Informed Decisions on Plant-Specific Changes to the Licensing Basis, Regulatory Guide 1.174 Rev 3, USNRC, Washington DC (2018)

[A–125] ARGENTINEAN NUCLEAR REGULATORY AUTHORITY, Regulatory Standards: Schedule of documentation to be submitted before the commercial operation of a Nuclear power reactor, AR 3.7.1 Rev 1, ARN, Buenos Aires (2002)

[A–126] ARGENTINEAN NUCLEAR REGULATORY AUTHORITY, Preliminary tests and starting of nuclear reactors power, AR 3.8.1 Rev 1, ARN, Buenos Aires (2002).

[A–127] INTERNATIONAL ATOMIC ENERGY AGENCY, Safety of Nuclear Power Plants: Design, SSR-2/1 Revision 1, IAEA, Vienna (2016)

[A–128] CANADIAN NUCLEAR SAFETY COMMISSION, Licensing Process for Class I Nuclear Facilities and Uranium Mines and Mills, REGDOC-3.5.1 Version 2, CNSC, Ottawa (2017)

[A–129] CANADIAN NUCLEAR SAFETY COMMISSION, General Nuclear Safety and Control Regulations SOR/2000-202, Ministry of Justice, Ottawa (2000).

[A–130] CANADA MINISTRY OF JUSTICE, Nuclear Fuel Waste Act, Ottawa (2002)

[A–131] CANADIAN NUCLEAR SAFETY COMMISSION, Pre-licensing Review of a Vendor's Reactor Design, REGDOC-3.5.4, CNSC, Ottawa (2018)

[A–132] PEOPLE'S REPUBLIC OF CHINA, Nuclear Safety Law, Beijing (2017)

[A–133] CHINA, Regulation of the People's Republic of China on the Supervision and Management for Civil Nuclear Installations, NNSA, Beijing, (1986)

[A–134] CZECH REPUBLIC, Construction Act, Act No. 183/2006 Coll., Prague (2006)

[A–135] CZECH REPUBLIC, Environmental Impact Assessment Act, Act No. 100/2001 Coll., Prague (2001)

[A–136] RUSSIAN FEDERATION, Administrative Procedures for the Public Service of Licensing Activities in the Field of Atomic Energy, Moscow (2014)

[A–137] OFFICE FOR NUCLEAR REGULATION, New Nuclear Power Plants: Generic Design Assessment Guidance to Requesting Parties, GDA-GD-006 Rev 0, ONR, Bootle (2019)

[A–138]   FEDERAL   ENVIRONMENTAL,   INDUSTRIAL   AND   NUCLEAR SUPERVISION SERVICE, Provision on Safety Review (Safety Analysis Review) Procedure of Nuclear Facilities and (or) Types of Activities in the Field of Use of Atomic Energy, Order No 160, Moscow (2014)

[A–139]   OFFICE FOR NUCLEAR REGULATION, New Nuclear Power Plants: Generic Design Assessment Technical Guidance, GDA-GD-007 Rev 0, ONR, Bootle (2019)

[A–140]   Office for Nuclear Regulation,Technical Inspections Guides - TIG, ONR, UK.

[A–141]   ARGENTINEAN   NUCLEAR   REGULATORY   AUTHORITY,   Regulatory Standards: Management system for safety in facilities and practices, AR 10.6.1 Rev 0, ARN, Buenos Aires (2020)

[A–142]   INTERNATIONAL ATOMIC ENERGY AGENCY, Leadership and Management for Safety, IAEA Safety Standards Series No. GSR Part 2, IAEA, Vienna (2016).

[A–143]   INTERNATIONAL ATOMIC ENERGY AGENCY, The Management System for Nuclear Installations, GS-G-3.5, IAEA, Vienna (2009)

[A–144]   CANADIAN   NUCLEAR   SAFETY   COMMISSION,   Conduct   of   Licensed Activities: Construction and Commissioning Programs, REGDOC-2.3.1, CNSC, Ottawa (2016)

[A–145]   CANADIAN NUCLEAR SAFETY COMMISSION, Maintenance Programs for Nuclear Power Plants, REGDOC-2.6.2, CNSC, Ottawa (2017)

[A–146]   CANADIAN STANDARDS ASSOCIATION, Management system requirements for nuclear facilities, CSA N286-12, CSA, Toronto (2012)

[A–147]   CANADIAN STANDARDS ASSOCIATION, Quality Assurance of Analytical, Scientific and Design Computer Programs for Nuclear Power Plants, CSA N286.7-16, CSA, Toronto (2016)

[A–148]   NATIONAL NUCLEAR SAFETY ADMINISTRATION, Implementation Rules of Regulations on the Safety Supervision and Management for Civilian Nuclear Installations, HAF001/01, NNSA, Beijing ()

[A–149]   FEDERAL   ENVIRONMENTAL,   INDUSTRIAL   AND   NUCLEAR SUPERVISION SERVICE, Rules for assessment of compliance of products, for which requirements related to safety in the field of the use of atomic energy as well as requirements for processes of its design (including surveys), production, construction, installation, adjustment, operation, storage, shipment, sale, utilization and final disposal are established, NP-071-18, Moscow (2018)

[A–150]   ROSTECHNADZOR/ROSATOM, On the temporary procedure for assessment of conformity of equipment, components, materials and semi-finished products supplied to the nuclear-powered ships and floating structures and support facilities, with mandatory requirements, Decision No. 00-03-10/641, Moscow (2014)

[A–151]   RUSSIAN FEDERATION, On Federal Environmental, Industrial and Nuclear Supervision Service, Decree No. 401 of July 30, Moscow (2004)

[A–152] OFFICE FOR NUCLEAR REGULATION, Supply Chain Management Arrangements for the Procurement of Nuclear Safety Related Items or Services, TAG NS-TAST-GD-077 Rev 5, ONR, Bootle (2019)

[A–153] HEALTH AND SAFETY EXECUTIVE, Health and Safety (Enforcing Authority) Regulations, HSE, Bootle (1998)

[A–154] SMALL MODULAR REACTORS REGULATORS' FORUM, Report on manufacturability, supply chain management and commissioning of Small Modular Reactors, INTERIM REPORT, SMRRF-MCCO WG, Vienna (2019).

[A–155] ARGENTINEAN NUCLEAR REGULATORY AUTHORITY, Site Assessment for Nuclear Power Plants, AR 10.10.1, ARN, Buenos Aires (2015)

[A–156] ARGENTINEAN NUCLEAR REGULATORY AUTHORITY, Limitación de efluentes radiactivos en reactores nucleares de potencia, AR 3.1.2, ARN, Buenos Aires (2002)

[A–157] ARGENTINEAN NUCLEAR REGULATORY AUTHORITY, Limitación de efluentes radiactivos en reactores nucleares de investigación, AR 4.1.2, ARN, Buenos Aires (2002)

[A–158] ARGENTINEAN NUCLEAR REGULATORY AUTHORITY, Limitación de efluentes radiactivos de instalaciones radiactivas Clase I, AR 6.1.2, ARN, Buenos Aires (2002)

[A–159] ARGENTINEAN NUCLEAR REGULATORY AUTHORITY, Resolution Nº 191/2013, ARN, Buenos Aires (2013)

[A–160] INTERNATIONAL ATOMIC ENERGY AGENCY, Radiation Protection and Safety of Radiation Sources: International Basic Safety Standards, IAEA Safety Standards Series No. GSR Part 3, IAEA, Vienna (2014).

[A–161] CANADIAN NUCLEAR SAFETY COMMISSION, Minimum Staff Complement, REGDOC-2.2.5, CNSC, Ottawa (2019).

[A–162] STATE OFFICE FOR NUCLEAR SAFETY, Requirements for safety assessment, Decree 162/2017 Coll., SÚJB, Prague (2017)

[A–163] THE COUNCIL OF THE EUROPEAN UNION, Establishing a Community framework for the nuclear safety of nuclear installations, Directive 2009/71/EURATOM (2009)

[A–164] THE COUNCIL OF THE EUROPEAN UNION, Establishing a Community framework for the nuclear safety of nuclear installations, Directive 2014/87/EURATOM (2014)

[A–165] French Public Health Code (Code de la santé publique), art. R.1333-8, France, https://www.legifrance.gouv.fr/codes/id/LEGITEXT000006072665/ (accessed 10 June 2021)

[A–166]   French Labour Code (Code du travail), art. R.4451-6-8, France, https://www.legifrance.gouv.fr/codes/id/LEGITEXT000006072050/ (accessed 10 June 2021)

[A–167]   FORMER NUCLEAR SAFETY COMMISSION OF JAPAN, Guide for Dose Target Value for commercial light water reactors, NSC Decision on 29 March 2001, Former NSC, Tokyo (2011)

[A–168]   FORMER NUCLEAR SAFETY COMMISSION OF JAPAN, Guide for the safety review of water cooled research and test reactors, NSC Decision on 29 March 2001, Former NSC, Tokyo (2011)

[A–169]   FEDERAL SERVICE FOR SURVEILLANCE ON CONSUMER RIGHTS PROTECTION AND HUMAN WELLBEING, Radiation Safety Norms (NRB-99/2009) SP 2.6.1. 2523-09, Moscow (2009)

[A–170]   FEDERAL SERVICE FOR SURVEILLANCE ON CONSUMER RIGHTS PROTECTION AND HUMAN WELLBEING, Basic Sanitary Rules of Radiation Safety, OSPORB-99/2010, Moscow (2010)

[A–171]   NUCLEAR ENERGY AGENCY, Status of Site-Level (Including Multi-Unit) PSA Developments, NEA/CSNI/R(2019)16, Organisation for Economic Co-Operation And Development (OECD) Nuclear Energy Agency (NEA), Committee on the Safety of Nuclear Installations (CSNI), Paris, France (2019).

[A–172]   INTERNATIONAL ATOMIC ENERGY AGENCY, Multi-Unit Probabilistic Safety Assessment, Safety Reports Series No. 110, IAEA, Vienna (in publication).

[A–173]   OFFICE FOR NUCLEAR REGULATION, Research Report on Multi-Unit Level 3 PSA – Executive Summary, ONR-1901-EX, Jacobsen Analytics, Bootle (2019)

[A–174]   OFFICE FOR NUCLEAR REGULATION, Probabilistic Safety Analysis, Nuclear Safety Technical Assessment Guide, NS-TAST-GD-030 Revision 7, Bootle, UK (2019)

[A–175]   INTERNATIONAL ATOMIC ENERGY AGENCY, Development and Application of Level 1 Probabilistic Safety Assessment for Nuclear Power Plants, IAEA Safety Standards Series No. SSG-3, IAEA, Vienna (2010).

[A–176]   FEDERAL ENVIRONMENTAL, INDUSTRIAL AND NUCLEAR SUPERVISION SERVICE, Requirements for the Content of Safety Analysis Reports for Nuclear Power Plant Units with VVER Reactors, NP-006-16, Moscow (2017)

[A–177]   SCIENTIFIC AND ENGINEERING CENTRE FOR NUCLEAR AND RADIATION SAFETY, Requirements for the Content of the Safety Analysis Report for Nuclear Power Plants with Fast Reactors, NP-018-05, Moscow (2005)

[A–178]   SCIENTIFIC AND ENGINEERING CENTRE FOR NUCLEAR AND RADIATION SAFETY, Requirements to the Content of a Safety Analysis Report for Nuclear Research Installations, NP-049-17, Moscow (2017)

[A–179]   INTERNATIONAL ATOMIC ENERGY AGENCY, Deterministic Safety Analysis for Nuclear Power Plants, Specific Safety Guide No. SSG-2 Rev. 1, IAEA, Vienna (2019).

[A–180]   NUCLEAR REGULATORY COMMISSION, Assessing the Technical Adequacy of the Advanced Light-Water Reactor Probabilistic Risk Assessment for then Design Certification Application and Combined License Application, Design Certification/Combined License DC/COL-ISG-028, Washington D.C., USA (2016)

[A–181]   INTERNATIONAL ATOMIC ENERGY AGENCY, External Human Induced Events in Site Evaluation for Nuclear Power Plants, Safety Guide No. NS-G-3.1, IAEA, Vienna (2012).

[A–182]   CANADIAN NUCLEAR SAFETY COMMISSION, Probabilistic Safety Assessment (PSA) for Reactor Facilities, REGDOC 2.4.2 Version 2, Ottawa (2014)

[A–183]   NUCLEAR SAFETY AUTHORITY, Determination of the Seismic Risk for the Safety of Surface Basic Nuclear Installations, Basic Safety Rule 2001-01, ASN, Montrouge (2001)

[A–184]   NUCLEAR SAFETY AUTHORITY, Risks related to industrial environment and transportation arteries, Basic Safety Rule I.2.D, ASN, Montrouge (1982)

[A–185]   NUCLEAR SAFETY AUTHORITY, Risks induced by projectile emission following the burst of a turbine-generator unit, Basic Safety Rule I.2.B, ASN, Montrouge (1980)

[A–186]   NUCLEAR SAFETY AUTHORITY, Risks induced by an aircraft crash (1980) Basic Safety Rule I.2.A, ASN, Montrouge (1980)

[A–187]   FEDERAL SERVICE FOR ENVIRONMENTAL, TECHNOLOGICAL AND NUCLEAR SUPERVISION, Accounting of External Natural and Man-Induced Impacts on Nuclear Facilities, NP-064-17, Moscow (2017)

[A–188]   OFFICE FOR NUCLEAR REGULATION, Technical Assessment Guide External Hazards, NS-TAST-GD-013 Revision 7, ONR, Bootle (2018)

[A–189]   DEPARTMENT OF ENERGY AND CLIMATE CHANGE, National Policy Statement for Nuclear Power Generation (EN-6), London, UK (2011)

[A–190]   OFFICE FOR NUCLEAR REGULATION, Technical Assessment Guide, Internal Hazards, NS-TAST-GD-014 Revision 5 ONR, Bootle (2019).

[A–191]   OFFICE FOR NUCLEAR REGULATION, Technical Assessment Guide, External Hazards, NS-TAST-GD-013 Revision 5, ONR, Bootle (2014).

[A–192]   WESTERN EUROPEAN NUCLEAR REGULATORS' ASSOCIATION (WENRA) (2014). Safety Reference Levels for Existing Reactors, WENRA Reactor Harmonisation Working Group (2014).

[A–193]   INTERNATIONAL ATOMIC ENERGY AGENCY, Protection against Internal Fires and Explosions in the Design of Nuclear Power Plants, Safety Guide No. NS-G-1.7, IAEA, Vienna (2004).

[A–194]   INTERNATIONAL ATOMIC ENERGY AGENCY, Protection against Internal Hazards other than Fire and Explosions in the Design of Nuclear Power Plants, Safety Guide No. NS-G-1.11, IAEA, Vienna (2004).

[A–195]   INTERNATIONAL ATOMIC ENERGY AGENCY, Protection against Internal Hazards in the Design of Nuclear Power Plants, Safety Standards Series No. SSG-64 (Under publication), IAEA, Vienna (2021).

[A–196]   INTERNATIONAL ATOMIC ENERGY AGENCY, Safety Aspects of Nuclear Power Plants in Human Induced External Events: General Considerations, Safety Reports Series No. 86, IAEA, Vienna (2017)

[A–197]   INTERNATIONAL ATOMIC ENERGY AGENCY, Safety Aspects of Nuclear Power Plants in Human Induced External Events: Assessment of Structures, Safety Reports Series No. 87, IAEA, Vienna (2018).

[A–198]   INTERNATIONAL ATOMIC ENERGY AGENCY, Safety Aspects of Nuclear Power Plants in Human Induced External Events: Margin Assessment, Safety Reports Series No. 88, IAEA, Vienna (2017).

[A–199]   INTERNATIONAL ATOMIC ENERGY AGENCY, Consideration of External Hazards In Probabilistic Safety Assessment For Single Unit And Multi-Unit Nuclear Power Plants, Safety Reports Series No. 92, IAEA, Vienna (2018).

[A–200]   INTERNATIONAL ATOMIC ENERGY AGENCY, Technical Approach for Multi-Unit Site Probabilistic Safety Assessment, Safety Reports Series No. 96, IAEA, Vienna (2019).

[A–201]   INTERNATIONAL ATOMIC ENERGY AGENCY, Application of the Safety Classification of Structures, Systems and Components in Nuclear Power Plants, TECDOC-1787, IAEA, Vienna (2016).

[A–202]   NATIONAL NUCLEAR SAFETY ADMINISTRATION, Safety Functions and Component Classification for BWR, PWR and PTR, Safety Guide HAD102/03, Beijing (1986)

[A–203]   FORMER NUCLEAR SAFETY COMMISSION OF JAPAN, Guide for Evaluation on Safety Design of water-cooled research and test reactors and its attachment Basic concept for classification on the importance of safety function of water-cooled research and test reactors, NSC Decision on 18 July 2001, Former NSC, Tokyo (2001)

[A–204]   AMERICAN NATIONAL STANDARDS INSTITUTE, Nuclear Safety Criteria for the Design of Stationary Pressurized Water Reactor Plants, ANSI/ANS 51.1, ANSI, New York, (1988)

[A–205]   INTERNATIONAL ATOMIC ENERGY AGENCY, Safety related terms for advanced nuclear plants, TECDOC 626, IAEA, Vienna (1991)

[A–206]   INTERNATIONAL ATOMIC ENERGY AGENCY, IAEA Safety Glossary: 2018 Edition, STI/PUB/1830, Vienna (2018)

[A–207]   INTERNATIONAL ATOMIC ENERGY AGENCY, Safety of Nuclear Power Plants: Design, NS-R-1, IAEA, Vienna (2000)

[A–208]   INTERNATIONAL ATOMIC ENERGY AGENCY, Considerations in the development of safety requirements for innovative reactors: Application to modular high temperature gas cooled reactors, TECDOC 1366, IAEA, Vienna (2003)

[A–209]   OFFICE FOR NUCLEAR REGULATION, Categorisation and Classification of Structures, Systems and Components Technical Assessment Guide, NS-TAST-GD-094, Revision 2, Bootle, UK (2019).

[A–210]   INTERNATIONAL ATOMIC ENERGY AGENCY, Design of Electrical Power Systems for Nuclear Power Plants, IAEA SSG-34, IAEA, Vienna (2016)

[A–211]   WORLD NUCLEAR ASSOCIATION, Safety Classification for I&C Systems in Nuclear Power Plants – Current Status and Difficulties, CORDEL Digital Instrumentation & Control Task Force, 2019 Revision, (2019)

[A–212]   WORLD NUCLEAR ASSOCIATION, Safety Classification for I&C Systems in Nuclear Power Plants: Comparison of Definitions of Key Concepts, CORDEL Digital Instrumentation & Control Task Force, 2020 Revision, (2020)

[A–213]   INTERNATIONAL ELECTROTECHNICAL COMMISSION, Nuclear Power Plants – Instrumentation, Control and Electrical Power Systems important to Safety – Categorisation of Functions and Classification of Systems, IEC 61226, IEC, Geneva, Switzerland, (2020)

[A–214]   U.S. NUCLEAR REGULATORY COMMISSION, Quality Group Classifications and Standards for Water-, Steam-, and Radioactive Waste Containing Components of Nuclear Power Plants, Regulatory Guide (RG) 1.26 Revision 5, Washington DC, (2017)

[A–215]   AMERICAN SOCIETY OF MECHANICAL ENGINEERS, ASME Boiler and Pressure Vessel Code (BPV Code), ASME, New York (2019).

[A–216]   CANADIAN NUCLEAR SAFETY COMMISSION, Accident Management, REGDOC-2.3.2 Version 2, CNSC, Ottawa (2015)

[A–217]   CANADIAN NUCLEAR SAFETY COMMISSION, Nuclear Emergency Preparedness and Response, REGDOC-2.10.1 Version 2, CNSC, Ottawa (2016)

[A–218]   CANADIAN STANDARDS ASSOCIATION, Fire protection for nuclear power plants, N293, CSA, Toronto (2012)

[A–219]   CANADIAN STANDARDS ASSOCIATION, General requirements for pressure retaining systems and components in CANDU nuclear power plants, N285.0, CSA, Toronto, (2007)

[A–220]   INTERNATIONAL ATOMIC ENERGY AGENCY, Basic Safety Principles for Nuclear Power Plants 75-INSAG-3 Rev 1, INSAG-12, IAEA, Vienna (1999).

[A–221]   INTERNATIONAL ATOMIC ENERGY AGENCY, Considerations in the development of safety requirements for innovative reactors: Application to modular high temperature gas cooled reactors, TECDOC-1366, IAEA, Vienna (2003)

[A–222]   INTERNATIONAL ATOMIC ENERGY AGENCY, Defence in Depth in Nuclear Safety, INSAG-10, IAEA, Vienna (1996)

[A–223]   INSTITUTE OF ELECTRICAL AND ELECTRONICS ENGINEERS, IEEE Standard for Application and Management of the Systems Engineering Process, IEEE 1220, IEEE, New York (2005)

[A–224]   U.S. NUCLEAR REGULATORY COMMISSION, Glossary, US NRC, Washington DC (2021), https://www.nrc.gov/reading-rm/basic-ref/glossary/defense-in-depth.html

[A–225]   CANADIAN STANDARDS ASSOCIATION, Requirements for support power systems of CANDU NPPs, CSA N290.5, CSA, Toronto (2016)

[A–226]   NUCLEAR SAFETY AUTHORITY, Resolution n° 2015-DC-0532 Relating to the Safety Report of Basic Nuclear Installations, ASN, Paris (2015)

[A–227]   USSR, Main Dimensions of Surface Ships and Vessels. Terms, Definitions and Symbols, Moscow, USSR (1980)

[A–228]   USSR, Structural Components of the Metallic Hull of Surface Ships and Vessels. Terms and Definitions, GOST 13641-80, Moscow (1980)

[A–229]   USSR, Operational Documentation for Vessels. Logbooks, GOST 19439.2-74, Moscow (1974)

[A–230]   USSR, Cast Deck and Board Hawsepipes. Specification, GOST 25056-81, Moscow (1981)

[A–231]   USSR, Deck Machinery and Hull Gear. Terms and Definitions, GOST 26069-86, Moscow (1986)

[A–232]   USSR, Surface ships and vessels. Method for calculation of landing and initial stability, OST 5.0099-74, Moscow (1974)

[A–233]   USSR, Technological preparation of the shipyard process. Terms and definitions, OST 5.0369-83, Moscow (1983)

[A–234]   KRYLOV STATE RESEARCH CENTRE, Design Documentation for Vessels. Rules for Development, Concurrence (Approval) and Endorsement, St. Petersburg (2001)

[A–235]   RUSSIAN FEDERATION, Federal Law No. 317-FZ on the State Atomic Energy Corporation Rosatom, No. 317-FZ, Moscow (2007)

[A–236]   NATIONAL NUCLEAR SAFETY ADMINISTRATION, Computer Based Safety Important System Software for Nuclear Power Plants, HAD 102-16, NNSA, Beijing (2016)

[A–237]   STATE OFFICE FOR NUCLEAR SAFETY, Management System Requirements, Decree No. 408/2016, SUBJ, Prague (2016)

[A–238]   NUCLEAR SAFETY AUTHORITY, Qualification Of Scientific Computing Tools Used In The Nuclear Safety Case, ASN Guide n°28, Montrouge (2017)

[A–239]   FEDERAL ENVIRONMENTAL, INDUSTRIAL AND NUCLEAR SUPERVISION SERVICE, Computer Testing of Declared Software Capabilities And Verification of Their Adequacy For Conducting Declared Calculations, Order No. 325, Rostechnadzor, Moscow (2018)

[A–240]   U.S. NUCLEAR REGULATORY COMMISSION, Verification, Validation, Reviews, and Audits for Digital Computer Software Used in Safety Systems of Nuclear Power Plants, Regulatory Guide 1.168, NRC, Washington DC (1997)

[A–241]   AMERICAN NATIONAL STANDARDS INSTITUTE, Documentation of Computer Software, ANSI/, ANSI, New York, (1995)

[A–242]   OFFICE FOR NUCLEAR REGULATION, Validation of Computer Codes and Calculation Methods, NS-TAST-GD-042, ONR, Bootle (2020)

[A–243]   U.S. NUCLEAR REGULATORY COMMISSION, Piping Benchmark Problems. Volumes I and II, NUREG/CR-1677, US NRC, Washington DC (1985)

[A–244]   CANADIAN NUCLEAR SAFETY COMMISSION, Stakeholder Workshop Report: Application of the Graded Approach in Regulating Small Modular Reactors, CNSC, Ottawa (2017)

[A–245]   CANADIAN STANDARDS ASSOCIATION, Risk management – Risk assessment techniques, CAN/CSA-IEC/ISO 31010-10, CSA, Toronto (2010)

[A–246]   CANADIAN STANDARDS ASSOCIATION, Risk-informed decision making for nuclear power plants, CSA N290.19, CSA, Toronto (2018)

[A–247]   OFFICE FOR NUCLEAR REGULATION, Ventilation, NS-TAST-GD-022, ONR, Bootle (2020)

[A–248]   OFFICE FOR NUCLEAR REGULATION, Nuclear Lifting Operations, NS-TAST-GD-056, ONR, Bootle (2018)

[A–249]   OFFICE FOR NUCLEAR REGULATION, Design Safety Assurance, NS-TAST-GD-057, ONR, Bootle (2017)

[A–250]   OFFICE FOR NUCLEAR REGULATION, Pressure Systems Safety, NS-TAST-GD-067, ONR, Bootle (2017)

[A–251]   NATIONAL NUCLEAR SAFETY ADMINISTRATION, Nuclear Power Plant Safety Requirements for Quality Assurance, HAF 003-1991, NNSA, Beijing (1991)

[A–252]   NUCLEAR SAFETY AUTHORITY, Specific Requirements to Be Met by EDF-SA at the Flamanville Nuclear site Regarding the Design and Construction of the Flamanville NPP, Resolution n° 2008-DC-0114, ASN, Montrouge (2008)

[A–253]   FEDERAL ENVIRONMENTAL, INDUSTRIAL AND NUCLEAR SUPERVISION SERVICE, Resolution of the Government of the Russian Federation No. 121 of February 14, 2012, Moscow, (2012)

[A–254]   FEDERAL ENVIRONMENTAL, INDUSTRIAL AND NUCLEAR SUPERVISION SERVICE, General Provisions for Ensuring the Safety of Nuclear Power Plants, NP-001-15, Moscow

[A–255]   ARGENTINEAN NUCLEAR REGULATORY AUTHORITY, Confinement system in nuclear power reactors, AR 3.4.3 Rev 1, ARN, Buenos Aires (2001)

[A–256]   ARGENTINEAN NUCLEAR REGULATORY AUTHORITY, Radiological Criteria Relating to Accidents in Nuclear Power Reactors, AR 3.1.3 Rev 2, ARN, Buenos Aires (2001)

[A–257]   FEDERAL ENVIRONMENTAL, INDUSTRIAL AND NUCLEAR SUPERVISION SERVICE, Rules for the Design and Operation of Confining Safety Systems for Nuclear Power Plants, (NP-010-16), Moscow (2016)

[A–258]   OFFICE FOR NUCLEAR REGULATION, Civil Engineering Containments for Reactor Plants, NS-TAST-GD-020 Revision 6, ONR, Bootle (2020)

[A–259]   ARGENTINEAN NUCLEAR REGULATORY AUTHORITY, The set of extinguishing systems must have at least two independent extinguishing systems with adequate diversity, each with sufficient negative reactivity as to cause the extinction of the reactor without the action of the other being necessary, AR 3.4.2, Rev. 1, ARN, Buenos Aires (2001)

[A–260]   CANADIAN NUCLEAR SAFETY COMMISSION, Design of New Nuclear Power Plants, RD-337, CNSC, Ottawa (2014).

[A–261]   FEDERAL ENVIRONMENTAL, INDUSTRIAL AND NUCLEAR SUPERVISION SERVICE, Nuclear Safety Rules for Reactor Installations of Nuclear Power Plants, NP-082-07, Moscow (2008)

[A–262]   ARGENTINEAN NUCLEAR REGULATORY AUTHORITY, Radiation Protection Requirements, AR 10.1.1, Rev. 3, ARN, Buenos Aires (2016)

[A–263]   NATIONAL NUCLEAR SAFETY ADMINISTRATION, Basic Standard for Protection Against Ionizing Radiation and Safety of Radiation Sources, GB18871-2002, NNSA, Beijing (2002)

[A–264]   NATIONAL NUCLEAR SAFETY ADMINISTRATION, Regulations on Radiation Protection of Nuclear Power Plants, GB6249-2011, NNSA, Beijing (2011)

[A–265]   UK STATUTORY INSTRUMENTS, The Ionising Radiations Regulations 2017 (IRR 17), HSE, Norwich (2017)

[A–266] HEALTH AND SAFETY EXECUTIVE, Working with ionising radiation, Ionising Radiations Regulations 2017, Approved Code of Practice and guidance, L121, HSE, Norwich (2018)

[A–267] HEALTH AND SAFETY EXECUTIVE, Reducing Risks, Protecting People, HSE's Decision Making Process, ISBN 0 7176 2151, HSE, Norwich (2018)

[A–268] OFFICE FOR NUCLEAR REGULATION, Radiological Shielding, TAST-GD-002. Rev 8. ONR, Bootle (2019)

[A–269] OFFICE FOR NUCLEAR REGULATION, Fundamental Principles, NS-TAST-GD-004 Rev 7, ONR, Bootle (2019)

[A–270] OFFICE FOR NUCLEAR REGULATION, Radiological Protection, NS-TAST-GD-038 Rev 9, ONR, Bootle (2019)

[A–271] OFFICE FOR NUCLEAR REGULATION, Criticality Safety, NS-TAST-GD-041 Rev 9, ONR, Bootle (2019)

[A–272] OFFICE FOR NUCLEAR REGULATION, Radiological Analysis Normal Operation, NS-TAST-GD-043 Rev 6, ONR, Bootle (2019)

[A–273] OFFICE FOR NUCLEAR REGULATION, Radiological Analysis Fault Conditions, NS-TAST-GD-045 Rev 6, ONR, Bootle (2019)

[A–274] OFFICE FOR NUCLEAR REGULATION, The Ionising Radiations Regulations 2017, NS-INSP-GD-054 Rev 7, ONR, Bootle (2019)

[A–275] OFFICE FOR NUCLEAR REGULATION, Numerical targets and legal limits in Safety Assessment Principles for Nuclear Facilities - an explanatory note, ONR, Bootle (2006)

[A–276] NUSCALE POWER, LLC, Mitigation Strategies for Loss of All AC Power Event, Technical Report (TR)-0816-50797, NuScale, Corvallis (2016)

[A–277] U.S. NUCLEAR REGULATORY COMMISSION, Severe Accident Risks: An Assessment for Five U.S. Nuclear Power Plants, NUREG/CR 1150, NRC, Washington DC (1990)

[A–278] ATOMIC ENERGY COMMISSION, Calculation of Distance Factors for Power and Test Reactor Sites, TID-14844 (ADAMS Accession No. ML021720780), AEC, Washington DC (1962)

[A–279] U.S. NUCLEAR REGULATORY COMMISSION, Accident Source Terms for Light Water Nuclear Power Plants, NUREG-1465, US NRC, Washington DC (1995)

[A–280] U.S. NUCLEAR REGULATORY COMMISSION, Alternative Radiological Source Terms for Evaluating Design Basis Accidents at Nuclear Power Reactors, Regulatory Guide (RG) 1.183, US NRC, Washington DC (2000)

[A–281]   CANADIAN STANDARDS ASSOCIATION, General requirements for nuclear emergency management programs, CSA N1600, CSA, Toronto (2021)

[A–282]   INTERNATIONAL ATOMIC ENERGY AGENCY, Arrangements for Preparedness for a Nuclear or Radiological Emergency, Safety Guide GS-G-2.1, IAEA, Vienna (2007)

[A–283]   INTERNATIONAL ATOMIC ENERGY AGENCY, Actions to protect the public in an emergency due to severe conditions at a light water reactor, EPR-NPP Public Protective Actions, Vienna (2013)

[A–284]   STATE OFFICE FOR NUCLEAR SAFETY, On details of ensuring radiation extraordinary event management, Decree No. 359/2016 Coll, SÚJB, Prague (2016)

[A–285]   STATE OFFICE FOR NUCLEAR SAFETY On Radiation Protection and Security of a Radioactive Source, Decree No. 422/2016, SÚJB, Prague (2016)

[A–286]   NUCLEAR REGULATORY AUTHORITYOF JAPAN, Guideline for Emergency Preparedness and Response, NRA Notice No 6 of 2019, NRA, Tokyo (2019)

[A–287]   FEDERAL ENVIRONMENTAL, INDUSTRIAL AND NUCLEAR SUPERVISION SERVICE , Standard content of the action plan for protection of personnel in the event of accident at nuclear power plant, NP-015-12, Rostechnadzor, Moscow (2013)

[A–288]   HEALTH AND SAFETY EXECUTIVE, Radiation (Emergency Preparedness and Public Information) Regulations 2019, REPPIR 2019, HSE, Norwich (2019)

[A–289]   ARGENTINEAN NUCLEAR REGULATORY AUTHORITY, Licensing of Class I facilities personnel, AR 0.11.1 Rev 3, ARN, Buenos Aires (2002)

[A–290]   ARGENTINEAN NUCLEAR REGULATORY AUTHORITY, Psychophysical fitness requirements for specific authorizations, AR 0.11.2 Rev 2, ARN, Buenos Aires (2002)

[A–291]   ARGENTINEAN NUCLEAR REGULATORY AUTHORITY, Retraining of personal of Class I facilities, AR 0.11.3 Rev 1, ARN, Buenos Aires (2001)

[A–292]   OFFICE FOR NUCLEAR REGULATION, Organizational capability, LC36, ONR, Bootle (2019)

[A–293]   ARGENTINEAN NUCLEAR REGULATORY AUTHORITY, Guarantees of no deviation of nuclear materials and materials, facilities and equipment of nuclear interest, AR 10.14.1, ARN, Buenos Aires (2007)

[A–294]   CANADIAN NUCLEAR SAFETY COMMISSION, Safeguards and Nuclear Material Accountancy, REGDOC-2.13.1, CNSC, Ottawa (2018)

[A–295]   INTERNATIONAL ATOMIC ENERGY AGENCY, International Safeguards in the Design of Nuclear Reactors, NP-T-2.9, IAEA, Vienna (2014)

[A–296] INTERNATIONAL ATOMIC ENERGY AGENCY, International Safeguards in Nuclear Facility Design and Construction, NP-T-2.8, IAEA, Vienna (2013)

[A–297] FEDERAL ENVIRONMENTAL, INDUSTRIAL AND NUCLEAR SUPERVISION SERVICE, Fundamental Rules on Accounting and Control of Nuclear Materials, NP-030-19, Moscow (2020)

[A–298] EUROPEAN UNION, The Nuclear Safeguards (EU Exit) Regulations 2019, EU Exit Regulations 2019, London (2019), https://www.legislation.gov.uk/uksi/2019/196/contents/made

[A–299] ARGENTINEAN NUCLEAR REGULATORY AUTHORITY, Requirements and procedures for physical protection, AR 10.13.1 Rev 1, ARN, Buenos Aires (2002)

[A–300] FRANCE, Defence Code (Code de la Défense); Paris, (2015), https://www.legifrance.gouv.fr/codes/texte_lc/LEGITEXT000006071307/2021-06-02/

[A–301] STATE OFFICE FOR NUCLEAR SAFETY, On security of nuclear installation and nuclear material, Decree No. 361/2016 Coll., SUBJ, Prague (2016)

[A–302] GOVERNMENT OF THE RUSSIAN FEDERATION, About approval of Rules of physical protection of nuclear materials, nuclear installations and Items of storage of nuclear materials, Resolution No. 456 of July 19, 2007 (amended in 2018), Moscow (2018)

[A–303] GOVERNMENT OF THE RUSSIAN FEDERATION, Resolution No. 646 of May 27, Moscow (2017)

[A–304] FEDERAL ENVIRONMENTAL, INDUSTRIAL AND NUCLEAR SUPERVISION SERVICE, Requirements for the Physical Protection of Vessels with Nuclear Reactors, Nuclear Service Vessels, Vessels Transporting Nuclear Material, and Floating Nuclear Plants, NP-085-19, Moscow (2019)

[A–305] OFFICE FOR NUCLEAR REGULATION, New nuclear reactors: Generic Design Assessment Guidance to Requesting Parties for the UK HPR1000, ONR-GDA-GD-001. Rev 4, ONR, Bootle (2019)

[A–306] OFFICE FOR NUCLEAR REGULATION, Categorisation for Theft, CNS-TAST-GD-6.1. Rev 1, ONR, Bootle (2020)

[A–307] OFFICE FOR NUCLEAR REGULATION, Target Identification for Sabotage. CNS-TAST-GD-6.2. Rev 1, ONR, Bootle (2020)

[A–308] OFFICE FOR NUCLEAR REGULATION, Physical Protection System Design (OS). CNS-TAST-GD-6.3. Rev 0, ONR, Bootle (2017)-Not Found

[A–309] OFFICE FOR NUCLEAR REGULATION, Effective Cyber and Information Risk Management, CNS-TAST-GD-7.1, ONR, Bootle (2020)

[A–310] OFFICE FOR NUCLEAR REGULATION, Protection of Nuclear Technology and Operations, CNS-TAST-GD-7.3 Rev 1, ONR, Bootle (2020)

[A–311]   OFFICE FOR NUCLEAR REGULATION, Physical Protection of Information, CNS-TAST-GD-7.4 Rev 1, ONR, Bootle (2020)

[A–312]   OFFICE FOR NUCLEAR REGULATION, Guidance on the Security Assessment of Generic New Nuclear Reactor Designs, CNS-TAST-GD-11.1. Issue 1.2, ONR, Bootle (2021)

LIST OF ABBREVIATIONS

ALARA	As low as reasonably achievable
ALARP	As low as reasonably practicable
ACS	Adjuster and Control System
ACOP	Approved Code of Practice
AMP	Ageing Management Programme
AMR F&D	Advanced Modular Reactor Feasibility and Development
ANO	Authorized Nuclear Operator
ANT	Advanced Nuclear Technologies
AOO	Anticipated Operational Occurrence
APS	Probabilistic Safety Analysis – PSA (*Análisis Probabilístico de Seguridad*)
ARN	National Regulatory Authority – ARN–Argentina (*Autoridad Regulatoria Nuclear*)
ASN	French Nuclear Safety Authority – ASN–France (*Autorité de Sûreté Nucléaire*)
ASME	American Society of Mechanical Engineers
ATWS	Anticipated Transient without Scram
AVR	German *Arbeitsgemeinschaft Versuchsreaktor*
BDBA	Beyond Design Basis Accident
BDBEE	Beyond Design Basis External Event
BLR	Basic Licensing Requirement
BSL	Basic Safety Level
BSO	Basic Safety Objective
BTC	Basic Technical Characteristics
BWR	Boiling Water Reactor
CBCS	Core Barrel Conditioning System
CANDU	Canada Deuterium Uranium (Canadian pressurized heavy-water reactor)

CC	Computer Code
CCS	Core Conditioning System
CDE	Core Damage Event
CDF	Core Damage Frequency
CDST	Core Damage Source Term
CEO	Chief Executive Officer
CNA	Atucha Nuclear Power Station (*Central Nuclear Atucha*)
CNE	Embalse Nuclear Power Station (*Central Nuclear Embalse*)
CNEA	National Atomic Energy Commission – Argentina
CNSC	Canadian Nuclear Safety Commission
COL	Combined Operating License
COTS	Commercial off the shelf
CP	Coated fuel Particle
CRD	Control Rod Drive system
CRO	Control Room Operator
CRSS	Control Room Shift Supervisor
CSA	Comprehensive Safeguards Agreement
CSA	Canadian Standards Association
CSC	Core Structure Ceramics
CVCS	Chemical and Volume Control System
DAC	Design Acceptance Confirmation
DBA	Design Basis Accident
DBE	Design Basis Event
DBST	Design Basis Source Term
DBT	Design Basis Threat
DC	Design Certification
DCA	Design Certification Application

DEPZ	Detailed Emergency Planning Zone
DiD	Defence in Depth
DIQ	Design Information Questionnaire
DEC	Design Extension Condition
DeP	Defence in Depth – DiD (*Defensa en Profundidad*)
DHRS	Decay Heat Removal System
DLOFC	Depressurized Loss of Forced Cooling
DNB	Departure Nucleate Boiling
DPP	Demonstration Power Plant
DRACS	Direct Reactor Auxiliary System
DSA	Deterministic Safety Assessment
EEAA	Absorber Element
ECCS	Emergency Core Cooling System
ECI	Emergency Coolant Injection System
EECC	Combustible Element
EHRS	Emergency Heat Removal System
EIMT	Examination, Inspection, Maintenance and Testing
EKP	Engineering Key Principles, as defined by UK Office for Nuclear Regulation
EOP	Emergency Operating Procedure
EPR	Emergency Preparedness and Response
EPR	European pressurized reactor
EPZ	Emergency Planning Zone
ESKOM	Electrical Utility in South Africa
EU	European Union
FHA	Fuel Handling Accident
FMEA	Failure Mode and Effects Analysis
FNPP	Floating Nuclear Power Plant

FOAK	First of a Kind
FP	Fission Product
FPOT	First-Plant-Only-Tests
FSS	First Shutdown System
GDA	Generic Design Assessment
GDC	General Design Criteria
KLI	Key Licensing Issue
KSyPP	Key Security Plan Principle
HAZOP	Hazard and Operability Study
HBSC	Human Based Safety Claim
HF	Human Factor
HFDS	High Official for Defence and Security – as defined by ASN-France
HFI	Human Factor Integration
HPB	Helium Pressure Boundary
HRA	Human Reliability Analysis
HTGR or HTR	High Temperature Gas Cooled Reactor
HTR-PM	High Temperature Gas Cooled Reactor Pebble-bed Module
HTTR	High Temperature Engineering Test Reactor
HVAC	Heating, Ventilation and Air Conditioning
I&C	Instrumentation and Control
ICRP	International Commission on Radiological Protection
IE	Initiating Event
IEC	International Electrotechnical Commission
IEEE	Institute of Electrical and Electronics Engineers
INET	Tsinghua University's Institute of Nuclear and new Energy Technology
INSAG	International Nuclear Safety Group

IRR	Ionising Radiations Regulations
IRSN	Institute for Radiological Protection and Nuclear Safety - France
ISI	In-Service Inspection program
IST	In-Service Testing
JOEN	*Jefe Operativo de Emergencias Nucleares*
LBE	Licensing Basis Event
LC	License Condition
LFR	Lead-cooled Fast Reactor
LLWR	Large Light Water Reactor
LOCA	Loss of Coolant Accident
LOFA	Loss of Flow Accident
LRF	Large Release Frequency
LWR	Light Water Reactor
MCCO	Working Group on Manufacturing, Construction, Commissioning, and Operations
MCR	Main Control Room
MS	Member State
MSLB	Main Steam Line Break
MLD	Master Logic Diagram
MOU	Memorandum of Understanding
MSR	Molten Salt Reactor
MUPSA	Multi-Unit Probabilistic Safety Assessment
NA-SA	Nucleoeléctrica Argentina S.A.
NEA	Nuclear Energy Agency (European Union)
NIA	Nuclear Installations Act
NIR	Nuclear Installations Regulations
NM	Nuclear Material

NNR	National Nuclear Regulator - South Africa
NNSA	National Nuclear Safety Administration - China
NOAK	Nth-of-a-kind
NPP	Nuclear Power Plant
NPT	Treaty on the Non-Proliferation of Nuclear Weapons
NRA	Nuclear Regulation Authority – Japan
NRC	Nuclear Regulatory Commission – United States
NRU	National Research Universal reactor
NSR	Nuclear Security Regulations
OECD/NEA WGRISK	OECD/NEA Working Group on Risk Assessment
OIL	Operational Intervention Level
ONR	Office for Nuclear Regulation – United Kingdom
OPEX	Operating Experience
OPZ	Outline Planning Zone
ORE	Occupational Radiation Exposure
OTS	Operational Surveillance Test
PAZ	Precautionary Action Zone
PLOFC	Pressurized Loss of Forced Cooling
PBMR	Pebble Bed Modular Reactor
PIE	Postulated Initiating Event
PIRT	Phenomena Identification and Ranking Table
PPB	Primary Pressure Boundary
PRA	Probabilistic Risk Assessment
PS	Prevention System
PSA	Probabilistic Safety Assessment
PSAR	Preliminary Safety Analysis Report

PSE	Probabilistic Safety Evaluation
PSI	Preservice Inspection
PWR	Pressurized Water Reactor
QA	Quality Assurance
QG	Quality Group
QM	Quality Management
QMS	Quality Management System
R&D	Research and Development
RAI	Request for Information, as defined by the US NRC.
RCCS	Reactor Cavity Cooling System
RCS	Reactor Cooling System
RCPB	Reactor Coolant Pressure Boundary
RE	Responsible Entity
REA	Rod Ejection Accident
REPPIR	Radiation (Emergency Preparedness and Public Information) Regulations
RGP	Relevant Good Practice
RHR	Residual Heat Removal
RHRS	Residual Heat Removal System
RP	Radiation Protection
RP	Requesting Party
RPR	Recipiente de Presión del Reactor
RPS	Reactor Protection System
RPV	Reactor Pressure Vessel
RSS	Rapid Shutdown System
SAFDL	Specified Acceptable Fuel Design Limit
SAP	Safety Assessment Principle
SA	Severe Accident

SAA	Severe Accident Analysis
SAR	Safety Analysis Report
SBD	Safeguards by Design
SBO	Station Black-Out
SC	Safety Case
SC-H	Safety Class - High
SC-M	Safety Class - Medium
SCT	Scientific Computing Tool
SFP	Spent Fuel Pool
SFR	Sodium-cooled Fast Reactor
SFR	Small Release Frequency
SG	Steam Generator
SGTF	Steam Generator Tube Failure
SGTR	Steam Generator Tube Ruptures
SIS	Safety Injection System
SM	Shift Manager
SMR	Small Modular Reactor
SMRs	Small Modular Reactors
SoDA	Statement of Design Acceptance
SRA	Safeguards Regulatory Authority
SSAC	State system of accounting for control of nuclear material
SSC	Structures, Systems and Components
SSE	Systems, Structures and Equipment
SSECR	Sistema de Seguridad de Extracción del Calor Residual
SSRP	Safety Standards and Regulatory Practices
SSS	Second Shutdown System
SÚJB	State Office for Nuclear Safety – Czech Republic

SyAP	Security Assessment Principle
TAG	Technical Assessment Guide
TIG	Technical Inspection Guide
TQC	Test, Qualification and Commissioning
TRA	Threat and Risk Assessment
TSO	Technical Support Organization
UHS	Ultimate Heat Sink
UPZ	Urgent Protective action planning Zone
V&V	Verification and Validation
VDR	Vendor Design Review
VLPC	Ventilated Low Pressure Containment
WENRA	Western European Nuclear Regulators Association
WG	Working Group

CONTRIBUTORS TO DRAFTING AND REVIEW

Ahmed, M.	International Atomic Energy Agency
Bester, P.	National Nuclear Regulator, South Africa
Cranston, G.	Nuclear Regulatory Commission, United States of America
Dudek, M.	Nuclear Regulatory Commission, United States of America
El Ghalbzouri, R.	French Nuclear Safety Authority, France
Fanielle, S.	International Atomic Energy Agency
Fong, M. C.	Canadian Nuclear Safety Commission, Canada
Horvath, K.	International Atomic Energy Agency
Li, B.	National Nuclear Safety Administration, China
Lisbona, D.	Office for Nuclear Regulation, United Kingdom
Masriera, N.	Autoridad Regulatoria Nuclear, Argentina
Miller, D.	Canadian Nuclear Safety Commission, Canada
Miranda, S.	International Atomic Energy Agency
Mooney, K.	International Atomic Energy Agency
Polyakov, D.	Rostechnadzor, Russian Federation
Santini, M.	International Atomic Energy Agency
Santos, M.	International Atomic Energy Agency
Scotto de Cesar, C.	International Atomic Energy Agency
Torano, P.	Autoridad Regulatoria Nuclear, Argentina
Zhu, L.	National Nuclear Safety Administration, China
Waldman, R.	International Atomic Energy Agency

Whitlock, J.   International Atomic Energy Agency

Wu, J.   National Nuclear Safety Administration, China

Consultants Meetings

Ottawa, Canada: 22–26 July 2019
Vienna, Austria: 22–26 June 2020, 7–11 December 2020

ORDERING LOCALLY

IAEA priced publications may be purchased from the sources listed below or from major local booksellers.

Orders for unpriced publications should be made directly to the IAEA. The contact details are given at the end of this list.

NORTH AMERICA

Bernan / Rowman & Littlefield

15250 NBN Way, Blue Ridge Summit, PA 17214, USA

Telephone: +1 800 462 6420 • Fax: +1 800 338 4550

Email: orders@rowman.com • Web site: www.rowman.com/bernan

REST OF WORLD

Please contact your preferred local supplier, or our lead distributor:

Eurospan Group

Gray's Inn House

127 Clerkenwell Road

London EC1R 5DB

United Kingdom

Trade orders and enquiries:

Telephone: +44 (0)176 760 4972 • Fax: +44 (0)176 760 1640

Email: eurospan@turpin-distribution.com

Individual orders:

www.eurospanbookstore.com/iaea

For further information:

Telephone: +44 (0)207 240 0856 • Fax: +44 (0)207 379 0609

Email: info@eurospangroup.com • Web site: www.eurospangroup.com

Orders for both priced and unpriced publications may be addressed directly to:

Marketing and Sales Unit

International Atomic Energy Agency

Vienna International Centre, PO Box 100, 1400 Vienna, Austria

Telephone: +43 1 2600 22529 or 22530 • Fax: +43 1 26007 22529

Email: sales.publications@iaea.org • Web site: www.iaea.org/publications

Printed and bound by CPI Group (UK) Ltd, Croydon, CR0 4YY

06/07/2026

02160600-0002